PRIMATE AGGRESSION, TERRITORIALITY, AND XENOPHOBIA

LIST OF CONTRIBUTORS

ORLANDO J. ANDY

IRWIN S. BERNSTEIN

C. R. CARPENTER

SHARON F. CATLIN

PIERRE CHARLES-DOMINIQUE

IRENÄUS EIBL-EIBESFELDT

M. Y. FAROOQUI

THOMAS P. GORDON

RALPH L. HOLLOWAY

LEWIS L. KLEIN

ARTHUR KLING

HANS KUMMER

ROSLYN MASS

MANFRED MAURUS

UELI NAGEL

B. C. PAL

THOMAS K. PITCAIRN

ROD PLOTNIK

FRANK E. POIRIER

ALISON RICHARD

ROBERT M. ROSE

JOHN PAUL SCOTT

M. F. SIDDIQI

M. W. SORENSON

CHARLES H. SOUTHWICK

HEINZ STEPHAN

ROBERT W. SUSSMAN

PRIMATE AGGRESSION, TERRITORIALITY, AND XENOPHOBIA

A Comparative Perspective

EDITED BY

RALPH L. HOLLOWAY
Department of Anthropology
Columbia University

ACADEMIC PRESS New York and London 1974

A Subsidiary of Harcourt Brace Jovanovich, Publishers

To Louise, Marguerite, and Eric

ACADEMIC PRESS, INC.
111 Fifth Avenue, New York, New York 10003

United Kingdom Edition published by
ACADEMIC PRESS, INC. (LONDON) LTD.
24/28 Oval Road, London NW1

Library of Congress Cataloging in Publication Data

Holloway, Ralph
Primate aggression, territoriality and xenophobia.

1. Primates–Behavior. 2. Aggressive behavior in animals. I. Title.
QL737.P9H76 599'.8'045 73-18954
ISBN 0–12–352850–X

PRINTED IN THE UNITED STATES OF AMERICA

CONTENTS

LEWIS L. KLEIN (77), Department of Anthropology, University of Illinois, Urbana, Illinois

ARTHUR KLING (361), Department of Psychiatry, College of Medicine and Dentistry of New Jersey, Rutgers Medical School, Piscataway, New Jersey

HANS KUMMER (159), Institute of Zoology, University of Zurich, Zurich, Switzerland

ROSLYN MASS (361), College of Medicine and Dentistry of New Jersey, Rutgers Medical School, Piscataway, New Jersey

MANFRED MAURUS (331), Max-Planck-Institut für Psychiatrie, Munich, West Germany

UELI NAGEL (159), Institute of Zoology, University of Zurich, Zurich, Switzerland

B. C. PAL (185), Department of Zoology, University of Kalyani, Kalyani, West Bengal, India

THOMAS K. PITCAIRN (241), Arbeitsgruppe für Humanethologie, Max-Planck-Institut für Verhaltensphysiologie, Enzianweg, West Germany

ROD PLOTNIK (389), Department of Psychology, San Diego State University, San Diego, California

FRANK E. POIRIER (123), Department of Anthropology, Ohio State University, Columbus, Ohio

ALISON RICHARD (49), Department of Anthropology, Yale University, New Haven, Connecticut

ROBERT M. ROSE (211, 275), Department of Psychosomatic Medicine, Boston University School of Medicine, Boston, Massachusetts

JOHN PAUL SCOTT (417), Center for Research on Social Behavior, Bowling Green State University, Bowling Green, Ohio

M. F. SIDDIQI (185), Department of Geography, Aligarh Muslim University, Aligarh, India

M. W. SORENSON (13), Division of Biological Science, University of Missouri, Columbia, Missouri

CHARLES H. SOUTHWICK (185), Department of Pathobiology, The Johns Hopkins University, Baltimore, Maryland

HEINZ STEPHAN (305), Neurobiologische Abteilung, Max-Planck-Institut für Hirnforschung, Frankfurt, West Germany

ROBERT W. SUSSMAN (49), Department of Anthropology, Washington University, St. Louis, Missouri

PREFACE

In planning this volume two topics interested me particularly—ones that I believed would contribute substantively to the growing field of primatology: (*1*) primate endocrinology and (*2*) a truly comparative work on primate aggression. Contributors were not selected on the basis of their particular viewpoints on aggression, for it was my expressed purpose that the volume should contain *no* particular theoretical orientation, but rather as many viewpoints as there were contributors, if possible. Accordingly, contributors were chosen on the basis of their substantive research in the area of aggression and/or primate behavior, whatever their own identities—anthropologists, ethnologists, behaviorists, naturalists—might be.

The authors whose efforts are represented here felt this approach worth while, and were willing to contribute their time and energy to produce a fairly holistic and unique volume of primate aggressive behavior in comparative perspective. I most sincerely thank each of the contributors for his time and effort in preparing these chapters.

This volume is not as complete as my initial ideal intentions would have had it, for the human side of aggression has not been as fully covered as originally intended. On the other hand, what is here is already unique in scope, and the inclusion of more material would have made either another volume or an unwieldy one imperative.

This is, as far as I am aware, the first comparative volume on primate aggressive behavior which offers a review of present knowledge at each major taxonomic level, and which reviews in depth endocrinological and neurological concomitants of aggression in primates. One might argue the veracity of including the tree shrews in the order Primates, but whatever their ultimate taxonomic affinities, a profile of their aggressive behavioral patterns, as presented by Dr. Sorenson's chapter, provides an interesting glimpse into the adaptive structure of a group possibly representing the ancestral condition from which primates did evolve. The inclusion of views on the endocrinological and neuroanatomical data is of particular interest to me, and I hope to others, for these provide the comparative basis against which the human species must ultimately be measured, and which form a substantive, empirical basis for further synthesis with genetic, behavioral, and evolutionary frameworks. These are two areas which have been pretty much ignored in the past, and which hopefully will spur on some much needed future research. Some of these data are presented for the first time in this volume.

Dr. Carpenter's chapter on aggressive episodes purposely has been left open-ended, so that the reader can assess for himself the areas of analogy and homology between widely separated animals and environments, and reach his own conclusions if he wishes. But indeed, each chapter could be similarly underlined, for each is unique in its own way.

Finally, I would like to express my particular appreciation to Mrs. Sue Gould, who played an invaluable role in managing the correspondence and coordination of this effort when I was on sabbatical leave during 1971–1972 in Africa, to Miss Roberta Britt for her typing assistance in the later stages of this book, and to Miss Kate McFeeley for her help in the indexing. I am grateful to the staff of Academic Press, for their encouragement, cooperation, and suggestions in putting this volume together.

INTRODUCTION

RALPH L. HOLLOWAY

Columbia University

The ideal volume on a comparative approach to primate aggressive behavior has yet to be written. Such a volume would start with a close examination of the definitions of aggression, their historical development, and an appraisal of their strong and weak points. Next, one would hope for a number of review articles that would systematically report and assess all available naturalistic and laboratory observations on *all* of the primates. Then, but not necessarily in this order, one would like to see special chapters that review all available knowledge of the anatomical and physiological bases of primate aggressive behavior, starting with the anatomical equipment of different primates (e.g., Wickler, 1967), such as sexual dimorphism in size, coloration, vocalization, strength, etc., then moving to the endocrinological side, and finally, the neuroanatomical and neurobiological aspects. Next, with all the data available, one could then have the various experts examine the spectrum of primate aggressive behavior and indicate their views about the place and nature of human aggression within that of the primates, and mammals, in general. Finally, each would critically summarize the hiatuses in their own research and suggest concrete areas needing further effort. Ideally, this might be repeated every 10 or so years.

HIATUSES

This volume has attempted to approach such an ideal conception and has failed, but not badly. There are some hiatuses, and these might be pointed out: (*1*) not all of the possible range of viewpoints are represented here; (*2*) not all of the laboratory experimental evidence is given; (*3*) there is no review of the external anatomical equipment utilized by primates when they are aggressive; (*4*) the ontogenetic development of aggressive behavior in both natural and laboratory colonies of primates is not fully covered; (*5*) the range of *human* aggressive behavior (in its own comparative framework, i.e., ethnological) is not covered here; (*6*) no grand synthesis of all primate aggression is presented that extracts each distinctive theoretical approach, weighs it against all others, and lists the strengths and weaknesses of each, and finally details the hiatuses and provides explicit suggestions for future research.

My own view of our task to understand fully our own species' aggressive behavior has not changed much since I wrote the paper, "Human Aggression: The Need for a Species-Specific Framework" (Holloway, 1968c). This does *not* mean, however, as some might wish to interpret it, that I regard comparative frameworks valueless, or that comparative knowledge of the order Primates or even the mammals *in toto* has nothing to offer the theorist, observer, or experimenter in coming to a closer understanding of human aggression in all of its fearsome manifestations. Still, the human animal is the only one we know, at least on this planet, that has undergone an evolutionary development resulting in a highly social animal using arbitrary symbols to reconstitute and define reality, modes of behavior, and which has such a uniquely organized brain. As animals, we share with our closest relatives (the great apes) much of our anatomy, physiology, and even behavioral attributes, except our unique ability to think and communicate in terms of arbitrary symbol systems.

Most anthropologists, at least physical ones, believe that the human animal is neither simply a *tabula rasa* written upon by cultural norms, nor a creature forever condemned merely to express its inherited "instincts" or innate dispositions. Those theorists who regard man as an evolutionary product, a "human" animal, but with no inherent biologically determined dispositions, have a peculiar logic and understanding of evolutionary forces. Those theorists who regard the human beast as simply one more animal, without unique qualities aside from external appearance, that simply acts out its genetic instructions, are also lacking in logic. We study primate aggression to understand what is and what is not shared by the human animal so that we may have a sane and substantive basis for recommending to society where its programs are inhuman, unjust, stupid, insane, and nonproductive. We can only truly know ourselves when we have a basis for comparisons. We only *know* ourselves to the extent that we can *predict* what our actions will be (behavioral, anatomical, physiological, cultural, etc.) given the *field* (*sensu* Lewin, 1937) of our existences.

Aggressive behavior is generally viewed in a pejorative sense; we wish to make it stop, and live in a society where we will not be prey to it. As most of the papers in this volume indicate, aggressive behavior is common to all primates, even the tree shrews, whatever their actual taxonomic placement. Aggressive behavior in micro- (and thus macro-) evolutionary perspective is eufunctional for the species; it helps to space out its members, thus working to maintain a best fit between population sizes and resources; mating dynamics, and the ability to maintain social control and cohesion within and between populations. It is hardly surprising then, that as Moyer (1968, 1969, 1971a,b) has so convincingly shown, there are many "kinds" of aggressive behavior, each with its own set of environmental conditions, cue stimuli, internal milieu, and neural circuitry (e.g., predatory, intermale, fear-inducing, irritable, territorial, maternal, and instrumental). In discussing the controversy resulting from the "looseness of the various definitions of drive," Moyer (1969, p. 104–105) notes:

> There are certain basic circuits in the nervous system. When they are active, certain complex behaviors occur. The problem for the student of behavior is to determine the variables, both internal and external, which activate and deactivate these circuits. There are certainly differences in the mechanisms for turning the basic neuronal circuits on and off. There are also remarkable similarities. The evidence seems to indicate that there are basic (in a sense, built-in) circuits for aggressive behavior just as there are for consummatory and sex behavior.

Here, for example, is a rich zone indeed for further research in an attempt to put aggressive behaviors and neural structures in a comparative framework. We have Moyer's (1969) aggression types, and now, thanks to the numerous researches of Stephan, Bauchot, and Andy (see references in Andy and Stephan, this volume), we have an excellent quantitative basis for comparing neural nuclei and circuits between primates. As far as I am aware, only Chance (1962) has attempted to speculate on the differences in aggression between primates on the basis of neural circuitry. Obviously, much more basic works of a neurophysiological kind are needed to work out aggressive events in the central nervous system (CNS) between neural centers and along neural fiber tracts before we can simply plug in our quantitative data and make predictions.

The major point I wish to make here is that, as Andy and Stephan's paper in this volume clearly demonstrates, the neuronal structures subserving aggressive behavior are quantitatively different in all of the primate species, indicating that each species has a unique organization of neural elements gained through natural selection and evolution, and that the human animal is no exception. Man is obviously an evolutionary product, and his CNS must, in some way, reflect the evolutionary forces that molded him. As I have pointed out in other contexts (Holloway, 1969, 1970), we cannot yet predict either his behavior from knowing the quantitative relations in his brain, or predict the quantitative relations from his behavior. This might be taken as an indication of where we stand scientifically with regard to our knowledge of brain and behavior. Indeed, this is why I insist that a species-specific framework is essential for understanding human aggression; indeed, any animal species. That is also why the comparative framework is indispensable.

It is precisely this area of neurobiology and endocrinology that holds forth so much promise, and in which this volume is particularly strong. I remember submitting in 1966 a paper on some theoretical speculations regarding human evolution that were based on sexual dimorphism and endocrinology in the primates. Fortunately (I think), it was rejected because at that time precious little comparative data on primate endocrinology (except in relation to ovulatory cycles) were available in terms of aggressive behavior. Obviously, as Rose *et al.* make clear, in this volume, there is a long way to go in gaining basic quantitative data on the comparative picture of primate endocrinology and aggressive behavior. The whole question of arboreal versus terrestrial primates in terms of quantity and thresholds of aggressive behavior and testosterone remains a moot one, as does our basic knowledge of individual variation in natural and experimental primate groups. Are the steroid hormones in ourselves and the great apes affected to any degree by seasonal changes? Are there any correlations between dominance, agonistic behavior, and thresholds, and hormones in individuals in primate populations? Have there been any evolutionary changes in hormonal target tissue interactions and aggressive thresholds in human evolution during the past 10 or 50 million years (e.g., Holloway 1967, 1968a)? What is happening in the ghettos, the subways, in undernourished and underdeveloped nations with extremely high population densities, in terms of stress, hormones, and thresholds of aggressive behavior? Do various cultural groups (e.g., the Yanomamo, Bushmen, New Guinean tribes) vary in any of these biological foundations?

The questions one may ask are almost endless, for here is an area of research that until now has been avoided, for both technical and moral reasons. Until we know the answers to these questions, and many others, we do not have the basic foundation of biological facts upon which to erect hypotheses and test them.

Surely the comparative neurological data, as well as that of the last 40 years of observing primates in the wild and captivity, show at least three important conclusions:

1. The quantitative as well as observational data on neuroanatomical differences in neural nuclei and fibers mediating agonistic behavior in primates indicate all are present and differ quantitatively, if not qualitatively, between species within the order Primates. *Homo sapiens* shares these structures and differs but quantitatively from other primates in this regard. There is no evidence to suggest that the neural development of these systems (i.e., the hypothalamus, amygdaloid complex, hippocampus, septum, anterior thalamus, cingulate cortex, etc.) and their associated fiber connections shows any dependence on environmental modification during growth, beyond possible adjustments in thresholds of facilitation and inhibition. (For example, see Moyer 1968, 1971a,b, for involvement of different aggression–mediating circuits and different modes of aggressive behavior.) To believe that this incredibly complex integration of centers and fibers has no relation to behavior, or an evolutionary significance and history, is naive. No, the centers are there, and do operate in man, and the quantitative data (e.g., Andy and Stephan, this volume) suggest a unique (species-specific) course for *Homo sapiens*.

It should not be overlooked that the central nervous system is an organ showing variability in structure, just as any other organ system. Our data on individual variations of structures in the human nervous system (or indeed, in any of the primate species) are wanting, and we can gauge the limits of our knowledge from recent compilations of Blinkov and Glezer (1968) on quantitative data for neural subsystems or Talairach *et al.* (1949) on the human thalamus. I have reviewed some of this data elsewhere (Holloway 1968b,c).

Beck and Gadjusek (1966) published some interesting preliminary data on variability of the septum in New Guinean natives,[1] which, if verified, should caution us to keep in mind the possibility of population differences in the structure of the CNS. These differences might have important significance for understanding human variability in different breeding isolates for behavior such as aggression or its thresholds.

At a minimum, we should accept the probability that different human groups (as well as different primate species) have different quantitative (if not neurochemical) relationships among the same structures, and

[1] Most of the sample was from brains of natives suffering from kuru.

that threshold differences in provoked aggression and temperamental differences might well exist between different populations as well as among their members.

Obviously, the ontogenetic development of reactivity of these brain nuclei and their interconnecting fiber systems occurs within a milieu of cultural conditioning that is perhaps more pervasive in terms of setting thresholds than are the neuroanatomical relations per se.

On the other hand, some of the current work on XYY males (see Shah, 1973 for review) indicates a genetic component on the Y chromosome influencing stature and testosterone and thus aggressive behavior through reduction of threshold. While suggestive, it would be premature to accept these conclusions without further testing and analysis.

2. Musculoskeletal organization, bodily stances, and facial expressions. Unfortunately, this volume does not include a review chapter on the similarities and differences between primate taxa relating to the repertoire of bodily and facial movements incorporated into agonistic interactions, although each of the taxon reviews (Section I, pp. 11–272) gives the repertoire for most. Analyses of these relationships may be found in Bertrand (1969) and Van Hooff (1967).

Homo sapiens of course shows the same basic expressive anatomy as most of the higher primates, and a number of ethologically oriented analyses (Blurton-Jones, 1971; McGrew, 1971; Eibl-Eibesfeldt, 1970, this volume) indicate a considerable degree of homology in structure, expressions, and behavioral functioning. One must suppose that neural controls of these assemblages are basically the same in organization. Intimidation and appeasement behavioral patterns exist in all primates (indeed, all mammals), and human evolution must surely have involved selection for components of these orientations utilizing the same basic organizations of facial, bodily, and neural patterns. Human beings, however, seem to have a proclivity for selectively attending to these gestures which goes beyond what data we have garnered from other primate societies. That is, human beings are capable of continuing extreme aggressive actions whether appeasement gestures are offered or not. Because of their cortical evolution and symbolic facility they can maintain hostilities over and beyond stimulus presentation and can extend aggressive impulses and channel energy to pursue aggressiveness to a wide variety of arbitrary symbolic clusters. Is this really loss of "control" or is it perhaps "super" control? In other words, the human animal has both a neural apparatus and cognitive structuring that permit cortical facilitation of circuit loops involving noncortical structures independent of the fear or appeasement gestures offered by the victim or victims, human or other animals. Extreme aggressive actions can be carried out by human actors on commands from other social actors without reference to either internal rage reactions or the appeasement/fear responses of their victims—witness My Lai, Biafra, Rwanda, or Europe during the 1930s and 1940s of this century. None of the papers in this volume, or elsewhere, based on nonhuman primate research, suggests any serious parallels to the range of human aggressive expression.

In sum, despite all of the similarity of nervous structure, facial and bodily expressions and stances between the human animal and his closest primate relatives, neither appeasement nor retreat need be inhibitive cue stimuli for further aggressive actions. Ideas and arbitrary symbols, cortically mediated, can drive aggressive behavior beyond that known for any other primate. We need to know much more about human abilities to alter figure–ground perceptions and how frustration and demagoguery combine to drive facilitation of aggressive behavior beyond the normal inhibitive influences of submissive and fear gestures and previous cultural conditioning.

3. Almost all of the papers in this volume attest to the extreme importance and significance of ecological variables in the frequency, strength, and duration of most primate aggression (e.g., Poirier, Southwick *et al.*, Nagel and Kummer, this volume).

In my 1968c article I stressed that both the human affectual and cultural milieus were generative of aggressive behavior; these are structured into human existence. Certainly, Bard's (1971) analysis of the high frequency of violent aggressive behavior within human families cannot be ignored, and the nonhuman primate data presented in this volume do not suggest the phenomenon is only human. Nonwestern societies are extremely variable in their aggressive thresholds within the group; most well-ordered societies have strong sanctions against the display of hostilities within the groups, particularly families. Xenophobia is defined simply as fear of strangers (Southwick, 1972) and as Livingstone (1968) has shown, is also quite variable in its expression among different cultures. I would submit that at least for adult humans, xenophobic responses are normative unless there has been strong cultural training and conditioning against it. Clannishness, or strong intragroup affiliation coupled with distantiation of other ethnic, religious, racial, or political groups, is an enforcing mechanism for continued xenophobias. The demagogue knows this fact only too well. It is all well and good to claim the human species as a single family, with intragroup variation not exceeding that of intergroup variation, but human groups *do* recognize themselves as units *different* in some ways from other groups, whether it be on a physically or culturally determined basis. *We and they*. Surely the roots for this basic perceptual style are very deep in our evolutionary history, and not easily overcome.

We can only assume some eufunctional basis for these kinds of perceptual stances, which are hardly human specific, but shared throughout most of the animal kingdom. We overcome this to some extent through cultural learning, and a pride in and positive value on tolerance and goodwill. But put stress on the system, and the age-old dispositions are dominant.

No, the human animal is not some *tabula rasa* upon which are written only cultural values, which has no biological evolutionary history, or scientifically describable dispositions to act out a number of neurobehavioral patterns shared with other animals, particularly other primates. The human animal has such dispositions indeed, and is all the more dangerous because of his numbers, social densities, and facilitative cerebral cortex, and his seeming need for strong external social control, be it through demagogues or others.

Add to this a last species-specific proclivity, the use of extrasomatic weaponry and tools, and we have introduced a new murderous logic into consideration. Man is fascinated by tools, by devices that make his labors easier, and yet which also hold the promise of power over others that is independent of physical strength, prowess, and courage. Tools have inherent logics in them. Sharp edges are for cutting, points for puncturing, triggers to be pulled, buttons to be pushed. These logics can be coupled to human action by commands of arbitrary puffs of air, with symbolic forms, whether a cue stimulus is present or not. An idea is sufficient.

I believe this volume will help to understand human aggression, both in terms of negative counterpoise, meaning those unique aspects of human adaptation discussed above, and positively with homologies between the human and other primates in aggressive and xenophobic behavior. *Homo sapiens,* the most dangerous animal currently making an evolutionary experiment, cannot understand let alone control himself without understanding his relationship to the rest of the natural world. Hopefully, this volume makes a contribution to that understanding.

REFERENCES

Bard, M. (1971). The study and modification of intra-familial violence. "The Control of Aggression and Violence" (J. L. Singer, ed.), pp. 154–164. Academic Press, New York.

Beck, E., and Gadjusek, C. (1966). Variable size of the septal nuclei in man. *Nature (London)* **210,** 1338–1340.

Bertrand, M. (1969). The behavioral repertoire of the stump-tail macaque. *Bibl. Primat.* No. 11. Karger, Basel.

Blinkov, S. M., and Glezer, I. I. (1968). "The Human Brain in Figures and Tables." Basic Books, New York.

Blurton-Jones, N. G. (1967). Some aspects of the social behavior of children in a nursery school. *In* "Primate Ethology" (D. Morris, ed.), pp. 347–368. Weidenfeld and Nicolson, London.

Chance, M. R. A. (1962). Social behavior and primate evolution. "Culture and Evolution of Man" (M. F. A. Montagu, ed.), pp. 84–130. Oxford Univ. Press, London and New York.

Eibl-Eibesfeldt, I. (1970). "Ethology: The Biology of Behavior." Holt, New York.

Holloway, R. L. (1967). The evolution of the human brain: Some notes toward a synthesis between neural structures and the evolution of complex behavior. *Gen. Syst.* **XII**, 3–19.

Holloway, R. L. (1968a). Cranial capacity and the evolution of the human brain. *In* "Culture: Man's Adaptive Dimension" (M. F. A. Montagu, ed.), pp. 170–196. Oxford Univ. Press, New York and London.

Holloway, R. L. (1968b). The evolution of the primate brain: Some aspects of quantitative relations. *Brain Res.* **7**, 121–172.

Holloway, R. L. (1968c). Human aggression: The need for a species-specific framework. *In* "War: The Anthropology of Armed Conflict and Aggression" (M. Fried, M. Harris, and R. Murphy, eds.), pp. 29–48. Natural History Press, New York.

Holloway, R. L. (1969). Some questions on parameters of neural evolution in primates. *Annals New York Academy of Science* **167**, 332–340.

Holloway, R. L. (1970). Neural parameters, hunting, and the evolution of the human brain. *In* "The Primate Brain" (C. R. Noback and W. Montague, eds.), pp. 299–309. Appleton, New York.

Lewin, K. (1936). "Principles of Topological Psychology." McGraw-Hill, New York.

Livingstone, F. B. (1968). The effects of warfare on the biology of the human species. *In* "War: The Anthropology of Armed Conflict and Aggression" (M. Fried, M. Harris, and R. Murphy, eds.), pp. 3–15. Natural History Press, New York.

McGrew, W. (1972). "An Ethological Study of Children's Behavior." Academic Press, New York.

Moyer, K. E. (1968). Kinds of aggression and their physiological basis. *Comm. Behav. Biol.* **2**, 65–87.

Moyer, K. E. (1969). Internal impulses to aggression. *Trans. N.Y. Acad. Sci.* (II) **31**, 104–114.

Moyer, K. E. (1971a). The physiology of aggression and the implication for aggression control. *In* "The Control of Aggression and Violence" (J. L. Singer, ed.), pp. 61–93. Academic Press, New York.

Moyer, K. E. (1971b). A preliminary physiological model for aggressive behavior. *In* "The Physiology of Defeat and Aggression" (B. E. Eleftheriou and J. P. Scott, eds.), pp. 223–263. Plenum Press, New York.

Shah, S. A. (1973). The XYY chromosomal abnormality. *In* "Aggression and Evolution" (C. M. Otten, ed.), pp. 37–60. Xerox College Publ., Massachusetts.

Southwick, C. H. (1972). Aggression among nonhuman primates. *Module* **23**, 1–23.

Talairach, J. *et al.* (1949). Recherches sur la coagulation thérapeutique des structures sous-corticales chez l'homme. *Rev. Neural.* **81**, 4–24.

Van Hooff, J. A. R. A. M. (1967). The facial displays of the catarrhine monkeys and apes. *In* "Primate Ethology" (D. Morris, ed.), pp. 7–68. Weidenfield and Nicolson, London.

Wickler, W. (1967). Socio-sexual signals and their intra-specific imitation among primates. *In* "Primate Ethology" (D. Morris, ed.), pp. 69–147. Weidenfeld and Nicolson, London.

COMPARATIVE BEHAVIORAL DATA

A REVIEW OF AGGRESSIVE BEHAVIOR IN THE TREE SHREWS

M. W. SORENSON

University of Missouri

INTRODUCTION

Aggression refers to actual or symbolic attack upon another individual and therefore includes all behavior that is more likely to lead to attack than to retreat (Stokes and Cox, 1970). According to Freud (1959) and Lorenz (1966), aggression builds up spontaneously and must be released. Lorenz includes certain external stimuli in what is attacked and, in this sense, agrees with Barnett (1967), Berkowitz (1962), Dollard *et al.* (1939), and Scott (1958), who state that aggression is internally motivated but dependent on appropriate external stimuli for expression. Alexander and Roth (1971), Myers (1966), and Southwick (1955) feel that increased animal density is the major cause of aggression.

Aggression is either intra- or interspecific and most likely is associated with one or more of the following: (*1*) competition for food, (*2*) defense of an infant by its parents, (*3*) a struggle for dominance or change in social status between two animals of near equal rank, (*4*) a redirection of aggression, (*5*) a failure to comply with signals, (*6*) the consort formation at estrus, and (*7*) changes in the internal biological state of the animal (Hamburg, 1971; Vernon, 1969).

Most fighting among animals is intraspecific and occurs in species with well-integrated societies. Group organization in social animals is based on sex pairing, territory, hierarchy, and leadership (Cloudsley-Thompson, 1965). Matthews (1964) divides intraspecific fighting into two kinds: ritual and overt. He doubts whether overt fighting normally occurs in nature. Lorenz (1966) states that aggression has survival value to the species, e.g., aggression spaces out social animals so that individuals are spread evenly over the available food and housing areas; aggression allows for selection of the strongest individual by rival fights; and aggression is used for protection of the brood. To have survival value, aggression must bring order into the social circle (George, 1966). Social organization of most animals is based in part on status hierarchies; the subordinates are dispersed and affect the rate of evolution (Christian, 1970). Dominant animals function to control birth rate and overpopulation, to integrate groups for defense against other groups, and to reduce intragroup aggression (van Kreveld, 1970). Collias (1953) states that a hierarchy is a stress-reducing device; however, Gartlan (1968) questions the concept that social dominance is a structuring mechanism. He points out that hierarchies are considered to reduce the amount of aggression, but that when hierarchies are most rigid, such as among animals in captivity, aggression is most common.

It seems that man has always known aggression but only recently has he understood its biological value to the species. With the tragic misuse of the environment and the tardy realization that overpopulation increases rivalry for the quality aspects of the environment, man now studies other animal societies to understand the inherent values of organized competition. Man realizes that he cannot continue to rely on technology to create new sources of food, new housing, and new comforts for newborn millions because the race between technology and the human gene pool is not a fair contest—man enjoys reproducing his kind more than he enjoys providing for them, and thus poverty and overpopulation merge to supply the raw materials for aggression.

The following review of aggression in the tree shrews (systematics according to Lyon, 1913) points out the ways these primitive primates organize their aggressive tendencies under captive conditions. It must be emphasized that tree shrews have not been studied carefully in their natural setting and that captivity increases animal density, restricts animal movement, and forces a familiarity between individuals not common to feral populations. Elliot *et al.* (1969) note that island groups of *Tupaia glis* have greater densities and smaller home ranges than do mainland groups (natural populations of *T. glis* in Thailand reach 15 to 30 individuals per acre; Morris *et al.,* 1967). Island forms are bolder but are less active than mainland animals. They also have larger adrenal glands and higher serum cholesterol levels than do their mainland counterparts living under less density. These crowded, island conditions are precisely those man is forcing on himself through overpopulation; it is this parallel that warrants this review.

STUDIES OF AGGRESSIVE BEHAVIOR IN TREE SHREWS

In 1827 Temminck cast the first stone at tree shrews when he called them "barbarous." Cantor (1846) remarked: "In a state of nature it (*Tupaia ferruginea*) lives singly or in pairs, fiercely attacking intruders of its own species. When several are confined together, they fight each other, or jointly attack and destroy the weakest [p. 189]." Hendrickson (1954) reports that it is impossible to keep two males of *T. glis* in the same cage because an aggressive attitude toward another individual, once begun, remains fixed and the animals must be separated or they will fight to the death. A male and one or more nonpregnant females may be housed together without conflict. Sprankel (1951, 1961a) also reports that individuals of *T. glis* are extremely aggressive and that family units are broken up by paternal aggression toward the young.

After observing eight specimens of *T. glis* in a 50-ft^2 outdoor enclosure over a 5-month period, Vandenbergh (1963) concluded that *Tupaia* is aggressive. Fighting occurs between adult males and between adult females, but not between adult males and females. Vandenbergh reports the deaths of three males and one female from wounds received in fighting. He also reports paternal aggression toward juvenile males. Draper (1963) cites 10 deaths from fighting among 13 individuals of *T. glis* in captivity. In his study of the behavior of 14 adults of *T. glis* in Puerto Rico, Kaufmann (1965) describes an aggressive episode between five males. One male and two females were placed in a 50-ft^2 outdoor cage and four males and two females later introduced one at a time into the enclosure. The resident male was the dominant animal and he harassed the introduced males continually. The introduced males were wounded and all died in a period of 2 days to 3 months. The resident male then remained peacefully with the females.

Studies of individuals of *Tupaia chinensis* in captivity by Hasler (1969a,b), Sorenson (1964, 1970), and Sorenson and Conaway (1966) indicate that this species is as aggressive as *T. glis*. Both males and females of *T. chinensis* attack individuals of other species without apparent reason. The dominant male is a despot who interrupts social activities among members of other species. Fights between either males or females of *T. chinensis* in which two animals "stand their ground" are unusual.

Hasler cites the deaths of four out of eight males as a result of aggressive encounters during the first 6 weeks of his study. Male–female aggression also occurs.

The behavior of members of *Tupaia longipes* is similar to that of specimens of *T. chinensis*. Fights between adult males, between adult females, among adult males and females, and between adult males and juvenile males are reported by Sorenson (1964) and Sorenson and Conaway (1964, 1966). Martin (1968) states that aggressive tendencies among members of *Tupaia belangeri* begin at about 3 months of age (juveniles first fight when they are 84 days old and strife occurs between parents and their offspring when the latter are 90 days old). Aggression in males coincides with the beginning of discoloration of the hair by secretions of the gular marking glands.

According to Sorenson (1970) and Thompson (1969), members of *Tupaia palawanensis* are less aggressive than are individuals of the preceding species. The largest male *T. palawanensis* is the most dominant animal in the colony and he maintains his rank by threatening, chasing, and biting the lower-ranked males. Male–female and female–female aggression also occurs, but is less intense than male–male aggression.

Unlike the despot male dominance found in members of *T. glis* and *T. chinensis*, males of *Tupaia montana* display codominance or mutual tolerance (Sorenson and Conaway, 1968). Most male–male and male–female aggression is associated with competition over estrous females. The more serious conflicts among members of *Lyonogale tana* are also caused by sexual competition or by hierarchy reversals. Sorenson (1964) reports a fight between two adult males that lasted over 1 hr and involved a semicontinuous chase of 73 circuits of the cage (a distance of approximately 2600 ft). These periods of hyperactivity leave the animals in a state of near exhaustion and activity is greatly reduced over the next few days.

Wharton (1950) states that individuals of *Urogale everetti* seldom fight in captivity but do much whimpering and squealing. They will bite when they are first handled, but they soon learn to trust their attendant, although they always vocalize a protest. Polyak (1957) also notes that outright squabbles or fights among members of *Urogale* do not occur if ample food is available.

There is meager information concerning aggression between members of the more arboreal species *Tupaia minor* and *Tupaia gracilis*. Sorenson (1964, 1970) and Sorenson and Conaway (1966) note that adults of these species seldom display overt aggression; chases are of low intensity and infrequently evolve into fights.

The preceding comments point out that aggression varies among species of tree shrews in captivity. Individuals of *T. glis* and *T. chinensis* are more aggressive than are members of *T. montana* and *Urogale everetti*. These differences in aggression are inherent in the species, as shown by studies of several species maintained together in captivity (Sorenson, 1964, 1970; Sorenson and Conaway, 1966), and are dependent on the number of animals in each cage, the size of the cage, and the naturalness of the surroundings. In the feral state, density seldom approaches conditions common to laboratory studies. Aggression becomes a negative force among animals in captivity; it cannot function to space out individuals; rather, it leads to physical and psychological stress detrimental to the species. It is not surprising, therefore, that animals devise ways to reduce fighting among themselves whenever they are either forced to or choose to live together in social systems.

THE DOMINANCE HIERARCHY

The structuring mechanism used by tree shrews in a social environment is usually the dominance hierarchy. The males of most species and the females of certain species maintain linear status hierarchies based on aggressive and agonistic displays. The most common displays are threat calls, threat postures, lunges, chases, and fights. Following establishment of a hierarchy, the overall level of aggression is reduced and rank is maintained by ritualized fighting patterns. Dominance is also subtly expressed by the displacement of subordinate animals from food and rest areas and by the mounting of subordinates by dominant animals.

Females of *Tupaia chinensis* have linear hierarchies, but among the males a single dominant animal assumes the role of a despot. The rank of this male is seldom reversed and he harasses all subordinate males. A similar situation is reported for individuals of *Tupaia glis* by Hendrickson (1954), Sprankel (1961a), and Vandenbergh (1963). Among members of *Tupaia montana*, social rankings include males and females, with the number one- and the number two-ranked males being codominant or mutually tolerant (Sorenson and Conaway, 1968). Males of *Tupaia longipes* have a linear hierarchy; a secondary female hierarchy is also evident but less stable than that of the males. Aggression between the number one- and two-ranked males is usually initiated by the second-ranked male whereas conflict between the number two and three animals is mutually induced. On occasions following serious male–male aggression, the defeated animal may become asocial and remain in out-of-the-way places for 2 or 3 days before slowly renewing interactions (Sorenson and Conaway, 1966). Dominance hierarchies also are evident among members of *Lyonogale tana*, *Tupaia minor*, and *Tupaia gracilis*.

When individuals of each sex of *T. chinensis*, *T. longipes*, and *Lyonogale tana* are housed together, intra- and interspecific hierarchies are formed. *T. chinensis* is dominant to both *L. tana* and *T. longipes*, and *T. longipes* is dominant over *L. tana*. When members of *T. gracilis* and *T. minor* are introduced into a cage containing individuals of *T. longipes*, each species accepts the others without aggression and intraspecific hierarchies are maintained (Sorenson, 1964).

Both males and females of *Tupaia palawanensis* form hierarchies. Thompson (1969) reports that in a population of 25 animals (5 males and 20 females), 12 females have equal standing, 1 male and 3 females are midway in the hierarchy, and 6 animals (1 male and 5 females) never assert themselves and are always the objects of threats and chases by the three highest-ranked males. The highest- and the lowest-ranked animals in the population die after a few months in captivity. Their positions are filled by animals next of rank and these in turn die after short periods. The result of the deaths of the highest- and lowest-ranked tree shrews is a stable population composed of the original center part of the hierarchy. This group displays a minimum of aggression. Whenever new animals are added to the stable group, the hierarchy breaks down, conflict ensues, and a new hierarchy is formed which includes members of the introduced group. Again, after 1 or 2 months, the most dominant and the most subordinate animals die.

HOSTILE BEHAVIOR

As noted, the dominance hierarchies of tree shrews are established by aggressive behavior but are maintained by less hostile episodes that reduce conflict. Hostile episodes associated with dominance hierarchies include auditory, visual, olfactory, and contactual patterns (Kaufmann, 1965).

Auditory Patterns

Members of *T. minor* are the least vocal of the species studied. They emit only two basic calls, interpreted as *warning* and *threat* calls. The *warning* call is a rapid, high-pitched chatter made with an open mouth as the animal crouches in a threat posture. This call occurs after chases between animals or other major disturbances in the cage. The *threat* call is made with an open mouth and sounds like a low-pitched bark. It is emitted during conflict by an animal that is cornered. Individuals of *T. gracilis* also have two basic calls. These calls are similar to those of *T. minor* except that the *warning* call is more rhythmic and melodic and is emitted with a semiclosed mouth. This call is elicited by aggression, cage disturbances, and in response to the *warning* calls of *T. minor.* Members of *T. gracilis* seldom emit *threat* calls. Those given are made with an open mouth and consist of two or three short, barking sounds (Sorenson, 1964, 1970).

Warning calls, *threat* calls, *rage* calls, and *fear* calls are emitted by members of *T. longipes. Warning* calls are most often heard and are made with a partially closed mouth and slight abdominal contractions. *Warning* calls are medium-pitched warbles which slowly decrease in volume. They are emitted after sudden noises or other disturbances and follow aggressive chases. Subordinate males call while the dominant male is chasing other subordinate animals. It is difficult to determine which animal is making the *warning* call because of a ventriloquistic quality of the call; however, the calling animal usually is perched high in the cage and the call is accompanied by tail flicks. *Threat* calls are made with an open mouth and resemble explosive barks and hisses. Often, the animal lunges forward immediately after calling. *Threat* calls are mostly responses to overt aggression by other tree shrews, although nonreceptive females emit *threat* calls when males approach them, and both sexes emit calls in defense of nest areas and after being handled. When *threat* calls fail to cause an aggressor to flee, the calls increase in intensity and become shrill, continuous screams indicative of *rage* calls. *Fear* calls are similar to *rage* calls but they are not continuous; rather, they are shrill, short cries emitted following a bite or other injury (Sorenson, 1964, 1970).

Individuals of *L. tana* seldom emit *warning* calls; when these animals are disturbed they run to cover and remain silent. By contrast, they do emit *threat* calls that are low-pitched guttural barks made with an open mouth in conflict situations. These calls also become shrill, staccato cries of *rage* similar to those calls of *T. longipes* (Sorenson, 1964, 1970).

According to Hasler (1969a), one of the few investigators to record and analyze the calls of tree shrews, individuals of *T. chinensis* have two distinctly aggressive calls. The *snort* call is a deep, explosive, coughlike sound delivered with an open mouth. *Snorts* are sometimes uttered repetedly and are characteristic of two situations: (*1*) aggression between males during which only the dominant male calls, and (*2*) on occasions when either sex is cornered in a nest box or is handled. The *squeal* call is an open-mouthed cry given by subordinate males and females during hostile encounters. The squealing of a chased animal is a continuous emission of five or more harmonics between 2 and 14 kHz.

Individuals of *T. palawanensis* have three calls associated with aggression. A *chatter* call is emitted by both sexes, but usually by the dominant male when he threatens a subordinate or when a chased tree shrew stops and threatens its pursuer. Initially, the call is of low volume with definite breaks between notes. If a low intensity call fails to thwart an intruder, it increases in volume and it is repeated as a series of piercing notes resembling *rage* calls. *Rage* calls are given by animals, with mouths open, during periods of stress. *Whistle* calls are pure notes with antiphonal qualities in the frequency range of 6.2 to 8.2 kHz. *Whistle* calls are given 1–3 min after a major disturbance in or out of the cage. These calls are interpreted as *warning* calls and sometimes they are repeated for 10–15 min (Thompson, 1969; Williams *et al.*, 1969).

Mountain tree shrews (*T. montana*) are very vocal although they have a smaller variety of calls than most species. *Warning* calls are pronounced; following an initial call by one animal, as many as five other animals may begin calling and continue to call for the length of the disturbance. During these vocalizations, the animals hold their tails upright and forward over their backs and flick them back and forth. *Warning* calls are emitted with an open mouth and strong abdominal contractions. Usually, *threat* calls are associated with sexual interactions. A proestrous female calls whenever males or females approach her. The dominant male calls whenever lower-ranked males come near estrous females. *Threat* calls are continuous, short chatters (Sorenson and Conaway, 1968). Pournelle (1954) describes the *warning* call of *T. montana* as a rolling-trill terminated by a sharp squeak. He also notes tail-flicking at the end of the call.

Each of the aforementioned species recognizes the distinct *warning*, *threat*, and *rage* calls of the other species and frequently the calls of animals in one cage elicit calls from animals in adjacent enclosures.

In his excellent study of the behavior of individuals of *T. belangeri* in captivity, Martin (1968) compares the interpretations of vocalizations of *T. belangeri* and of *T. glis* reported by Andrew (1964), Hofer (1957), Kaufmann (1965), Kuhn and Starck (1966), and Sprankel (1961a). Martin lists a *chirp* call (probably "Schnalzen" of Sprankel and chirp of Kaufmann) given by males at boundary disputes and between males where neither is dominant and little or no actual contact occurs. This call is emitted with a closed mouth and usually is accompanied by tail-flicking. *Squeak* calls ("gellende, abgehacktes Geschrei," "Keckern," "langgezogene, spitze Schreie" of Sprankel and squeal call of Kaufmann) are drawn-out, high-pitched calls denoting extreme fear or anger by defeated opponents and sometimes are accompanied by a submissive posture. Kaufmann did not distinguish the position of submission (reported by Sprankel) and states that both attacking and attacked animals squeal. Hofer states that some *squeaks* ("Keckern and abgehacktes Geschrei" of Sprankel) contain elements of threat. Explosive *snorts* ("platzendes, schnarchendes Gerausch" of Sprankel, "Schnarren" of Kuhn and Starck, and snort of Kaufmann) are vocalizations given by adults with open mouths during fights with conspecifics, and after the animals are disturbed in nest boxes. Kaufmann reports that *snorts* are emitted as alarm calls during hostile encounters, chiefly by subordinate individuals. Andrew states that the *snort* call of *T. glis* is a "sharp call" indicating threat.

Liat (1968) also notes that individuals of *T. glis* have warning and threatening calls. The latter are described as harsh, chattering noises used to frighten away intruders. Cantor (1846) reports two calls of *T. ferruginea:* "A short peculiar tremulous whistling sound . . . marks their pleasurable emotions . . . while the contrary is expressed by shrill protracted cries [p. 189]."

Visual Patterns

TAIL FLICKING

Almost all authors report the occurrence of tail flicking by tree shrews. The tail is held upright, curled slightly forward, and flicked back and forth over the back. Kaufmann (1965) states that this pattern is typical of conflict situations and is used almost equally by dominant and subordinate animals. He implies that tail flicking indicates a state of nervous tension. Sorenson (1964) also notes tail flicking among members of five species of tree shrews in conflict situations and during periods of excitement.

Autrum and von Holst (1968) report that tree shrews ruffle their tail hairs in response to social and nonsocial stimuli in their environment. Ruffling is determined by adding the single intervals of ruffling during a continuous 12-hr period and expressing these intervals as a percent of the total observation time (%SST, i.e., "Strauben der Schwanzhaare"). The authors state that noise, the number of individuals in the cage, the degree of subordination of an animal, and so forth, affect the %SST. Under 20% SST animals lose weight and the growth rate is slow; between 20% and 50% SST adult females display male copulatory behavior, females fail to mark their offspring, and the young are eaten by members of the group; between 50% and 80% SST females become infertile and males retract their testes; above 80% SST animals die within a 10-day period.

In a later report, von Holst (1969) explains that the tail hairs of tree shrews normally lie flat but when tree shrews are disturbed, the arretores pilorum muscles are activated by fibers of the sympathetic nervous system and the tail hairs are erected. The tail then has a bushy appearance. The %SST is therefore a measure of stress, e.g., ruffling occurs when animals are placed in a new environment. After the tree shrews become familiar with their surroundings, ruffling is almost exclusively caused by the presence of other animals. As much as 90% SST is seen when a subordinate tree shrew is confronted by a dominant tree shrew, even though fighting seldom occurs.

Raab (1971) attempts to separate social and psychic interactions of tree shrews via measuring the amounts of serotonin (5-HT) and 5-hydroxyindoleamine (5-HIAA) in discrete brain areas following physical and visual stress. Raab exposes male tree shrews twice daily to experienced male fighters for a 2-min period. The defeated males and a control group of males are then left in visual contact with the fighters for 12 hr a day as a means of stress. Raab states that the stressed animals lose body weight and show elevated amounts of 5-HT and 5-HIAA in the septal and frontal areas of the cortex.

PRESENTING

Sorenson (1964, 1970) and Sorenson and Conaway (1966) report presenting among males and females of *Lyonogale tana* and *Tupaia longipes* during initial periods in captivity. The animal presenting lifts its tail directly overhead and positions its anal area a few inches in front of the other animal who assumes the posture of an open-mouthed aggressor. The aggressor then either licks or ignores the perineum of the presenting tree shrew. On occasion, two animals will present and then turn to face each other and display threat postures. Presenting nearly always represents subordinance. After animals are separated into species groups, a dominance hierarchy is formed and presenting declines.

Hasler (1969a) describes the introduction of an adult male *T. chinensis* into the cage of two young males 7 months of age. The three animals became hyperactive and then each of the juveniles presented to the adult who licked their anal areas for several minutes but did not attack them. Again, presenting seems to indicate subordinance; however, Kaufmann (1965) reports that presenting of the perineum by females to males of *T. glis* is an invitation to mount and not necessarily a sign of submission (of 28 full mounts recorded, 24 were immediately preceded by presenting).

POSTURES

Sorenson (1964) and Sorenson and Conaway (1966) state that intra- and interspecific recognition of dominance postures is pronounced. Animals will respond from distances of up to 10 ft. Members of *T. longipes* display a threat posture in which the animal either sits or crouches with its neck stretched forward, its head tilted slightly upward and its mouth open, its body held rigid, and its hind legs positioned to allow a forward lunge. This posture usually is assumed in area defense by subordinate animals. Hasler (1969a) reports that prior to fighting, members of *T. chinensis* face one another in a crouch, 1 or 2 ft apart, for a period of up to 2 min. The animals remain motionless and stare at one another. Kaufmann (1965) states that individuals of *T. glis* have two distinct postures: an elongated posture and a withdrawn posture. The first posture is typical of an aggressive individual where the animal stretches its body, with its nose thrust forward and its head either level or depressed, sometimes touching the floor. The withdrawn posture is common to intimidated individuals who crouch with their heads withdrawn and depressed, touching their forepaws. Usually their mouths are closed. This posture is a defensive posture and not a ritualized submissive posture that functions to interrupt fights.

MOUTH OPEN OR SHUT

Kaufmann (1965) states that members of *T. glis* open their mouths and expose their teeth during chases and during all chittering except that of very low intensity. On some occasions the open mouth pattern is used during hostility without other patterns, indicating that it can function independently as an aggressive signal. A closed mouth posture is typical of conflict behavior and is indicative of subordination. Hasler (1969a) reports that sometimes a chased tree shrew stops, faces its pursuer with an open mouth, and emits *squeal* calls. The open-mouthed stance is a submissive posture, for it usually stops aggression. However, other authors report open-mouthed stances by tree shrews emitting *threat* calls and just prior to lunging at their opponents.

LUNGES

Lunges by tree shrews of all species function as aggressive signals. A short, quick forward lunge in which the animal moves no more than 1 ft is typical of conflict situations among individuals of *T. longipes*, *T. minor*, *T. gracilis*, *T. montana*, and *L. tana* (Sorenson, 1964; Sorenson and Conaway, 1966), *T. chinensis* (Hasler, 1969a; Sorenson, 1964), *T. belangeri* (Martin, 1968), and *T. glis* (Kaufmann, 1965; Sprankel, 1961a; Vandenbergh, 1963). A lunge often is accompanied by an open-mouthed posture and the animal being lunged at usually retreats. Kaufmann (1965) notes that snapping and air biting by members of *T. glis* also function as signs of aggressiveness and often are closely associated with lunges.

Olfactory Patterns

CHEST RUBBING AND ANAL DRAGGING

Scent marking by secretions of gular and abdominal glands is common among most species of tree shrews studied. Sprankel (1961a,b) reports the presence and use of scent glands in the skin of the chest and throat in members of *T. glis*. Vandenbergh (1963) cites chest rubbing immediately after aggressive encounters by males and females of *T. glis* and Kaufmann (1965) states that members of *T. glis* vigorously rub their chins, their throats, and especially their chests back and forth on objects in their cage. The males frequently have a sticky yellow fluid present in the midventral line from the chin to the abdomen. Rubbing occurs during hostile encounters (20 out of 30 observed instances) and is done by the dominant male. Kaufmann further states that males and females drag their rumps on branches while in a sitting position. Rump dragging occurs in conjunction with grooming, in hostile episodes, and when a male is introduced into a new pen.

Martin (1968) states that the gular and the abdominal glands of members of *T. belangeri* are used for marking nest boxes, food trays, and offspring. He notes that the secretions of the gular gland of the female are colorless and nonadhesive. Thompson (1969) reports anal dragging along limbs, logs, and nest boxes after defecation by individuals of *T. palawanensis*. He interprets such behavior as either cleaning or marking behavior. Hasler (1969a) reports that only the dominant male *T. chinensis* exhibits marking behavior. This male sometimes alternates chinning (rubbing of the throat and sternum) with aggressive chases of lower-ranked males and with mounting attempts of estrous females. Both Hasler (1969a) and Schatzman (1971) point out that secretions of the gular and abdominal glands are under the control of androgens.

URINE MARKING

Sprankel (1961a) cites two methods of urine marking by males of *T. glis:* (*1*) the male dribbles urine as he moves back and forth along branches, and (*2*) the male urinates in one spot, then stamps his feet up and down in the puddle, presumably impregnating the soles of his feet and therefore leaving scent traces wherever he walks. Sprankel proposes that territoriality is maintained by males using secretions of their interramal glands and urine trailing as scent markings. Kaufmann (1965) agrees that males of *T. glis* dribble urine on branches, but he does not report the dancelike treading of males in puddles of urine. Hasler (1969a) describes urine dribbling by all males of *T. chinensis* in his colony, but he finds no correlation between urine dribbling and other behaviors. Thompson (1969) reports that males of *T. palawanensis* urinate on limbs and logs in captivity. During urination the male is very quiet; he sits motionless with his ventral surface close to the log and does not tread in his urine. Neither Hasler nor Thompson report territoriality among their animals in captivity although there are areas in the enclosures to which specific individuals retreat and places where certain tree shrews go to groom and to rest. Urine marking of these rest areas is not reported.

Sorenson (1964, 1970) and Sorenson and Conaway (1966) report few indications of territorial behavior among tree shrews in captivity. Dominant males of *T. chinensis* seem to prefer certain rest areas and will eject one or more resting animals from these areas. Members of *T. minor*, housed with specimens of *T. montana*, use only a restricted part of their cage and defend that area against the other species. In Borneo, members of *T. montana* live in discrete groups of about 12 individuals separated from other small groups by distances of approximately ½ mile. Boundary defense is not reported.

Martin (1968) states that urine, fecal, gular, and abdominal gland secretions are used by males of *T. belangeri* to mark intercage boundaries. Agonistic encounters are associated with marking sites. Martin reports that scent marking functions in four ways: (*1*) to reassure the animal within its home range, (*2*) to deter conspecifics from entering the home range, (*3*) to mark paths, and (*4*) to deter other species.

Contact Patterns

MOUNTING, GENITAL LICKING, AND ANAL NUZZLING

These forms of contact behavior occur mostly during sexual episodes in which the female becomes hostile and attempts to repel the male. Kaufmann (1965) cites 44 sexual episodes between males and females of *T. glis* and reports that hostile acts by females interrupted 21 interactions. In response to an aggressive female, the male may lick his penis or rub his penis along a branch. Aggression by the female often closely follows anal nuzzling by the male. Kaufmann also reports hostile acts in 16 of 22 cases of homosexual activity between females.

Conaway and Sorenson (1966) and Sorenson (1964, 1970) report anal nuzzling, genital licking, and other forms of contact behavior among members of several species of tree shrews. The authors state that following of estrous females by males of *T. longipes* is intense, and that males keep their noses as close as possible to the perineum of the females. Whenever a female stops, a male either tries to mount her or pushes her

rump upward, forcing her into lordosis. Sometimes the male grooms her middorsally and bites her along her nape and shoulder. If the female fails to accept the male, he becomes aggressive and tries to force copulation. Such behavior results in male–female fighting and unless the female is fully receptive, she repels the male by use of either open-mouthed threat postures and threat calls or bites. The dominant male often chases and bites subordinate males when they try to copulate with estrous females.

Snedigar (1949) reports that among members of *Urogale*, aggressive behavior is associated with sexual behavior and the female is dominant and the male "hen-pecked."

BOXING

Only Kaufmann (1965) and Sorenson and Conaway (1968) report observing boxing behavior among tree shrews. Kaufmann states that twice during hostile episodes one *T. glis* hit at another with a forepaw. Sorenson and Conaway describe male–male and male–female boxing between members of *T. montana*. The boxing animals stand upright on their hind legs near one another and thrust their forefeet toward the head of the other animal. Several thrusts are made, although striking of the other tree shrew seldom occurs. Boxing is independent of social rank and seems to reduce overall aggression.

BITING AND GRASPING

Tree shrews grasp and bite one another during aggressive episodes of high intensity. Sorenson (1964, 1970) describes fights between males of *T. longipes* in which the animals face off at about 18 inches, crouch in a threat posture, move side to side or circle each other, and then lunge forward to bite one another. Most bites are superficial and occur as the animals roll over and over on the floor of the cage. Hasler (1969a) reports that severe biting of the rump and the base of the tail occurs when the dominant male *T. chinensis* overtakes a subordinate male during a chase. Biting by females is uncommon and involves only a few nips on the rump area.

Kaufmann (1965) describes a hostile encounter between males of *T. glis* following introduction of a strange male into a pen housing a dominant resident male. At first, both animals moved nervously about, displaying tail flicks and emitting *chirp* calls. The resident male then closely approached the intruder; they faced each other repeatedly in elongated postures with their mouths open. The resident male was relatively quiet and held his head low, he circled the intruder and lunged at his underside. The intruder held his head low and with his mouth open, he faced his attacker. The resident rubbed his chest on branches, and finally he grasped and bit the intruder. Both animals rolled over on the floor chittering and squealing. The intruder broke free and fled. This sequence was repeated several times until the resident male constantly chased the intruder and bit him at every opportunity. Most wounds were on the scrotum, hind legs, rump, and base of the tail and these areas often developed large infected sores.

SUMMMARY AND CONCLUSIONS

A few generalizations can be made concerning the aggressive behavior of tree shrews in captivity. First, sociality and aggression vary among species; some species are nearly asocial whereas others demonstrate polygamous family groups with paternal dominance. Second, dominant males vary in dominance expressions; however, most male aggression is characterized by chases and bites. Third, male–male, male–female, and female–female aggression occurs between adults of most species and paternal aggression occurs in some species. Fourth, female aggression is displayed mostly by lunges, threat calls, and threat postures. Fifth, the overall level of aggression is increased by increasing the ratio of males to females, by increasing the total number of animals, and by decreasing either the amount of space or the quality of the space. Sixth, territoriality is evident in some species and probable in others. Seventh, group organization is achieved and overall aggression is lowered by the establishment and maintenance of status hierarchies. Therefore, as Cloudsley-Thompson (1965) points out, well-integrated societies are based on sex pairing, territoriality, hierarchy, and leadership.

Man still carries his animal heritage of group territoriality, and this coupled with his cultural ability has prolonged his life to a crowded state. His capacity to make and to use tools allows him to direct his aggression toward distant foes and reduces the appeasement of injury (Tinbergen, 1968). Without appeasement, fighting among humans becomes disruptive and leads to fatalities not common to societies of lower animals. As Craig notes in 1921, ". . . fighting is not sought nor valued for its own sake . . . animals fight to remove the presence or interference of another, not to kill him [p. 267]."

I question the comparison of human and nonhuman behavior, and yet man so desperately needs examples to help him realize that he remains part of a vast gene pool—a gene pool in which aggression is given a positive role during phylogeny. Since man cannot escape aggression, he must use it as evolution intended: to ensure individual distance among social animals. The proper use of aggression requires organization, and, thus, men unite and elect leaders; leaders form groups and contrive laws; laws become the framework of governments; governments build nations; and nations guarantee men a suitable niche in the environment.

As nations expand, their borders push against those of their neighbors and individuals fence their territorial lines. International conflicts occur, national boundaries change, and the defeated masses migrate into new habitats to begin anew the cycle of national expansion. With time, however, the amount and the quality of the unused habitat depreciates and the existing space becomes saturated with people and their by-products. As man again strives to space out his kind, he finds no new lands to tread upon and thus aggression can no longer play a positive role. Man is at the crossroads—either he limits his numbers by reducing human births or he allows aggression to assume the negative role of genocide.

REFERENCES

Alexander, B. K., and Roth, E. M. (1971). The effects of acute crowding on aggressive behavior of Japanese monkeys. *Behaviour* **39**, 73–90.

Andrew, R. J. (1964). Displays of the primates. *In* "Evolutionary and Genetic Biology of Primates" (J. Buettner-Janusch, ed.), Vol. 2, pp. 227–309. Academic Press, New York.

Autrum, H., and von Holst, D. (1968). Sozialer "Stress" bei Tupajas (*Tupaia glis*) und seine Wirkung auf Wachstum, Korpergewicht und Fortpflanzung. *Z. vergl. Physiol.* **58**, 347–355.

Barnett, S. A. (1967). On the hazards of analogies between human aggression and aggression in other animals (A review of "On aggression" by Konrad Lorenz). *Sci. Amer.* **216**, 135–138.

Berkowitz, L. (1962). "Aggression: A Social Psychological Analysis." McGraw-Hill, New York.

Cantor, T. (1846). Catalogue of mammalia inhabiting the Malayan Peninsula and Islands. *J. Asiat. Soc. Bengal* **15**, 171–279.

Christian, J. J. (1970). Social subordination, population density, and mammalian evolution. *Science* **168**, 84–90.

Cloudsley-Thompson, J. L. (1965). "Animal Conflict and Adaptation." Dufour, Philadelphia, Pennsylvania.

Collias, N. E. (1953). Social behaviour in animals. *Ecology* **34**, 810–811.

Conaway, C. H., and Sorenson, M. W. (1966). Reproduction in tree shrews. *In* "Comparative Biology of Reproduction in Mammals" (I. W. Rowlands, ed.), *Symp. Zool. Soc. London No. 15* pp. 471–492. Academic Press, New York.

Craig, W. (1921). Why do animals fight? *Int. J. Ethics* **31**, 264–278.

Dollard, J., Doob, L., Miller, N., Mowrer, O., and Sears, R. (1939). "Frustration and Aggression." Yale Univ. Press, New Haven, Connecticut.

Draper, W. A. (1963). Laboratory maintenance of the tree shrew (*Tupaia glis* Diard 1820). *Lab. Primate Newsl.* **2**, 1–2.

Elliot, O., Wong, M., and Shearman, C. E. (1969). Serum cholesterol of Malayan tree shrews. *Primates* **10**, 97–100.

Freud, S. (1959). Why war? Letter to Professor Einstein. *In* "Collected papers of Sigmund Freud" (J. Strachey, ed.), Vol. 5. Basic Books, New York.
Gartlan, J. S. (1968). Structure and function in primate society. *Folia Primatol.* **8,** 89–120.
George, J. (1966). Why do animals fight? *Audubon* **68,** 18–20.
Hamburg, D. A. (1971). Psychobiological studies of aggressive behaviour. *Nature (London)* **230,** 19–23.
Hasler, J. F. (1969a). Behavior of *Tupaia chinensis* in captivity. M.A. Thesis, Univ. of Missouri, Columbia, Missouri.
Hasler, J. F. (1969b). Agnostic behavior in captive tree shrews. *J. Colo.-Wyo. Acad. Sci.* **6,** 58.
Hendrickson, J. R. (1954). Breeding of the tree shrew. *Nature (London)* **174,** 794–795.
Hofer, H. (1957). Uber das Spitzhornchen. *Natur Volk* **87,** 145–155.
Kaufmann, J. H. (1965). Studies on the behavior of captive tree shrews (*Tupaia glis*). *Folia Primatol.* **3,** 50–74.
Kuhn, H. J., and Starck, D. (1966). Die Tupaia-Zucht des Dr. Senckenbenbergischen Anatomischen Institutes. *Natur Mus.* **96,** 263–271.
Liat, L. B. (1968). Distribution of the primates of west Malaysia. *In* "Recent Advances in Primatology," *Proc. Int. Congr. Primat.* Vol. II, pp. 121–130. Karger, Basel, Switzerland.
Lorenz, K. (1966). "On Aggression." Harcourt, New York.
Lyon, M. W. (1913). Treeshrews: An account of the mammalian family Tupaiidae. *Proc. U. S. Nat. Mus.* **45,** 1–188.
Martin, R. D. (1968). Reproduction and ontogeny in tree-shrews (*Tupaia belangeri*), with reference to their general behavior and taxonomic relationships. *Z. Tierpsychol.* **25,** 409–529.
Matthews, L. H. (1964). Overt fighting in mammals. *In* "The Natural History of Aggression" (J. D. Carthy and F. J. Ebling, eds.), pp. 23–32. Academic Press, New York.
Morris, J. H., Negus, N. C., and Spertzel, R. O. (1967). Colonization of the tree shrew (*Tupaia glis*). *Lab. Anim. Care* **17,** 514–520.
Myers, K. (1966). The effects of density on sociality and health in mammals. *Proc. Ecol. Soc. Aust.* **1,** 40–64.
Polyak, S. (1957). "The Vertebrate Visual System." Univ. of Chicago Press, Chicago, Illinois.
Pournelle, G. (1954). Taming the tree shrew. *Zoonooz* **27,** 3.
Raab, A. (1971). Der serotoninstoffwechsel in einzelnen hirnteilen vom *Tupaia* (*Tupaia belangeri*) bei soziopsychischem Stress. *Z. vergl. Physiol.* **72,** 54–66.
Schatzman, J. A. (1971). Effects of testosterone propionate in tree shrews. M.A. Thesis, Univ. of Missouri, Columbia, Missouri.
Scott, J. P. (1958). "Aggression." Univ. of Chicago Press, Chicago, Illinois.
Snedigar, R. (1949). Breeding of the Philippine tree shrew, *Urogale everetti* Thomas. *J. Mammal.* **30,** 194–195.
Sorenson, M. W. (1964). The behavior of tree shrews in captivity. Ph.D. Thesis, Univ. of Missouri, Columbia, Missouri.
Sorenson, M. W. (1970). Behavior of tree shrews. *In* "Primate Behavior: Developments in Field and Laboratory Research" (L. A. Rosenblum, ed.), Vol. I, pp. 141–194. Academic Press, New York.

Sorenson, M. W., and Conaway, C. H. (1964). Observations of tree shrews in captivity. *Sabah Soc. J.* **2**, 77–91.

Sorenson, M. W., and Conaway, C. H. (1966). Observations on the social behavior of tree shrews in captivity. *Folia Primatol.* **4**, 124–145.

Sorenson, M. W., and Conaway, C. H. (1968). The social and reproductive behavior of *Tupaia montana* in captivity. *J. Mammal.* **49**, 502–512.

Southwick, C. H. (1955). The population dynamics of confined house mice supplied with unlimited food. *Ecology* **36**, 212–225.

Sprankel, H. (1959). Fortpflanzung von *Tupaia glis* Diard 1820 (Tupaiidae, Prosimiae) in Gefangenschaft. *Naturwissenschaften* **46**, 338.

Sprankel, H. (1961a). Uber Verhaltensweisen und Zucht von *Tupaia glis* Diard 1820 in Gefangenschaft. *Z. Wiss. Zool. Abt. A* **165**, 186–220.

Sprankel, H. (1961b). Histologie und biologische Bedeutung eines jugulo-sternalen Duftdrusenfeldes bei *Tupaia glis* Diard 1820. *Verh. Deutsch. Zool. Ges. Saarbrucken* 198–206.

Stokes, A. W., and Cox, L. M. (1970). Aggressive man and aggressive beast. *Bioscience* **20**, 1092–1095.

Temminck, C. J. (1827). "Monographies de Mammalogie," Vol. I (Quoted by Lyon, 1913, *Proc. U. S. Nat. Mus.* **45**, 1–188).

Thompson, P. J. (1969). Behavior of tree shrews (*Tupaia palawanensis*) in captivity. M.A. Thesis, Univ. of Missouri, Columbia, Missouri.

Tinbergen, N. (1968). On war and peace in animals and man: An ethologist's approach to the biology of aggression. *Science* **160**, 1411–1418.

Vandenbergh, J. G. (1963). Feeding, activity and social behavior of the tree shrew, *Tupaia glis*, in a large outdoor enclosure. *Folia Primatol.* **1**, 199–207.

van Kreveld, D. (1970). A selective review of dominance-subordination relations in animals. *Genet. Psychol. Monogr.* **81**, 143–173.

Vernon, W. M. (1969). Animal aggression: Review of research. *Genet. Psychol. Monogr.* **80**, 3–28.

von Holst, D. (1969). Sozialer Stress bei Tupajas (*Tupaia belangeri*). Die Aktivierung des sympathischen Nervensystems und ihre Beziehung zu hormonal ausgelosten ethologischen und physiologischen Veranderungen. *Z. vergl. Physiol.* **63**, 1–58.

Wharton, C. H. (1950). Notes on the Philippine tree-shrew *Urogale everetti* Thomas. *J. Mammal.* **31**, 352–354.

Williams, H. W., Sorenson, M. W., and Thompson, P. J. (1969). Antiphonal calling of the tree shrew *Tupaia palawanensis*. *Folia Primatol.* **11**, 200–205.

AGGRESSION AND TERRITORIALITY IN NOCTURNAL PROSIMIANS

PIERRE CHARLES-DOMINIQUE

Museum National d'Histoire Naturelle

INTRODUCTION

Aggression, viewed in the context of social behavior and ecological factors, contributes to population equilibria among all the vertebrates. It manifests itself among gregarious primates both between neighboring groups and, more importantly, within the group, each individual defending a hierarchical position. Thus "socialization," as much as encephalization or the evolution of the hand, lies among the principal characteristics of the order. Among nocturnal prosimians, the most primitive primates in their behavior as well as in their morphology, there is no gregariousness; each individual leads a solitary existence.

The "social" higher primates have often been contrasted with the solitary lower primates. While the existence of complex social relations is clearly evident among the former, it is less so among the latter, which despite their solitary existences, do communicate with each other. Such communication is generally accomplished by vocalizations and scent-marking when the animals are out of visual contact. It is therefore better to speak of *gregarious* and *solitary* primates: All are social, but their means of communication are different.

Among solitary primates, aggression, like social relationships, is not manifested in the same way as it is in gregarious forms. It is for this reason that nocturnal prosimians warrant particular attention. The biology of these animals is almost unknown; nonetheless, what is known of those species studied reveals a certain unity in their social structures.

Galago demidovii, Perodicticus potto (a galagine and a lorisine, representing both major groups of African lorisids) and *Lepilemur mustelinus* (a Malagasy lemur belonging to the Lemuridae) provide examples showing how social and aggressive relations in each species can adapt to ecological circumstances. The major characteristics of the social organization of all three are similar, and are outlined below.

Adults occupy fixed territories, the arrangement of which somehow

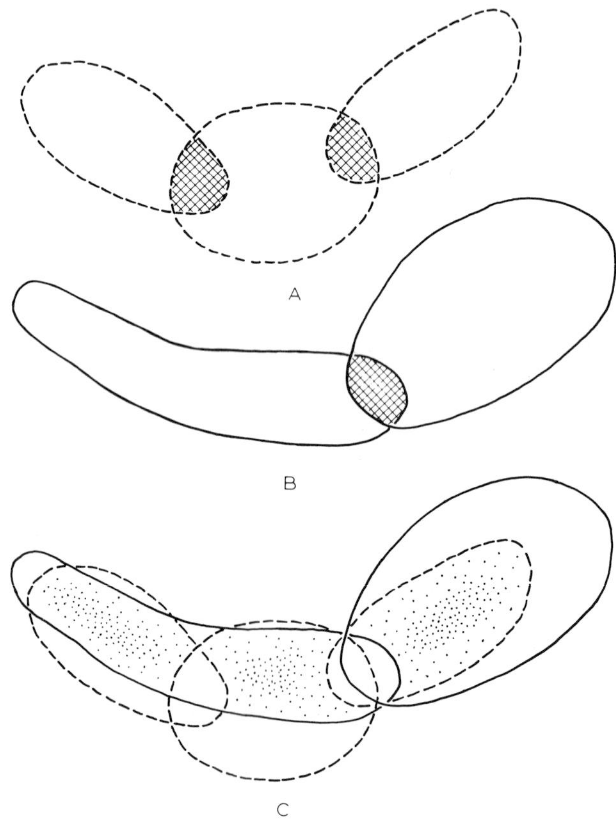

Fig. 1 Diagrammatic representation of the territories of *Galago demidovii* and *Perodicticus potto.* A: Female territorial boundaries (dotted line), showing zones of overlap. B: Male territorial boundaries (continuous line), also showing zones of overlap. C: Male and female territorial boundaries. Hatched areas: Zones of overlap where aggressive territorial contacts take place between males and between females. Dotted areas: Zones of social contact between males and females.

represents interindividual relationships (Figs. 1 and 3). Each female lives in a territory bordering or slightly overlapping those of other females. The same holds between males, but male territories overlap with those of females to an extent varying more or less with the physical strength of the animal concerned. The largest males occupy territories overlapping with those of one or several females with which they hold permanent social relationships, independent of breeding seasons. Small males do not permanently inhabit areas overlapping with those of females; sometimes excluded, at other times tolerated by the large males, they often wander at random. The young stay with the mother until well after weaning. At puberty, which occurs rather late, males leave the maternal territory and roam around, visiting other females; this is the time at which the young male is most likely to become involved in confrontations with already established large males. Pubescent young females are much more sedentary and live near the maternal territory. According to both species and individual preference, they may retain social contact with the mother over a long period of time, to the point, among galagos, of forming "matriarchies" (Charles-Dominique, 1972).

In such a social system, the defense of territories assumes a major importance, and it is at their territorial boundaries that individuals most often behave in an aggressive manner. In accord with their solitary existences, such behavior most often passes unnoticed. Indeed, an active system of defense involving the direct repulsion of intruders seems difficult for nocturnal animals occupying vast territories in a forest, where visibility is greatly reduced. However, each individual knows his immediate neighbors, with whom he came into contact at the time of adoption of his territory. An equilibrium is rapidly established in which individuals visit their boundaries, where they make their presence known while holding their neighbors at bay. When one of them disappears, an immediate readjustment of territories takes place (such events are rare, occurring about once a year). The first stage of territorial defense thus consists of an individual's announcing his presence and identity, in a quasi-permanent way, to known neighbors who stay where they are as long as he makes himself known. This signaling by various behaviors (marking, vocalization, etc.) is very frequent in contact zones, but may also be elicited by conditions of high emotional stress, for instance, the presence of a predator, an unusual event, or any conflict situation. These different communication behaviors correspond to a first degree of aggression directed toward neighbors. We shall see in the following three examples how they can be adapted to the particular ecological conditions of each species.

ECOLOGICAL ADAPTATIONS OF AGGRESSIVE BEHAVIOR

Galago demidovii

This small tropical and equatorial African species lives in thickets entwined with lianas. The long tarsus of galagos permits them to run and jump rapidly, to escape easily from predators, and to exploit large areas. *Galago demidovii* feeds primarily off insects (70%), the remainder of its diet comprising fruits and gum. Since its prey is dispersed through the forest, this galago has to range widely to fulfill its nutritional requirements. In consequence, its territories are large for such a small species (60 gm body weight); the average area is 1.8 ha for the male (maximum 2.7 ha 500 m in length) and 0.8 ha for the female.

The galago employs two methods of informing other individuals of his identity and presence in his territory: vocalizations and urine marking.

VOCALIZATIONS

Galago demidovii has a large repertoire of vocalizations; one of them, the most powerful, is used at moments of great excitement. Facing a predator, following a fight with another individual or even spontaneously in certain parts of his range, the galago repeats his vocalization, sometimes for almost an hour, at a frequency of 1 or 2 a second. In captivity, it is easy to recognize each individual by his vocalizations; moreover, laboratory experiments presently in progress show that galagos adopt different behaviors when listening to others, whether rivals or not. In the forest, they can be heard by a human observer from between 50 and 100 m away; they are probably audible to their own species even further away.

URINE MARKING

A well-known facet of galago behavior is the dropping of urine onto the sole of a hind foot and wetting the digits of the hand of the same side by rubbing them together. The extremities of the opposite side are then moistened in the same way. This urine washing of the extremities, which enables the animal to scent-mark the path of his travel, has been interpreted as a means of facilitating the return home (Eibl-Eibesfeldt, 1953; Sauer and Sauer, 1963). But field observations have shown that galagos, like the other lorisids, seldom use the same route twice except to places to which there is but a single passage, and when they are perfectly familiar with their territory, in which case each zone is visited every 1–3 days. Urine marking, as among most mammals, must be related to territorial definition, and the washing of the feet seems to be an arboreal adaptation related to the diffusion of the scent. When one realizes the multitude of possible itineraries through the dense thickets of the equatorial forest, it is evident that another individual will more easily recognize a trail than a spot. In captivity, *Galago demidovii* normally marks from four to eight times an hour, but the introduction of a novel object, the passage of another galago, or the presence of unfamiliar markings considerably increases this frequency. Moreover, urine is more frequently deposited on the dividers separating the cages.

The identifying vocalization of a galago is the analog of that of a bird occupying a territory; however, whereas the bird has visual landmarks allowing him to delimit his territory with great precision, the galago has to define his by olfactory signs (urine). In order to inform each other of their presence, two galagos must have partially overlapping territories (1/5 to 1/10 among *Galago demidovii*); thus they have not so much boundary lines as boundary zones. The behavior of galagos in zones of overlap is not well known; trapping seems to show that they avoid each other.

Perodicticus potto

This lorisine, weighing about 1 kg, eats fruits, gum, and insects, and covers large territories (average: 12 ha among males, 7.5 ha among females). The potto possesses major adaptations, both morphological and behavioral, for unobtrusive movement. Its tail is reduced and it never jumps; rather, it moves slowly, in perfectly smooth movements, without disturbing the vegetation. In contrast to the galagines, which use jumping and rapid locomotion in escaping from predators, the potto becomes still at the slightest danger (Charles-Dominique, 1971). These ecological adaptations have profound consequences; the potto (like *Arctocebus*, the other Africa lorisine) does not have an identifying vocalization like that of the galagines.[1] It is easy to understand that such a cry is incompatible with a system of locomotion based on unobtrusiveness; thus only scent-marking is used by neighboring pottos for communication.

In contrast to *Galago demidovii*, which wets its extremities for marking, *Perodicticus potto* deposits urine directly, lowering its hindquarters until its penis or clitoris is just touching the support (vibrissae increase the effective length of the penis and clitoris). Without interrupting its progress, it thus drops onto the support a trail of urine 1 or 2 m in length. In

[1] I know of only three vocalizations in the repertoire of the potto: a "tic" rich in overtones, which allows the mother and the young to find each other in the morning (a vocalization used also by the male in moments of courtship); a cry of distress related to pain; and a cry of anger related to fighting.

captivity, the potto marks most often on branches of large diameter, less frequently on narrower twigs. This behavior, interpreted by Seitz (1969) as a means of marking a route to facilitate the return journey, seems to me instead to be related to territorial marking.[2] The same author, observing that urine is deposited most often along the dividing partitions of cages, would equally accept the latter interpretation.

This system of depositing urine seems to be well adapted to the life of the potto, which, to cross from tree to tree, has to use alternately the large branches and their forks, then the twigs carrying foliage (Charles-Dominique, 1971). While there are many alternative routes from place to place along fine branches, the choice of routes along large branches is more restricted. In leaving his scent on big branches, the potto therefore affords his neighbors many more opportunities to detect him while moving around. Moreover, while most of the urine deposited in a single place will fall to the ground, laid down in the form of a trail, the urine is much more effectively used.

The potto is slow and never jumps, thus providing a distinct advantage in field studies. Between 1967 and 1972 I was able to follow quite closely a small population of pottos living around the field station in Gabon that had been set up by the Centre Nationale Recherches Scientifiques.

In the vicinity of the laboratory at Makokou the secondary forest is broken by trails and clearings which frequently oblige the pottos to follow complicated routes to get from one island of vegetation to another. In order to facilitate observation, at the beginning of the study I isolated certain wooded zones by cutting down a number of trees, and then connecting these zones by long lianas which acted as "bridges." It then became easy to monitor the passage of pottos by the use of electrical contacts connected to pen recorders and to warning devices; alerted to the passage of a potto by an electric signal, it was possible, if one hurried, to arrive in time to identify the individual animal, distinguishable 10 m away by a system of marks.

Year by year, this small population underwent modification (in individuals as well as in territories), and in 1969 the situation became particularly favorable for studying the relationship between two neighboring males (Fig. 2). Their territories overlapped precisely in the area of the laboratory, and two liana "bridges," with their electric contacts, apprised me of any movements along them. Convinced that I would at last witness interactions between two neighboring males, I set up blinds and reversed my activity rhythm to be ready at any hour of the night. After 2 months of waiting, the two males had still not encountered each other: Should

[2] As in *Galago demidovii,* the potto only rarely takes the same route twice; experiments carried out in the forests of Gabon (Charles-Dominique, 1971) show that in the case of complex itineraries, only memory allows this species to choose the optimum route. Moreover, there is no definite "home," any thicket sufficing.

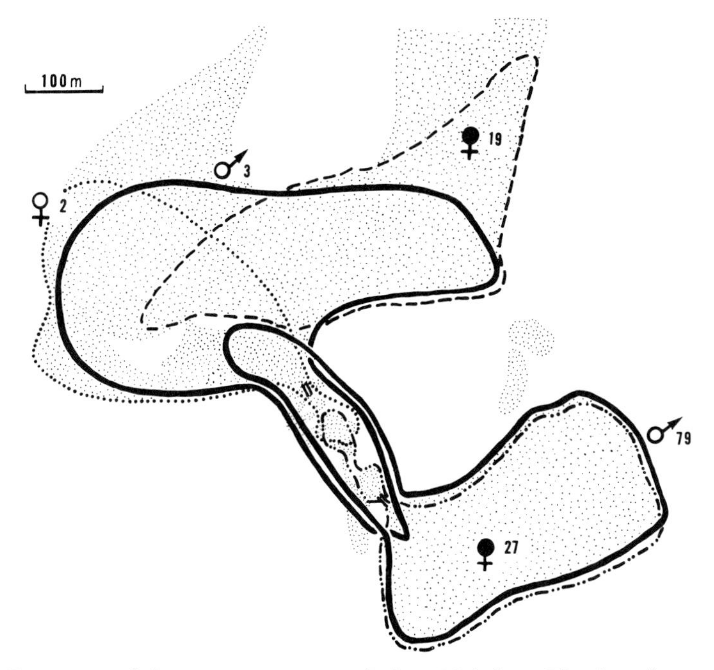

Fig. 2 Territories of the various pottos studied at Makokou. The dotted area represents the forest, while the two symbols = denote the liana bridges on which were placed the electric contacts. ♂ 79 and ♀ 27 form one pair, ♂ 3 and ♀ 2 the other. Note the communal area where the males and females of the two different pairs avoid each other.

one arrive while the other was in occupation of the border zone, he would retreat about 50 m away. The two neighboring females, which had a similar territorial arrangement, likewise avoided each other in the frontier zone, which nonetheless they visited at intervals of from 1 to 2 days. This avoidance behavior also exists between neighboring males and females; however, in contrast, members of the same couple often join up in the border zone. The further an animal penetrates into the border zone, the less he is in his exclusive territory and the more he is in that of his neighbors; the territorial limit therefore assumes the nature of a gradient.

Lepilemur mustelinus

This nocturnal Malagasy lemur is almost exclusively folivorous; in the south of the island where we have observed it between September and October (Charles-Dominique and Hladik, 1971), it ate only leaves and flowers. This dietary specialization has given rise to extensive physiological and behavioral adaptations: Of the 11 hr of its nocturnal activity, only 75 min are devoted to feeding, the animal merely browsing from time to time on the foliage or flowers available almost everywhere in its territory.[3] A very small portion of the forest suffices for each individual, and this species can attain an enormous population density—300 animals per square kilometer. *Lepilemur mustelinus leucopus,* a little smaller than the potto (600 gm) occupies a territory 40 times as small as that of the latter: 0.18 to 0.3 ha, about 50 m across, as opposed to between 7 and 12 ha 300 to 500 m in diameter. In so restricted a territory, *Lepilemur* devotes nearly all its time to surveying its boundaries, which are strictly defined (Fig. 3). Neighboring animals position themselves at their territorial boundaries, 2 or 3 m apart, and sometimes thus watch each other for close to an hour; from time to time one of them will utter a cry which immediately provokes his partner into answering. These vocal duets happen most often, moreover, when visibility is best; bright moonlight increases their frequency by a factor of 3.

Visual signals are equally important; the two animals facing each other sometimes begin to turn their heads rapidly from side to side, as if shaking them in negation, a display found in all aggressive situations (threat, presence of an enemy, etc.). During these confrontations, the antagonists also adopt certain postures which mimic the first phase of jumping; sometimes they bound vigorously and noisily onto a nearby branch, a form of behavior which recalls slightly the tree-shaking of monkeys. When one of them moves close to the borders of his territory, the other may follow him, observing him all the while. Such confrontations happen almost exclusively between males; they hardly ever occur between females, and never between a female and a male, whose territories overlap as among other nocturnal prosimians (Fig. 3).

Lepilemur may deposit urine on the branches, but it seems that they most frequently rely on vocal and visual signals in defining and guarding their territories. This is made possible only because of the tiny size of their territories which are easy to survey directly, and because of the amount of time available to individuals (12.5% of activity devoted to feeding, 87.5% to the policing of territories). These conditions, exceptional among primates, are due exclusively to the highly specialized diet of this species.

AGGRESSION IN A NATURAL ENVIRONMENT AND CAPTIVITY

Galago demidovii, Perodicticus potto, and *Lepilemur mustelinus,* in spite of certain differences in their means of communication, show a

[3] This study, carried out in collaboration with C. M. Hladik, showed that 70% of the food intake of *Lepilemur* consisted of leaves and flowers of *Alluaudia ascendens* and *Alluaudia procera.* These two species constituted 65% of the local vegetation, the remaining 35% consisting of the foliage of bushes and lianas.

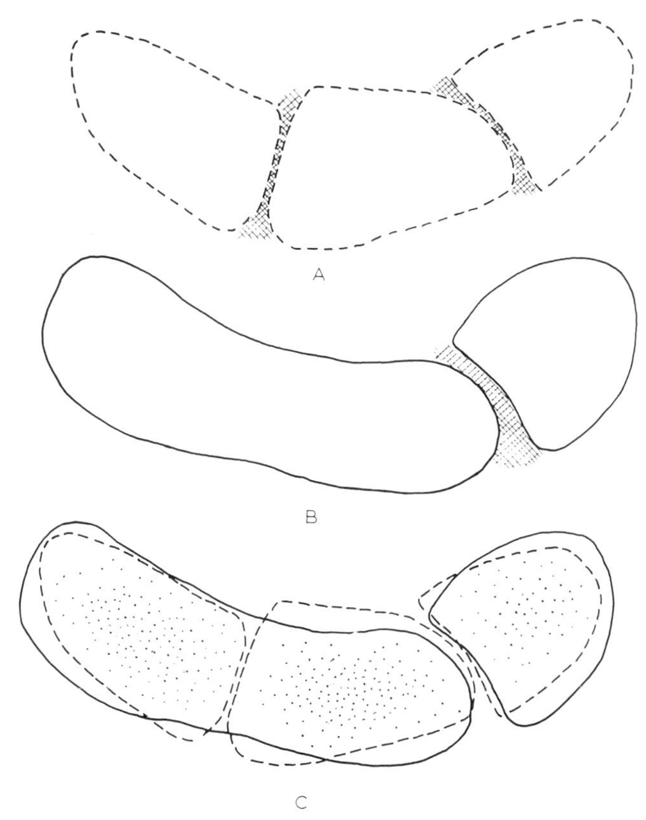

Fig. 3 Diagrammatic representation of the territories of *Lepilemur mustelinus*. A: Female territorial boundaries (dotted line). B: Male territorial boundaries (solid line). C: Male and female territorial boundaries. Hatched area: Zone of aggressive territorial contacts. Dotted area: Zone of male–female social contacts. Note that male and female territories do not overlap; this situation, unique among nocturnal prosimians, is due to the peculiar requirements of territories which the animals can survey directly (visual and auditory signals).

unity in their social structure: The territories, similarly arranged, are defined, surveyed, and defended. *Perodicticus potto* and *Lepilemur mustelinus* are extreme cases due to ecological specializations; *Galago demidovii* seems most generalized in these respects. *Euoticus elegantulus, Galago alleni* in Gabon, *Galago senegalensis, Galago crassicaudatus* in South Africa (Sauer and Sauer, 1963; Bearder, 1974), *Microcebus murinus* in Madagascar (Martin, 1972), also communicate by means of auditory and olfactory (urine-marking) signals; their social organization resembles that described for *Galago demidovii.*

In the vertebrates, the concept of territory was first defined among birds, then extended to mammals, which explains many current controversies. *A territory is that area occupied by an animal or group of animals, and which is defended against intrusion by others of the same species.* When, for a given species, it is not certain that an area is defended, the term "home range," whose definition is less restricted, is preferable. This definition of territory is rather theoretical, for in the field fights are rarely witnessed; however, the nocturnal prosimians can bear the marks of combat: teeth-torn ears; mangled or stumpy tails; broken fingers set crookedly; various scars; gouged eyes, etc. These wounds can only be attributed to intraspecies combat between adults, since not one of 152 pubescent individuals of several species examined (30 *Arctocebus calabarensis,* 15 *Perodicticus potto,* 20 *Euoticus elegantulus,* 22 *Galago alleni,* 60 *Galago demidovii,* 5 *Lepilemur mustelinus*) bore scars; had they done so, wounds in all the animals might have been attributed to simple "accidents." Traces of serious wounds, from two to four times as numerous among males as among females, are nevertheless rare among adults: 21% in *Arctocebus calabarensis* ($n = 32$), 23% in *Euoticus elegantulus* ($n = 34$), 20% in *Galago alleni* ($n = 20$), 13% in *Galago demidovii* ($n = 60$), 30% in *Lepilemur mustelinus* ($n = 13$). (These figures, however, represent only minima; certain traumas could have escaped notice.) In a secondary forest population of *Galago demidovii* followed for 4 years, out of 21 animals later rechecked (11 after 1 year, 3 after 2 years, 2 after 3 years, 5 after 4 years), only one showed evidence of fighting after he was first observed. All this shows that, taking into account the life span of nocturnal prosimians, fighting is rare in the natural environment.

On the other hand, observation in captivity would suggest that fighting is frequent: When two males (or two females) are kept together in a small cage, they are easily provoked into fighting. Rapidly, one of the antagonists becomes the aggressor, and the other the pursued; the latter, lacking space, cannot escape from the aggressor, which can then wound him gravely, even fatally, if they are not separated in time. In the forest, the same animals would never reach this point, since the weaker has the opportunity for flight. In the case of confrontation, there is normally an aggressor (the individual defending his territory) and an attacked (the trespasser) which, in certain cases, can become the aggressor; only then is there a fight. The dentition of nocturnal prosimians is well developed, in particular the canine: 4.5 mm in *Euoticus elegantulus,* which weighs 300 gm; 5.0 mm in *Arctocebus calabarensis,* which only weighs 200 gm; 8.0 mm in *Perodicticus potto,* which weighs 1 kg, etc. Anyone who has handled these animals knows how serious a bite they can inflict, and it is conceivable that too much fighting might be detrimental to the species. Happily, behavioral mechanisms permit territorial defense without the necessity of fighting. These mechanisms essentially involve flight and avoidance.

Avoidance behavior consists of fleeing from a neighbor's territory without having to face him: The animal rapidly recognizes the scent marks of the occupant, and leaves. This avoidance behavior is especially clear when animals are released in the forest, distant from the place of capture. If released in an area already occupied by an animal of the same sex, a new individual leaves rapidly, and establishes himself in an unoccupied zone. In the opposite case, he will establish himself on the spot, as long as the local vegetation fulfills his ecological requirements. Such experiments were carried out 15 times on *Perodicticus potto,* 7 on *Euoticus elegantulus,* 6 on *Galago alleni,* 3 on *Galago demidovii,* and once on *Lepilemur mustelinus.* In 13 instances of introduction into areas already occupied, there were 11 avoidances, 2 fights, and once the intruder was chased away.

When an animal is released in an unknown area already occupied by several individuals, he may wander around for several days before finding an unoccupied zone. While searching, he will enter occupied territories, if only because he is driven by hunger (all lose weight during this period), and encounter the occupants. The usual reaction in such circumstance is flight, always downwards. The animal bolts straight down, sometimes reaching the ground. This reaction of downward flight is observed in all instances of confrontation without fighting; sometimes the weaker animal merely descends or passes beneath a branch to placate the aggressor (Fig. 4).

In Madagascar we introduced an adult male *Lepilemur* into the territory of another adult male; the stranger jumped to the base of a trunk, gained 1 m in height, then froze. Wanting to witness an encounter between this animal and the territory's occupant, we tried in vain to push him upward; despite our efforts, he insisted on staying close to the ground, under the eyes of the occupant, who watched him without interfering. Later, the animal we had introduced fled, and we suppose he got lost, for he never returned to his territory, which was situated only 25 m away.

In some cases the occupant of the territory is aggressive in spite of the apparent submission of the intruder: In Gabon, where I had introduced *Euoticus elegantulus* into the forest surrounding the station, we saw a female chasing another, newly arrived, away. The female intruder plunged to the ground, followed by the occupant. The two went through our house, and chased off into the forest! In these extreme cases of aggression, the antagonist no longer pays attention to the environment, and loses all notion of danger, to the extent that the observer can approach closely

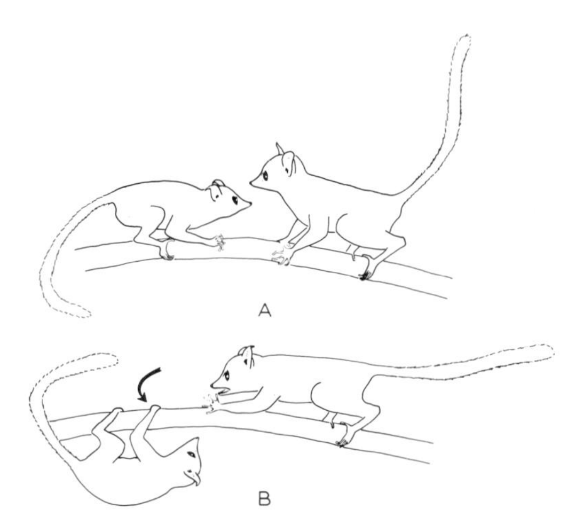

Fig. 4 Meeting of two *Galago demidovii*. That on the right is dominant and the other subordinate. A: First stage of submission (ears folded, lowered posture). B: A stronger degree of submission, in which the threatened animal protects itself beneath a branch.

without alarming the animals. They are then much more vulnerable to predators, which might occasionally take advantage of the situation (it is easy, by imitating the distress call of a galago or a rodent, to attract small carnivores). One thus realizes the number of mechanisms which operate to eliminate undesirable fighting, not only to avoid the risk of wounds, but also to cut down on the opportunities offered to predators.

Wishing to study the social and aggressive relationships of *Galago demidovii* in captivity, I recreated territories as observed in nature by providing cages with compartments intercommunicating by trapdoors. Each territory comprised two compartments which could be isolated; this permitted the establishment of communication with or without physical encounters. First (Fig. 5A), I allowed a male, "♂ a," the chance to enter an empty compartment never occupied or marked by another: "♂ a" went through rapidly, explored and urine marked it. Then, (Fig. 5B), I allowed males to enter an empty compartment which had already been occupied by another male "♂ b," which was meanwhile confined in the next cubicle. The "♂ a" generally hesitated a great deal before entering, put his head through the door, and then backed away; even when he entered, he nevertheless remained nervous and immediately returned to his own cubicle. Generally, the situation developed abnormally if it was allowed to continue too long, and "♂ a" stayed longer and longer in the territory of his neighbor; animals in captivity lack space, and more readily enter the empty territory of a neighbor than they would under natural conditions. If then the door separating "♂ a" from "♂ b" was withdrawn (Fig. 5C), "♂ a" fled back to his territory, frequently pursued by "♂ b." Sometimes, because of the lack of space, "♂ a" jumped up and down in all directions, knocking against the partitions of the cage, which scared his antagonist to such an extent that both males fled from each other, bouncing brutally against the sides of the cage, and risking knocking themselves unconscious. In other cases, the intruder was rapidly cornered by the occupant of the territory, who would grip and bite him.

When the same experiments were carried out with adult females (Fig. 5A, D, E), it was observed, first, that these animals hesitated less than males to trespass upon empty neighboring territories, and that, in cases of encounter, the reactions were more varied: According to the individual, they would (*1*) lick themselves for a long time, (*2*) lick themselves briefly, threaten each other, then separate (flight), (*3*) fight immediately. To understand these variable reactions, one must realize that it is of course possible that special social relationships may exist between cer-

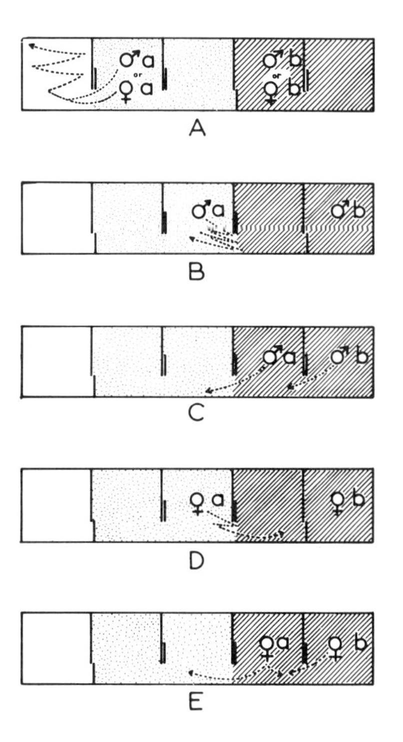

Fig. 5 Reconstruction of a series of experiments carried out in captivity on *Galago demidovii*. Only the compartments used in the experiments are shown here; the entire setup comprised 12 intercommunicating compartments on three levels. Hatched areas: Compartments usually occupied by ♂ b and ♀ b. Dotted areas: Compartments normally occupied by ♂ a and ♀ a. Clear areas: Normally unoccupied compartment.

tain adult females which share the same territory. The aggressive reactions of the females are therefore more subtle than those of the males; this agrees with the fact that, among all species studied, males bear two to four times as many scars due to fighting than do females.

During aggressive contacts, the hands are always used. Generally a slap, accompanied by a threatening call suffices to make one flee. This behavior, which exists among the galagines, the cheirogaleines, and

Fig. 6 Defensive posture of *Lepilemur mustelinus.* The hand ready to administer a slap, by comparison with the gesture of a boxer, has given the species the name "sportive lemur." The galagines and cheirogaleines also adopt this posture, but less spectacularly.

Lepilemur, is responsible for the latter's common name of "sportive lemur" (Fig. 6). During fighting itself, the hands are used extensively, the aggressor trying to grip his antagonist, which in his turn uses his hands to ward him off.

Among most nocturnal prosimians, the very large ears are precise direction finders. During fighting the animals fold them completely to protect them; the eyelids are likewise narrowed, reducing the vulnerable surface of the eye. The loser, trying to escape from the aggressor, makes characteristic calls of distress which presumably exert an inhibitory influence on the winner, who lets go quickly.

Young animals compete neither with themselves nor with adults. In *Galago demidovii,* after weaning, young females follow their mothers alone, while young males not only follow their mothers, but also the dominant male, whom they may accompany into the territories of his females. Essentially, aggressive interactions only exist between adult males and adult females. When a female occasionally attacks a male, it is invariably because he is courting her, and she is not ready to accept his advances. She simply rebuffs him, without attempting to expel him from her territory, her aggressiveness ceasing as soon as he retires several meters away.

It seems that there are certain differences between the aggression of males and females; whereas the female defends only the area she inhabits, the male defends not only his own territory, but also those of his female or females.

TERRITORIAL DEFENSE BY THE MALE

Field observations made on different nocturnal prosimian species show that, in the absence of females, males tolerate each other's presence much more. Their aggressiveness manifests itself above all when they are defending the territories of females. This aggressiveness does not exist during the period of courting, but only from the time when a female accepts a male.

Galago demidovii

Among this species, the large males drive the smaller males to the periphery of the females' territories (the peripheral males weigh an average 60 gm, as opposed to the average 70 gm of central males). In the population studied in Gabon, three small males formed a small group sharing the same territory; they were not aggressive among themselves, since they were often found together in the same traps. The following year, one of them had established himself on the territory of a couple of females, and was no longer associated with other males.

Lepilemur mustelinus

After the experimental removal of an adult male who occupied the territories of several females, his immediate neighbors moved into his old territory. Two of them, who were courting one female each, would pass within a few meters of each other without exhibiting aggression. As soon as they returned to their old limits, however, they would confront each other again and exchange vocalizations.

Perodicticus potto

After we introduced two females to the experimental area at Makokou, two males visited them every day to court them. These two males would "ignore" each other even when they had approached to within 10 m. One of them died accidentally on touching a live electric wire, but the other was accepted by his female after 3 months of courting. I then released another male, who was immediately attacked by the occupant.

This competition in defense of the territories occupied by females excludes the weaker males. We have seen in *Galago demidovii* that the smaller animals were driven to the population periphery. In *Perodicticus potto,* established males weigh 1.15 kg on average, as opposed to the 1.0 kg of wandering adult males. *Microcebus murinus* of Madagascar (Martin, 1972) shows the same phenomenon: A smaller number of large males live on the territory of females and the others, smaller, live on the periphery.

TERRITORIAL DEFENSE AMONG FEMALES

Essentially, females defend an area large enough to feed themselves and their young, and it seems that territorial size is adapted to the nutritional needs of the animal concerned. In the south of Madagascar, *Lepilemur mustelinus leucopus* feeds mainly during the lean season (southern winter) on the flowers of two species of *Alluaudia,* which appear in succession. During our study, 65% of the flowers had been eaten by the end of October (the flowers are not eaten after the onset of fruiting) and in certain areas where one of the two species of *Alluaudia* was lacking, there was always a species to "replace" it. These were *Xerosicyos perrieri* or *Salvadora angustifolia,* rarer species which were at the time the only ones bearing leaves.

It is much more difficult to estimate the availability of food in equatorial forests, but in Gabon, during the long dry season, the prosimians noticeably lose weight; examination of their digestive tracts also shows a diminution in the weight of food ingested (Charles-Dominique, 1972).

The competition which takes place between females, because of the peculiarities of their aggressiveness, allows the maintenance of territorial limits, but with larger or better areas for the more vigorous females. In *Galago demidovii*, for example, the average territorial size is 10,000 m^2 for females which weigh over 55 gm, and 7000 m^2 for lighter ones. In certain exceptional circumstances, when food becomes scarce, one may suppose that, because of competition, the most vigorous females are at a selective advantage.

In the social system of nocturnal prosimians, aggressiveness, tempered by certain inhibitory behaviors which reduce the chances of fighting, acts in many ways:

1. Females may feed in a sufficiently large area which they know perfectly, and which they can therefore exploit in a profitable way. The stronger among them are favored relative to the weaker, which allows natural selection.

2. Among the strongest, a small number of males share the territories of females. The more females a male controls, the less he uses the food resources on the territory of each one. This "economical" system, which also assures a sexual selection useful to the species, drives the supernumerary males into less favorable areas uninhabited by females.

In contrast to the higher primates among which, because of direct contacts, aggressiveness contributes to the establishment of a social hierarchy within the group, nocturnal prosimians mark and defend individual territories which, according to their size and positions, give each animal a social status. Since the nocturnal prosimians are the most primitive, it may be supposed that their social organization represents an ancient situation among the primates; moreover, we can also find this type of organization among several primitive families of carnivores, rodents, ungulates, etc.

The principal characteristic of the diurnal primates (the more evolved lemurs and higher primates) is gregariousness: a greater or lesser number of individuals moving together over a communal territory. Aggression may therefore take on a social character (fighting between neighboring groups), but it may also play a role in the establishment of the hierarchy within the group. It is probable that the different social structures of gregarious primates, in which aggression takes on a different significance, derive from the social system of the nocturnal prosimians, among which the defense of individual territories is paramount.

ACKNOWLEDGMENTS

The editor would like to express particular thanks to Dr. Ian Tattersall and Mrs. Christine Lémery for translating this paper from the original French.

REFERENCES

Bearder, S. (1974). Ecology of Bushbabies, *Galago senegalensis* and *Galago crassicaudatus,* with some notes on their behaviour in the field. *In* "Prosimian Biology" (G. A. Doyle, R. D. Martin, and A. C. Walker, eds.). Duckworth, London.

Charles-Dominique, P. (1971). Eco-éthologie des Prosimiens du Gabon. *Biol. Gabon* **7** (2), 121–228.

Charles-Dominique, P. (1972). Ecologie et vie sociale de *Galago demidovii* (Fischer 1808; Prosimii). *In* Behaviour and ecology of nocturnal prosimians, *Beih. Tierpsychol.* **9,** 7–41.

Charles-Dominique, P., and Hladik, C. M. (1971). Le Lépilémur du Sud de Madagascar: écologie, alimentation et vie sociale. *La Terre Vie.* **25** (1), 3–66.

Eibl-Eibesfeldt, I. (1953). Eine besondere Form des Duftmarkieren beim Riesen Galago, *Galago crassicaudatus* (E. Geoffroy 1812). *Saugetierk.* **1,** 171–173.

Martin, R. D. (1972). A preliminary field-study of the Lesser Mouse Lemurs. *In* Behaviour and ecology of nocturnal prosimians. *Beih. Tierpsychol.* **9,** 43–89.

Sauer, E. G. F., and Sauer, E. M. (1963). The South West African Bushbaby of the *Galago senegalensis* group. *J. S. W. Afr. Sci. Soc.* **16,** 5–36.

Seitz, E. (1969). Die Beteudung geruchlicher Orienteirung Beim Plumploris *Nycticebus coucang* Boddaert 1785 (Prosimii, Lorisidae). Z. *Tierpsychol.* **26,** 5–36.

THE ROLE OF AGGRESSION AMONG DIURNAL PROSIMIANS [1]

ROBERT W. SUSSMAN
Washington University

ALISON RICHARD
Yale University

INTRODUCTION

In this paper we consider aggression among three diurnal species of Malagasy lemurs: *Lemur fulvus rufus, Lemur catta,* and *Propithecus verreauxi.* These species were studied in the course of two separate 18-month field studies in Madagascar (*L. catta* and *L. f. rufus* by Sussman, 1969–1970; *P. verreauxi* by A. Richard 1970–1971).

The concept of aggression is ill-defined and has been variously used to denote a spectrum of interactions between individuals, between conspecific groups, and between members of different species (see Hinde, 1970). In this paper, we shall refer to as aggression or agonistic behavior those interactions which tend to increase the distance between conspecific individuals or groups and which may lead to the infliction of physical harm on one or more of the participants. It is not our intention to formulate an all-inclusive definition of these terms or to consider all aspects of

[1] The study conducted by RWS was supported in part by the National Institute of Mental Health Predoctoral Research Fellowship and Training Grant #F01-MH-46268, by Duke University Biomedical Support Grant NIH-5S05FR07070-05, and by a Duke University Graduate Fellowship. The study by AR was supported in part by Royal Society Leverhulme Award, Explorer's Club of America, the Boise Fund, the Society of the Sigma Xi, a NATO Overseas Studentship, the John Spedan Lewis Trust Fund for Advancement of Science, and the Central Research Fund of London University.

aggression among the lemurs. Our specific aim is to discuss agonistic behavior, as delimited above, in an ecological context instead of in a purely social setting, and to examine differences in the role played by such behavior in the ecology of the three species studied.

The approximate distribution of *Lemur catta*, *Lemur fulvus rufus*, and *Propithecus verreauxi* on Madagascar is shown in Figs. 1 and 2. The crude nature of these estimates should be emphasized. Only the distribution of *L. catta* has been documented in any detail (Sussman, 1972). Furthermore, even these approximate distributions become daily more inaccurate as progressive felling of forests and hunting of prosimians continue throughout large areas of Madagascar.

Certain associations can be made between the distribution of these species and the distribution of vegetational zones. *Lemur fulvus rufus* is found in forest areas which provide a more or less continuous canopy level. In contrast, *Lemur catta* is found in habitats ranging from the rich, deciduous forests of the southwest, through the arid *Didierea* forests that cover much of the south of the island, to the low-profile brush and scrub vegetation which replaces the *Didierea* or deciduous forests in some areas. Finally, *Propithecus verreauxi* is found in many of the deciduous forests west of the central plateau and in the southern *Didierea* forest, but there is no record of this species living in brush and scrub vegetation.

In the section that follows, the nature of intergroup and intragroup spacing and the ecology of each species are discussed in some detail. Briefer reference is made to agonistic behavior during the mating season. In the concluding section, an attempt is made to integrate our findings and thereby to give some insight into the functional importance of agonistic behavior in the ecology of these species.

PROPITHECUS VERREAUXI

A detailed 18-month study was conducted on four groups of *Propithecus verreauxi*. Two of the groups were sampled from a population living in semideciduous forest in the northwest of the island, and two from a population living in the *Didierea* forest in the south. The age and sex composition of these groups is given in Table 1.

Ecology

NORTHERN STUDY AREA

The northern study area (16° 35′ south latitude and 46° 82′ east longitude) is situated in a forest to the west of the forestry station at Ampi-

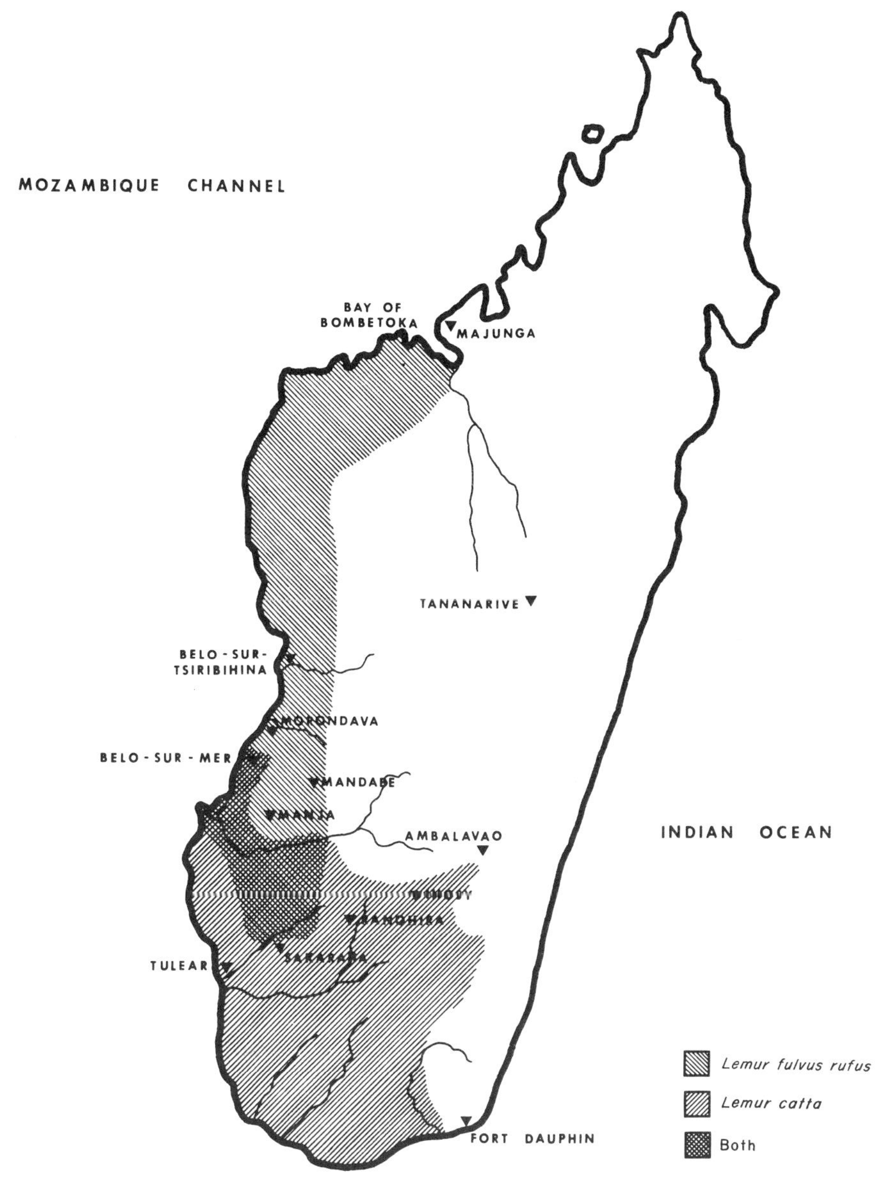

Fig. 1 Distribution of *Lemur fulvus rufus* and *Lemur catta*. Populations are not continuous within these areas but are found only where suitable primary vegetation exists.

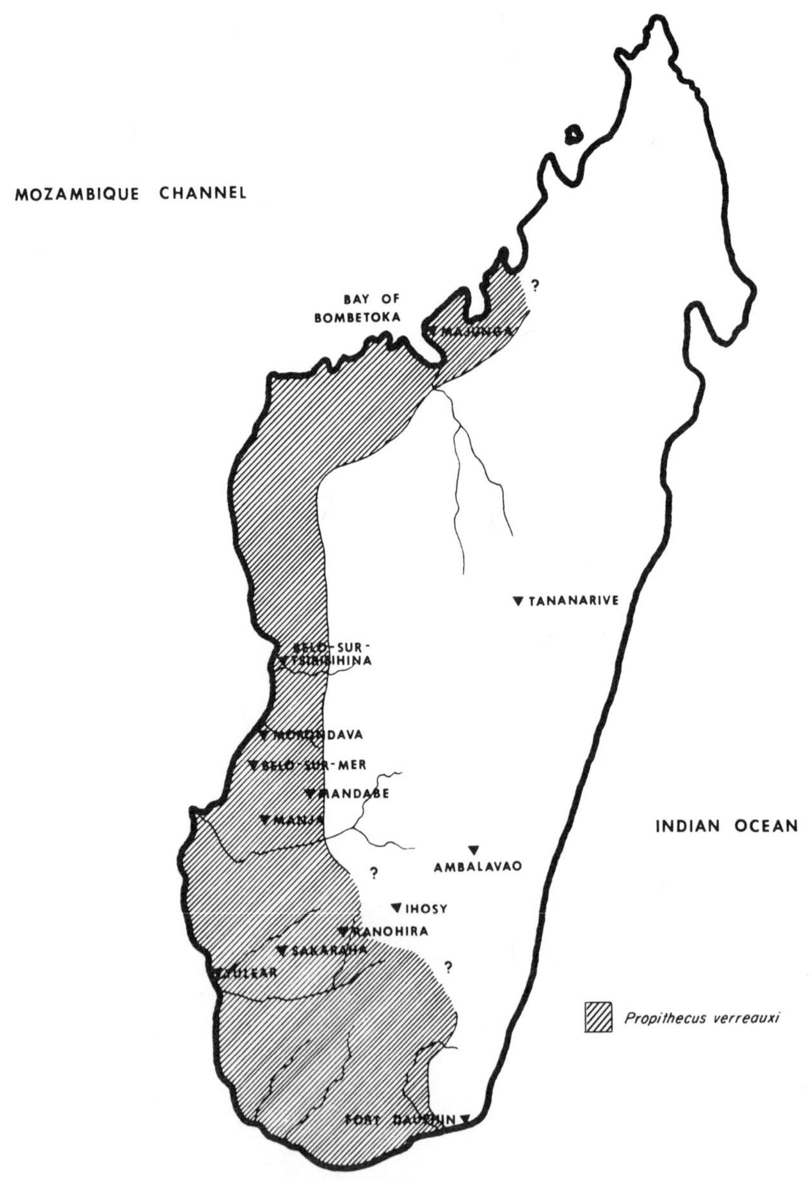

Fig. 2 Distribution of *Propithecus verreauxi*. Populations are not continuous within these areas but are found only where suitable primary vegetation exists.

TABLE 1 **Composition of Groups of *Propithecus verreauxi* When Observations Were Begun in July 1970 (North) and September 1970 (South)**

Group	Adult male	Adult female	Subadult[a]	Juvenile[a]	Infant	Total
I	2	5	0	0	0	7
II	1	1	1	1	1	5
III	2	2	1	1	2	8
IV	1	2	1	0	2	6
	6	10	3	2	5	26

[a] All subadults and juveniles were males.

joroa, in Reserve Nationale No. 7 (Fig. 3). It is about 100 km from the west coast of the island.

Soil in the hilltop study area is extremely sandy, and the small stature of trees is probably a reflection of the poverty of the soil as well as of the age of the forest. Apart from the forest's low profile, the usual characteristics of a very young secondary succession (Richards, 1966) are absent. This suggests that the forest has not been disturbed in the immediate past. Most trees are between 3 and 13 m in height, with emergents rarely exceeding 21 m. Both deciduous and evergreen trees are present, the former losing their leaves at the onset of the dry season in May.

Seasonal changes in climate are clearly defined: Heavy rain begins to fall, particularly at night, in the middle of October. Rains continue until the end of March, but little or no rain falls between March and October. Mean annual rainfall in this area is 1600 mm. Maximum and minimum temperatures recorded in the study area were 39°C and 14°C.

The forest contained six species of prosimians in addition to *Propithecus verreauxi coquereli: Lemur fulvus fulvus, Lemur mongoz mongoz, Microcebus murinus, Cheirogaleus medius, Lepilemur mustelinus ruficaudatus,* and *Avahi laniger occidentalis.*

SOUTHERN STUDY AREA

The southern study area, 24° 84′ south latitude and 46° 50′ east longitude, is situated in *Didierea* forest 1 km from Hazafotsy. This is about 1500 km from the northern study area and 60 km from the south coast.

The forest is dominated by xerophytic vegetation and particularly by species of the Euphorbiaceae and Didiereaceae families. Over 80% of the plant species in this forest are endemic, although the overall number of species present is lower here than in the north. Many species exhibit water-conserving adaptations, such as extreme reduction of leaf size or huge-girthed "bottle" trunks. Most trees are less than 13 m high. It ap-

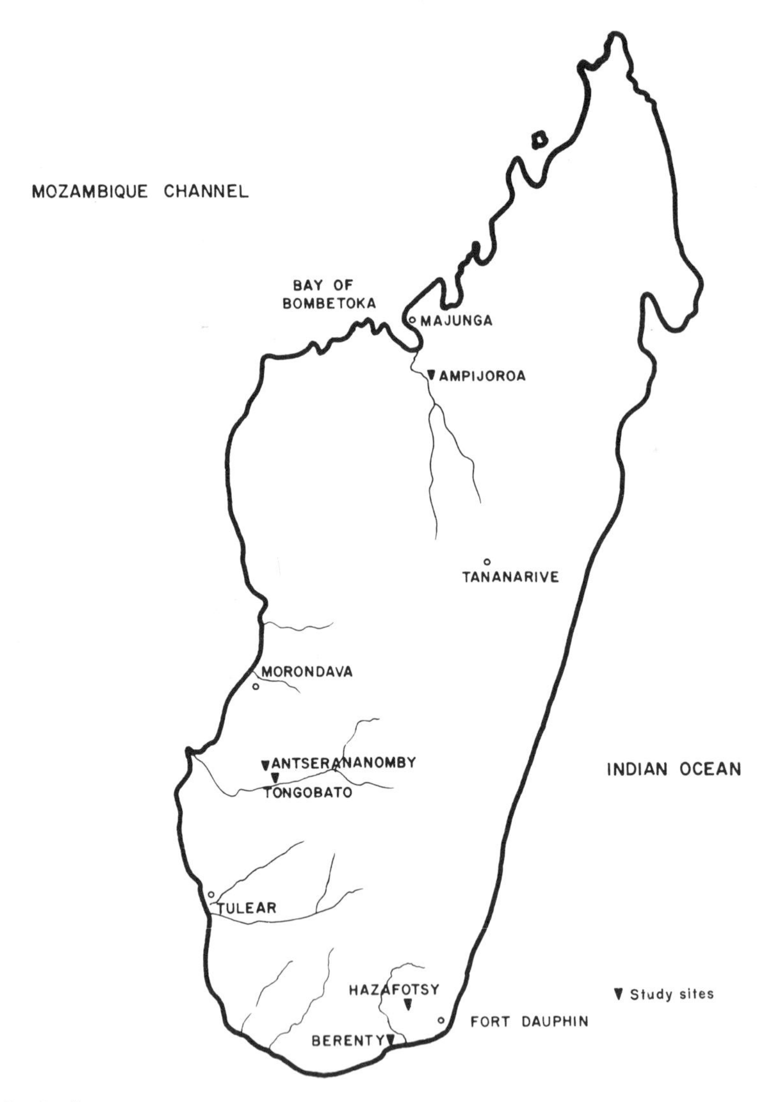

Fig. 3 Study sites.

pears that, in this area anyway, there is a critical size beyond which two of the most abundant *Alluaudia* species present, *A. procera* and *A. ascendens,* become top heavy, with the entire branching superstructure ultimately being torn off the trunk by its own weight. As in the north, many trees in this forest are deciduous and shed their leaves at the beginning of the dry season.

The contrast between seasons is greater in the south and the changeover from one season to the other is more abrupt than in the north. Almost all rains fall in January, February, and March, with a virtual drought for the rest of the year. Periodically, the rains fail altogether, as in 1972. Mean annual rainfall in this area is 600 mm. Maximum and minimum temperatures recorded in the study area were 40°C and 8°C.

Only two species of prosimians other than *Propithecus verreauxi verreauxi* were seen in the southern forest: *Microcebus murinus* and *Lepilemur leucopus*. Vocalizations of *Lemur catta* were heard on 1 day, apparently from animals moving rapidly through the forest.

Diet and Utilization of Habitat

In both study areas, groups spent considerable amounts of time at all levels of the forest, including the ground. These levels represented only height above the ground: In neither study area was the vertical composition of the forest sufficiently uniform to correlate each level with a characteristic vegetational structure. The southern groups tended to spend more time at lower levels than the northern groups. Factors contributing to this difference were the overall lower stature of the southern forest and the fact that most of the taller trees in the south are *Alluaudia ascendens* and *Alluaudia procera*. The extremities of the thin, spiny, vertical branches of these species provided no regular or firm support for locomotion and were a source of food only in September. Animals were rarely seen high on these branches at any other time.

The diet of *Propithecus verreauxi* in both study areas was composed of leaves, shoots, flowers, and fruit. The relative importance of these items varied seasonally: In the dry season, for example, a large percentage of the diet of animals in both regions consisted of adult leaves. In contrast, in the wet season, fruit was the most important constituent. These fluctuations in the composition of diet were probably associated with fluctuations in the availability of various foods.

A limited analysis of the vegetation in both study areas (Richard, 1973) indicates that most tree species occur rarely in each forest and only a few species are abundant. Although groups in both areas spent a high percentage of their *total feeding time* eating food from tree species which occur frequently in the forest, they did not rely totally on abundant tree species as food sources. Even among the groups' most commonly eaten foods, there was a wide range of variation in the abundance of the food species. Furthermore, rarely occurring tree species constituted almost two-thirds of the *total number* of plant species eaten in each area.

Intergroup Encounters

In the north, the study groups lived in home ranges of approximately 6 ha. These ranges overlapped extensively. Areas of exclusive use were widely scattered and often located on the periphery of the home range (Fig. 4). It is possible that they were an artifact and simply areas in

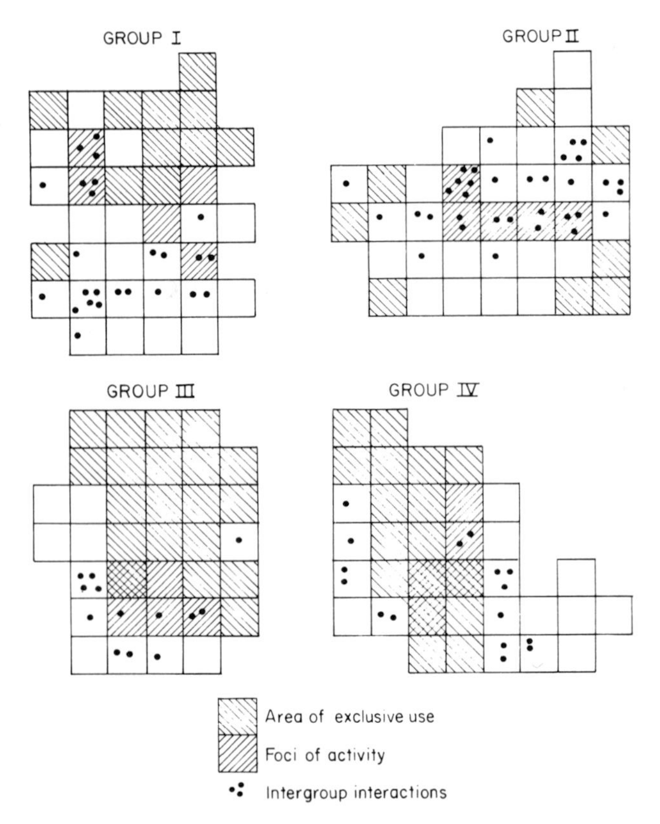

Fig. 4 *Propithecus verreauxi.* Areas of exclusive use, foci of activity, and the locations of intergroup interactions within the home range of each study group.

which other groups were not seen, rather than areas into which other groups did not go.

Intergroup interactions occurred throughout the home range of each group, except in the areas of exclusive use. It is, therefore, improbable that the groups were defending these areas or the boundary of some other well-defined area during encounters.

Only 18 out of the 59 encounters observed in the north culminated in a ritualized "battle" composed of ". . . leaping, staring, and scent-marking with very low growls or in silence . . . everything depends on a fast, formal pattern of movement, each animal occupying sections of tree rather than opposing individuals of the other troop. . . . [Jolly, 1966, pp. 49–50]." Most of these 18 "battles" occurred with reference to a particular food source in which one of the groups was feeding prior to the encounter. In most other instances, after reciprocal staring, some scent-marking and growling, the groups moved away from each other toward the geographical center of their home range. The frenetic activity of the "battle" was completely absent.

In the south, the study groups lived in home ranges of about 5 ha, but overlap between the ranges was minimal. In contrast to the scattered distribution of areas of exclusive use in the north, the central part of the home range of each group in the south formed one exclusively used block. Thus it appeared that the groups had clear borders to their area of exclusive use, defined by periodic agonistic encounters in the narrow zone of overlap with neighboring groups (Fig. 4).

Only 29 encounters were observed in the south outside of the mating season, but all except 9 culminated in "battles." These encounters took place whenever two groups came into close proximity along the "boundaries" of their home ranges and were not related to a specific ongoing activity.

Available data are limited, and interpretations of these regional differences must thus be tentative. However, it is postulated that in the north, each group required its total home range in the course of a whole year because of the varying distribution and small size of important but scattered sources of food. At any one time, however, the total food available within the home range was probably in excess of the group's immediate requirements and overlap between groups could be extensive. In the south, food probably became a critical limiting factor toward the end of the dry season, and at that time the whole home range was necessary to support one group. The round-the-year territoriality found in the area may be an adaptive response to this minimum foraging area requirement that operates at times of greatly reduced availability of food.[2]

Intragroup Encounters

An animal that displaced, threatened, bit, or cuffed another animal in its own group was considered to be the aggressor in an agonistic en-

[2] At Berenty, as Jolly (1966) suggests, behavioral as well as ecological requirements are probably determinative of home range size; in this area of high population density, crowding rather than food may be a limiting factor, for the large, closely concentrated food sources may permit theoretical reduction of home range size per group beyond that compatible with behavioral requirements.

counter. Aggression was responded to by submissive gestures (Fig. 5). These included baring the teeth, rolling up the tail between the hind legs, hunching the back, and giving "spat" vocalizations (see Jolly, 1966). After such an encounter, the submissive animal always moved away from the aggressor. It should be noted that this repertoire of aggressive and submissive behavior differed from that seen in agonistic encounters between groups.

In all groups, most agonistic encounters occurred in a feeding situation, and spatial displacements without reference to a food source were rarely seen. The only other context in which agonistic behavior was commonly seen was when animals tried to handle a mother's newborn infant: The mother would cuff and bite the approaching animal to prevent it from gaining access to her infant.

Fig. 5 Adult male *Propithecus verreauxi* displaying submissive gestures.

The frequency of agonistic behavior varied considerably between the four groups, but not consistently between areas. Agonistic encounters occurred more frequently in the wet season than in the dry in all four groups. Most animals contributed to this increase, and in not one was a decrease in frequency recorded. It is probable that the increase in the wet season was partially associated with the increase in general activity and, specifically, in time spent feeding. However, the frequency increased more in some group members than in others. This may have been partly due to the presence of infants in groups in the dry season and not in the wet season. It is likely that the mother's protectiveness toward her newborn infant caused her to have a high frequency of agonistic encounters in the dry season. This factor being absent in the wet season, her frequency did not increase proportionately as much as that of other group members.

It would be an oversimplification to see all the social relationships of *Propithecus verreauxi* solely in the light of a simple linear dominance hierarchy. Unidirectionality of agonistic encounters and displacements in the groups could be used to define a hierarchy in each group, but there was no consistent correlation between the rank of individuals in a hierarchy established on this criterion and their ranks in hierarchies established according to other criteria such as the frequency of aggression, the direction and frequency of grooming, or preferential access to females during the mating season (see the following section).

For these reasons, the clear-cut hierarchy existing with respect to access to food was called a "feeding hierarchy." In all four groups, the highest ranking animal in this hierarchy was an adult female. High rank in the feeding hierarchy was not necessarily a function of sex, however.

Agonistic Encounters during the Mating Season

Behavioral changes that occurred in *Propithecus verreauxi* during the mating season have been described in detail elsewhere (Richard, 1973; 1974), and only a brief resumé is given here.

Flushing of the vulva of a female in one of the southern study groups in mid-January 1971 coincided with a sudden, significant increase in the following activities in that group: scent-marking by adult males, intragroup agonistic encounters, intergroup encounters, and "roaming" by adult males. This last category describes the frequent excursions made by adult males, alone or in pairs, deep into the home ranges of other groups.

Copulation was observed only during the first week of March 1971, although the associated increase in various behaviors (see earlier) had been present throughout the preceding month. During the week when mating occurred, there was some degree of breakdown in group structuring: This was marked by the appearance of aggression directed at males with a high rank in the feeding hierarchy by males ranking low in that hierarchy. This aggression was not ritualized and involved fierce fighting which frequently resulted in severe wounds. It is postulated that some males with priority of access to food outside of the mating season (i.e., nonmating season dominant males) extended this priority to include access to females in their own and other groups during the mating season. After the mating season, these males remained in the groups in which they were "dominant" before the mating season. Males who were subordinate prior to the mating season could gain priority of access to females during the mating season by fighting and ousting the resident nonmating season "dominant" male in his own or another group. These males then stayed in that group to become nonmating season dominant males. The mating season in *Propithecus verreauxi* appeared to act as a social catalyst, a period when the status quo of the previous year was abandoned and a readjustment of the social structure and, specifically, of the feeding hierarchy occurred.

LEMUR FULVUS RUFUS

Groups of *Lemur fulvus rufus* were studied in a number of forests in the southwest of Madagascar and the behavior of these groups was compared to that of sympatric and allopatric populations of *Lemur catta*. The age and sex composition of the groups of *L. f. rufus* studied is given in Table 2.

Ecology

The range of *Lemur fulvus rufus* extends throughout forested areas of western Madagascar, from near Majunga in the north to the Fiherenana River in the south. It is the only subspecies of *Lemur fulvus* found in this region. An intensive study of this species was carried out in two forests just north of the Mangoky River (21° 46′ south latitude, 44° 7′ east longitude): Antserananomby, in which *L. f. rufus* and *Lemur catta* coexist; and Tongobato, in which *L. f. rufus* is found without *L. catta*. These forests are 10 km apart and were probably once part of a continuous forest range.

Both Antserananomby and Tongobato are primary deciduous forests made up of a *Tamarindus indica* (kily) consociation. Other rarely occurring species of trees are scattered throughout the forests but are found most frequently on the forest edge. Kily trees make up a closed canopy about 7 to 15 m high throughout the forest. Rising above the closed

TABLE 2 Composition of Groups of *Lemur fulvus rufus*

Name of group	Adult		Juvenile		Infant	Total	Adult sex ratio
	Male	Female	Male	Female			M:F
Antserananomby							
AF-1	4	6	1	1	0	12	1:1.50
AF-2	4	5	1	0	0	10	1:1.25
AF-3	4	3	0	1	0	8	1.33:1
AF-4	2	5	2	1	0	10	1:2.50
AF-5	4	6	0	2	0	12	1:1.50
AF-6	3	2	0	0	0	5	1.50:1
AF-7	3	4	1	1	0	9	1:1.33
AF-8	2	2	0	0	0	4	1:1.00
AF-9	2	2	1	0	0	5	1:1.00
AF-10	4	5	2	0	0	11	1:1.25
AF-11	5	7	1	2	0	15	1:1.40
AF-12	4	4	1	0	0	9	1:1.00
Totals	41	51	10	8	0	110	41:51 = 1:1.24
Means	3.42	4.25	.83	.66	0	9.17	
Tongobato							
TF-1	5	8	0	0	4	17	1:1.60
TF-2	2	3	1	0	1	7	1:1.50
TF-3	4	5	0	1	2	12	1:1.25
TF-4	3	4	0	0	1	8	1:1.33
TF-5	3	2	0	1	1	7	1.50:1
Totals	17	22	1	2	9	51	17:22= 1:1.29
Means	3.40	4.40	.20	.40	1.80	10.20	
Overall totals	58	73	11	10	9	161	58:73 = 1:1.26
Overall means	3.41	4.29	.64	.59	.53	9.47	

canopy are a number of scattered tall trees or stands of trees which form an emergent layer about 15–20 m high. A subordinate tree layer, 3–7 m high, consists of certain species of smaller trees and the saplings of the taller trees.

The forest at Tongobato is surrounded by cultivated fields and degraded vegetation. The forest at Antserananomby is bordered on the east by the Bengily River, a tributary of the Mangoky River. The Bengily contains no water during the dry season. To the north and northwest, the continuous canopy forest is replaced by primary brush and scrub vegetation. This vegetation reaches the height of the subordinate tree layer in the continuous canopy portion of the forest. There is no dominant tree species but only an association of many codominant species. The tallest trees are the baobab (*Adansonia madagascariensis*) and a few scattered *Diospyros* sp., *Acacia* sp., and *Tamarindus indica.* Most of the trees have thin trunks and are spaced closely together. The underbrush is thick with lianas, euphorbiaceous bushes, and thorny bushes. It is extremely difficult to move about or to see for any distance in this type of vegetation.

Mean annual rainfall in the southwest is around 700 mm. However, the rainfall is seasonal. It rains from late December to March, but little or no rain falls between April and November. Maximum and minimum temperatures recorded in the study area were 40°C and 10°C.

The forest at Antserananomby contained five species of prosimians in addition to *Lemur fulvus rufus* and *Lemur catta: Propithecus verreauxi verreauxi, Lepilemur mustelinus ruficaudatus, Phaner furcifer, Cheirogaleus medius,* and *Microcebus murinus.* In addition to *Lemur fulvus rufus,* the following species were sighted at Tongobato: *Propithecus verreauxi verreauxi, Lepilemur mustelinus ruficaudatus,* and *Phaner furcifer.*

Diet and Utilization of the Habitat

Lemur fulvus rufus spent over 90% of its time in the closed canopy or the adjacent strata of the forest. Over 85% of the travel of the groups took place within the closed canopy. Animals rarely came to the ground and groups were never seen in areas of the forest which would have necessitated terrestrial locomotion. *L. f. rufus* was never seen in brush and scrub vegetation.

The home ranges of the groups which were mapped during the period of Sussman's study were small—approximately 1.0 ha. However, only rough estimates could be made of the home ranges and it is likely that the extreme limits of the ranges were not known. Furthermore, it was not difficult to force a group out of the area in which it was usually seen if the animals were sufficiently frightened. Therefore, the home ranges of groups of *L. f. rufus* (Figs. 6 and 7) may be more accurately thought of as equivalent to core areas. It is likely that these core areas are not stable and change seasonally and over time. The groups move little within the period of a day; day-ranges were recorded between 125 and 150 m. The population density was very high at both forests, averaging about 1000/km^2.

The restricted use of vertical and horizontal space by *Lemur fulvus rufus* can be related to the very restricted diet of this species. The diet of *L. f. rufus* was composed of leaves, shoots, flowers, and fruit, but consisted of only a few species of plants. Leaves of the *Tamarindus indica* made up a very large proportion of the diet. In the dry season, at Ant-

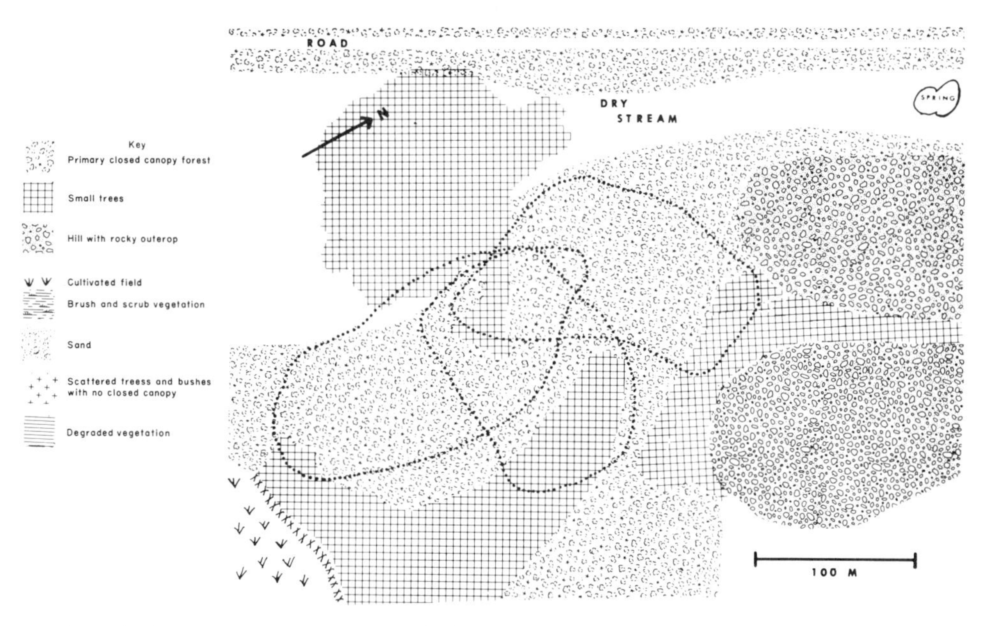

Fig. 6 Tongobato. Home ranges of three groups of *Lemur fulvus rufus*. This map does not include all of the groups seen in this area.

serananomby, approximately 75% of the observed feeding was done on mature kily leaves. The amount of fruit, flowers, and young leaves eaten depended upon the season and the distribution of trees other than the dominant kily tree within the forest. Groups supplemented a diet consisting of kily—leaves, fruit, and flowers—with parts of those species of trees which happened to be within their small home range. At both Tongobato and Antserananomby, only three species of plants made up over 80% of the diet, with kily accounting for 50% and 75% of the diet respectively. It should be noted, however, that populations of *L. f. rufus* are found in forests which are not dominated by kily trees and that they are not specifically dependent upon kily.

Intergroup Encounters

The home ranges of groups of *Lemur fulvus rufus* were very small and overlapped extensively (Figs. 6 and 7). The borders of the ranges were not defended, but in most cases groups maintained spatial separation by means of frequent vocalizations. These vocalizations begin with gutteral grunts and grade into high-pitched rasps. They were often given when groups were moving and evoked similar vocalizations from neighboring

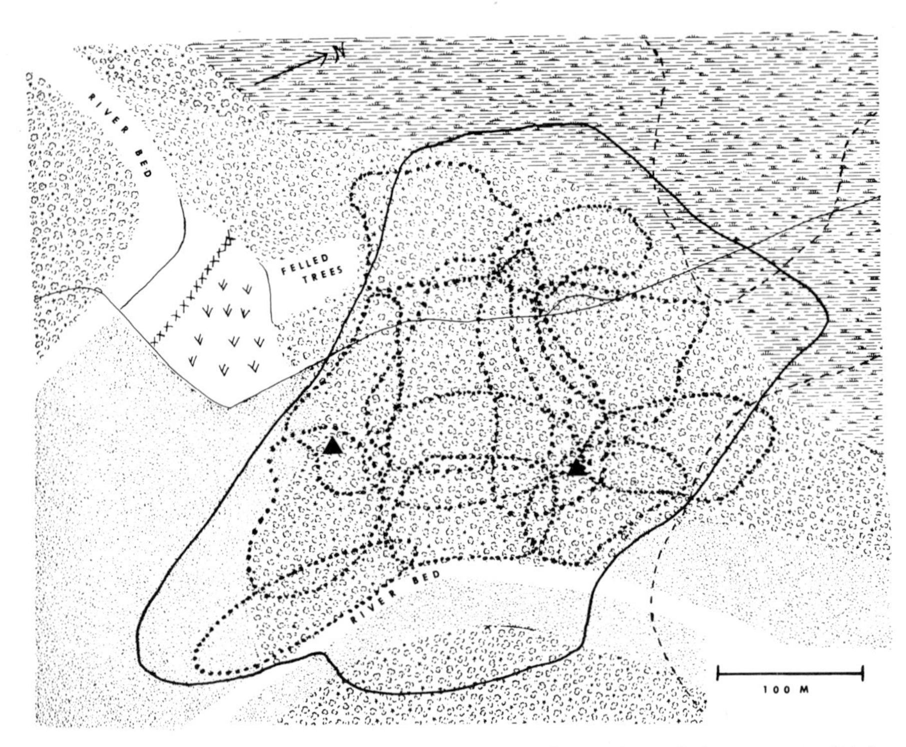

Fig. 7 Antserananomby. Home ranges of *Lemur fulvus rufus* and *Lemur catta.* (▲): *Ficus soroceoides* tree; (——): *Lemur catta* (study group); (-----): *Lemur catta* (peripheral groups); (·····): *Lemur fulvus rufus.* (See Key in Fig. 6.)

groups. When agonistic group encounters did occur, they most often developed when groups were moving or when one group surprised another as it entered a tree in which the first group was feeding or resting.

During intergroup encounters, the animals from each group faced each other in adjacent trees or branches, grunted and rasped, rhythmically wagging their tails back and forth and changing positions frequently (Fig. 8). Most encounters ended when one of the groups retreated and moved off in another direction. A chase sometimes developed, but never lasted long. For the most part, marking did not occur during an actual encounter. After the two groups separated, however, the members of both groups often marked the branches of the trees in which they settled. Both male and female *Lemur fulvus rufus* mark anogenitally; males also mark with their foreheads.

The frequency of agonistic group encounters depended upon the season and was highest when new leaves and fruit were forming on the kily trees. At these times, many groups of *Lemur fulvus rufus* would concentrate more toward the center of the forest where the density of the kily trees was highest, and several groups would often feed, sun, or rest in adjacent trees. In fact, more than one group was frequently found in branches on different sides of the same tree.

At other times of the year, the groups remained further apart. They did, however, come together to feed in particular species of trees in which there was ripe fruit and which occur rarely throughout the forest. At such times, two or three groups of *Lemur fulvus rufus* could often be found feeding on different branches of one of these trees. For example, at Antserananomby there were two large trees (*Ficus soroceoides*) within the closed canopy portion of the forest which began to fruit during the driest part of the year (in July). Although there were only two of these trees in the forest (Fig. 7), at least 6 of the 12 groups fed on them.

Thus, by not defending particular territorial boundaries and by having highly overlapping ranges, *Lemur fulvus rufus* could not only exploit the evenly distributed products of the kily trees, but a number of groups could utilize the rarely occurring species of trees. Since the density of the population was very high and groups were often in close proximity,

Fig. 8 Posture of an adult *Lemur fulvus rufus* during an agonistic intergroup encounter.

agonistic group encounters were very likely important factors in maintaining the coherence of groups of *L. f. rufus.* This particular pattern of group spacing thus allowed many groups to utilize resources in close temporal and spatial proximity while the integrity of the group was maintained.

Intragroup Encounters

Agonistic encounters between members of the same group were infrequent—approximately 0.10 per hour—and usually occurred when an animal approached or tried to sit in contact with or groom another. A few occurred while animals were attempting to gain access to waterholes in trees. During agonistic encounters, one animal would simply lunge at or attempt to grab or slap the other and many interactions ended with a brief chase. Most encounters were quite mild, seldom involving more than momentary physical contact. Harrington (1971), however, in a field study of olfactory communication and social behavior of *Lemur fulvus fulvus,* saw two fights in which animals fell from the trees while holding onto each other. Sometimes a grunt accompanied a threat and in the most intense encounters, high squeaks were given. Neither Harrington nor Sussman could discern any dominance hierarchy in groups of *Lemur fulvus.*

Agonistic Encounters during the Mating Season

The mating season was marked by an increase in the frequency of scent-marking, and a number of copulations or attempted copulations were observed. However, in contrast to *Propithecus verreauxi* and *Lemur catta,* agonistic encounters in groups of *Lemur fulvus* did not increase in frequency or intensity during the mating season.

LEMUR CATTA

Intensive studies were carried out on three groups of *Lemur catta.* One group was located at Antserananomby, where it coexisted with groups of *Lemur fulvus rufus;* the other two groups were found in a similar deciduous forest in the south of Madagascar. The age and sex composition of the three groups is given in Table 3.

Ecology

The range of *Lemur catta* extends throughout the south and southwest of Madagascar as far north as the Morondava River. *L. catta* is the only

TABLE 3 Composition of Groups of *Lemur catta*

Name of group	Adult Male	Adult Female	Juvenile	Infant	Total	Adult sex ratio M:F
Antserananomby						
AC-1	7	8	4	0	19	1:1.14
Berenty						
BC-1	4	5	2	4	15	1:1.25
BC-2	7	6	3	4	20	1.17:1
Totals	11	11	5	8	35	11:11 = 1:1.00
Means	5.50	5.50	2.50	4.00	17.50	
Overall totals	18	19	9	8	54	18:19 = 1:1.05
Overall means	6.00	6.30	3.00	2.70	18.00	

diurnal species of lemur which inhabits the dry brush and scrub forests of the south and southwest and utilizes the dry, rocky, and mountainous areas in the southern portion of the central plateau where only patches of deciduous forest remain.

The forest at Antserananomby is described above. The second forest in which *Lemur catta* was studied is Berenty in the south of Madagascar (Fig. 3). This forest has been described by Jolly (1966). It is a protected reserve which consists of about 100 ha of gallery forest and 0.5 ha of *Didierea* forest, although part of the gallery forest (perhaps 30 ha) is degraded. The whole forest is surrounded by sisal fields, so that it forms an island of natural habitat isolated by terrain which the lemurs cannot cross. The reserve has existed for about 25 years.

The vegetation at Berenty differs only slightly from that at Antserananomby and Tongobato. The gallery forest is dominated by a continuous canopy of *Tamarindus indica.* Other large trees include some species which are not found in the other two forests. The main structural difference between Berenty and the other forests is the relative absence of an emergent layer. There are no primary brush and scrub regions in the 9 ha of the forest that were studied by Sussman, although there are many areas in which the vegetation is scattered and the trees do not form a continuous canopy.

Maximum and minimum temperatures recorded in the study area were 39°C and 15.4°C. The mean annual rainfall in this area is 600 mm. During 1970, however, the rainfall at Berenty was the lowest recorded in the past 10 years (Charles-Dominique and Hladik, 1971). The species of prosimians at Berenty besides *Lemur catta* are: *Propithecus verreauxi verreauxi, Lepilemur mustelinus leucopus, Microcebus murinus,* and *Cheirogaleus medius.*

Diet and Utilization of the Habitat

Lemur catta utilized all of the vertical strata of the forest. For the most part, this species moved and traveled on the ground, rested during the day in low trees, rested at night in the closed canopy, and fed in all of the forest strata. At both study areas the animals spent most of their time on the ground. When groups of *L. catta* traveled, approximately 70% of this activity was performed on the ground. The home range of the group studied at Antserananomby was 8.8 ha (Fig. 7), while those of the groups studied at Berenty were about 6 ha (Fig. 9). The population density was estimated to be 215/km^2 at Antserananomby and 250/km^2 in

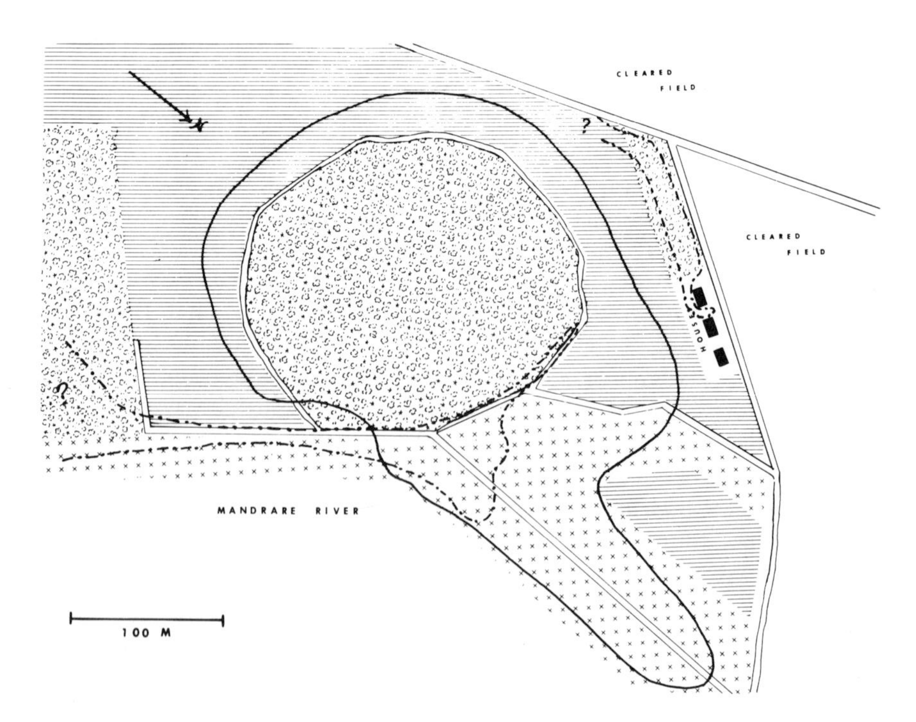

Fig. 9 Berenty. Home range of *Lemur catta* group BC-1 and borders of the home range of *Lemur catta* group BC-2. (═══): road; (——): group BC-1; (·—·—·): group BC-2. (See Key in Figure 6.)

the study area at Berenty.[3] Day-ranges were approximately 950 m. The group of *L. catta* at Antserananomby spent 58% of the daylight hours outside of the portion of the forest with a continuous canopy, even though this represented only 30% of the total home range of the group. The animals spent much of their time foraging in the brush and scrub vegetation and in other areas adjacent to the continuous canopy portion of the forest.

The diet of *Lemur catta* was composed of leaves, shoots, flowers, and fruit. The diverse vertical and horizontal ranging pattern of this species was associated with a varied diet. A group of *L. catta* would usually utilize one part of its home range for about 2 or 3 days and then change to another part. In a period of 7–10 days the group visited most of its total range. The constant surveillance of a relatively large home range allowed groups of *L. catta* to exploit a number of different plants which had a patchy distribution over a wide area: Trees that were blossoming or that had fruit could be located by the group and utilized. In this way, the group of *L. catta* at Antserananomby exploited food sources in the brush and scrub vegetation that were not available to *Lemur fulvus rufus* and some resources in the portion of the forest with a continuous canopy that were available to only a few groups of *L. f. rufus*. Thus, during the dry season at Antserananomby, *L. catta* had access to and ate considerably more fruit than *L. f. rufus*.

Intergroup Encounters

At Berenty in 1963–1964 (Jolly, 1966) and at Antserananomby, each group of *Lemur catta* maintained almost exclusive use of its relatively large home range, and boundaries of the home ranges of adjacent groups overlapped only slightly. In general, distance was probably maintained between groups by terrierlike barks that were exchanged at various times throughout the day.

Only one intergroup encounter was seen at Antserananomby during approximately 200 hr of observation. In this encounter, when the groups noticed each other, they both immediately changed directions and moved away from the mutual border. Jolly (1966) saw only five encounters between adjacent groups in 400 hr of observation. In four of these encounters the two groups avoided each other. In one encounter a male from

[3] In 1963–1964, Jolly (1966) estimated the population density of *L. catta* in the entire reserve at Berenty to be 320/km^2. The difference may be due to the fact that the groups studied by Sussman lived on the periphery of the reserve.

one group attempted to mate with a female of another and was chased away by the males of her group.

At Berenty in 1970, a different pattern of intergroup interaction emerged. The two groups studied by Sussman shared a large portion of their home ranges (Fig. 9). Both extensively used the area of overlap, but were found in this area at different times. Four encounters were seen in 105 hr of observation. In two of these the groups moved off in opposite directions. The other two encounters involved agonistic interactions between the two groups. These consisted of quick spats, chases, genital and brachial marking, and tail waving displays between several members of the two groups (Fig. 10). In another part of the forest in the same year, four groups studied by Jolly (1972) had less than 50% of their ranges to themselves. In her study area, the number of animals had increased from 41 in September 1963, to 58 in September 1970, and the number of groups had doubled. In 54 hr of observation she saw 13 group encounters, 9 of which involved agonistic interactions between two groups or individual members of two groups. Many of these encounters lasted as long as 10 min and, in some, vocalizations were used which were usually given as alarm calls to hawks.

Near the Mangoky River and in many areas of the south and southwest of Madagascar, the populations of *Lemur catta* range widely in largely undisturbed vegetation. The greatest proportion of these areas are dry brush and scrub or desertlike *Didierea* forests. The ranging and foraging pattern of *L. catta* may be an adaptation to these arid environments in which the resources are sparse and unevenly distributed. At Berenty, the animals are restricted to the protected reserve because of the insular condition of the forest. It is possible that the population at Berenty is increasing and the animals are retaining their ranging and foraging pattern at the expense of exclusiveness of their home range. The result may be an extensive overlapping of home ranges and a "time-plan" (Jolly, 1972) sharing of resources at the cost of increasing time and energy spent in intergroup agonistic encounters.

Intragroup Encounters

Agonistic encounters between members of the same group are frequent in *Lemur catta*. Outside of the mating season there are, on the average, approximately 1–2 per hour (Jolly, 1966, 1972). The characteristics of these encounters are variable and can range from a stare or slap to a

Fig. 10 *Lemur catta* from different groups genital marking during an agonistic intergroup encounter.

highly ritualized scent-marking display. Scent-marking includes genital marking of branches by both males and females and palmar marking of branches, marking of the tail, and tail waving by males only (see Jolly, 1966). Although scent glands are involved in these ritualized gestures, the display itself is undoubtedly a visually oriented signal. Agonistic encounters outside of the mating season often ended in chases but very seldom involved physical contact. The animals have not been seen to inflict injuries during these encounters. The behavioral repertoire is the same for both intragroup and intergroup agonistic encounters.

Jolly (1966) studied interactions between individual members of one group at Berenty in 1963–1964. She found a clear linear dominance order among the males and a fairly clear one among the females. She considered an animal to be dominant over another "if it won consistently in spats and other aggressive encounters. In troop 1, in 1964, it was possible to rank all the animals in linear order, each animal being defeated by those above it and defeating those below it, although the exact rank of all the females was not certain [Jolly, 1966, p. 104]."

Associated with this well-defined dominance hierarchy is the tendency to peripheralize subordinate males. As a group moves from place to place, the females, juveniles, and dominant males usually move together and ahead of the subordinate males. The subordinate males also tend to feed and rest together, often being joined by some of the juveniles. Thus the group is frequently divided into subgroups which separate while moving, foraging, or resting. In most cases, the entire group does not feed or rest in one tree, and the females and dominant males get first choice of feeding and day-resting sites. In brush and scrub regions, where the resource units are small and scattered, the large group often divides into a number of subgroups and spreads out over a wide area. However, the membership of the subgroups has not yet been analyzed.

In 1970 at Berenty, two groups were studied (one by Jolly, 1972, and one by Sussman) which were more fragmented. In both of these groups, two of the adult males were restricted to the periphery of the group and were threatened or chased whenever they approached the main body of the group too closely (within about 5 m). Sometimes these males moved entirely alone, but they usually followed the group and entered feeding trees after the rest of the group had left. This intense peripheralization of subordinate males may be due to a shortage of food and/or overcrowded conditions.

In a study still in progress, however, Budnitz (personal communication) has observed two males from two different groups leave their groups and attempt to enter others. In both cases the animals were not low ranking and were not being driven out of their original groups. These males were not immediately allowed into the new group nor were they allowed back into their original groups. Whether this is a normal pattern in groups of *Lemur catta* which may allow gene exchange to occur between adjacent groups or whether it is due to unstable social structure resulting from high population density at Berenty, or to other unknown factors must await further comparative studies.

Agonistic Encounters during the Mating Season

The following account is based on observations made by Jolly during the mating season in 1964 (Jolly, 1966). The frequency and intensity of agonistic encounters increased during the mating season. Agonistic encounters rose in frequency to 21 per hour during the week of mating and were high in the weeks immediately preceding and following mating. The proportion of scent-marking displays rose and the animals were also seen jumping at each other, grappling with their hands while on their hind feet. Jolly called these interactions "jump fights." In such fights the animals would grab at each other, often making contact and attempting to bite one another. This type of encounter was seen only during the mating season. By the end of the mating season all five males in the group studied had wounds 4–5 cm long on their thighs, arms, or sides. It is probable that these injuries were acquired during jump fights.

During most of the mating sequences observed by Jolly, the male was harassed by other members of the group and had to chase or fight them before he could actually mount the female. Most of this harassment was done by adult males. In some cases the male had to chase his rivals up to 20 times between approaches to the female. In four instances, juveniles or females approached a pair during mating and flung themselves onto the male, cuffing and clawing. Matings actually culminated apart from the main body of the group, because each time the male dismounted to chase, the female would move further away from the group.

Access to females during the mating season was not related to an animal's position in the dominance hierarchy during the nonmating season. One of the subordinate males accomplished half of the observed matings in 1964. In fact, subordinate animals successfully challenged and chased dominant ones and the dominance hierarchy essentially broke down. However, the original hierarchy of the group was reestablished after the mating season.

CONCLUSIONS

We have described some aspects of aggressive behavior among these prosimian species and attempted to put them into an ecological and social context. The question now remains: Is there anything we can call a pattern of aggression among diurnal prosimians? Charles-Dominique (this volume) has emphasized similarities in the nature of aggression among the nocturnal prosimians. The parallel between the three diurnal species described in this paper is less marked: Even at a general level of social organization there are limited criteria by which the behavioral taxonomist could establish relatedness. A few features are common to all three: a relatively small home range; relatively small groups; a very short breeding season; and the extensive use of olfactory communication. Beyond that, however, in behavioral/ecological terms, each species has found a very different "solution" to the "problem" posed by the environment. All three appear to exploit efficiently the habitat in which they live, but in all three the means of exploitation are different. Associated with this variation in the nature of exploitation is a markedly different pattern of aggression in each species.

In much of its range, *Lemur catta* lives in extremely arid environments where food resources are sparse and patchy. Animals forage over extensive areas at all levels of the forest and on the ground and live in relatively large home ranges. This foraging pattern gives groups access to widely scattered food resources. Distance is usually maintained between groups by frequent vocalizations and avoidance. At Berenty, the pattern of foraging is maintained, but the groups share large parts of their home ranges. This leads to a time-plan sharing of some of the resources and an increase in the number of agonistic encounters between groups.

Lemur fulvus rufus spends most of its time in the continuous canopy of the forest and moves little throughout the day. Groups have very small home ranges. These ranges overlap extensively, and borders between groups are not defended. A very high proportion of feeding is done on the dominant species of tree in the forest, but many groups may also share rare, seasonally available resources. Intergroup aggression in this species is not specifically over food resources, but rather serves to maintain group integrity.

Propithecus verreauxi was studied in two highly contrasting areas and revealed two different patterns of group spacing which, it is held, were closely related to ecological factors. In the northern study area, specific food sources were small, widely scattered, and only seasonally available, although overall, food was abundant. Groups lived in overlapping ranges and shared resources. Encounters were frequent between groups but, in most instances, not overtly hostile. As in *Lemur fulvus rufus,* the function of interactions appeared to be the maintenance of group integrity rather than the defense of a particular area. In the south, food was sparse and it is likely that, at least at some times of the year, the food present in any single home range was sufficient to feed only one group. In contrast to *Lemur catta,* in this study area, groups of *P. verreauxi* spaced themselves by defending borders rather than by means of vocalization and avoidance.

The pattern of intragroup aggression appears to be closely integrated with that of intergroup aggression in the three species. In *Lemur fulvus rufus,* the absence of competition for food resources between groups and the extensive overlap between ranges are mirrored within the group by a very low frequency of agonistic behavior and apparent absence of hierarchical structuring. *Lemur catta* has a higher frequency of agonistic behavior and there is a clear hierarchy within the group. It is suggested that this is related to establishing priority of access to food resource units of limited size, and that it also facilitates the division of the group into foraging subgroups. Similarly, groups of *Propithecus verreauxi* have a linear hierarchy with relation to food; as with *L. catta,* this is probably a function of the limited size of resource units rather than of overall limited availability of food in the north. In the south, during periodic extreme droughts, preferential access to limited resources may be crucial for survival.

Each of the species studied has an extremely short mating season, lasting between 1 and 2 weeks. However, each shows very different patterns of aggression within this period. Although the frequency of marking increases in *Lemur fulvus rufus* during the mating season, there is no appreciable change in the frequency or intensity of agonistic encounters. In contrast, *Propithecus verreauxi* and *Lemur catta* exhibit a striking increase in both the frequency and intensity of aggression within the group. In fact, encounters at this time may result in severe physical harm to the participants. In both species, this increase in aggression is associated with a breakdown in the internal structuring of the group: This is manifested by previously subordinate males directing aggression at dominant males and gaining access to estrous females. However, where in *L. catta* the end of the mating season is accompanied by a reversion to the premating season group structure, in *P. verreauxi* the mating season appears to act as a social catalyst: Previously subordinate males may emerge from the mating season with high rank. This period is also associated with an increase in the frequency of intergroup aggression in *P. verreauxi,* something not found in *L. catta.*

In conclusion, we postulate that each of these species has a species-specific pattern of aggression integral to its social structure and ecology. However, within the constraints of each pattern, there is evidence of considerable variability associated with different ecological conditions. Thus, patterns of aggression can be meaningfully understood only within a total context of species-specific patterns of social organization and local population responses to specific ecological conditions.

REFERENCES

Charles-Dominique, P., and Hladik, C. M. (1971). Le *Lepilemur* du sud de Madagascar: écologie, alimentation et vie sociale. *La Terre et la Vie* **25**, 3–66.

Harrington, J. E. (1971). Olfactory Communication in *Lemur fulvus.* Ph.D. Thesis, Duke Univ.

Hinde, R. A. (1970). "Animal Behavior," 2nd ed. McGraw-Hill, New York.

Jolly, A. (1966). "Lemur Behavior." Univ. of Chicago Press, Chicago, Illinois.

Jolly, A. (1972). Troop continuity and troop spacing in *Propithecus verreauxi* and *Lemur catta* at Berenty (Madagascar). *Folia Primat.* **17**, 321–334.

Richard, A. (1973). Social Organization and Ecology of *Propithecus verreauxi* Grandidier. Ph.D. Thesis, Univ. College, London.

Richard, A. (1974). Patterns of mating in *Propithecus verreauxi verreauxi. In* "Prosimian Biology" (R. D. Martin, G. A. Doyle, and A. C. Walker, eds.). Duckworth, London.

Richards, P. W. (1966). "The Tropical Rain Forest," 4th ed. Cambridge Univ. Press, London and New York.

Sussman, R. W. (1972). An Ecological Study of Two Madagascan Primates: *Lemur fulvus rufus* (Audebert) and *Lemur catta* (Linnaeus). Ph.D. Thesis, Duke Univ.

AGONISTIC BEHAVIOR IN NEOTROPICAL PRIMATES

LEWIS L. KLEIN

University of Illinois

INTRODUCTION

There are 64 recognized species of New World monkeys, customarily grouped into 16 genera, 7 subfamilies, 2 families, and one superfamily, the *Ceboidea* (Napier and Napier, 1967). All New World species are believed to be related by descent from a single type of extinct prosimian or primitive monkey which flourished in the Americas during the Oligocene contemporaneously with the taxa ancestral to present-day African and Asian monkeys and apes. (For further detail see Simons, 1972.) Despite their monophyletic origin, however, neotropical primates are a diverse phylogenetic lineage, especially compared to the Old World monkeys (Schultz, 1970). They range in size from less than ½ lb pygmy marmosets to animals frequently as heavy as 20 lb (e.g., wooly and spider monkeys), and several striking morphological specializations distinguish the various taxonomic divisions (e.g., the redevelopment of claws on all but one of the digits of marmosets, the *Callithricidae,* and the development of prehensile tails in five of the *Cebidae* genera). Although few field studies have been completed to date, the information so far available suggests that the morphological diversity is reflected in significant behavioral variability as well, including both positive and negative types of social behavior. In this chapter some of the differences in hostile social interactions are described, and related where possible to methodological, ecological, and behavioral problems of wider scope. Discussion is limited to those neotropical primates for which reports of behavioral field studies of substantial duration (a minimum of 3 months) of natural populations were available. These include the following: the mantled howler, *Alouatta villosa* (Carpenter, 1934, 1965; Collias and Southwick, 1952; Altmann, 1959; Bernstein, 1964; Chivers, 1969; Richard, 1970); the red howler, *Alouatta seniculus* (Klein, 1972; Neville, 1972a,b); the spider monkeys, *Ateles geoffroyi* (Carpenter, 1935; Eisenberg and Kuehn, 1966); *Ateles belzebuth* (Klein, 1972); the capuchin monkey, *Cebus capucinus* (Oppenheimer, 1968); the titi monkey, *Callicebus moloch* (Mason, 1966, 1968); and the squirrel monkey, *Saimiri* spp. (Thorington, 1967, 1968; Baldwin and Baldwin, 1971, 1972). Observations of agonistic behavior in spider monkeys, the primary subjects of the author's research (Klein and Klein, 1971; Klein, 1972) will provide the major focus of the comparisons to be discussed.

PROBLEMS OF DEFINITIONS AND METHODS

A major problem of cross-taxa behavioral comparisons is defining categories in a manner that facilitates, rather than impedes, the task of subsequently relating discovered differences and similarities to other types of variables, e.g., ecological, ontogenetic, and physiological processes. Deciding what categories or parameters of comparison will be fruitful beforehand or exclusively by definition is risky. However, definitions are always either implicitly or explicitly made. Our decision on the problem of defining agonistic behavior was to (*a*) try to avoid as much as possible basing a definition on some hypothetically unitary motivational entity, (*b*) avoid simply treating all avoidance behavior as agonistic, and (*c*) avoid narrowly equating agonistic interactions solely with physical fighting. On the other hand, we also wanted to make maximum use of the observational material collected in the natural forest habitats of New World primates. However, obtaining lengthy and uninterrupted behavioral sequences so that all the relevant behavior can be considered is unfortunately almost impossible. As a consequence, in the analysis of the 627 hr of *Ateles belzebuth* field protocols we collected, rather than rely upon a simple definition, we broadly considered agonistic (*a*) all interactions that resulted in conspicuous rapid evasion by one or more of the conspecific participants; (*b*) interactions in which nonreciprocal biting, pulling, and pushing occurred, even though flight by one or more of the participants was not always observed; and (*c*) interactions during which the performance of specific conspicuous behavioral patterns which frequently preceded active evasion by conspecifics on other occasions occurred. Included for the latter reason were incidents of conspecifically oriented branch shaking, charging toward, lunging or running at, and growling and coughing vocalizations. In many of the cases in which these behavioral patterns were observed *not* to precede evasion or avoidance, they nevertheless appeared to elicit in the animal being oriented to signs of either disturbance, e.g., defecation, urination, screeching, or a reciprocation of similar behaviors, e.g., growling or charging toward.

For purposes of relating this behavior to other parameters, both social and ecological, we have also made some general and tentative analyses of situations in which agonistic behavior might be expected to occur, even though the more conspicuous or unambiguous indicators such as those listed above were not necessarily seen. These situations include encounters between spider monkeys known to be relatively unfamiliar with one another, between spider monkeys and potential predators, and observations of animals feeding in relatively close proximity. Discussion based on these observations has been included in an attempt to discern what influence situational contexts might have upon the expression and frequency of agonistic behavior as specified earlier, and on behavioral patterns which may represent more subtle, less intense, or more ambiguous expressions of it. In spider monkeys these would include social spacing and a number of behavioral patterns whose relationship to aggression appear to be more complex and indirect, such as branch marking, salivation, and several types of loud vocalization. It should be clear in the ensuing discussion when reference is being made to situations presumed to have an influence on aggression in distinction to actually observed agonistic incidents.

DISCUSSION

Ateles belzebuth

FREQUENCY AND TYPES OF AGONISTIC INTERACTIONS

Agonistic interactions between spider monkeys of 1 or more years of age were subdivided into three categories: *supplantations, altercations,* and *confrontations.* Observed in addition were four anomalous cases in which juveniles appeared to totally ignore a clear-cut aggressive action (biting or pushing) by an adult animal trying to get at a specific source of food, several instances of apparently pain-producing play between young animals, and a few episodes of physical restraint occurring in a weaning context.

Considered supplantations were instances of rapid location changes by one animal in a direction away from an approaching animal with neither disturbance vocalizations by the animal moving away nor "aggressive" vocalizations or gestures by the animal approaching. Thirteen such interactions were noted. Since supplantations were inconspicuous, it is assumed that a large number were overlooked in the field. In many instances, even in a captive situation, supplantations between *Ateles* may be difficult to distinguish from what we call delayed avoidance, an interaction in which the departure of one animal is delayed until some time after the arrival of another animal with, again, no gestures or vocalizations of disturbance or aggression, and no symptoms of distress. The relationship of delayed avoidance to supplantation is complex, and at this time not entirely clear.

Classified as altercations were interactions in which obvious rapid evasion, avoidance, flight and/or symptoms or signals of distress or aggressive intent were known to have occurred. Sixty altercations between spider monkeys were observed under field conditions. Loud vocalizations and noisy flight meant that these interactions were conspicuous and only very occasionally overlooked when they occurred between animals within 100 yd of the observers.

Classified as confrontations were interactions in which no clear-cut immediate avoidance or evasive activities were performed by the participants, but reciprocally oriented patterns of aggression and disturbance were. Confrontations relative to the other types of agonistic interactions were drawn out, lasting as long as an hour. Few, if any confrontations were missed if they occurred within 200–400 yd of the observers. Although confrontations were actually observed on only 10 occasions, at least another 20 such interactions were presumed to have occurred, but at distances of $\frac{1}{4}$ to $\frac{3}{4}$ mile from the observers.

Observed confrontations did not include physical contact between antagonists, but 7 of the 61 cases of altercations did involve a degree of physical contact such as quick nips, hair pulling, pushing or slaps. None of these instances resulted in externally apparent wounds. One case was observed of a male with a fresh 4-inch thigh wound in circumstances which strongly suggested that it had resulted from an unseen but heard confrontation between males of different social groups. In addition, one other male at the study site was badly scarred about the mouth. Wounding and scarring of a similar sort are known to occur in fights between *Ateles* males kept in captivity. Although other spider monkeys in our study population were seen with fresh cuts, swellings, etc., there were no substantial reasons for believing they were more likely the result of fighting than falling or insect bites scratched open.

CONSPECIFIC PARTICIPANTS IN AGONISTIC INTERACTIONS

The age, sex, and number of participants in agonistic interactions offer a potentially interesting basis upon which to compare primate taxa. Essential for these comparisons, however, are reliable estimates of the sex and age composition of the study population, and of the frequency and duration of their visual and spatial contacts with one another. At our Colombian study site (approximately 3 square miles within a much larger continuous forest), social interactions occurred among a population of *Ateles belzebuth* estimated to consist of 10 adult males, 30 adult females, 11 juveniles, and 6 infants, who in turn could be assigned to 3 separate and mutually exclusive social groups. Observed nonagonistic interactions almost always occurred between members of the same social group. However, since the members of any one spider monkey social group were rarely assembled together at a single location (for further information see Klein, 1972), the number of possible participants in face-to-face encounters was usually much smaller. Over the year, the median number of animals within 200 yd of one another (subgroup size) was 3.5 independently locomoting spider monkeys (i.e., animals approximately 1 year or older). Agonistic interactions tended to occur more frequently, however, when the animals were in visual contact with more than the usual number, a not too unexpected phenomenon. Approximately 85% of observed agonistic incidents occurred when the animals were assembled in groups of four or more, and slightly more than 50% when in subgroups of six or more.

Although most agonistic interactions were dyadic, coalitions—instances in which two or more animals simultaneously threatened or attacked others—were frequent. Thirty percent of the 60 altercations included an attack by two or more animals against a single individual. An interesting aspect of these coalitions was the extremely high degree of mutual physical contact between cooperating animals, frequently including intensive arm and tail wrapping (Klein and Klein, 1971). Aggressive behavior would frequently cease if this physical contact was broken and begin again only when it was renewed. Similar coalitions, in addition to a highly increased probability of the occurrence of mutual pectoral sniffing between cooperating animals—an important intragroup tension-reducing social signal—occurred in 10 of the 11 confrontations between contending animals of different social groups. Supplantations were almost always dyadic.

A conspicuous element of all observed agonistic interactions was the disproportionate frequency of instances of males aggressing against females, both in coalitions and in dyadic encounters. In contrast, females aggressing against males were never observed. Of the 52 altercations in which we were able to identify the participants with respect to sex and age, 50% involved males against females with the males in all cases the victors. Significant was the fact that several of the more intense of these interactions, as judged by intensity of vocalizations and vigor of flight, involved a female bearing a new infant. Coalitions of two or more males aggressing against single females were 7% of the total number of altercations, and an additional 21% involved two or more females arrayed against a single female. In only one case was a male observed to form a coalition with a female in an encounter against a single female. The remaining altercations (15%) were between two adult females. Adult male contra adult male altercations, except possibly in one partially observed interaction, were not observed, although one instance of an adult male supplanting a subadult or large juvenile male was noted, and 7 of the 10 observed confrontations involved males of different groups arrayed against one another.

The group membership of participant animals appeared to have a significant bearing on the outcome and nature of the agonistic interaction. Almost all altercations and supplantations occurred between members of the same social group, accounting perhaps for the brevity of the interaction, and the clarity with which "winners" and "losers" could be observed. Hostility in these cases appeared to dissipate over a short time. "Losers" usually desisted from either reentering the tree from which they had been forced out by another animal or coalition of animals, or remained in a position very much closer to the ground, a sort of safety area where losers of altercations frequently fled to.

Most fleeing animals maintained a constant face-to-face orientation with their antagonists. Signals functionally qualifying as appeasement, i.e., gestures facilitating or accompanying approach to a threatening animal, did not appear to be part of the behavioral repertoire of *Ateles*.

Confrontations involved animals from differing social groups. These interactions, in contrast to intragroup altercations, were usually prolonged and their outcomes considerably less decisive. Confrontations usually ended with a deflection of travel direction, or pull back by the members of one of the groups involved. Which of the two groups withdrew appeared to depend mostly upon the number of male animals from each social group actually at the site of the encounter, but the total number of animals of both sexes and the location of the meeting may also have had an effect.

The clearest results occurred when males of one group encountered an all-adult female subgroup of the other group. In these cases (two were observed), the female reactions were prolonged ook-barking bouts (as long as an hour) with very gradual withdrawal. Ook-barks when performed in these and other circumstances, mostly intertaxa contacts, were usually made by stationary animals. In contrast, the presence of males from both groups usually meant that reciprocal charging, sustained growling, and branch shaking would occur. Females sometimes participated in these displays, but their involvement always seemed to subside at a much more rapid rate than that of the participating males. Thereafter their immobility and inconspicuousness during these prolonged and active confrontations was generally striking relative to their normal participation in social interactions. Although growling, chasing, and charging were predominantly male activities when both males and females were present, encounters between females of two different groups were seen in which the aggressive patterns normally performed by the males were performed by several of the females. Similar performances of male-frequent behaviors less obviously aggressive (e.g., whoops and scent-marking) by females, in the absence of males, were known to have been performed on other occasions as well.

Behavioral patterns were seen at confrontations which did not usually occur in the other types of face-to-face agonistic interactions described. These included loud vocalizations (whooping and screaming—see Klein, 1972) and scent-marking of tree trunks and branches. Whoops were given during or shortly after 7 of the 10 confrontations. Scent-marking, in which secretions of the sternal glands mixed with saliva were rubbed onto trees, was observed three times during or shortly after confrontations, and on two other occasions when vocalizations indicated that intergroup contacts may have taken place shortly beforehand. Whether the performance of these particular behavioral patterns in association with confrontations reflects specifically distinct stimuli, specific types of motivation, the duration of the agonistic encounter, or some combination of all three is difficult to determine from the data collected. The effects of duration of excitement itself, i.e., long, drawn-out sequences of ambivalent and unresolved agonistic responses, on the occurrence of behavioral patterns so obviously related on the one hand to what is frequently referred to as "territorial" behavior, and on the other to patterns of response known to be controlled by the autonomic system, e.g., salivation and other types of glandular secretions, perhaps ought to be stressed, since a possible tie between duration of autonomic excitement and the performance of this functional class of gestures has so rarely been considered.

CONTEXT OF AGONISTIC BEHAVIOR

Agonistic interactions between spider monkeys occurred in many activity contexts; however, considering the proportion of daylight time actually spent in various activities (approximately 20% feeding, 65% resting, 15% moving—see Klein, 1972 for further detail), a relatively large percentage of the agonistic interactions appeared to occur when at least one of the participants was feeding: 40% of the supplantations and 23% of the altercations. All but two of these episodes occurred when spider monkeys were feeding on trees with crowns less than 60 ft in diameter and about 60% occurred when the animals were feeding on substances restricted to an area less than 25 ft in diameter. The association of feeding with an increased probability of agonistic interactions in *Ateles* may have been more subtly reflected in spacing phenomena as well. Specifically, (*a*) the number of simultaneously feeding spider monkeys observed within the same tree was frequently less than the number of individuals which had been traveling together as a single subgroup; (*b*) consistently fewer spider monkeys were observed to be simultaneously feeding than resting simultaneously in similarly sized and even identical trees when they were not bearing ripe fruit; (*c*) spider monkeys appeared reluctant to rest in certain trees when they bore ripe fruit although they would enter these same trees to feed and in some instances these same trees were known to be preferred resting sites at other times; (*d*) feeding adult spider monkeys generally appeared to be evenly spread out within fruiting trees, and a change of location by one animal was usually followed by an immediate change by another, reestablishing the former interanimal distances; and (*e*) spider monkeys were frequently observed waiting on the periphery of a fruiting tree, entering only after another individual had exited. (For further detail, see Klein, 1972.)

Agonistic interactions and agonistic behavioral patterns were also relatively frequent events at encounters between members of the same social group who had been traveling apart for periods of days or hours. Altercations occurred at 14% of these subgroup mergers, accounting for an additional 13% of all the spider monkey intragroup fights or chases observed. Threats or chasing redirected toward other species of nearby primates were also recorded at five subgroup mergers: on two occasions to nearby howler monkeys, and on three occasions to the observers.

INTERTAXA AGONISTIC INTERACTIONS

Agonistic behavioral patterns directed at animals of other taxa were not limited to the above-noted instances, and spider monkeys were seen threatening and behaving defensively in the presence of other species with whom they made both frequent and infrequent contact. For example, 63 contacts between howler (*Alouatta seniculus*) and spider monkeys were observed in the course of the field study. Interactions clearly agonistic were observed on 10 of these occasions. In all but three of these, the spider monkeys appeared to force the howlers into a retreat. The three exceptions involved either a female or juvenile spider monkey and an adult male howler, but similar outcomes with similar participants were variable. Several incidents of female *Ateles belzebuth* threatening and successfully supplanting male howlers were also observed (for further detail see Klein, 1972).

As between spider monkeys of the same social group, agonistic interactions between spider and howler monkeys frequently appeared to be food related. (Klein, 1972; Richard, 1970, p. 255.) Eight of ten of these altercations and supplantations occurred at or in trees with medium or small crown widths (less than 60 ft in diameter) bearing ripe fruit. In contrast, the crowns of trees in which individuals of both taxa were observed simultaneously feeding on ripe fruit (15 cases), with a single exception, were larger than 60 ft. The single exception resulted in a precipitous exit from a fruiting *Hyeronima* tree by two adult female howlers and a rapid retreat to the periphery of the tree by an adult male howler following a series of agonistic vocalizations by one or more members of a subgroup of spider monkeys composed of two adult females and two juvenile females. When feeding simultaneously on other substances in the same tree (e.g., decaying wood and leaves), proximity did not appear as likely to provoke conspicuous agonistic interactions.

Spider monkeys approaching howler females with infants, seemingly attracted by the infant, were observed on three occasions. In all instances the howler females behaved defensively, either moving away or shielding their infants by turning their backs. Eisenberg and Kuehn (1966) observed a female spider monkey carrying an infant howler for several days on Barro Colorado Island, and Richard (1970) reports a similar incident.

In the Colombian field site spider monkey contacts with capuchins were slightly less frequent (53 occasions), and considerably shorter in duration than with howlers. Only three relatively inconspicuous agonistic interactions were seen. In two cases, *Cebus apella* were observed rapidly doubling back after having moved into a tree in which a spider monkey was already situated, and an adult male spider monkey was once observed shaking branches while staring in the direction of oncoming *C. apella*. Several more clear-cut instances of aggression have been described by Eisenberg and Kuehn (1966), and Oppenheimer (1968) between *Ateles geoffroyi* and *Cebus capucinus* on Barro Colorado Island

(B.C.I.), with the spider monkeys consistently gaining the advantage.

Our data on feeding space and priority of entry into fruiting trees suggests a similar tendency. Although the diets of both taxa consisted of many of the same types of fruit (see Hladik and Hladik, 1969; Klein, 1972), and although individuals of both taxa were observed feeding successively in fruiting trees with crowns of less than 60 ft diameter, foraging groups of *C. apella* appeared very reluctant to enter such a tree if a spider monkey was already in it. Simultaneous feeding of *C. apella* and a spider monkey in a tree with crown diameter of less than 60 ft was observed on only one occasion.

Encounters between spider monkeys and potential predators at times elicited behavioral patterns similar to those observed during confrontations and altercations between conspecifics. These included ook-barking, growling, charging, and branch shaking. The intensity and duration of these reactions appeared to depend more upon the potential predator's location when first seen by the spider monkey, and its subsequent behavior in relation to its potential prey, i.e., whether it approached, moved away, or remained stationary, than on its taxa. This conclusion is largely based upon an observed range of spider monkey reactions, from almost complete indifference, to flight, to approach, in response to the same type of predator, the tyra (*Eira barbara*), depending upon its behavior and whether it was in the trees or on the ground.

One of these interactions suggested to us that branch shaking could be an effective method of upsetting the balance of a predator generally considered quite agile in an arboreal milieu. A male spider monkey pulling and shaking lianas and branches produced several near falls in a pair of tyras trying to avoid him after we had arrived on the scene. Conceivably, branch shaking, a very frequent occurrence during conspecific altercations and confrontations, could at times be serving a similar but perhaps more subtle function in these situations as well. Thus branch shaking in other situations, e.g., in response to the presence of terrestrial observers, and in other primate taxa in response to some types of social tension (see the following, and Hall, 1963) may be interpretable as either a generalized and/or ritualized expression of an immediately instrumental aggressive act, particularly likely to be performed by those arboreal animals for whom balance and agility must play as important a role as strength and speed in determining relative fighting ability.

COMPARISONS WITH SPIDER MONKEY POPULATIONS IN OTHER LOCATIONS

Few data have been published on the agonistic behavior of *Ateles* sp. in other locations or habitats. Carpenter (1935) observed a "few instances" of presumed fighting between males. No mention was made of male aggression against females, nor group affiliation of the contending males. Aggressive coalitions between adult males, a conspicuous aspect of confrontations and altercations in our experience, were presumably alluded to in the following sentence: "When a group is disturbed, the males, including those in the male subgroupings, rush toward the place of disturbance and under these conditions, rudimentary types of cooperation are to be observed [Carpenter, 1935, p. 180]."

Eisenberg and Kuehn (1966) reported several instances of agonistic interactions between the members of the small released group of *Ateles geoffroyi* on B.C.I., in addition to several instances of aggressive interaction between some of these animals and howlers, *Alouatta palliata*, and capuchins, *Cebus capucinus*. Richard (1970) observed four supplantations of howlers by spider monkeys in a fruiting *Spondias* tree, and the result of an apparent howler infant kidnapping. She also observed an adult male chasing a female and possibly inflicting a wound. It was suggested that this type of interaction might be a behavioral abnormality (p. 255). As noted above, similar interactions occurred between male and female spider monkeys in a natural habitat, and in a previous paper (Klein and Klein, 1971) the frequent occurrence of single male and male coalition attacks upon females was reported for a large colony of *A. geoffroyi* at the San Francisco Zoo.

Recently, more data on this aspect of the social behavior of the zoo colony have been collected, using an improved sampling technique. The data confirm our initial impressions that the patterns of participation in agonistic interactions in the zoo colony were similar to those in a natural population (Rondinelli and Klein, in preparation). All activities and interactions of six sexually mature males and six females were recorded in separate daily 10-min sample intervals for 2 months. In a period of approximately 350 hr, 34 agonistic interactions were observed in which at least one of the sampled adults was a participant during its sampling period. In 24 of these cases males attacked or chased females; 50% of these interactions were dyadic, the remaining 12 were coalitions of two or more males arrayed against a single female. Only three incidents of males aggressing against other adult males were observed; all were male coalitions against single males. No female was seen to attack an adult male, and only two male–female coalitions against a single female were noted. The remaining five agonistic interactions scored included four dyadic interactions between females, and one case of a female coalition attacking a single female. Chi-square values for the number of male attacks on females, and the frequency with which these attacks involved coalitions of two or more animals on the winning side were all significant at a level greater than 0.001.

Alouatta

At a simple level, virtually all observers of howler monkeys, *Alouatta* spp., have commented on the rarity of conspecific agonistic interactions and aggression (Carpenter, 1934, 1965; Altmann, 1959; Bernstein, 1964; Richard, 1970; Neville, 1972a,b), and our field data are corroborative. In comparison to spider monkeys, conspicuous agonistic interactions occur less frequently among howler monkeys, despite the fact that in all habitats sampled so far the actual number of howler monkeys in continuous and close contact with one another is usually greater than the comparable number of spider monkeys. During approximately 70 hr of observation we observed only three clear-cut cases of intragroup agonistic interactions; Neville reports 41 cases in 603 hr of observation, and Bernstein one case in 221 hr. However, the task of comparing the agonistic interactions of *Ateles* with those of *Alouatta* perforce raises issues which are of considerably broader behavioral scope than just the number of agonistic interactions observed. In this light some of the aspects underlying the impressive consensus concerning howler pacificity deserve closer attention, particularly the following five points. In the main these points suggest that there are rather narrow limits beyond which generalization concerning peacefulness and aggression based exclusively upon the absolute number of occurrences are relatively meaningless, at least at this time. First, there are good reasons for believing that some of the criteria used to distinguish agonistic interactions from other types of social interactions in other primates, including spider monkeys, are less appropriate for howler monkeys. For example, altercations between spider monkeys were interactions made conspicuous as a consequence of rapid evasion, flight, and the utterance of vocalizations associated with attacks. Faced with a necessity of relying heavily upon scoring criteria such as silent facial expressions, few if any such altercations would have been scored unambiguously under field conditions. It is of some interest to note, then, that a frequently silent facial grimace has been used as one of the key indicators of an aggressive action (Carpenter, 1934; Altmann, 1959; Neville, 1972b). In contrast, no important facial gesture in feral spider monkeys characteristically occurs without an accompanying vocalization. And while bursts of rapid movement are characteristic of the behavior of spider monkeys of both sexes and most ages in many types of contexts, howler monkeys are among the slowest and most lethargic of monkeys, and despite their famous "roars," relatively silent primates (Klein, 1972; Richard, 1970).

Second, although the absolute number of aggressive actions and defensive reactions recorded for howlers per hours observed may be considerably lower than for spider monkeys, a ratio of conspicuous agonistic to conspicuous friendly activity (e.g., grooming), would probably result in at least a somewhat different picture considering the overall low rate at which conspicuous social interactions of any type occur between howler monkeys (Klein, 1972; Richard, 1970). Plausible arguments tracing both the inactivity and relative silence of howler monkeys can be developed around the fact that the major components of their diet are noncompetitive, high-bulk, low-energy foodstuffs requiring a relatively lengthy digestive process (Klein, 1972).

Third, although the information available on the causation of male howler roaring is only of the most elementary descriptive type and virtually nothing is known concerning its ontogeny or physiological correlates, the existing data have suggested to many that their function and possible motivation may be aggressive (Carpenter, 1934; Altmann, 1959). Since the roars of group males are among the most frequently performed of howler vocalizations, inclusion of these events in a general category of agonistic acts would result in a significant upward reassessment of aggressive potential, at least in males.

Fourth, although certain vocalizations, e.g., the "squeaking door screech" (Neville, 1972b), and a single facial expression (bared lips) may have been noted frequently enough by several investigators in temporal and sequential association with abusive physical contact, chasing, clear-cut avoidance, and supplantation to justify considering them agonistic in function and probable motivation, functional and motivational interpretations of less frequently recorded gestures and vocalizations have not been attempted, despite the fact that the few data available do suggest the possibility of agonistic functions for at least some of them. In this category could be placed grunting, whimper, and oddle vocalizations (Altmann, 1959; Bernstein, 1964), sham eating, a rapid stripping of leaves with mouth and hands (Carpenter, 1934), and possibly play-fighting. For example, conspicuous sham feeding, although mentioned infrequently, has been observed performed by adult males in competitive, disturbing, and somewhat ambiguous situations, directed toward human intruders, at nongroup males, attractive but noncopulating females, and at times prior to play (Carpenter, 1934; Klein, pers. obs.). It does not appear to occur in other circumstances.

A major obstacle to understanding sham feeding and some of these other gestures is the current unavailability of captive colonies where intensive studies might be feasible. Moreover, since the discovered relevancy of behavioral acts often results in a more systematic recording of their occurrence, functional interpretations of greater numbers of communicatory acts might again force a reassessment of the consensus concerning howler pacificity. (For a possible analogy see Klein, 1971 and Klein and Klein, 1971.)

Finally, hesitation is certainly warranted by Chivers' (1969) observations that 15% of the adult males in the groups he followed were either scarred or seen with freshly torn lips, and his interpretation that these injuries were most likely to have been the result of intragroup fighting. Neville (1972a) reports a similar percentage of adult and subadult males with scars, torn ears, and missing fingers. His arguments that these were caused by nocturnal predatory attacks are unconvincing, particularly on the matter of why such wounds should be evident in adult and subadult males, but not in females. The possibility that serious fighting is radically inhibited by the presence of a human observer in an animal whose reaction to mildly stressful situations is frequently extreme passivity and relative inactivity deserves some serious attention.

PARTICIPANTS

Although the average size and composition of howler monkey groups varies between habitats, perhaps between species, and with changes in population density, behavioral observations have been limited in almost all cases to areas where cohesive and stable groups of 8 to 20, usually consisting of more than one adult male and a two to four times greater number of adult females and immatures, are the norm. Since conspicuous occurrences of intragroup agonistic interactions have been reported only rarely, however, clear patterns of differential age–sex participation or roles in most types of agonistic interactions are difficult to discern. Carpenter (1934, 1965), for example, repeatedly emphasizes the pacific nature of intragroup relationships between and among all age–sex classes and most cases of intragroup aggression have been treated as isolated, nonsystematic, or atypical events. In this vein Collias and Southwick (1952), for instance, noted seven aggressive interactions (lunging, backing away, bared teeth, and cackling vocalizations) between the adult females of a single group—possibly the same pair over a 17-day period. They considered that it may have been the result of an unusually high ratio of males to females in that particular group that year (one adult male and nine adult females). Bernstein (1964) reported a single instance of aggression, a male attack on a female, and Richard (1970) noted a short physical confrontation and chase between two adult males in a six-male troop. The most systematic sample of intragroup (*A. seniculus*) agonistic interactions of several types and degrees of intensity are provided by Neville (1972b). Of 41 incidents, 7 are reported to have occurred between adult males and 16 between females. In this particular study population (Hato Masaguaral) the ratio of sexually mature males to females was approximately equal. Participants in the remaining cases are either unspecified, although the illustrative protocols indicate that at least a few occurred between an adult male and females, or involved younger animals. It will be recalled that Chivers (1969) interpreted the occurrence of scars and wounds on 15% of the adult males of the groups he observed to be the result of intragroup fighting.

Besides aggressive interactions presumably related to weaning, several instances of agonistic interactions between fully adult males and younger animals have been recorded, including a serious attack by an adult male on a young infant (Collias and Southwick, 1952), interference with play interactions (Carpenter, 1934; Neville, 1972b), and aggressive play between juveniles (Carpenter, 1934). Agonistic interaction between fully adult and younger but sexually mature males has sometimes been tentatively offered as an explanation of the occurrence of extragroup males and the greater number of females than males in most of the censused populations, despite the fact that actually very few such interactions have been reliably described from first-hand observations.

Exceedingly vigorous response by clan males to some extra-group males is one of the clearest patterns to emerge from the published observations. An extended series of such interactions was reported to have occurred repeatedly over a 4-month period between a specific extragroup male and four clan males (Carpenter, 1934). Behavior noted at these encounters was extended roaring, branch shaking, and vigorous locomotory movements by the clan males, and evasive hiding movements and "feeble barks" by the group follower. On one occasion the extra-group male was seen with wounds. At times, female as well as male group members appeared to react vocally to the extragroup male's presence. However, in addition to the exceedingly vigorous response some clan males make to the proximity of apparently nongroup males, other males appear to join groups freely with few indications of hostility directed against them (Carpenter, 1934; Neville, 1972b). With the existing data, it is impossible to determine what distinguishes the males reacted to in this way and those which appear to move freely into and out of groups. Carpenter's position that males reacted to strongly at first eventually become fully habituated to the original clan males is not fully supported by his own observations.

INTERGROUP AGONISTIC INTERACTIONS

Visual contact between howler monkeys of different groups occurs infrequently relative to vocal contact. Most reports indicate that group males roar at an average rate of between once a day or once every 2 days (Carpenter, 1934; Chivers, 1969; Altmann, 1959). And since these calls can be heard at distances greater than at least $\frac{1}{2}$ mile, presumably in most areas several groups are in daily auditory contact. In contrast, the highest estimate of intergroup visual encounters (Altmann, 1959) is about once every 6 days per group. The best descriptions of what may happen at these encounters are still those of Carpenter (1934). Somewhat paradoxically, as described, the most consistent result of these visual contacts are roaring bouts between the males of each group, similar to those occurring in response to audible stimuli. In some cases, roaring is all that appears to happen, although there are indications that the males of each group in at least some cases draw closer to one another than usual, suggestive of a degree of cooperative aggressive tendencies. In contrast to spider monkeys, however, little if any mutual contact between aggressive partners appears to occur despite the absence of the vigorous locomotory charges spider monkeys would usually perform under similar circumstances; likewise neither salivation nor behavior suggestive of scent-marking has been observed in howlers in these circumstances.

Only a few of the accounts of encounters between howler groups mention more conspicuous types of agonistic behavior such as branch shaking, charging, and back bowing; and female participation appears to be limited to occasional supporting vocalizations, barks according to Carpenter (1934), wail-like sounds according to Altmann (1959). Only Chivers (1969) has suggested that the roars given in these visual confrontations differ from those exchanged when the groups are out of sight of one another.[1]

Reported durations of intergroup visual encounters appear to be uniformly less than 1 hr, and in this respect, at least, interactions between groups may be less intense than the similar but nonreciprocal responses evoked by the proximity of extragroup males.

Although roars and other forms of threatening behavior which occurred at intergroup encounters were originally suggested to be the result of incursion into and the defense of territorial boundaries (Carpenter, 1934; Altmann, 1959), later work suggests that considerable overlap in range patterns occurs between adjacent groups, and boundary areas are not stable. What appears to provoke an immediately mutual active reaction

[1] Distinctions noted were those of amplitude and "participation by animals of all classes."

is a perceptible incursion into a consistently shifting area of about 150 yd surrounding a group's actual location (group space) following an imperceptible approach (Chivers, 1969). This can occur as a result of extensive movements following morning roars, or movements on days in which roaring does not occur or cannot be heard, e.g., during thunderstorms. Vocally advertised proximity of 150 to 220 yd appears to result in mutual avoidance rather than active confrontation. One hundred and fifty yards may represent a critical distance on B.C.I. It falls between 50 and 100% of several estimates of average daily travel distances (Altmann, 1959; Carpenter, 1934; Chivers, 1969). Vigorous and energy-demanding defensive reactions are apparently more likely to occur only when the possibility of chance encounters becomes highly, but at this time, unspecifiably, probable.

Insofar as contrasts between howler and spider monkeys can be drawn at this point, the most salient ones are negative. There is for instance, little evidence of special and systematic agonistic interactions occurring between males and females, and there are no indications of any clear intragroup type of social facilitation, coalition formation, or aggressive contagion occurring during the infrequent intragroup agonistic interactions, despite the occurrence of coordinated roaring between males in reaction to the proximity of extragroup males and other howler groups.

INTERTAXA AGONISTIC INTERACTIONS

Interactions between howler monkeys and other animals, including other primate taxa, may on occasion be agonistic, resulting in avoidance, chases, mild threats, etc. Based upon both published descriptions and our field experience, typical howler reactions to the proximity of animals ranging from about one-third their body size and larger, appear to be some version of lethargic wariness. Stimuli sufficient to provoke full-blown roaring or barking bouts do not occur frequently, and appear to require either the persistent approach of staring and unfamiliar human observers, attacks or lunges by animals within trees, and/or exceptionally loud or startling noises or events, e.g., gunshots, the crack of a suddenly breaking large branch, loud and abrupt bird calls, thunder, etc. Altmann (1959) suggests that the category of vocalizations alternately referred to as either roars or howls should be subdivided into roars and barks, the former predominant during intergroup encounters, the latter occurring in response to disturbances by humans.

Except for feeding (see the following) there are few data concerning specific conditions predisposing intertaxa agonistic interaction. In some cases, threatening or avoidance reactions by *Alouatta* to the approach of individuals of other taxa appears to depend upon the number of adult howler males present.

CONTEXT OF AGONISTIC INTERACTIONS

Most observers of howler monkeys have commented on the absence of any overt indication of food-related competition, feeding priorities, or feeding space (Altmann, 1959; Bernstein, 1964; Carpenter, 1934, 1965). Although the emphasis may turn out to be somewhat one-sided, for example, Neville (1972b) notes a few cases of food or food site appropriations and supplantations, the lack of a clear-cut relationship between feeding and agonistic interactions appears to be another major contrast between howler and spider monkeys, and perhaps capuchins as well (see below). In our opinion, the unimportance of feeding-induced aggression is more likely to be a function of the unique characteristics of the usual *Alouatta* diet (mature leaves and figs) than the result of hypothetical superabundancies of food products in neotropical forests (Klein and Klein, in press; c.f. Carpenter, 1965).

The insignificance of feeding competition within howler groups may not, however, extend to intertaxa agonistic interactions since many such contacts appear to occur as a consequence of mutual attraction to fruiting trees (Klein and Klein, 1972). At our study site, several instances of agonistic interactions between spider and howler monkeys, and clear approach inhibitions were traceable to feeding in medium and small crowned trees bearing ripe fruit. However, within larger trees howlers frequently have been observed feeding in proximity to many other taxa without incident both by ourselves and others (Klein and Klein, 1973; Carpenter, 1934). However, most of the recorded instances of agonistic feeding interactions between howlers and other primate taxa suggest that the reactions of the howlers are usually defensive following direct approaches by individuals of other genera (Richard, 1970; Collias and Southwick, 1952; Eisenberg and Kuehn, 1966; Klein and Klein, 1973).

Despite prevalent interpretations to the contrary, one of the more consistent situations in which *Alouatta* intragroup agonistic interactions appear to occur is competition for access to sexually receptive females. For example, Carpenter's (1934) protocols of sexual behavior indicate that certain types of interactions interpretable as agonistic, e.g., supplantations and inhibition of approach, regularly precede copulations in groups consisting of more than one adult male. In 5 out of 10 cases (observations numbered 34 to 43) there were indications of mating priorities clearly dependent upon intermale competition, not female preferences. And although the other five protocols did not positively indicate overt male–male competition, they represented either very brief observations, and/or observations of sexual relations between a single male–female consort pair. The latter is a situation consistent with what might be expected where stable male dominance hierarchies exist as well as with the overt competition described in the other protocols. Neville (1972b) reports similar events. The occurrence of extragroup males and the agonistic reactions to some of them by group males may be a related phenomenon.

The well-known difficulty of determining relative dominance between howler males, and in fact establishing the very existence of dominance hierarchies themselves, may be more a function of the relatively low levels at which howlers interact and react than to anything else. And this, as has been noted earlier, may be relatable to characteristics of their feeding ecology.

Callicebus

Mason (1966, 1968) has published the only existing extended behavioral reports on a natural population of titi monkeys, *Callicebus moloch.* His principal study site was a small and isolated relict forest (about 17 acres) in the Colombian llanos which contained 28 titi monkeys socially and spatially divided into nine groups. Eight of the nine groups consisted of a single pair of adult animals in addition to a maximum of two additional immature animals. No other taxa of primate appeared to be permanent residents of this particular 17-acre plot.

Mason recognized two major types of agonistic interactions: confrontations and intragroup agonistic interactions. Minimal descriptive information of agonistic interactions between individuals of the small adult male–adult female offspring unit is given. Presumably spatial supplantations, fighting, chasing, and active avoidance were rarely observed. The few cases noted involved chasing or biting between adults and subadults in three of the nine groups, and an unspecified degree of "mild aggression" between an adult pair subsequent to a female's consorting with a male from a neighboring group. Mason suggests, however, that the relatively more prevalent aggression between subadult and adult animals may be the reason two subadults eventually disappeared. Situational factors possibly affecting the frequency of these apparently rare instances of intragroup conflicts were not noted and aggressive responses toward predators, other taxa, or the observer were not mentioned. In our own very brief contacts with *Callicebus moloch* and *torquatus* in Colombia, the only responses to our presence noted were flight or visual investigation. The impression that intragroup aggression is rare is consistent with Moynihan's impression (1966) that *Callicebus* are less frequently and intensely aggressive than *Aotes* in captivity.

Mason described no conspicuous behavioral acts in the natural population, aside from avoidance, which appeared to remotivate or terminate intragroup agonistic interactions. Moynihan (1966) describes one vocalization, a combination trill-whistle, one facial expression, teeth-baring, and several cases of a subadult jumping onto the back of an aggressing adult, all of which sometimes apeared to terminate agonistic interactions without withdrawal. As described, however, the latter performance appears to be an artifact of captive rearing and the vocalizations and facial expression are difficult to interpret functionally, considering the physical and behavioral limitations imposed by the small cages.

In contrast to intragroup agonistic interactions, confrontations, "an agonistic display elicited by and directed toward neighboring groups . . . in which calling, rushing and chasing were prominent elements [Mason, 1968, p. 215] appears to occur frequently. Twenty intergroup confrontations in 12 days were reported to have occurred between a sample of three adjoining groups. Confrontations lasted from less than 5 min to more than 30 min and several types of conspicuous behavioral patterns frequently associated with confrontations appear *not* to have been seen or heard in other contexts. These would include moaning, arch postures, piloerection, tail-lashing and chest rubbing. Physical contact and relatively close proximity between the male and female of each pair was also noted as a conspicuous feature of these intergroup confrontations. Physical contact between antagonistic animals is described as both rare and brief when it was observed to occur, and Mason reports seeing no *C. moloch* with wounds, torn ears, or scars.

Mason thinks intergroup confrontations were neither random nor the result of an attraction to specific food sources, and suggests that the motivation may be intrinsic, with lowered thresholds occurring during the early morning hours, when in the proximity of nongroup animals, and at specific locations (1968). Mason does not specifically mention any encounters between groups which did not result in confrontations although they may have occurred.

It will be recalled that some data (Chivers, 1969) indicated that one of the events partially responsible for at least several close at hand visual–vocal confrontations between two groups of howler monkeys following mutual rapid agonistic aproach was intergroup proximity of 150 yd or less resulting from inconspicuous feeding and traveling movements of the preceding day apparently made mutually conspicuous by the morning roaring bouts. It was also suggested that the proximity eliciting this type of agonistic interaction in howlers was not too discrepant from the best estimates of *Alouatta* daily travel (Carpenter, 1965). Implied was the fact that the minimum distances maintained without engaging in relatively active and energy-consuming displays were related to the distances generally covered in the course of normal daily feeding and moving between feeding locations and sleeping sites. Superficially, a similar hypothesis appears to be irrelevant to *Callicebus* confrontations which apear to occur regularly in small stable sectors of each group's home range, but it is interesting to note that the considerably more frequent active confrontations between *Callicebus* groups occurs in animals whose daily path length (approximately 550 m) is considerably greater than the diameter of its annual home range (less than 100 m), a situation not as characteristic of howlers on Barro Colorado Island (Chivers, 1969) nor of our study site (Klein and Klein, in press). Although by itself this possible relationship cannot explain the development of territorial behavior in *Callicebus moloch,* it does appear to me to have possibilities for explaining why it has developed in a classical way in that taxa, and not in *Alouatta.* On the basis of the sketchy information so far available on the subject, both genera should be able to find a sufficient variety and amount of their specific dietary requirements within sharply circumscribed and defendable areas in many types of forest habitats throughout their respective ranges in tropical America, but perhaps the relevant dietary differences have not yet been recognized, e.g., relative importance of fig fruit for howlers, camouflaged and cryptic insects for *Callicebus.*

Saimiri

FREQUENCY AND TYPES OF AGONISTIC INTERACTIONS

Of all the neotropical primates, squirrel monkeys appear to have been studied the most intensively, frequently, and by more investigators than any of the others. Curiously, this research interest has not yet appeared to spark either the desire, funding, and/or personnel necessary to carry through a long-term study in a reasonably intact natural habitat. Baldwin and Baldwin (1972) and Thorington (1967, 1968) currently hold the distinction of having watched squirrel monkeys for the longest period of continuous time (10 weeks) at locations within their range of natural distribution. Baldwin's site was in western Panama, the subject species *Saimiri oerstedi;* Thorington's site, east of the Andes in the Colombian llanos, his subjects *Saimiri sciureus.* Both studies were made in forests seriously affected by both past and ongoing human activities.

Interactions considered agonistic and most frequently observed by the Baldwins in Panama were highly vocal chases. Less frequently noted were what appear to be a succession of consecutive supplantations called by the authors "semi-chases," as well as 13 "mild threats" which included pushing–pulling, lunging, and sham biting usually accompanied by a spitting vocalization. Genital displays, a frequent and conspicuous aspect of the behavior of *Saimiri* in captive conditions[2] and treated by some researchers as an exceptionally important threat display, were observed only twice by Baldwin and Baldwin and six times by Thorington. In addition to penile displays, Thorington notes an unspecified number of fights and chases.

In common with the other taxa reviewed in this chapter, it is difficult on the basis of existing material to make any meaningful assessment of either relative or absolute frequency of agonistic interactions. The problems again are ones of weighing the effects of nonrandom and noncontrolled sampling, different behavioral criteria, probabilities dependent upon the number of animals in contact with one another and the observer at any one time, overall rates of social interaction, as well as a general paucity of observational data from natural habitats.

Our impressions based on personal but unsystematic field observations of *Saimiri* were that over the year we were more likely to see or hear an intragroup chase between squirrel monkeys per contact hour than any comparable agonistic event within either a howler or spider monkey group. The contrast with *Cebus apella* was less marked. Specifically, at our Colombian field site we observed six intraspecific fights or chases fully, and heard or partially saw an additional 41 during a cumulative contact period of 55 hr with groups composed of about 30 independently locomoting animals. Baldwin and Baldwin (1972) indicate an even higher rate of occurrence in a Panamanian group of 23 animals sampled during their breeding season. "Chases were the most common social behavior observed. During some hours they occurred as frequently as 20 discrete bouts/hr, and at other times as infrequently as 1 bout/hr [pp. 174–175]."

[2] The extent to which penile displaying should or can be relied upon as an index of threat behavior should probably receive more critical attention than it has. It does, however, appear reasonably certain that relative changes in the frequency with which adult squirrel monkeys at Monkey Jungle and in laboratory situations perform closed position penile displays correlates rather well with changing frequencies of fights and chases on a seasonal and shorter term basis (Baldwin, 1968; Ploog and MacLean, 1963; Ploog *et al.*, 1963). The proposed threat function of the display is also strengthened by observations that a rank order correlation exists between frequency of displaying and an intermale probability of chasing and winning fights. Fights are also reported to be preceded and/or associated with penile displays. The status of additional motor patterns, e.g., urine washing and kicking, branch shaking (Baldwin, 1968) and arr and yap vocalizations (Winter, 1968) is less certain, but probably similar. For purposes of this review, however, the functional interpretation of these displays is not a major one because very few of them have either been observed or noted in a field situation.

CONSPECIFIC PARTICIPANTS IN AGONISTIC INTERACTIONS

Reported group size for natural populations of *Saimiri* sp. counted under conditions in which animals were not fleeing from observers vary from 10 to 300 animals (Baldwin and Baldwin, 1971). Although Baldwin and Baldwin suggest that group size correlates highly with area of undisturbed forest, our data are not supportive. Groups composed of more than 40 independently locomoting squirrel monkeys were not seen by us either in the undisturbed and extensive forests along the southern boundary of the Colombian national park La Macarena, nor in the smaller 30–60-acre forest island relicts in the inhabited and cleared regions of the Colombian llanos. Group size in all areas in Colombia where we saw *Saimiri,* in both altered and unaltered forest, ranged between approximately 20 and 40 animals. The evidence on group size and undisturbed *Saimiri* habitat as of now suggests at best only a correlation between group size and some factor related to geographical latitude, since groups in altered and unaltered Amazonian regions appear to be from 3 to 10 times as large as those in altered and unaltered forests further north.

Information on characteristic group composition is limited by the small number of groups censused and the failure to apply unambiguous morphological criteria to the task of distinguishing adult from "subadult" males in field and semi-field conditions.[3] However, the within-group ratios in censused natural populations of females with body and canine length fully developed to similarly fully developed males (approximately 30-month and older animals—Long and Cooper, 1968) are: 6:5, 7:5 (Baldwin and Baldwin, 1972), and 5:3 (Thorington, 1968). The number of immature animals (less than 36 months) in the above reports appear to be slightly less than double the number of fully adult females (3 years or older).

Baldwin and Baldwin's (1972) Panamanian observations indicate clearly that mature male *Saimiri* as a class are much more frequent participants in intragroup agonistic interactions than any other age–sex

[3] The criteria used by Baldwin (1969) to make this particular age–class distinction appear to be based on behavioral patterns, testicular size, and presumed ability to develop into the "fatted" condition (in areas other than Panama). The difficulty is that these masculine characteristics appear to be both (*a*) seasonal, and (*b*) highly affected by social stimuli. Thus they may be partially independent of chronological age and maturational stages. Discussion of behavior characteristics, including agonistic behavior, emphasizing differences between adult and subadult males may then be tautological. It might be better in the future to limit class distinctions between full longitudinally and dentally developed males to those features which have actually been perceived: "fatted," fully fatted, actively spermatogenic, etc., rather than to a presumed but unverified chronological development.

group. All chases and "semi-chases" observed (and these were reported to be the most commonly observed types of social interaction) included an adult male participant. In contrast to chases in *Ateles,* however, the most frequent of this type of interaction appeared to be coalitions of females and juveniles chasing after (or away) single adult males. It was indicated that these coalitions could include as many as 10 animals; not mentioned are either physical contact or special signals between coalition members. Most approaches by males to females were considered by the Baldwins to be sexual rather than aggressively motivated, and if only one male and female were involved, they indicate "the interaction usually ended without chases or disturbance [Baldwin and Baldwin, 1972, p. 175]." It was implied that approached juveniles and infants, although likely to react, were unlikely to chase an adult male unless joined by a female. However, at least one case in which a juvenile appeared to successively supplant two adult males was observed. Although no chases or "mild threats" other than two penile displays were observed to have been exchanged between the two adult males most frequently in proximity to the majority of other group members, one or both of these males were seen chasing a third "peripheral" male. The frequency of these chases was not reported.

The apparently very few observations of agonistic interactions between the two feral males during the breeding season contrasts remarkably with the situation described in detail for Monkey Jungle (Baldwin, 1968, 1971) and said to have been observed in the much larger troops observed very briefly in Peru during the mating season (Baldwin and Baldwin, 1971). However, the existing data suggest strongly that the very high incidence (presumably reflecting high individual rates) of penile displays and agonistic social interactions at Monkey Jungle may be an artifact of captivity in which the usual intensive foraging by *Saimiri* for insects may have been reduced by provisioning. Of course, the total observed frequency of such interactions would be expected to be greater in groups of larger size, regardless of habitat.

A proportionately high frequency of agonistic interactions between adult males usually proximate to adult females and seven to nine extra-group "bachelor males" has also been reported for Monkey Jungle (Baldwin, 1971; DuMond, 1968). The existence of independent groups of male *Saimiri,* however, has never been reported for areas inhabited by native populations of squirrel monkeys.

Agonistic interactions considered less intense by Baldwin and Baldwin (1972) were infrequently noted during their Panamanian study. No specific breakdown of the participants is given, but it is stated that in none of these cases were adult males involved. At Monkey Jungle at least infrequent agonistic interactions between all age–sex classes, including mothers and infants, have been noted (Baldwin, 1971).

INTERGROUP AGONISTIC INTERACTIONS

Specific interactions between feral squirrel monkeys belonging to different social groups have not been observed. Baldwin and Baldwin (1972), however, did describe a coordinated zig-zag pattern of travel by two troops which may have represented a series of active unilateral or reciprocal avoidances. DuMond (1968) describes the release of extragroup animals into the Monkey Jungle colony. The results indicated that adult females were readily accepted and integrated rapidly and successfully with little or no agonistic behavior, while adult males were intermittently aggressed against by both colony females and adult males, depending upon season and sexual condition.

INTERTAXA AGONISTIC INTERACTIONS

Most described interactions between *Saimiri* and individuals of other taxa, either comparable or larger in size, indicate a neutral or clearly defensive–evasive reaction by the squirrel monkeys. A few exceptions, however, have been described, and ought to be noted, along with some comments on their special spatial relationship with some species of capuchin monkeys.

On occasion, individual or small numbers of squirrel monkeys have been observed making tentative approaches to larger stationary animals (to distances of 10 or so feet) while vocalizing loudly and successively assuming fixed visual orientations from differing locations and angles. Responses of this sort have been observed given in response to human observers (Klein, pers. obs.; Baldwin and Baldwin, 1971), dogs (Baldwin and Baldwin, 1972), and to a tayra (Thorington, 1968). A loud raucous call directed at a nearby potentially threatening uakari was noted by DuMond (1968) and a similar call directed at an insectivorous hawk, *Harpagus bidentatus*, with a recently caught katydid was noted by Klein and Klein (1973). Baldwin and Baldwin (1972) observed two squirrel monkeys pluck tail feathers from a sleeping nocturnal bird; Thorington (1968) reported a supplantation of a toucan, and Winter (1972) reported a burst of aggressive vocalizations in a small free ranging group in response to herons and sudden movements of an observer. Actually, the relative paucity of conspicuously aggressive responses to potentially dangerous and/or competitive taxa is perhaps the most significant aspect of *Saimiri* intertaxa behavior, considering the seemingly high levels of intragroup chasing and fighting.

A special relationship between *Saimiri sciureus* and small groups of *Ce-*

bus apella (in Colombia between 5–12 animals) has been noted by several observers (Klein and Klein, 1973; Baldwin and Baldwin, 1971). Where sympatric, the two primate species appear to spend about 50% of the daylight hours in close spatial associations that extend beyond simply feeding in trees bearing ripe fruit. There is, nevertheless, little evidence of mutually friendly interactions, and some indications of at least occasional food-related chases and supplantations. In all reported instances individual capuchins appear to have been the winners of these interactions. It is of some interest, therefore, that in the area of Panama studied by the Baldwins, spatial proximity is reportedly infrequent between sympatric groups of *Saimiri oerstedi* and at least one group of 27 *Cebus capucinus*. The Baldwins (1972) report that squirrel monkeys responded to the presence of *C. capucinus* with marked avoidant and fearful reactions. Perhaps the key to this avoidance is the much larger capuchin groups in this area.

CONTEXT OF AGONISTIC INTERACTIONS

Feeding. The bulk of the squirrel monkey's diet, generally considered to be a mixture of food and insects (Fooden, 1964; Thorington, 1967, 1968; Klein, 1972), is usually obtained while the animals are actively mobile. For example, in one sample period while feeding on fig fruit, the longest period a squirrel monkey remained at a single location was 6 sec; the simultaneous comparable durations for *Ateles* and *Cebus* were 30 and 12 sec. When feeding on insects, stationary intervals for foraging *Saimiri* are usually even briefer (Klein, 1972). Baldwin and Baldwin (1972) report for *S. oerstedi* an average of one stationary minute per each large item observed eaten with a maximum of 2 min on one occasion. The more commonly observed pattern, however, "consisted of eating while moving from branch to branch and engaging in almost continuous manual and visual investigation [1972, p. 171]." Baldwin and Baldwin explain this mobility as a consequence of the small size of most of the objects eaten by the squirrel monkeys and the presumed scarcity of insects and other foods in that particular forest. Other interpretations are possible (see, e.g., Thorington, 1967; Klein, 1972; and Klein and Klein, in press), and more compatible with the fact that (*a*) groups of *Saimiri* frequently return to the same fruiting tree several times in the course of a short period (sometimes in less than an hour), (*b*) the average duration of stationary feeding bouts at any one location usually does not exhaust all the ripe fruit obtainable there, and (*c*) the fact, as exemplified above, that other taxa when feeding on identical items are considerably less mobile.

In either case the normal pattern of feeding while moving would make it extremely difficult to score agonistic interactions, such as supplantations, in a feeding context even if they were occurring. And food items as a precipitator of fights and chases would be both difficult to recognize and relatively unimportant to the animals as a consequence of their exaggerated run and grab pattern of foraging, i.e., there would generally be nothing to defend, hoard, or steal away. The same active feeding pattern may also be responsible for the fact that some laboratory researchers (Ploog and MacLean, 1963; Castell and Heinrich, 1971; Green, *et al.*, 1972), but not all (Plotnik *et al.*, 1968) appear to have had difficulty in measuring consistent feeding or drinking priorities correlated with other measures of social dominance.

At the 4-acre Monkey Jungle where provisions are primarily obtained from two specific feeding platforms, the qualitative descriptions leave the reader with the impression that agonistic interactions under these conditions are rather rare, at least in comparison to what would be expected in similar circumstances for some macaques.

> It might be expected that the close quarters of the feeding area would pose an unnatural social stress in the colony as certainly would be expected among macaques or baboons. However, there has been no observed evidence of competition for food or space at the feeding stations. The youngest animals eat contentedly beside adult males with no observable tension [DuMond, 1968, p. 94].

However, the problem of determining what effect if any feeding competition has upon the frequency of agonistic interactions in *Saimiri* relative to other contexts, if not relative to other taxa, has not been determined. Baldwin's (1971) quantitative data did not include social interactions occurring when the animals were feeding at the provision sites.

> The behaviors and individual distances of the animals were atypical when they were at these [feeding, L.K.] stations; therefore, no observations made on animals at these stations have been included in the calculations or descriptions of this or previous reports [Baldwin, 1971, p. 30].

Although rarely noted, agonistic interactions in a feeding situation have, however, been noted in the field. Thorington notes "jockeying for position" when in a fruiting tree (1967, p. 182), and Baldwin and Baldwin (1972) despite stressing the lack of aggression when these monkeys are feeding on certain fruits do mention "attempts at food stealing [p. 173]." Our field impressions were that *Saimiri* were at least sometimes competing with one another, primarily through supplantations, more for insects and locations likely to produce insects, e.g., within epiphytic bromeliads than for fruit. More conspicuous conspecific agonistic interactions clearly occurring in these contexts were seen only once.

Seasonality. One of the better known results of the investigations conducted at Monkey Jungle was the discovery of a marked seasonal increase in male–male and male–female agonistic interactions associated with seasonal mating, the onset of spermatogenesis, and an easily perceivable increase in fat deposition and pelage "fluffiness" around the shoulders and upper torso of some adult males (DuMond and Hutchinson, 1967; DuMond, 1968; Baldwin, 1968). These events have been called the "fatted" male phenomena. Although the original research (DuMond and Hutchinson, 1967) clearly demonstrated that the physiological changes associated with the "fatted" condition must also have been occurring in the natural habitat areas (Iquitos, Peru) in which the *Saimiri* were being trapped, the behavioral field data published to date (Baldwin and Baldwin, 1971) inadequately support the hypothesis that the seasonal development of these secondary sexual characteristics correlate with significant increases of agonistic behavior in native populations. Thorington's (1968) Colombian observations were made from January through most of March, encompassing part of the birth season, and Baldwin and Baldwin's (1972) 10-consecutive-week Panamanian study encompassed just the breeding season. Moreover, they report that at least one of the indices of seasonal sexual changes, the deposition of noticeable amounts of subcutaneous fat, did not appear to be characteristic of the two adult males of the group they observed, nor, as noted above, did they observe frequent agonistic behavior between the two group males. They attribute this apparent discordance to a food supply judged to be inadequate primarily on the grounds that the amount of time they observed *Saimiri* engaged in foraging activities appeared surprisingly great. However, considering the lack of normative data on *Saimiri* activity levels in any other natural environment, either during a mating or nonmating season, their interpretation strikes me as exceedingly tenuous and premature.

Curiously, in another report Baldwin and Baldwin (1971) reject the possibility that seasonal effects could account for differences in individual rates of social interaction, including the frequency of agonistic interactions, between Colombian and Amazonian populations of *Saimiri* in favor of an hypothesis involving group size, peer play group sizes, and socialization processes. Unfortunately the rejection is based on the erroneous assumption that the groups observed in Colombia and the Amazonian regions between June through September 1969 were both sampled during their respective breeding seasons (Baldwin and Baldwin,

1971, pp. 54–55). Contrary to Baldwin's belief, however, the *Saimiri* infants he observed in June 1969 were probably no older than 4 months, as *Saimiri* births in the areas of Colombia he surveyed were clustered between late February and early April in 1965 (Thorington, 1968, and pers. comm.), and in our Colombian site, approximately same longitude but 2° further south, in April and May in 1968 (Klein, 1971). There is good reason to believe that these birth seasons remain relatively constant from year to year (DuMond and Hutchinson, 1967; Klein, pers. obs.). Working from an estimated gestational period of 165–170 days (Rosenblum, 1968), periods of intensive mating in both these areas should generally fall sometime between September and February. Baldwin's Colombian observations were made primarily in June, a month not part of the mating season in 1964, 1967, nor 1968. On the other hand, his observations in Amazonia during July and August do correspond to an estimated peak of sexual activity in the regions of Iquitos and Leticia (DuMond and Hutchinson, 1967; Rosenblum, 1968).

The very limited data we recorded on *Saimiri* agonistic interactions at our study site support the possibility that active and vocal fighting and chasing are relatively more frequent at those times when conception is likely. Approximately 70% of the *Saimiri* fights and chases we recorded (50 cases) over 11 months occurred in that 3-month period (December–February). Regular and marked seasonal changes in the incidence of agonistic interactions may be one of the major contrasts between *Saimiri*, and the other neotropical taxa discussed in this chapter.

APPEASEMENT GESTURES AND MATING SUCCESS

As described for Monkey Jungle, while most agonistic interactions between adult males and females appear to be just the result of males approaching females, serious fighting and chasing between troop males appeared primarily to be the result of closed penile displays in which at least one of the displayed to and contacted animals did not remain stationary and in a huddled, restlike position (Baldwin, 1968; DuMond, 1968). This immobile rest position appears to be the only behavioral pattern of adult *Saimiri*, besides perhaps an "affectionate purr" vocalization, noted by DuMond (1968), which might functionally qualify as an appeasement display. The immediate causes of the initial penile display are rather mysteriously said to be "general excitement" resulting from an "aversive or threatening stimulus, either social or not [Baldwin, 1968, p. 289]."

One of the anomalies in the material published from Monkey Jungle is the fact that the relatively stable penile display hierarchy appears to have no additional social implications.

> What is so interesting about the display-dominance hierarchy phenomenon is that the dominance standing of a male seemed to have virtually no implications in its social life within the colony when the males were not actively participating in a display episode. The most submissive male, Scar, would eat with absolute composure side by side with Pink, and during sexual interactions Scar at times was more actively involved than any of the other more dominant males [DuMond, 1968, p. 120].

> During the three month mating season, I observed seven changes in the ranking of the four males, the changes occurring after fights in which some of the males were often seriously injured. But rank in the hierarchy did not correlate with privileges to food or sex (or anything else that I could observe) [Baldwin, 1968, p. 311].

Repeated emphases on this particular anomaly (see, e.g., Jolly, 1972) ought, I think, to be tempered by a full recognition that the degree of excitability recorded for these animals (the most likely explanation of the lack of copulatory success by the dominant animals) is likely to be one of those social parameters most affected by artificially high densities and decreased amounts of time spent in foraging activities. Moreover, a careful reading of Baldwin's and DuMond's reports indicates that one of the most active participants in penile displays, a male not observed to successfully mate in the breeding season of 1967, was the only adult male troop member in July 1966 (the three other adult male troop members were introduced thereafter) and was very likely to have been the progenitor of the approximately 30 offspring born between June and August 1966.

Cebus

FREQUENCY AND TYPES OF AGONISTIC INTERACTIONS

The best source on the social behavior of a natural population of capuchin monkeys is the doctoral dissertation of Oppenheimer (1968) based on material collected during an 18-month study of *Cebus capucinus* on Barro Colorado Island. Therein the capuchins are described as generally peaceful animals. This characterization appears to have been based on two types of observations: (*a*) physical wounding as a direct consequence of conspecific agonistic interactions in a noncage situation was rare (i.e., one case was noted), and (*b*) conspicuously aggressive acts probably constituted only a small percentage of any animal's total social interactions and/or accounted for only a small percentage of the animal's daily time. However, since the latter possibilities are not quantified, rigorous comparisons with other primate taxa are precluded. On the other hand, the general impression of peacefulness is consistent with Bernstein's (1965, p. 221) conclusions that "The cebus [*C. albifrons*, L.K.] were also intermediate with respect to rhesus and gibbons in regards to agonistic responses. . . ." Unfortunately, the data published by Bernstein (1965, p. 216; Bernstein and Mason, 1963, p. 456) from the series of controlled comparisons between similarly sized and composed groups of *Cebus albifrons* and *Macaca mulatta* do not unambiguously illustrate this particular purported difference. The inconsistencies again appear to be tied into the dual problems of scoring criteria and judgments concerning base lines to be used when assessing relative frequency.

Somewhat contrary to the above conclusions, our experiences have left us with a strongly held impression that of all the neotropical primates we have observed either in the field or handled as captives, the capuchins are most like the macaques and baboons with respect to pattern and frequency of agonistic interactions. In turn, these impressions are based upon the following. (*a*) Supplantations and rigid patterns of stable spacing when feeding appear to be more characteristic of capuchins than any of the other neotropical taxa. Neither Oppenheimer, Bernstein, nor ourselves, however, have collected the types of quantitative information necessary to check these particular impressions. (*b*) In our experience with capuchins (*Cebus apella*), both captive and feral, the frequency of participation in interactions classifiable as altercations appeared to be more frequent than in any of the other taxa with which we have had first-hand experience. This is not meant to imply that such interactions account for most of their social interactions or are engaged in for any great amounts of time, and in this sense are compatible with Oppenheimer's and Bernstein's assessments. (*c*) The development of specialized visual gestures of threat, appeasement, and submission appear more highly developed in *Cebus* sp. than in any of the other neotropical taxa discussed in this review. This personal impression of specialization is based upon the number of clearly and easily discriminable agonistic gestures and displays which occur in agonistic contexts. Their exaggerated, repetitive, and sustained nature contrasts considerably with gestures made in similar contexts by *Ateles* sp. Oppenheimer (1968) e.g., cites the following as indicative of attack: scalp and moderate horizontal lip retraction with an open mouth (open mouth threat), bouncing, branch shaking, and tooth grinding. More problematic in function, but probably associated with attack is also urine rubbing and at least occasionally penile displaying in *C. apella* (Klein, pers. obs.). Most striking are the number of discriminable signals frequently given by capuchins in response to an attack, or the above threats which appear to have a conciliatory effect upon the aggressing animal without requiring either active evasion or instrumentally defensive reactions. Conspicuous gestures fulfilling these functions appear not to be given by *Ateles.* Head shakes, tail-rump waggles, and grins (a closed mouth with marked horizontal lip retraction) appear to function in this manner for *Cebus capucinus.* For both *Cebus apella* and *Cebus albifrons* I would add tooth chattering or lip smacking (Bernstein, 1965), and for *Cebus apella,* head clenching (Klein, pers. obs.).

PARTICIPANTS

The pool of potential participants in any agonistic interaction at any one time can include only those animals in actual contact. In this respect, *Cebus* contrasts with the other taxa so far discussed. The two largest of the 10 *Cebus capucinus* groups censused by Oppenheimer (1968) consisted of 15 animals, including two infants. The remainder (Oppenheimer is not too clear on this point) appeared to fall between the maximum number and seven animals. All but one of these groups (a group in the process of splitting and reforming) contained at least one fully adult male and at least several consisted of two of these adult males (estimated 8+ years old). There also appeared to be a substantial population of nongroup isolated males of 4 years and older. All groups contained three or four adult females (4+ years) and one group was known to have five adult females. The remaining group members were juveniles of both sexes. Groups of *C. capucinus* larger than this, however, have been observed on B.C.I. (Enders, 1935) and in Panama (Baldwin and Baldwin, 1972).

Cebus apella groups at our Colombian study site ranged in size from 6 to 12 independently locomoting animals. Median group size was 8. Thorington (1967) reports groups of 6 and 7 on the Colombian llanos. Although not entirely clear, the descriptions of group cohesion in natural populations of *Cebus* sp. seem to imply that either visual or aural contact is maintained between all group members throughout most of the daylight hours and consequently most intragroup agonistic interactions, excluding supplantations, would probably be perceived by all or most group members. In this respect they contrast with *Ateles.*

For purposes of comparison, Bernstein's (1965, 1966) group of captive *Cebus albifrons* was originally composed of 12 animals: 3 adult males, 6 females, and 3 immatures, and had the run of a large outdoor cage. Social contact between colony members was probably continuous except for occasional removal and certain experimental manipulations.

Specific quantitative information from all studies on participation in agonistic interactions is minimal. Concerning intragroup altercations and disputes, Oppenheimer (1968) indicates that "Juveniles and adult females were responsible for most of the disputes" and

> The old male or males of the group usually did not interfere with agonistic interactions of other members of the group, unless they were in some way directly involved. The adult male might hit an animal that is disturbing him . . . or might give a quick warning bite . . . but did not make open-mouth threats to the other members of the group, and rarely made agonistic vocalizations within the group [pp. 119–120].

He also indicates that agonistic behavior was not observed being directed toward single nongroup males. The picture is somewhat inconsistent with Bernstein's (1965) observation that the largest male was involved in agonistic episodes more often than other group members (p. 219). However, Bernstein's most serious series of fights involved an adult female with infant who twice had to be removed from the colony cage. Her antagonists were not specified.

Capuchin intragroup agonistic interactions sometimes involve more than two animals simultaneously. A third animal assisting or backing one of the original participants appears to be the most frequent of these more extensive interactions (Oppenheimer, 1968; Bernstein, 1965). However, actual quantitative information on the frequency, regularity of occurrence, and number of individuals involved has not been reported. In our experience these coalitions are less frequent, predictable, and inclusive than in spider monkeys, but in turn appear to be considerably more frequent than in *Alouatta*. This difference exists in spite of the fact that the number of animals in close contact with one another in capuchin groups is usually more than the number customarily in visual and aural contact in an *Ateles* group. Suggestive of an additional contrast with spider monkeys is the fact that coalitions between capuchin adult males have not been noted and this difference appears to apply to intergroup as well as intragroup agonistic interactions. Only coalitions between adult females, and adult females and juveniles directed against other females or juveniles are reported by Oppenheimer. Bernstein, however, reports that the aid of the largest male was frequently enlisted in his captive situation, another seeming contrast between his and Oppenheimer's observations. Bernstein gives no indication, however, of whether agonistic coalitions occurred between the several adult males in his study group.

Cebus coalitions when they do occur are similar to those of *Ateles*, and contrast with *Saimiri* in that the mutually oriented members of an agonistic coalition are frequently in close physical contact despite vigorous bodily movement. Both Oppenheimer (1968) and Bernstein (1965) describe an "assistance position" which involves extensive bodily contact, frequently with the arm of one animal thrown over the shoulder of the other.

INTERGROUP AGONISTIC CONTACTS

Encounters between members of distinct social groups of capuchins have been described by Oppenheimer (1968). Most of these are implied to be at least partially agonistic even when consisting exclusively of vocal and auditory exchanges. The descriptions suggest that such interactions were generally reciprocal, and the most conspicuous and active participants were usually the larger adult males from each group. Encounters appear to occur most frequently at the border of each group's approximately $\frac{1}{3}$ square mile home range and Oppenheimer interprets this agonistic behavior as territorial defense.

Several interesting qualifications of this general picture are also provided by Oppenheimer. At least one of the adult females in a temporarily maleless group was known to behave aggressively at contacts with members of another group. Her vocalizations and charges, however, appeared *not* to have been reciprocated by the adult male of the group of which she had been a former member, *nor* did they appear to have been specifically oriented at him. Participation in these agonistic intergroup confrontations by group members in addition to the largest male occasionally occurred. Oppenheimer suggests that the degree of this participation was in part a function of the location of contact. Intergroup contacts well within the "territorial boundaries" of one group appeared to provoke more group members, primarily adult females, into more patently aggressive responses such as charging the intruding conspecifics. Cooperative coalitions between the adult males of any single group regardless of the location of intergroup contact appeared to be a rare event; only one case was described (Oppenheimer, 1968). In contrast is the report that the adult male spider monkey released on B.C.I. frequently appeared to follow one of the control male capuchins during capuchin intergroup confrontations, a phenomenon which I take to support the earlier suggestion that propensities to form aggressive coalitions are considerably greater in spider than capuchin monkeys.

INTERTAXA AGONISTIC INTERACTIONS

Agonistic interactions between *Cebus* sp. and other nonhuman primates have been described relatively frequently in the existing field reports, and those interactions involving *Ateles, Alouatta,* and *Saimiri* have been outlined above. The reports suggest that capuchins are usually the more conspicuously aggressive participants in such interactions, and, with the exception of their interactions with *Ateles,* capuchins appear also to be the usual "winners" even when howler monkeys, a considerably larger primate, are their primary antagonists. Attention was also given to the possibility that the special spatial relationship between *Cebus apella* and *Saimiri sciureus* may in part be based upon this aggressive potential, allowing the capuchins, where they occur in small enough numbers, to habitually exploit to their own advantage the insectivorous foraging techniques of *Saimiri.*

Their apparent exceptional agonistic relationships with *Ateles* merit some additional comment. On B.C.I. where spider monkeys were recently introduced and where, when Oppenheimer's observations were made, there were still few adults, social interactions between spider monkeys and the members of certain *Cebus capucinus* groups were frequent. With respect to agonistic interactions, two types of observations are of special interest. (*a*) In the more serious instances of agonistic behavior, *Ateles* are almost always the winners, and (*b*) Oppenheimer reports that spider monkeys frequently appeared not to react to capuchin threats and in some cases not even to aversive physical attack. In contrast, at our study site, a habitat in which conspecific and allospecific relationships had had considerably more time to stabilize, capuchins (*C. apella*) generally appeared to avoid interactions and close spatial proximity with *Ateles,* particularly in competitive situations, and social interactions of all types were limited in scope (Klein, 1972). For example, although both taxa ate many of the same types of fruit, and were observed feeding simultaneously from the same tree if its crown was larger than 60 ft in diameter, *C. apella* were highly unlikely to enter fruiting trees of smaller dimensions if spider monkeys were already in them, even though they were seen entering similar or even the identical ones when *Ateles* were either not present or shortly after their departure. Moreover, neither play nor grooming interactions were observed to occur between these particular taxa, although both types of interactions occurred between *Ateles* and *Alouatta.* In time I expect that the capuchins of B.C.I. will begin more assiduously to avoid contacts with spider monkeys, particularly in competitive situations. The development will probably quicken as the spider monkeys, particularly the male, begin to form close associations and coalitions with the increasing number of maturing conspecifics.

Few clear-cut agonistic interactions between capuchins and potential predators other than man were noted by Oppenheimer (1968). Reactions seen ranged from active evasion to vigorous threats. With the exception of two types of vocalizations, a series of gyrrahs and single gyrrahs (the latter sometimes evoked by birds flying overhead), specific behavioral patterns observed during these types of intertaxa encounters—including those with the observer—were similar to those seen at conspecific intergroup encounters: branch shaking, branch breaking, staring, head bobs, and occasional tooth grinding. On many of these occasions the most active participants appeared to be juvenile capuchins, occasionally backed by a fully adult male. At other times, only the single fully adult male was observed threatening.

Our experiences with *C. apella* in Colombia are, in general, consistent with the above. Early in the study, either one or two adult animals, and after several months, immatures more frequently were the animals who tended to noticeably and vigorously react to our presence. These reactions generally included repetitive chuck vocalizations, stares, eye tuft raising, head bobbing, and sometimes branch shaking and breaking. Branch breaking and dropping was, at times, a very conspicuous component of these interactions. Although we were the frequent recipients of branch shaking displays by spider monkeys and occasionally had objects dropped or dislodged close to us by howler and wooly monkeys, we observed nothing in the behavior of these other taxa comparable with respect to apparent intent and instrumentality in their behavior.[4]

Another interesting taxa difference in these types of interactions, and again consistent with conspecific agonistic reactions, was the limits on aggressive coalitions in capuchins compared to spider monkeys. Threatening capuchins, if backed by conspecifics at all, were usually conspicuously assisted by only one other animal, despite the numbers actually present, and in those cases described by Oppenheimer (1968) this backing animal was frequently the largest male. In contrast, aggressive reactions by unhabituated spider monkeys to ourselves often involved the active participation of most adult males present, and in the absence of males, adult females. Since some of these reactions were prolonged, the size of these coalitions increased over time, as did several of the observed spider monkey intergroup confrontations (see earlier discussion). The facilitation of intragroup, intergroup, and intertaxa aggressive responses thus appears to be either more highly developed or less inhibited in spider monkeys than in capuchins. If forced to distinguish between the two alternative processes at this time, without additional data, I would favor it being less inhibited in spider monkeys since the evidence generally suggests that carefully measured aggressivity and its reciprocal behavioral patterns—

[4] For example, on one occasion an adult male capuchin while vocalizing moved toward us a distance of about 10 ft and stopped at a large (8 ft long, 12 inch diameter) hanging dead tree limb. Suspended by his tail and pushing outward with his hind limbs, the capuchin pushed the branch in an arc toward us. It broke off and fell to the ground several feet short of our location. The capuchin returned immediately to his original position about 10 ft away.

submission, appeasement, protected threats, etc.—are more characteristic of capuchins than any of the other neotropical taxa dealt with in this review. Perhaps related is the fact that aggressive responses by spider monkey coalition members are generally similar, while the capuchin coalition dyads frequently have a clear-cut element of complementarity: e.g., one capuchin threatens with its mouth open while the assisting animal with its arm around and over its partner generally keeps its mouth shut (Oppenheimer, 1968). The complementarity includes vocalizations as well—in one case a capuchin giving a series of chucks in response to our presence switched immediately to growl screams upon being touched by his partner. The partner in turn began producing the chucks (Klein, pers. obs.).

At our study site capuchins were known to react to the presence of several potentially dangerous animals other than man. They were heard giving bark-chuck vocalizations to a cayman (*Caiman sclerops*) swimming up a usually dry creek bed under trees in which they were foraging, and a repetitive chuck vocalization was given as a great black hawk (*Buteogallus* sp.) flew away from a 40 to 50-ft liana-enclosed tree within which a group of *Cebus apella* had been stationary and silent for at least 5 prior min.

On several occasions we were told by local people that capuchins (both *Cebus albifrons* and *Cebus apella*) would physically attack domestic dogs. We heard no reports of this type concerning other New World primates.

Bernstein (1966) presented several test situations to his captive colony of *C. albifrons* which were designed to provoke defensive reactions. The patterns of behavior elicited appear to be consistent with those observed by Oppenheimer and myself. A snake presented in a transparent box produced three series of unspecified types of threats, primarily from immature animals, at times backed by the largest adult male and an adult female, while the more serious stimulus of a human walking within the cage and swinging a stick elicited threats almost exclusively from the largest male.

CONTEXT OF AGONISTIC BEHAVIOR

Lack of specific and quantitative information makes it almost impossible to deal systematically with the effects of situational contexts on the frequency or form of capuchin agonistic interactions. Of that which appears to be known, Oppenheimer notes that all contacts between stable *Cebus capucinus* groups appeared to precipitate some form of conspicuous aggressive behavior; it was frequently his first indication of another group's presence. The location of contact appeared to affect the reciprocity of these intergroup interactions, i.e., whether they resulted in a mutual confrontation or a one-sided chase (see earlier discussion). With respect to intragroup agonistic interactions, only the following is offered by Oppenheimer. "Usually the disputes arose when an adult had something, such as food or an infant, or could provide something, such as grooming, that was attractive to the juvenile or other adult [1968, p. 120]." The implication that disputes might sometimes occur as a consequence of a refusal or interruption of grooming is interesting. If true, it appears to be a rather unique aggression-eliciting stimulus, but more data are needed.

Our own observations of free-ranging *C. apella* add only a little, as we actually saw very few intragroup agonistic interactions clearly (five cases). Three of these, however, were definitely food related and in combination with food-related agonistic interactions with other taxa (*Alouatta, Saimiri,* and guans, *Pipile cumanensis*), and the fact that small fruiting trees (crown diameter of less than 25 ft) invariably contained only a single adult capuchin at any one time led us to believe that competition over food had an important effect upon the frequency of agonistic interactions in capuchin monkeys (Klein, 1972).

This competition was of a different kind than that seen in *Ateles* in that the more eclectic diet of capuchins appeared to allow all members of a group to eat or forage simultaneously on a variety of items despite concentrations of specific edibles. Food competition in *C. apella* relative to *Ateles* thus appeared to be based upon "quality," not the quantity of frequently limited amounts of ripe fruits. Elsewhere (Klein, 1972) we have suggested that this type of competition in contrast to that more prevalent where daily diet regularly consists of just a few items clustered in single trees is one of the bases for an emphasis on controlling agonistic interactions with conspicuous appeasement and submissive behavior rather than simple avoidance.

The above is not supported by Bernstein's observations on captive *Cebus albifrons*. However, the effect food competition might have had upon either the frequency of fights, threats, or supplantations appears not to have been investigated directly. Bernstein (1965) reports

> Observations following the addition of fresh food to the compound did not reveal a consistent "priority to incentives" . . . even for the largest male. Individuals frequently ate side by side and either of a pair might take food from the other. Only on rare occasion was this followed by flight or aggression. In addition, the youngest infant reached into the mouths of many group members and explored the contents with impunity [p. 218].

Bernstein also notes that "Exchange of food between individuals has been seen in both gibbon and cebus groups . . . [p. 221]." Unfortunately, this particular aspect of Bernstein's study was not reported quantitatively and the citing of infants or young animals occasionally appearing to "take" food from an older animal is interesting, but not too surprising since it should be expected that infants, of all the animals in a group, would be least likely to evoke either competitive or aggressive responses in adults by these behaviors.

In comparison to Bernstein's study, the result of a series of experiments by Plotnik *et al.* (1968) with "adolescent" male *C. albifrons* as subjects did demonstate a consistent association between feeding competition and frequency of aggression. Maintained at a body weight of 90% normal, the frequency of aggressive responses—pushing, pulling, biting, chasing—was increased by about 50% when food was presented in their home cages (2½ ft x 3 ft x 3 ft), and about 200% when food was presented in a much smaller shuttle box apparatus. Although the experimental conditions were designed to intensify aggressive responses, they do support the field observations suggesting that food competition may have a significant effect upon agonistic interactions and aspects of social spacing.

Very few data are available on sexual behavior in natural populations. Oppenheimer (1968) observed two copulations and one copulation attempt between feral *C. capucinus*. During these interactions there were no indications of agonistic behavior between sexual partners nor harassment by nearby onlooking animals. Bernstein (1965, p. 219), however, suggests that a sexual attraction might have been responsible for a considerable amount of threatening behavior between two captive males and a female. Although unambiguous sexual behavior was not seen, the female did abort during this period.

An association between colony formation and a high frequency of conspecific agonistic responses was one of the clearest positive relationships between aggression and situational context in Bernstein's observations (1965, 1966). Several hours after the release of the animals, the frequency of agonistic responses appeared to peak at a rate approximately 20 times the subsequent 50-day average. Rapid group formation of this type, of course, would probably not occur in a more natural habitat.[5] The introduction of several types of novel inanimate and nonedible stimuli to the same caged group did not appear to affect agonistic behavior at all (Bernstein, 1966).

SUMMARY

Comparisons between primate taxa can be made in many ways and for several purposes. In this review of agonistic behavior, we have chosen to

[5] Oppenheimer (1968) documents the splitting and formation of a new *C. capucinus* group on B.C.I.

stress parameters which we believe to be (*a*) those about which at least some reliable information for New World primates is available; (*b*) those which we believe to be relatively unaffected by situational features (e.g., see above for comparisons between captive and natural populations of *Ateles* spp.); (*c*) those factors which from our viewpoint appear to be potentially related to taxa-specific ecological and physiological adaptations; and (*d*) therefore, potentially relatable to specific evolutionary processes.

Minimized, although not ignored, were formal differences and similarities in the repertoire of agonistic social gestures and vocalizations. Although for some neotropical primates, detailed descriptive studies of the signaling behavior of captive specimens have been published (see, e.g., Moynihan, 1964, 1966, 1967, 1970), described differences are at the moment difficult to relate to ecological and social processes except at the grossest of levels, e.g., nocturnal versus diurnal.

Also minimized were comparisons between taxa based on the frequency of agonistic interactions, despite the importance such comparisons have played in traditional approaches to questions of aggressive behavior (Carpenter, 1942; Hall, 1964; Washburn and Hamburg, 1968; Jolly, 1972). In our opinion, comparisons based heavily on frequency, given the present state of the evidence, are akin to comparing oranges with apples. Few such comparisons are based upon similar and realistic criteria applicable across taxa, and few, if any investigators have been concerned with frequency of agonistic interactions relative to activity levels, positive social interactions, spatial dispersion, and/or group size. In this review, frequency data were included to (*a*) indicate the extent of the data upon which generalizations were made, and (*b*) to point to those situational features, both unique and common, which in each taxa appeared to raise the probability of certain types of aggression.

Spider monkeys, *Ateles* spp., the principal subject animal of the author's research, provided the pivotal point around which agonistic behavior in the other neotropical primates was compared. The following characteristics of spider monkey agonistic behavior were noted.

1. Despite poor visibility, most agonistic interactions, with the exception of supplantations, were conspicuous. Spider monkeys involved in agonistic interactions are almost always vocal, and frequently charge, chase, leap, shake branches, urinate and defecate.

2. Observed agonistic interactions frequently involve more than two animals and both social contagion and coalitions are striking. The latter almost always involves extensive and persistent physical contact between cooperating participants.

3. Intragroup agonistic interactions frequently occur between adult males and adult females. In a significant proportion of these interactions, seen both in captive colonies and in a natural habitat, more than one male was observed chasing or attacking single females. Cooperative coalitions of adult females arrayed against males have not yet been observed, although coalitions including males and females against single females do occur.

4. Intragroup agonistic interactions between adult males appear to occur very rarely. Only one such case was observed under field conditions, and few cases have been observed in stabilized captive colonies. The occasional fighting between males observed in captivity often results in serious wounding.

5. The behavioral repertoire of spider monkeys includes no gestures which regularly appear to appease or redirect the activities of threatening conspecifics. Agonistic interactions are usually terminated by avoidance or escape.

6. Visual encounters between individuals of differing social networks appeared to reliably precipitate lengthy agonistic interactions. Reciprocal aggressive responses appeared to occur, however, only if males of different groups were at the scene simultaneously. Behavior noted at intergroup encounters included whooping, screaming, squealing, chasing, and branch shaking. Although these acts were also observed in other contexts, including encounters between spider monkeys and other primate and nonprimate taxa, motor patterns in which secretions of the sternal glands mixed with saliva were repeatedly rubbed onto tree limbs were more prominent at these intergroup confrontations than at any other time under natural conditions.

7. Feeding, relative to other activities, appeared to increase the probability of agonistic interactions and affect interanimal and intragroup spatial relationships.

8. The probability of agonistic encounters also increased at those times when an encounter occurred between members of the same social group who up to then had been traveling apart for a period of days or hours. Merging subgroups and frequent isolation of adult animals is a relatively unique characteristic of the fission–fusion social organization of *Ateles*.

9. Encounters between spider monkeys and other taxa, including potential predators, sometimes elicited agonistic behavioral patterns similar to those observed during conspecific interactions. The nature and duration of these reactions appeared to depend upon the potential predator's location when first seen by the spider monkey and its subsequent behavior in relation to its potential prey.

Relative to spider monkeys, the following contrasts were noted in howler monkeys.

1. Conspicuous and clear-cut cases of intragroup aggression were infrequent. However, it proved difficult to determine the relevance of this consistent finding in light of the fact that (*a*) all types of social interactions were generally low, and (*b*) the motivational and functional significance of many of the howler monkeys' most conspicuous action patterns are unknown, or a matter of debate.

2. Vocal reactions appear to be the major, sometimes only response made to stimuli to which spider monkeys and other primates would react with more conspicuous facial, postural, locomotory, or manual means.

3. Although patterns of intragroup participation in agonistic interactions have not been delineated, the information available indicates that male–male agonistic interactions have been most frequently observed.

4. To date, there are no observations suggesting that within-group agonistic behavior results in either aggressive or defensive coalitions, or the social facilitation of aggressive behavior.

5. Sustained vocal and sometimes more active reactions to nongroup males, described for *Alouatta,* have not been observed in spider monkeys, partly perhaps because the existence of permanently peripheral or nongroup male spider monkeys has not yet been verified.

6. The distinction between intra and intergroup agonistic interactions is more marked for howler than spider monkeys because sustained and loud vocalizations have not yet been reported to occur either during or immediately after howler intragroup agonistic interactions.

7. Feeding activities in howler monkeys do not appear to increase the probability of intragroup hostilities, and as reported so far, appear to have little effect upon intragroup spatial proximity.

8. There are some indications in the published literature that the probability of intragroup hostility is increased as a consequence of intermale sexual competition and/or interference.

Relative to spider monkeys, the following major contrasts were noted in the agonistic interactions of titi monkeys (*Callicebus moloch*).

1. Agonistic interactions between sexual immatures and mature adults are reported to be the most frequent type of intragroup agonistic interaction.

2. Agonistic interactions between a single pair of adult male and female group members have been reported to occur only in a sexual context and under apparently exceptional circumstances.

3. Intergroup confrontations between *Callicebus* appear to be very frequent, conspicuous, regular, and predictable, occurring at stable boun-

dary areas, and either limited or markedly affected by time of day. Intergroup confrontations between adult *Ateles*, although similarly conspicuous as a consequence of intense and prolonged vocalizations, and several types of active display, including on at least some occasions branch marking, do not fit as neatly into the functional category of territorial defense and advertisement.

4. Cooperative agonistic coalitions have been reported to occur only between the male and female adults of a single pair during their intergroup encounters.

Relative to spider monkeys, the following major contrasts were noted in the agonistic interactions of squirrel monkeys.

1. Evidence for marked seasonal changes in the frequency of agonistic interactions between sexually mature males and between adult male and adult female squirrel monkeys. Peaks in frequency appear to correspond with breeding seasons.
2. The occurrence and aggressive success of coalitions of squirrel monkey females and juveniles in response to approaches by conspecific adult males.
3. The infrequent occurrence of conspicuously aggressive behavioral patterns in response to primates of other taxa, and potential predators.

Although systematic quantitative data on many aspects of capuchin agonistic interactions are lacking, observations suggest that they contrast with those of *Ateles* in at least the following ways.

1. Supplantations and rigid spatial patterns, rather than avoidance, appear to be more characteristic types of agonistic interactions among capuchins.
2. The development of specialized visual gestures of threat, appeasement, and submission appears more highly developed in *Cebus* spp. than in spider monkeys.
3. Adult male capuchins appear to be relatively infrequent participants in active intragroup altercations relative to adult female capuchins.
4. Cooperative aggressive coalitions, particularly between adult males, are not as readily formed by capuchins during conspecific agonistic interactions, although they do occur.
5. Feeding competition in capuchins, in comparison with spider monkeys, appears to be based more directly upon degrees of food quality rather than quantity. It was suggested that this type of feeding competition, in contrast to the situation in which essential food items are limited in variety and clustered in single trees, is one of the bases for the development of conspicuous appeasement and submissive behavior.

REFERENCES

Altmann, S. A. (1959). Field observations on a howling monkey society. *J. Mammal.* **40**, 317–330.

Baldwin, J. D. (1968). The social behavior of adult male squirrel monkeys (*Saimiri sciureus*) in a seminatural environment. *Folia Primatol.* **9**, 281–314.

Baldwin, J. D. (1969). The ontogeny of social behaviour of squirrel monkeys (*Saimiri sciureus*) in a seminatural environment. *Folia Primatol.* **11**, 35–79.

Baldwin, J. D. (1971). The social organization of a semifree-ranging troop of squirrel monkeys (*Saimiri sciureus*). *Folia Primatol.* **14**, 23–50.

Baldwin, J. D., and Baldwin, J. I. (1971). Squirrel monkeys (*Saimiri*) in natural habitats in Panama, Colombia, and Peru. *Primates* **12**, 45–62.

Baldwin, J. D., and Baldwin, J. I. (1972). The ecology and behavior of squirrel monkeys (*Saimiri oerstedi*) in a natural forest in Western Panama. *Folia Primatol.* **18**, 161–184.

Bernstein, I. S. (1964). A field study of the activities of howler monkeys. *Anim. Behav.* **12**, 92–97.

Bernstein, I. S. (1965). Activity patterns in a *Cebus* monkey group. *Folia Primatol.* **3**, 211–224.

Bernstein, I. S. (1966). Analysis of a key role in a capuchin (*Cebus albifrons*) group. *Tulane Stud. Zool.* **13**, 49–54.

Bernstein, I. S., and Mason, W. A. (1963). Activity patterns of rhesus monkeys in a social group. *Anim. Behav.* **11**, 455–460.

Carpenter, C. R. (1934). A field study of the behavior and social relations of howling monkeys. *Compar. Psychol. Monogr.* **10**(2).

Carpenter, C. R. (1935). Behavior of red spider monkeys in Panama. *J. Mammal.* **16**, 171–180.

Carpenter, CR. (1942). Societies of monkeys and apes. *Biological Symposia* **8**, 177–204.

Carpenter, C. R. (1965). The howlers of Barro Colorado Island. *In* "Primate Behavior: Field Studies of Monkeys and Apes" (I. DeVore, ed.), pp. 250–291. Holt, New York.

Castell, R., and Heinrich, B. (1971). Rank order in a captive female squirrel monkey colony. *Folia Primatol.* **14**, 182–189.

Chivers, D. J. (1969). On the daily behaviour and spacing of howling monkey groups. *Folia Primatol.* **10**, 48–102.

Collias, N. E., and Southwick, C. H. (1952). A field study of population density and social organization of howling monkeys. *Proc. Amer. Phil. Soc.* **96**, 143–156.

DuMond, F. V. (1968). The squirrel monkey in a seminatural environment. *In* "The Squirrel Monkey" (L. A. Rosenblum and R. W. Cooper, eds.), pp. 88–145. Academic Press, New York.

DuMond, F. V., and Hutchinson, T. C. (1967). Squirrel monkey reproduction: the "fatted" male phenomenon and seasonal spermatogenesis. *Science* **158**, 1067–1070.

Eisenberg, J. F., and Kuehn, R. E. (1966). The behavior of *Ateles geoffroyi* and related species. *Smithsonian Misc. Collect.* **151**(8).

Enders, R. K. (1935). Mammalian life histories from Barro Colorado Island, Panama. *Bull. Mus. Comp. Zool., Harvard College* **78**(4), 385–502.

Fooden, J. (1964). Stomach contents and gastrointestinal proportions in wild-shot Guianan monkeys. *Amer. J. Phys. Anthrop.* **22**, 227–232.

Green, R., Whalen, R. E., Rutley, B., and Battie, C. (1972). Dominance hierarchy in squirrel monkeys (*Saimiri sciureus*). Role of the gonads and androgen on genital display and feeding order. *Folia Primatol.* **18**, 185–195.

Hall, K. R. L. (1963). Tool-using performances as indicators of behavioral adaptability. *Current Anthropol.* **4**, 479–494.

Hall, K. R. L. (1964). Aggression in monkey and ape societies. *In* "The Natural History of Aggression" (J. D. Carthy and F. J. Ebling, eds.), pp. 51–64. Academic Press, New York.

Hladik, A., and Hladik, C. M. (1969). Rapports tropiques entre vegetation et primates dans la foret de Barro Colorado (Panama). *La Terre et la Vie* **1**, 25–117.

Jolly, A. (1972). "The Evolution of Primate Behavior." Macmillan, New York.

Klein, L. (1971). Observations on copulation and seasonal reproduction of two species of spider monkeys, *Ateles belzebuth* and *A. geoffroyi. Folia Primatol.* **15**, 233–248.

Klein, L. (1972). The ecology and social organization of the spider monkey, *Ateles belzebuth.* Unpublished Ph.D. thesis, Univ. of California, Berkeley.

Klein, L., and Klein, D. (1971). Aspects of social behavior in a colony of spider monkeys, *Ateles geoffroyi. Int. Zoo Yearbook* **11**, 175–181.

Klein, L., and Klein, D. (1973). Observations on two types of neotropical primate intertaxa associations. *Amer. J. Phys. Anthrop.* **38**, 649–654.

Klein, L., and Klein, D. Social and ecological contrasts between four taxa of neotropical primates (*Ateles belzebuth, Alouatta seniculus, Saimiri sciureus, Cebus apella*). The Hague: Mouton and Co. (In press.)

Long, J. D., and Cooper, R. W. (1968). Physical growth and dental eruption in captive-bred squirrel monkeys, *Saimiri sciureus* (Leticia, Colombia). *In* "The squirrel monkey" (L. A. Rosenblum and R. W. Cooper, eds.), pp. 193–205. Academic Press, New York.

Lundy, W. E. (1954). Howlers. *Natur. Hist.* **63**, 128–133.

Mason, W. A. (1966). Social organization of the South American monkey, *Callicebus molloch:* a preliminary report. *Tulane Stud. Zool.* **13**, 23–28.

Mason, W. A. (1968). Use of space by Callicebus groups. *In* "Primates: Studies in Adaptation and Variability" (P. C. Jay, ed.), pp. 200–216. Holt, New York.

Moynihan, M. (1964). Some behavior patterns of platyrrhine monkeys. I. The night monkey (*Aotus trivirgatus*). *Smithsonian Misc. Coll.* **146**, 1–84.

Moynihan, M. (1966). Communication in the titi monkey, *Callicebus. J. Zool.* **150**, 77–127.

Moynihan, M. (1967). Comparative aspects of communication in New World primates. *In* "Primate Ethology" (D. Morris, ed.), pp. 306–343. Aldine, Chicago, Illinois.

Moynihan, M. (1970). Some behavioral patterns of platyrrhine Monkeys. II *Sanguinus geoffroyi* and some other tamarins. *Smithsonian Contrib. Zool.* No. 28.

Napier, J. R., and Napier, P. H. (1967). "A Handbook of Living Primates." Academic Press, New York.

Neville, M. K. (1972a). The population structure of red howler monkeys (*Alouatta seniculus*) in Trinidad and Venezuela. *Folia Primatol.* **17**, 56–86.

Neville, M. K. (1972b). Social relations within troops of red howler monkeys (*Alouatta seniculus*). *Folia Primatol.* **18**, 47–77.

Oppenheimer, J. R. (1968). Behavior and ecology of the white-faced monkey *Cebus capucinus* on Barro Colorado Island. Unpublished Ph.D. thesis, Univ. of Illinois, Urbana.

Ploog, D. W., and MacLean, P. D. (1963). Display of penile erection in squirrel monkeys (*S. sciureus*). *Anim. Behav.* **11**, 32–39.

Ploog, D. W., Blitz, J., and Ploog, F. (1963). Studies on social and sexual behavior of the squirrel monkey (*Saimiri sciureus*). *Folia Primatol.* **1**, 29–66.

Plotnik, R., King, F. A., and Roberts, L. (1968). Effects of competition on the aggressive behavior of squirrel and cebus monkeys. *Behaviour* **32**, 315–332.

Richard, A. (1970). A comparative study of the activity patterns and behavior of *Alouatta villosa* and *Ateles geoffroyi*. *Folia Primatol.* **12**, 241–263.

Rondinelli, and Klein. An analysis of social spacing and related behaviors in a captive colony of spider monkeys, *Ateles geoffroyi*. (In preparation.)

Rosenblum, L. A., and Cooper, R. W. (1968). "The Squirrel Monkey." Academic Press, New York.

Rowell, T. (1972). "Social Behaviour of Monkeys." Penguin, Baltimore, Maryland.

Schultz, A. H. (1970). The comparative uniformity of the Cercopithecoidea. *In* "Old World Monkeys" (J. R. Napier and P. H. Napier, eds.), pp. 39–51. Academic Press, New York.

Simons, E. L. (1972). "Primate Evolution. An Introduction to Man's Place in Nature." Macmillan, New York.

Thorington, R. W., Jr. (1967). Feeding and activity of *Cebus* and *Saimiri* in a Colombian forest. *In* "Progress in Primatology" (D. Starck, R. Schneider, and H. J. Kuhn, eds.), pp. 180–184. Fischer, Stuttgart.

Thorington, R. W., Jr. (1968). Observations of monkeys in a Colombian forest. *In* "The Squirrel Monkey" (L. A. Rosenblum and R. W. Cooper, eds.), pp. 69–85. Academic Press, New York.

Washburn, S. L., and Hamburg, D. (1968). Aggressive behavior in Old World monkeys and apes. *In* "Primates" (P. C. Jay, ed.), pp. 458–478. Holt, New York.

Winter, P. (1968). Social communication in the squirrel monkey. *In* "The Squirrel Monkey" (L. A. Rosenblum and R. W. Cooper, eds.), pp. 235–253. Academic Press, New York.

Winter, P. (1972). Observations on the vocal behavior of free-ranging squirrel monkeys (*Saimiri sciureus*). *Z. Tierpsychol.* **31**, 1–7.

COLOBINE AGGRESSION: A REVIEW

FRANK E. POIRIER
Ohio State University

INTRODUCTION

This review has been difficult to complete because there has been little previous attempt to correlate existing information on aggression. On the surface, the topic of aggression seems the victim of overkill, but then there is scarcely a single definition fitting everyone's purposes. Much has recently been written about human aggression, but little of it is of help when dealing with the nonhuman primate literature. This is especially true when one works with a primate group like the colobines which has been separate from the hominid phylogenetic line for approximately 20 million years or more (Sarich, 1970; Simons, 1970).

Colobines are Asian and African primates sharing a number of common features. Colobines are primarily arboreal, all are basically leaf-eaters, almost all their infants possess a natal coat (hair coloring distinguishing them from the adults), most exhibit two types of group composition—male and multiple-male bisexual groups with a number of females. In bisexual groups females predominate over males and in many species there are all-male groups living on the periphery of bisexual groups. Most colobines exhibit territorial behavior, most show a reduced amount of aggression compared to other Old World monkeys, and there is minimal sexual dimorphism. While there is both intra- and interspecific variability regarding troop size, a modal troop ranges between 20–30 animals, much smaller than is common for most savannah-dwelling monkeys.

The colobines reviewed herein represent the total number which have been studied. According to Napier and Napier, 1967, the colobines can be divided into the two genera of *Colobus* and *Presbytis*. The African colobines, the genus *Colobus*, are further subdivided into 3 subgenera, 5 species, and 41 subspecies. The Asian colobines, the genus *Presbytis*, are subdivided into 4 species groups, 14 species, and 84 subspecies. Of this total the paper includes data shown in Table 1. However, most of

TABLE 1 **Sources for Data**

Genus	Investigators
	Africa
Colobus	
Olive colobus	Booth (1957)
Black and white colobus	Marler (1969), Schenkel and Schenkel-Hulliger (1967)
Red colobus	Marler (1970), Nishida (1972)
	Asia
	India
Presbytis	
North Indian common langur	Jay (1962, 1963a,b, 1965), Mohnot (1971a,b), Vogel (1970)
South Indian common langur	Sugiyama (1964, 1965a,b, 1967), Yoshiba (1968)
Nilgiri langur	Poirier (1967, 1968a,b,c,d, 1969a,b,c, 1970a,b,c, 1971, 1972), Tanaka (1965)
	Ceylon
Presbytis	
Ceylon gray langur	Beck and Tuttle (1972), Ripley (1967)
	Malaysia
Presbytis	
Purple-faced leaf monkey, lutong	Bernstein (1968), Furuya (1961–1962)

our information deals with the common langur or hanuman monkey of north and south India which has been studied by various investigators (i.e., Jay, 1962, 1963a,b, 1965; Mohnot, 1971a,b; Sugiyama, 1964, 1965a,b, 1967; Vogel, 1970; Yoshiba, 1968). The few scattered reports of some of the less readily observable colobines were of little value for this review. The other major Indian study deals with the Nilgiri langur (Poirier, 1967, 1968a,b,c,d, 1969a,b,c, 1970a,b,c, 1971, 1972; Tanaka, 1965). Leaf-eating monkeys have also been studied in Ceylon (Ripley, 1967; Beck and Tuttle, 1972). They have supplied some valuable information on territorial behavior. Bernstein's (1968) study of the purple-face leaf-eaters of Malaysia has been gleaned for relevant information (see also Furuya, 1961–1962). African colobines are poorly represented as we know little about them (Booth, 1957; Marler, 1969, 1970; Nishida, 1972; Schenkel and Schenkel-Hulliger, 1967; Ullrich, 1961).

AGGRESSION DEFINED

We have yet to arrive at and agree upon a working definition of primate aggression. Some authors have purposely avoided the term as undefinable (i.e., Kummer, 1967; Vowles, 1970). There is considerable disagreement among those attempting a definition. We need an inclusive definition—but one not too broad, specific enough to fit the general behavioral data, but general enough to be applicable cross-specifically. We need a definition making the concept of aggression meaningful in discussing complex behavioral patterns, many of which have an unknown genesis, even in an immediately observable situation. Numerous definitions were reviewed and pondered, yet none significantly increased the understanding of colobine aggression, its causes, context and intent. As Southwick (1972, p. 2) states, "Like many basic concepts, aggression is moderately easy to recognize but hard to define." Johnson (1972) discusses this dilemma in great detail.

One of the major problems when discussing aggression is ascription of intent. Berkowitz (1965) considered intention to injure as a prime requisite for labeling a behavior "aggressive." However, Buss (1961) among others, argues that ascription of intent is empirically useless in animal studies, and among nonverbal humans. Another basic problem is whether or not to restrict aggressive behavior to purely social situations, e.g., can certain behaviors directed toward inanimate objects or simply into space be labeled aggressive? Some authors (Washburn and Hamburg, 1968) include predation under the rubric of aggression. And some, for example Carthy and Ebling (1964), would include self-inflicted injury, such as suicide. Additional confusion arises from the interchangable, often undefined, use of substitute terms.[1]

Having read numerous works which increased the confusion as to

[1] One of the most frequently used substitutes is the term "agonistic." Agonistic behavior, according to Scott (1966), is a "behavioral system composed of behavior patterns having the common function of adaptation to situations involving physical conflict between members of the same species." This is widely accepted to refer to the full range of aggressive and defensive behaviors associated with conflict.

what the term aggression really means, I will simply present aspects of colobine behavior which others, including myself, have termed aggression. My definition is simplistic; I have made little effort to find intent in the various acts. Instead, the literature on colobines was read and all material was extracted which indicated that animals were acting in a stressful situation. My category of aggression includes such behaviors as actual attack patterns, actual infliction of injury, various attack and threat patterns associated with dominance, nondirected and redirected (or displacement) behaviors stemming from stressful situations, attack and threat behavior (even though ritualized) associated with intertroop behavior, repulsing of an infant by its mother during the later stages of weaning, infanticide by adult males (found among some common Indian langurs during group changes), and threat behavior directed toward predators or objects "perceived" as threats. I neglected the whole category of aggression as it occurs in play because we lack decent criteria to separate "playfighting" from outright aggression.

The relation between sexual and aggressive behavior is unclear. There is no question that among colobines, and among some other nonhuman primates, there is a rise in tension among the males as the females enter estrus.[2] In fact, there is increasing evidence that colobine intermale aggression rises with the simple presence of females. There are strong suggestions that one of the factors producing one-male colobine troops is a rise of intermale tension during the breeding season.

One of our major problems is that our observations and subsequent reporting of aggressive behavior are dependent upon our witnessing behavioral patterns. Because of restrictions imposed by the field situation,[3] we are forced to depend mainly upon external indicators of aggressive behavior. As methods develop and are refined, a more prudent and perhaps productive course would be measurement of hormonal changes and other forms of physiochemical, psychodynamic monitoring. Evidence is rapidly accumulating which shows that testosterone and androgen strongly influence male and intermale aggressive behavior (i.e., Beeman, 1947; Beach, 1965; Hamburg, 1970, 1971; Levy and King, 1953; Michael and Zimpe, 1970; Moyer, 1969; Scarf, 1972; Seward, 1945; Sigg, 1968;

[2] There is also some evidence suggesting increasing irritability among females. For a discussion of the rise of aggression as it is related to rhesus sexual behavior, see Wilson and Boelkins (1970).

[3] It must be remembered that almost all the information on colobines derives from naturalistic studies. Leaf-eating monkeys do poorly in captivity, thus there is little opportunity for controlled laboratory testing such as is feasible with macaques and many other nonhuman primates. This will continue to be a major obstacle in delimiting many features of colobine behavior.

Suchowski, 1968, among others). Recent experimental studies of aggression suggest that the physiological bases of various manifestations differ in origin (Moyer, 1971). Some kinds of colobine aggression are qualitatively different than others.

Aggression is a multifactorial behavioral expression strongly influenced by numerous external and internal variables. Field studies are primarily concerned with delimiting instances of external variables; discerning internal variables is still beyond our grasp. Most would agree that in addition to immediate internal and external influences, experiential factors such as socialization (Poirier, 1972, 1973a) and personal life experiences also play a marked role in determining an animal's aggressive tendencies. However, because of the short time span of most field studies, we are still in an unfavorable position to deal with these. A number of longitudinal studies are clearly needed if we are to understand anything but the most immediate sources of aggression. None of the studies referred to here fulfill this criterion; at best we have a limited perspective.

To summarize, this paper presents an interactionist approach to the study of aggression. It assumes that aggression is influenced by internal and external factors, as well as by learning, the particular social group and its socialization modes, and the econiche.

REVIEW OF FACTORS INFLUENCING AGGRESSION

Ecological Variables

Scott (1958) lists a number of ecological causes of aggression among which are temperature, food supply, space, and such changes in the physical environment as atmospheric pressure. To this we might add availability of water and, because colobines are arboreal primates, availability of sleeping trees. The effect of some of these factors is easier to assess than others. There is no evidence that shifts in atmospheric pressure affect the incidence of colobine aggression, and such is unlikely to be of major import. Evidence concerning temperature change is equivocal; however, there may be a subsidiary effect between temperature variation as it relates to the seasonality of resources and the birth and mating seasons.

Both availability and location of food supply may affect colobine aggression. This, along with the space parameter, will be discussed in the section on intertroop aggression. However, there seems to be no relationship between availability and location of water supply and colobine aggression. Colobines derive most of their water from the succulent leaves which are their dietary mainstay. For example, in about 1250 hr of observation on Nilgiri langurs (Poirier, 1970b), drinking behavior was witnessed only five times, licking water from a tree after a rain four times, and licking water from one's hair twice. Most colobines pass several months without visiting a water source. Thus, at least under the "normal" ecological conditions in which most observers have worked, water supply is not a factor inducing aggression. There is no aggression in north India, where Jay (1965) reports that groups are spaced to take advantage of artificial water sources used for irrigation, when different common langur groups come together to drink from the same water source.

Seasonal and Periodic Expression of Aggression

The evidence relating to seasonal variance in expression of colobine aggressive behavior is equivocal because most observers have not analyzed their data with an eye toward this variable. In reading through the literature, however, there is rather clear evidence of seasonal increases in aggression. The most facile explanation for a seasonal increase in aggression is that during the season(s) with the most daylight there is simply more time available for behavioral interaction; at dark most primates peacefully settle down for the night. The more daylight the greater the chance for behavioral interaction, and therefore the greater the possibility for aggressive behavior. This, however, is not the full story.

Although the idea was once contested, we now know that most nonhuman primates show a periodicity in birth and mating behaviors. Colobines are no exception; in fact, the relation between mating behavior and aggression is more clearly shown among north and south Indian langurs than other primates. There is a strong possibility that the all-male groups living on the periphery of many colobine bisexual groups result from intermale aggression, especially as expressed during the mating season. Within Nilgiri langur multiple-male troops, for example, there is a discernible rise of intermale aggression during the mating season.

The most striking case of periodic aggressive behavior is found in Sugiyama's (1965b, 1967) reports of south Indian common langurs and Mohnot's study (1971a) of north Indian common langurs. Both authors found an association between mating behavior and increased antagonism between males of all-male and bisexual troops. The antagonism usually continues until one of the males of the all-male group joins the bisexual troop, whereupon he kills many or most of the resident infants and drives out the other males. Sugiyama suggests that this extreme expression of aggression (infanticide) is periodic and occurs about once every 3 or 4 years on the average.

TYPES OF COLOBINE AGGRESSION

Interspecific Aggression

There are few reports of interspecific aggression among colobines and the argument over whether or not predation should be considered aggressive behavior is irrelevant. In some areas in which they have been studied, for example, north and south India and in Malaysia, colobines overlap with other nonhuman primates. In north India Nilgiri langurs and rhesus macaques overlap at higher altitudes. In Malaysia, the lutong and crab-eating macaque overlap. In most cases interaction between the species is peaceful and one of mutual disinterest. If a tension-producing situation should arise, for example, if macaques and langurs both arrive at a water hole simultaneously or if macaques arrive later, the langurs simply move away. One of the reasons for this peaceful interaction is likely due to the fact that colobines compete with few other nonhuman primates for food and other requisite components of the econiche.

There have been reports of colobines acting aggressively toward carnivorous predators, domestic dogs, and humans. In most cases, however, their first response is to flee rather than face the interloper. A most interesting contrast is visible in south India where bonnet macaques will flock above a dog or human and threaten it for various lengths of time. A langur will most likely flee and hide in a tree. The langur's modus operandi is to flee to a tree and remain unobtrusive.

Interesting reports have been made of colobines aggressively attack ing birds. Poirier (1967) reports, for example, that adult Nilgiri langurs occasionally attack crows, whereas infants often play with them. Bernstein (1968) reports that lutongs occasionally rush at owls, although it is not clear whether this is play or aggressive behavior. In any case, interspecific aggression is of minor importance and appears to be defensive rather than offensive.

Intertroop Aggression

One of the main arenas for colobine aggressive behavior is intertroop battles. Until quite recently few primatologists would accept the fact that many nonhuman primates frequently engage in territorial behavior. Territorial behavoir is now seen as a major component of an arboreal primate's life, and colobines are no exception. Except for Jay's study on common Indian langurs, colobine territorial behavior has been reported from India, Malaysia, Ceylon, and Africa. While there is a good deal of discussion as to a definition of territory (i.e., Bates, 1970), colobines roughly fit the classic pattern of defending some part of the home range by fighting or aggressive gestures from conspecifics (Burt, 1943).

A very interesting feature of colobine intertroop encounters is the fact that they have readily available means of avoiding such contact. As arboreal animals occupying upper story vegetation which provides a relatively unobstructed view of surroundings, and as possessors of loud, sonorous vocalizations, colobine groups could rather easily avoid contact. Nevertheless, contact is frequent. Colobines maintain troop separation by one or a combination of the following: variable movement patterns, the male whoop vocalization, and male vigilance behavior.

The major troop location signal is the whoop vocalization. Among Nilgiri langurs the whoop is a sonorous, booming male vocalization audible over long distances which serves to locate troops in their respective home ranges. The vocalization is usually given in two different circumstances—upon rising in the morning when it serves to locate troops within their respective ranges, and during territorial battles. Similar vocal patterns have been described for gibbons (Carpenter, 1940), howlers (Carpenter, 1934), and *Callicebus* (Mason, 1966). An interesting feature of these calls is that although they function as troop separating mechanisms, they also elicit approach and contact with neighboring troops. Marler (1968) suggests that eliciting contact is a prelude for distance-increasing "hostile displays."

Colobine intertroop relationships assume various forms including peaceful feeding in proximity, peaceful withdrawal of one troop, or quite often, exchange by the males of aggressive visual and/or vocal signals, and occasionally chasing and actual physical contact. It cannot be positively determined what influences the mode of troop interaction. No doubt there is some association with local environmental conditions and historical troop relationships. Troop size may also be a factor; e.g., males of smaller troops may avoid males of larger troops. Except among Ceylon langurs (Ripley, 1967), females seldom participate and are passive observers of intertroop battles. Among Nilgiri langurs, even when he enters an opposing male's home range, a male avoids the females and infants. Territorial battles are primarily undertaken by leader males; however, among Ceylon langurs lesser males may become involved.

There are recurrent patterns common to colobine intertroop encounters. Time of day seems to be one; among Nilgiri langurs (Poirier, 1968d, 1969c) altercations are most likely to occur between 09:00 and 12:00 and from 14:00 to 16:00 hr. These are all intervals of major movement and/or feeding. Among Nilgiri langurs there are also monthly peaks of antagonism occurring in January, March and April. January immediately follows the winter birth peak, March and April immediately precede the spring peak. Colobine intertroop encounters rarely extend beyond various types of threat behaviors; except for Ceylon langurs overt fighting seldom occurs. The most common behavior patterns are loud vocalizations, rapid, frenetic movement through the trees, raining dead branches to the ground (both the vocalizations and movement produce loud noise, and the latter produces a striking visual display in the more elaborately colored members of the colobinae), and what I have labeled "ritual chasing" among Nilgiri langurs. In the latter, males of opposing troops chase one another in and out their respective home ranges, but they rarely make physical contact. There is repeated jockeying back and forth until the males come to a standstill at some point of balance in their respective home ranges from where they threaten each other and eventually return, vocalizing loudly, to their groups. Territorial encounters may last over an hour.

The patterns most common to colobine intertroop encounters deserve brief analysis. The loud vocalizations serve primarily to locate troops, and do not seem to be used in active defense against intrusion. Excitement is high during this stage, which includes tremendous leaping and running through the tree tops, as is evidenced by frequent defecation and urination. Another indication of high excitement and/or tension is the fact that males may have penile erections. The rapid chasing back and forth across home range boundaries is interesting in two respects: (*1*) animals seldom enter another troop's core area, and (*2*) physical contact is rare. Using an explanation R. Spitz (1969) offered in a very different situation, we might suggest that chasing provides the male with a compromise between his aggression and his "super ego." Chasing serves to dissipate aggression, the lack of physical contact ensures that no animal will be injured.

We are still unsure of the underlying motives of colobine territorial encounters. Colobine intertroop encounters are not means for extending ranges, but are means whereby ranges are maintained. We know that spatial location of troops is involved, and increasing evidence suggests that the home range core areas are what is being defended. Although a prime function of colobine territorial behavior seems to be spacing of males and indirect prevention of overfeeding and overcrowding, there is insufficient evidence to relate it directly to either the breeding and rearing or the food and shelter type of mammalian territorial behavior.

There is some evidence among certain Nilgiri langur troops which I studied (Poirier, 1968d) of a direct relation between food supply and intertroop aggression. Nilgiri langur home ranges overlap considerably in some areas, useage of overlapping areas being determined by variable movement patterns which usually maintain troop separation. However, seasonally various plants lose their value as food sources. One such plant, the *Acacia* tree, is a favorite food source of Nilgiri langurs, and when *Acacia* stands found within overlapping home range areas contain fruits while other trees in the home range lack them, these trees are often focal points for territorial battles. Males of adjacent troops display and chase each other for long periods about these trees.

One-Male Troops

Although colobines are found in multi-male bisexual social groups, one-male troops are predominant in many species. A number of factors account for one-male troops, not the least of which are lack of predation and the capacity for rapid arboreal escape. Colobine males play little role in protecting troop members from danger; each animal's prime defense is its own ability to flee. Another factor which may account for one-male troops is the amount of male–male aggression. There is increasing evidence that excess adult male colobines leave or are driven from bisexual troops. Colobine males live peacefully together outside bisexual troops, in all-male groupings; however, there is increasing male hostility in the presence of females. There may be a positive attraction between males which breaks down in the presence of females. Sugiyama's (1967) and Yoshiba's (1968) data on south Indian common langurs and Kummer's (1968) data on hamadryas baboons suggest a similar interpretation.

Chance and Jolly (1970) using Jay's information on common langurs, suggest that there is a high level of aggression between adult and subadult males; mainly low-level aggression occurs between adult males and adult females and overlaps with reciprocal grooming behavior. Adult males and adult females groom, but are not groomed by juveniles. There is very active reciprocal grooming between adult females, infants, and subadult females, but little aggression. The authors conclude that the scope of social relations suggests an aggressive bond between the adult and subadult males and a mixed type of bond linking female–infant groups to the adult males.

As an example of the above proposition, several trends become apparent when one compares the social relationships of adult males and adult females in one- and multi-male Nilgiri langur troops (Poirier, 1969b, 1970a,b). Comparing the adult male with the adult female dominance hierarchy, it is clear that adult male dominance interactions are more frequent and severe. Taking as an example one of the multiple-male, multiple-female troops, adult male–adult male aggressive encounters totaled 84% of the adult–adult dominance sequences, or one interaction per 7.7 hr of observation. On the other hand, adult female–adult female aggressive encounters occurred just once per 39 observation hours. Over time, the existence of more than one adult male seems to prove disruptive to the troop. In multiple-male Nilgiri langur troops, subordinate adult males direct their hostile attention outside the male hierarchy at animals, such as infants and juveniles, that are not disposed to retaliate. Such aggression could prove injurious, especially since the leader male does not protect younger animals. Over time, the presence of more than one adult male in a troop could prove detrimental to younger troop members. Considering one multiple-male troop, the third and fourth males in the male dominance hierarchy directed 88% of their aggression outside the dominance hierarchy. The fourth-ranked male directed 57% of his aggression at an infant-2.

The inability of Nilgiri langur males to socially interact peacefully appears in grooming behavior. Grooming, a behavior suggestive of positive social bonds between its participants, is more frequent among adult females than among adult males. Duration of a grooming bout is a major differentiating feature; female–female bouts last longer than either female–male or male–male sequences. These data may reflect a relative inability of adult males to establish and maintain mutual contact when compared to adult females.

Furthermore, the one effective mechanism for dissipation of tension, redirection of aggression at younger, subordinate animals, would be harmful to the group and in the long haul dysgenic for the species.

Troop Change—Introduction of New Members

One of the most interesting, but hardest to comprehend, features of colobine life is the phenomenon of troop change, especially the introduction of new males to a bisexual troop. Although the pattern varies with the species, and perhaps with each troop according to historical factors, troop change is either nonaggressive (among Nilgiri langurs Poirier, 1969b, 1970b) or very aggressive (among north and south Indian common langurs). It is to the latter which we now turn.

When Sugiyama (1965a,b, 1967) and subsequently Yoshiba (1968) first reported infanticide among common langurs in south India, the incident seemed to be an isolated case due perhaps to special local circumstances. Now, however, similar behavior has been reported by Mohnot (1971a) from northwest India and may also occur in Ceylon. It is clear that infanticide is not an isolated phenomenon but rather, a widespread colobine response to yet-undetermined conditions. The social changes which Japanese primatologists documented for the south Indian langurs at Dharwar are frequent and dramatic intertroop encounters are a daily occurrence there. In fact, they are so frequent that troops can be ranked respective to one another in a hierarchy.

While encounters between bisexual group males seldom lead to physical contact, those between males of bisexual and all-male troops are a different matter. Relationships between males of bisexual and all-male troops are characteristically aggressive. Japanese researchers documented more than 10 major social changes in 2 years of observation, in every case these followed contact between all-male and bisexual troops. Changes were of several kinds, but they most always included a complete shifting of the troop leadership, subsequent division of the troop, killing of all the infants and driving out of juvenile males, and copulation between the new leader males and troop females.

A number of questions are raised by this behavior: Why are the infants killed, why does this behavior occur in some areas and not in others, and is there a wider function for such behavior than is immediately apparent? Infant killing by the new males in the process of social change may be connected with increasing levels of sexual excitability. Both Mohnot and Sugiyama report that peak sexual activity followed troop reconstructions. However, troop changes noted among Nilgiri langurs (Poirier, 1969b) showed no increase in sexual behavior. We do not know the reason(s) for this difference, although, the Nilgiri langur study may not have been conducted long enough to witness such events.

Other reasons have been suggested for infant killing. Mohnot (1971a) suggests that the simultaneous sexual excitement and tension upon battling other males might be the stimulus behind infant killing. Mohnot notes that a male's postkilling behavior is markedly relaxed. He feels that males lack the means of aggression dissipation common to females, such as embracing and mouth sniffing, and thus killing of infants is a tension releaser. Yoshiba (1968) suggests that because the infant is an unknown new stimulus the new troop leader is more aggressive to infants born prior to his arrival than those born subsequently.

Itani (1972) advances a third reason for infant killing—that it prevents incest. In one-male troops the probability of father and daughter incest would be 100% after the adult male resided for a period of 4 years in a bisexual troop because he has sexual access to all females, and 4 years is the requisite time needed for a female to gain sexual maturity. (This assumes that no psychological avoidance mechanism exists mediating against father–daughter mating.) According to Sugiyama's data the leader male of a bisexual troop retains his position for about 3 years.

Incest between himself and his daughter, who is born by the first mating between himself and her mother after he assumed control of the troop, could be completely avoided because he is driven from the troop prior to his daughter's reaching sexual maturity. Itani further notes that if the male's reign is longer than 3 years, and if that male is subsequently replaced by another, then some of the infants killed in the process might be products of incestuous matings between father and daughter. Frequent infanticide reduces the effects of incest.

One final question is why infant killing is found in some areas and not in others. What causes this social difference among colobines? The answer is yet to be found, but Sugiyama (1967) points to population density as one reason. The high population densities at Dharwar lead to males deserting troops and forming all-male groups. However, the lack of unoccupied ground makes it difficult for them to establish a new home range. According to Sugiyama, this leads to frequent attack on bisexual troops by "dissatisfied" males and blocks the increase of troop size. Similarly, demographic data from northwest India suggests that high population densities may be a contributing factor. Jolly (1972) notes that the mother's rather cavalier abandonment of the infant once the male bites it ensures a higher death rate.[4] There are parallels among mice when pregnant females exposed to the odor or presence of a strange male abort and immediately come into estrus. This could be advantageous to the male and female if it reduces inbreeding. It might also be advantageous if the male is inclined to act aggressively toward a female with which it has not mated, to the extent that she might lose an infant she already carries.

Other data support the contention of a positive relationship between aggression and ecological conditions. Overt aggression between olive baboon troops is rare in Nairobi Park where population density is low (DeVore and Hall, 1965); considerably more aggression occurs between hamadryas baboons in western Ethiopia where troop sizes and population densities are much higher (Kummer, 1968). Even among hamadryas, differences occur; in eastern Ethiopia densities are higher than in western Ethiopia where troops are smaller and more dispersed. Western Ethiopian hamadryas exhibit less aggression. However, this generalization of a relation between rising aggression and increasing population density did not hold in Gartlan and Brain's (1968) study of vervet monkeys. Their analysis added another dimension—that it is not population density per se which contributes to aggressive behavior, but rather population density in relation to environmental resources.

[4] Mohnot notes that some females tend to their wounded infants.

We are still a long way from fully understanding the mechanics involved, but infanticide as practiced by some colobine males is surely one of the more interesting cases of primate aggression. It certainly bears further research.

Sexual Behavior and Aggression

There is no doubt that there is some relationship between sexual behavior and colobine aggression; in some species it is more direct than in others. In the stressful situation of group change the relationship is quite vivid; however, even in the normal course of events sexual behavior evokes some aggressive behavior. Jay (1965) mentions that during the short consortships formed by north Indian common langurs, the consort pair may be harassed by less dominant males, which run about, bark, threaten, and slap at the consort male. The harassing male directs his aggressive behavior almost exclusively at the consort male. Vogel's (1970) study of common langurs in northwestern India likewise reports harassment during sexual behavior. Vogel notes that there is much "sexual jealousy" among males, especially leader males who may aggressively interrupt the copulations of subordinate males.

The Japanese note that south Indian common langur consort males are not harassed by subadults or juveniles in an ordinary bisexual troop. However, in an experimental troop formed by three males of an all-male group and adult females from more than two bisexual troops, copulations between the leader male and a female were harassed by subordinate males. When a male group joins with a bisexual troop, copulations between troop females and nontroop males are harassed by males of the all-male group. Yoshiba (1968) notes that adult and subadult females of bisexual troops often harass consort males. Sometimes estrous females harass the consort pair as do nonestrous adults and subadults.

Again the question: Why is there harassment and aggression during copulatory behavior among some langurs and not others? We lack an answer.

Aggression and Mother–Infant Behavior

The major concern here is the aggression directed from a mother to her infant and from an infant to its mother as it occurs during the weaning process, that is, as mother and infant begin to separate from one another. During weaning there is a gradual transition of the close mother–infant relationship from one of virtually continuous physical attachment and codirected attention, through several transitional stages, to one of ultimate independence and separate functioning of the mother and infant (Kaufman and Rosenblum, 1969). Weaning encompasses the physical and emotional rejection of an infant by its mother, who, although she was once the major source of comfort, warmth, and food, is now hostile and denying. The severity of rejection depends upon the mother's and infant's temperaments, species-specific behavior, gender of the infant, mother's dominance status, number of previous births a female has experienced, and econiche.[5] Multiparous females may wean their infants with less effort than primiparous younger females (Jay, 1962). Dominant females seem to have less trouble weaning their infants, who seem less reluctant to leave, then do subordinate females (DeVore, 1963). Male infants are weaned earlier and treated more roughly during weaning than are females (Jensen *et al.*, 1962; Itani, 1959).

Little is known about the mechanisms of colobine weaning behavior; however, some patterns do emerge. Although we do not know if males are weaned earlier or if dominant females wean earlier than subordinates, we do know that at later stages in the weaning process there is a good deal of aggressive interaction between a mother and her infant and occasionally between a mother and other females. Jay (1965) notes that among common langurs the severity of rejection varies among females. A rare female may be very positive in her rejection and after a few weeks strike the infant whenever it approaches. Jay feels that a mother's temperament and the amount of previous mothering experience influence the weaning process.

During the last month of weaning, the common langur mother persists in her rejection of the infant. The mother's tension increases and she often flees from the infant. She might bite her infant, but the infant is not visibly hurt. The infant's tension increases during the later weaning stages and it strikes and screams at its mother. Eventually, by about 12–14 months, the infant is weaned.

Nilgiri langurs wean early (Poirier, 1968c, 1969b, 1972, 1973a), by 9 weeks infants begin to take solid foods, facilitating weaning as the mother no longer needs to constantly satisfy the infant's nutritional requirements. Once the break is initiated, the mother continually leaves the infant who is forced to find ways through the trees to reach her. The infant chases the mother, catches her, and loses her again, all the while it screams loudly and may slap out at branches, etc. This highly charged process develops the infant's abilities of self-locomotion as well as teaching it the proper routes through the trees.

[5] For an exposition of this last point, see Chalmers (1972).

In the earliest weaning stages a Nilgiri langur mother alternates between accepting and forcefully rejecting her infant; advances toward the mother might stimulate a positive hugging or negative slapping response. The infant is caught in an emotional dilemma, never knowing what response to expect from a once warm and friendly object. The infant's tension is visibly high. As maternal solicitude wanes, maternal behavior appears which actively encourages dissolution of the dyad. The two most prevalent forms of separation behavior are nipple withdrawal and punitive punishment. During the final weaning stages the mother refuses with greater frequency and harshness the infant's attempts to nurse. A mother frequently loses patience with a persistent 10- or 11-month-old infant's attempts to nurse. She will slap and threaten the infant who reacts with violent muscle spasms, frenetic movement, and loud screaming. Although a mother may slap and bite her infant, there is never any serious infliction of injury. An infant's outbursts often disturb nearby animals who may threaten the infant and/or mother. The mother's persistent rejection initially increases rather than decreases the infant's nursing attempts. By 1 year the stressful weaning process is completed and the mother and infant establish a new life-long relationship.

A concentrated study of the weaning habits of various primates may reveal interesting answers. Different authors have suggested that there are relationships between the method and duration of weaning behavior and adult life (Poirier, 1972, 1973a). Mead (1935) remarked on the association of adult aggressiveness with rough and abbreviated nursing in contrast to the gentle and prolonged nursing in a more cooperative tribe. Heath (in an unpublished study) found a significantly higher degree of aggressiveness in nine early weaned rats compared to nine rats remaining with the mother. Finally, Anthoney (1968) suggests that there may be a link between the amount of grooming and length of the weaning period. Since grooming is an important behavioral ameliorant of aggression, this leads to other possibilities. For example, is there less grooming in relatively nonagressive species, and more grooming in relatively aggressive species? Is there more social and "psychological" strain among species exhibiting minimal grooming? I have previously suggested that one reason for the loose social structure of Nilgiri langur troops is the lack of such troop cohesive behaviors as grooming (Poirier, 1969b).

SUMMARY OF COLOBINE AGGRESSION

The following summarizes the major characteristics of colobine aggression. (*1*) For various reasons, some of which are most likely linked to the socialization process and the arboreal niche, with its lack of predation, colobines are not as aggressive as many other nonhuman primates. However, with high population densities and crowding even the peaceful colobine becomes aggressive. (*2*) Colobines exhibit some seasonality in aggressive behavior. A rise in aggression can be related to the seasonal fruiting and flowering of certain favored food sources when adjacent troops contest the right to feed from scarce sources, and during the mating season. (*3*) There is minimal interspecific aggression directed from colobines to other animals, primates and nonprimates. Generally the colobine pattern is peaceful coexistence with other animals. (*4*) Much of colobine aggressive behavior, and certainly some of its most dramatic instances, are witnessed between rather than within troops. Territorial behavior is a factor of colobine life. (*5*) Among some colobines, extreme aggression is witnessed between males of bisexual and all-male groups. When members of the latter invade a bisexual troop, infanticide and group leadership change is common. (*6*) Colobines evidence a link between sexual behavior and aggression. (*7*) While the mother–infant dyad is generally peaceful, aggression characterizes the weaning process where it helps sever one form of relationship and establish another. It is necessary for a female to rid herself of one dependent before accommodating another. (*8*) Females appear to be the stimulus of much intermale aggression. Male–male interactions in all-male groups are relatively peaceful; in multiple-male bisexual troops they are often aggressive. Male relationships seem to be strained in the presence of females. Males seem to lack aggression-dissipating mechanisms, such as grooming, common to females. (*9*) There is a sexual gradient in aggressive expression; males are more aggressive than and dominate females. (*10*) Aggressive behavior varies according to age as well as sex. There are certain periods in an animal's life when aggression is more common than in others, i.e., during weaning and during subadulthood when an animal begins to move into the adult hierarchy. In this instance aggression is more prevalent for males than females. This seems to be related to their adult roles (Poirier, 1973a). (*11*) Aggressive behavior which does occur entails minimal physical contact; most aggression occurs as threat behaviors (see for example Poirier, 1970a,c). An animal is seldom injured during aggressive interactions, for tactile signals such as biting, slapping, and wrestling seldom occur. (*12*) Most aggressive interactions are between two animals; alliances are seldom formed whereby a number of individuals attack another. (However, north Indian common langurs may form alliances whereby a number of females attack another.) (*13*) Colobines possess a number of appeasement and submissive gestures, postures, and vocalizations serving to interrupt attack sequences. (*14*) Colobines possess in the communicative repertoire behaviors, such as embracing or hugging, which dissipate the tension of an aggressed animal.

Southwick (1972) notes that there are minimally three categories of primate aggression: (*1*) species with a low frequency of aggressive behavior under natural conditions, (*2*) species with a high frequency of ritualized aggressive communication, but with a low frequency of overt attack, and (*3*) species with a high frequency of overt attack. Under this scheme colobines fall into categories 1 and 2. As concerns intragroup aggression, colobines are 1 and as concerns intergroup aggression they are 2.

DISSIPATION OF AGGRESSION

The Communication Repertoire

One of the major means whereby aggression is dissipated or avoided is through proper usage of the communication matrix. As do all animals, colobines possess a number of vocal, gestural, and postural communicative cues signaling appeasement, submission, and "I mean no harm," which, when used in the proper circumstances, substantially reduce the amount of aggressive behavior (Poirier, 1970a,c). Jay (1963) notes that 80% of common langur threat behavior is accomplished in a nonviolent manner using subtle forms of threat communication. Should the communication system fail to stop aggression, other communicative acts exist which reestablish a peaceful relationship among the participants. There is little lag in these communicative acts; if an aggressive signal is given, an appeasement signal is almost immediately returned. There appears to be little internalization of the situation, or better said, forethought about what the proper response should be.

A troop's dominance hierarchy can be determined once one recognizes the flow of dominant and subordinate signals. There is rather consistent agreement among colobines concerning such behaviors. The most common dominant signals include grinning, staring, biting air, slapping the ground, lunging, chasing, bobbing the head, and mounting another animal. Submissive gestures include presenting the hindquarters, looking away, running away, turning one's back to another animal, and being mounted. Depending upon the tension inherent in the situation, these may be accompanied by any one of a number of vocal cues. Subtle pauses and hesitations are gestures most often employed. Also very common is the maintenance of physical space. The higher the animal's position in the dominance hierarchy, the wider the personal space it controls which a less dominant animal may not enter without first clarifying its intent.

Displacement activities are another category of social communication serving to dissipate aggression. Some situations arouse conflicting tendencies during which animals manifest behaviors not obviously pertinent to the situation at hand, but by their performance these behaviors may prevent an animal's being aggressed against. The most common colobine displacement acts are scratching and sham or symbolic feeding. Sham or symbolic feeding is most common among males during moments of tension, especially during territorial battles. Scratching occurs in similar conflict situations; animals occasionally interrupt threat sequences to scratch themselves. It is uncertain how these behaviors function to dissipate aggression, but the sudden lull in activity and shift to a non-aggressive pattern serves to break the intensity of the situation, and the animals turn and leave.

Maintenance of Dominance Hierarchy

The maintenance of a stable dominance hierarchy is a very important method whereby primates and many other animals reduce the incidence of aggression. It may be more than coincidence that males, who are the most aggressive, are the easiest to rank; females, who are generally less aggressive, are harder to rank. The same has been found for a human nursery school group (Knudson, 1971). Dominance rankings markedly affect social integration and group control, and communicative patterns such as described earlier are essential aspects of troop integration. Societies based on the dominance principle might be expected to be anxious, for an obvious defect is the possibility of abuse. However, fighting is rare when the dominance hierarchy is well defined. The usual consequence of a rank ordering is minimization of social disruption; stable dominance orders supplant continuous fighting and excessive punishment by relative tolerance. Aggression usually accompanies instability.[6] But, it is often impossible to discern whether aggression caused the social instability, or resulted from it. An animal's position within the hierarchy subtly influences many aspects of its life, i.e., what and when to eat, who to approach and avoid, and where to sit, sleep, and move. However, contrasting with many other nonhuman primates, among colobines it does not relate to access to estrous females.

The dominance hierarchy dissipates aggression by placing each animal

[6] Observations supporting this have been made by Carpenter (1942) on the Cayo Santiago rhesus, by Zuckerman (1932) and Hall (1964) on captive baboons, and Southwick (1970) on captive rhesus.

in a known relationship with another. Since each animal knows its position vis-à-vis others and acts accordingly, and as long as each stays in its place, there is minimal disruption. Aggression is most apt to occur when the order is upset, when the leader male is displaced or when subadults are moving up the dominance hierarchy. One of the stormiest periods in the young primate's life (including the human animal) is when the subadult male begins to move into the adult hierarchy. The male experiences frequent tension and aggression until he establishes his position in the troop's hierarchy.

To have a stable colobine social troop, one lacking frequent aggressive output, one of two situations should prevail: (*1*) either the troop must be a one-male troop or (*2*) in multiple male bisexual troops the males must be arranged in a stable dominance hierarchy. The case for females is less clear. Jay (1965) and Yoshiba (1968) state that common langur females lack stable hierarchies. Schenkel and Schenkel-Hulliger (1967) agree for colobus. However, Poirier (1970a) states that Nilgiri langur females can be rank ordered.

Ritualized Behavior

Although labeling the chasing back and forth across home range "boundaries" during intertroop battles as ritualized behavior will pain some readers, it is clear that this pattern of male chasing male and seldom mixing is an important means of dissipating aggression. Furthermore, the rapid leaping about and movement in the treetops common to colobine intertroop communication may also be a ritualized display serving to dissipate aggression. Undoubtedly, if the energy spent during these activities was spent instead in direct physical confrontation, great physical harm might occur. (Although if physical contact which resulted in harm occurred, then one could turn the argument about and say that all the excitement produced by the displays and chasing was released through violence!) Ritualized aggression conserves energy, reduces chances of injury and death, but yet achieves the adaptive values of spacing and natural selection (Southwick, 1972). Jolly (1972) notes that it is uncertain "whether one should argue that the 'general' level of aggression is low in these animals because many of their ritualized battles seem perfectly calm to the naive observer, or whether one should argue that the battles are highly aggressive and that they somehow take it out on their neighbors instead of behaving aggressively within the troop." [7] But,

[7] From A. Jolly, "The Evolution of Primate Behavior," Macmillan Publishing Co., Inc., New York, p. 178. © 1972 by Macmillan Publishing Co., Inc.

since we cannot discern an animal's inner feelings, the question is not likely to be readily answered.

Grooming

A major method whereby most primates dissipate aggression is grooming behavior. Nonhuman primate grooming is a highly interdependent activity serving a biological and social function. In most species grooming is highly reciprocal and is the basis of much social conditioning. Grooming aids social integration; it depends upon a previous state of positive conditioning; and it enhances and strengthens social relationships (Carpenter, 1940). Grooming is a very important tactile cue frequently observed during aggressive sequences for it functions to reduce tension between animals involved in aggression. Among Nilgiri langurs (who actually groom little), 45% of all grooming bouts occurred after an aggressive episode (Poirier, 1967, 1970a). The most frequent response to the invitation of grooming is general relaxation of tension and adoption by the recipient of a posture inviting further grooming. Grooming indirectly favors a bond by diverting the recipient to an action—invitation to further grooming—incompatible with aggression (Marler, 1965). This is perhaps the reason for its occurrence in potentially aggressive situations.

Other tactile cues which apparently reduce tension have been recorded during colobine aggressive encounters. A most interesting pattern exhibited by the Nilgiri langur is labeled embracing. During threat sequences subordinates often seek rather than avoid physical contact with dominant animals. This is especially true of adult female, adult male–juvenile, and adult male–infant-2 encounters. Typically the subordinate runs to, embraces, and grooms the dominant animal. Juveniles and infants often chase adult males and force themselves to the male's chest before their tension subsides.

The Nilgiri langur embrace pattern is characteristic of other colobines, for example the common langur (Jay, 1965) and the black and white colobus (Schenkel and Schenkel-Hulliger, 1971). The embracing pattern seems to mimic the mother–infant embrace–clinging posture (Poirier, 1968a, 1970c). Corresponding behaviors in other species have likewise been interpreted as regressions to infantile behavior (Andrew, 1964; Bolwig, 1957; Kummer, 1967a,b). The original function of the maternal behavior, to protect the infant which is embraced or carried, is present in the dominance embrace behavior. Embracing reduces the probability of attack by a dominant animal and serves to soothe the subordinate's tension. The addition of grooming to embracing behavior is especially interesting since grooming helps alleviate tension. Recourse to embracing in aggressive situations definitely has a stress-reducing effect.

A final tactile gesture serving to alleviate tension was labeled the "comforting touch" among Nilgiri langurs (Poirier, 1970a,c). Depending upon the context, touching among Nilgiri langurs is either a mildly assertive gesture given by a dominant to a subordinate animal, or is a signal that an aggressive situation has terminated and results in gradual tension reduction. Occasionally rather than embrace an animal, a dominant Nilgiri langur simply places its hand upon the lower status animal, whereupon tension almost immediately decreases. This gesture is most often used by animals with a wide dominance gap, for example, between adult males and juveniles.

Socialization

In this section we explore the relationship between the socialization process and adult aggressive behavior. The major factor negating a full understanding of whatever relationship may exist stems from the fact that there are few long-term field studies where individual genealogies are known. Since this is as true for colobines as for most other nonhuman primates, we must rely on comparative data as an indication of what might be found if data existed.

Most nonhuman primates live within social groups where they are socialized and where they learn the life-ways of preceding generations. Within the social group an animal learns to express its biology and adapt to its surroundings. Differences in primate societies depend not only upon biology, but to a great extent upon the circumstances in which an individual lives and learns. The composition of the social group, and the particular balance of interanimal relationships, determines the nature of the social environment within which a youngster learns and matures.

One of the most exciting lines of research in socialization studies deals with gender differences in the rearing process. There is a direct relation between the socialization and learning processes of infant males and females and their subsequent adult roles (Poirier 1972, 1973a). Although most of our data derives from captive macaque colonies, it is instructive to review the material in light of what it may ultimately reveal about colobines and other primates. Goy (1968) and his associates suggest, on the basis of high hormonal levels circulating in the blood of newborn rhesus, that during fetal or neonatal life hormones act in an inductive way on the undifferentiated brain to organize certain circuits into male and female patterns. These hormones may act to produce behavioral patterns. For example, hormonal influences at a critical developmental period may affect later sensitivity to certain stimuli. Such a situation may account for the varying reactions of males and females to infant natal coats. Hormones may also act by reinforcing some behavioral patterns over others. Whereas a female may derive pleasure from hugging an infant to its chest, a male may derive more pleasure from the large muscle movements and fast actions involved in play-fighting and aggressive behavior (Lancaster, 1972).

Gender differences in behavior are partly developed by the dynamics of group social interaction, for example by learning role patterns. Social roles are not strictly inherited (Benedict, 1969), for laboratory studies suggest that primates without social experiences lack marked sexual behavioral differences (Chamove *et al.*, 1967). Studies of the mother-infant interactional dyad clearly show that there are sexual differences in the development of independence from the mother, i.e., among pigtail (Jensen *et al.*, 1968) and rhesus macaque (Mitchell and Brandt, 1970) infants. Such early differences reflect trends witnessed later in life (Poirier, 1972, 1973a).

A major feature differentiating male and female primates is the amount of aggressive output; males are usually more aggressive than females. Mothers of male infants are more punishing and rejecting than mothers of females, who tend to be restrictive and protective of their daughters. From the very beginning, male and female infants are treated differently, due perhaps to the fact that male infants behave differently and that the mother reacts to this.

Developmental studies of laboratory-reared rhesus elucidate some interesting points regarding male aggression. A rhesus mother threatens and punishes her male infant at an earlier age and at more frequent intervals than her female infant, which she restrains, retrieves, and protects (Mitchell, 1969; Mitchell and Brandt, 1970). The infant male's characteristic predisposition toward rougher play and rougher infant-directed activity is subtly supported by the mother's behavior, and through observations of other mothers and infants (Mitchell *et al.*, 1967). Vessey (1971) reports for the provisioned Cayo Santiago colony that male rhesus infants receive more punishment from the mother than do females.

Similar data have been generated by studies of feral baboons. Ransom and Rowell's (1972) study shows that as early as 2 or 3 months there are consistent differences in the development of the mother–infant relationship and peer interactions. By the time of the transitional period to the juvenile stage (1–2 years), sexual differences in play, frequency of initiation and withdrawal, and duration and roughness of play bouts are present. During the next 4 or 5 years young males interact mostly with each other on the group's periphery and are generally avoided by their mothers and other females. Young females, on the other hand, maintain close proximity to adult females and attendant adult males.

In a recent study of nursery school children, Knudson (1971) found similar gender-related differences in play and aggressive behavior. Boys engage in a higher total frequency of dominance behavior than girls. Secondly, although boys and girls show more physical than verbal dominance behavior, girls engage in a significantly higher proportion of verbal dominance. Knudson also found that the frequency of rough-and-tumble play for boys is significantly higher than for girls and that it was easier to establish dominance rankings for boys than for girls.

Where does the above highly suggestive line of reasoning lead us in considering colobine aggression? Lacking longitudinal and controlled studies, our information is only suggestive; however, there are some interesting points. First, in contrast to what we know about macaques and baboons, dominance status among colobines seems to be mainly acquired (that is, attained through one's own efforts) rather than derived (that is, influenced by the mother's ranking). Imanishi (1960) introduced the concept of identification from psychoanalysis to explain the fact that Japanese macaque infants with a dominant mother tend themselves toward dominance. Infants of higher ranking mothers identify with troop leaders; they eventually become leaders. Infants of lower ranking mothers are unable to identify; they become peripheral members or desert the troop (Itani *et al.*, 1963). Studies of several Japanese macaque troops show that in paired competition for food, successful monkeys are often infants of higher ranking mothers (Kawai, 1958a,b; Kawamura, 1958). Koford's (1963) and Sade's (1965, 1967) reports on the rhesus of Cayo Santiago indicate that adolescent sons of the highest-ranking females hold a high rank in the adult male hierarchy.

As concerns a female colobine, intratroop dominance activity is not a major feature in her life and her status (if it can even be determined) is seldom apparent in her relations with other troop members. It is therefore unlikely that a mother's status has any measurable effect upon that of her infant. Most likely a mother's status is less influential for the infant's development than is her temperament; irritable mothers may have less confident infants.

Another important feature of colobine socialization, clearly separating the process from that of other nonhuman primates, is the pattern of infant transference from one female to another. Unlike most monkeys, colobine mothers allow other females to hold and carry their newborn. Although there are differences as to time of commencement and termination, colobine infants are often passed around to other females from soon after birth. This pattern may continue for the first few months of life. In particular contrast to some macaques and baboons, every colobine infant has free access to every other infant, and females of all ranks have free access to all infants. Swapping of infants may be one of the roots of the nonaggressive colobine society.

No doubt variation in the socialization process accounts for many differences in primate behavior, including incidence of aggression. Eimerl and DeVore (1965) compared the social systems and socialization of colobines and baboons, noting that while patterns of physical and social development are similar, resulting adult behaviors differ. Different ecological pressures do not fully explain these differences. However, they become clear if we look at the unique occurrences and experiences that influence the infants' respective lives.

Arboreality and Aggression

Compared with many nonhuman primates, colobines are particularly lethargic and nonaggressive, at least in intratroop daily life. While comparative lack of intratroop aggression is related to many factors, there is also the possibility that it is related to the colobine's arboreal life-way. Although it still is not proven,[8] there are suggestions that the more arboreal a species, the smaller the amount of intratroop antagonism. DeVore (1963, p. 314) suggested that "intragroup aggression and agonistic behavior decrease by the degree to which a species is adapted to arboreal life." He further noted a correlation between expression of sexual dimorphism and dominance behavior, suggesting that terrestrial life leads to increased predation which leads to increased morphological specialization of the male for group protection. This leads to increased sexual dimorphism, with males becoming larger and more aggressive than females as an adaptation for group protection.

If there is truth to the above arguments, and there is evidence in their favor, one would expect to find minimal intragroup aggression and sexual dimorphism among arboreal forms. Both DeVore's hypotheses tend to be confirmed by present colobine data.

There may be other relationships between arboreality and lack of aggression. For example, an arboreal leaf-eating group (provided it is eclectic in its dietary patterns and includes fruits, buds, insects, and berries in the diet, as most colobines do), may have a greater potential food supply

[8] Thus arguments stating that the aggressiveness supposedly inherent to *Homo sapiens* and their ancestors is related to terrestrial life should be viewed as only suggestive.

than a terrestrial foraging group. The rather unrestricted food supply afforded by an arboreal existence could reduce potential aggressive outbreaks over food supply. This is merely suggestive and not conclusive. We do know, however, that leaf-eaters, probably because they derive less nourishment from the leaves forming a major part of the diet, spend a greater proportion of their day in slow feeding than do many terrestrial omnivorous forms. Since colobine feeding bouts are slow and peaceful, there is a reduction in time left for other activities, i.e., aggression. This argument may be spurious; however, colobine dietary patterns should not be overlooked as being one of many factors mediating against high aggressive outputs.

MAN AND THE MONKEY

Introduction

The final subject to be addressed is what are the relationships between colobine and human aggression? Are there any similarities or dissimilarities from which we can learn? When moving from nonhuman to human primate, a new set of rules comes into play, and it is hardly possible to simply take data generated from studies of one and apply them to the other. Many popular works on human aggression contain references to findings from animal behavior studies thought to be germane to the human situation. Some authors consider human aggression an innate component of human nature. A more direct attempt to equate human and nonhuman primate aggression is Russell and Russell's book *Violence, Monkeys and Man* (1969). A central problem in the study of behavioral evolution is that modern primates are not equivalents of hominid ancestors. There is considerable debate as to how useful current behavioral data are for reconstructing the hominid past. The resolution to the problem is not simple, and may vary according to the question broached.

Similarities

There are situational similarities producing colobine and human aggression. In this age of population control we are used to hearing that high population densities breed aggression and crime. Among colobines, population densities increase the likelihood of aggression, i.e., infanticide among some common langurs. Another similarity appears in the incidence of aggression against strangers. It may be a characteristic of most social group-living animals that outsiders are the focus of hostility. Whether such outwardly directed aggression is a long-term cohesive force, as some popularizers have claimed, is yet to be proven. Even if it were, to champion hostility at "outsiders" seems a fruitless and ultimately expensive way to cement group solidarity.

There are other tantalizing parallels. The fact that females exhibit more verbal aggression than males is a pattern established for colobines and for some human nursery groups. Another interesting point is usage of tactile appeasement gestures; physical contact is a major comfortant from aggression in both human and nonhuman primates. Analogies can be made between human and nonhuman primate tactile cues in stress and conflict situations, as indicators of social status, in sexual behavior, and in the mother–infant relationship (vide also Frank, 1957; McGrew, 1972; Poirier, 1968a; Ribble, 1943; Spitz, 1965).

An interesting point may be made concerning increased colobine aggression as the mother–infant bond is broken and later as the male begins to move into the adult dominance hierarchy. Is there a human correlate? Is there a greater amount of mother–infant (and because they are human, father–infant) tension as the human child breaks the family bond? Is the so-called generation gap a reflection of a biological trend whereby adolescents break away from their parents? A certain amount of aggressive output may be requisite for the adolescent youngster to work its way into adult culture. Human children are normally closely controlled by their families and parents, and under such conditions, fighting of any serious sort is repressed or reduced to teasing and similar mild expressions. However, at adolescence the human child begins to be free of family control, and stays relatively free of similar controls until marriage and the formation of new family relationships. During this transitional period from adolescence to adulthood we observe the maximum expression of aggression. Among humans rates of aggression are higher for males than for females. Crime rates are usually two to five times higher for males than for females. Males are more likely to exhibit delinquency in aggressive behavior while females are more often involved in sexual acts. This probably reflects hormonal differences, but also reflects the social tradition of longer family control over females (Scott, 1962).

Contrasts

Now what of the contrasts? In what ways do human and nonhuman primate aggression differ? For millions of years hominids have possessed extrasomatic (nonbodily) means of producing aggression; hominids made and used tools as weapons of violence. Except for the possible brandishment of sticks and other objects in intimidation among chimpanzees and gorillas, use of weapons clearly separates human from nonhuman primate aggressive behavior. The use of extrasomatic objects in aggressive behavior both limits and extends the amount and types of human aggressive behavior.

A major characteristic differentiating human and nonhuman primate aggression may be internalization of aggression. There is little time lag between threat and punishment in nonhuman primates. On the other hand, humans can (and frequently they do) internalize their aggression, releasing it years after the stimulus producing aggressiveness occurred. This is unlikely for nonhuman primates. They lack the complex neuronal structure, especially in the association areas of the brain allowing this. Increase in the human brain of the association areas probably results from new selection pressures combined with the evolution of more complex forms of social institutions, and is probably connected with evolution of language.

The possession of symbolic communication, of language, is another factor separating human and nonhuman primate aggression. We language-bearing animals have learned to substitute words for weapons and to use them as weapons (Hill, 1970; Spitz, 1969). Much of our lives involve linguistic contacts, and many activities appearing as motor acts in lower species appear linguistically in *Homo* (Freeman, 1964). Human language plays a major role in aggression, courtship, and greeting behaviors. The specific vocabulary is incidental to underlying behavior motivations. There is a category of human aggression labeled verbal displaced aggression. Such behavior includes gossip, grumbling, and scandal.

With increasing institutional complexity, itself related to increasing neuronal complexity and social demands, humans can vent aggression in new ways. No nonhuman primate can express aggression through sports, witchcraft, voodoo, sorcery, and games like chess. Human aggression can be more indirect than nonhuman primate aggression and can be expressed in many more ways.

Tantalizing Leads

Are there any leads or directions which can be developed from an analysis of nonhuman primate aggression? The following, although based on colobine data, would seem to hold true for most nonhuman primate behavior. To control human aggression we must learn to recognize and effectively use appeasement gestures. No doubt we possess appeasement behaviors, such as some of our greeting ceremonies which include smiling, kissing, embracing, handshaking, submissive lowering of the head, etc. But such signals are good only at close range. Perhaps what is needed is some meaningful cross-culturally recognized, long-distance appeasement signal. The use of other appeasement behaviors, such as grooming, might also be helpful. Analogies of primate grooming (e.g., stroking, hugging, scratching) are currently permissible in some social situations, perhaps they should be universally acceptable in any tension situation.

The colobine data suggest that it is not simply population density which is our single greatest source of aggression, but that it is population density relative to scarce resources. An emphasis on the dangers of overpopulation is certainly warranted, but not to the exclusion of stressing the importance of accessibility to resource materials. Under dense population conditions much aggression could be avoided if enough food, water and other necessities were available.

Social uncertainty, i.e., group instability, introduction of a new member, or loss of a leader, is a very potent generator of aggression. During unstable social periods traditional modes for dissipating aggression, like the dominance hierarchy, are voided, as is predictability within the system. It has been suggested (by Gilula and Daniels, 1969) that current examples of human violence and the factors encouraging it reflect our vacillation between the anachronistic culture of violence and the perplexing culture of continual change. Demands for especially rapid change are potentially dangerous because change activates a tendency to return to older, formerly effective, coping behaviors. Social disruption rooted in social/political change tends to increase violence as a coping strategy.

We must recognize that primates (especially when young) learn patterns of aggressive behavior. Some studies on social learning of aggressive behavior in young children show that they learn aggressive patterns with remarkable facility by viewing aggressively acting models. Monkey and ape youngsters mimic their peers and adults, human children do likewise. Human children readily learn aggressive behavior from imitating models in play, such behaviors persist in the child's repertoire for months after even a single exposure. Imitation is even greater if the model has been rewarded for its aggression (Hamburg, 1971b). One method of controlling human aggression would be to limit the amount of aggressive behavior a young child sees and has the opportunity to imitate. During adolescence, aggression is apt to be prevalent for family control is weakening, the social order is being tested, and the individual is personally in an unstable life situation trying to satisfy demands from both the adult and peer society. It would be wise to closely monitor the adolescence period to insert as much stability as possible into the life order.

CONCLUSION

Far from being all bad, that is, maladaptive, aggression is an adaptive mechanism enabling primates to meet a variety of environmental circumstances, to structure their social lives in functional groups, and to survive more satisfactorily in competitive communities (Southwick, 1972). Washburn and Hamburg (1968, p. 462) describe primate aggression as "an essential adaptive mechanism . . . and a major factor in primate evolution." Hall (1964, p. 52) stresses the same theme, noting that "controlled aggressiveness . . . is a valuable survival characteristic in that it ensures protection of the group and group cohesion." Some aggressive behavior is obviously adaptive if used in the right situation and with moderation.

Maybe our immediate ancestors were territorial and aggressive; however, evidence from comparative primate studies is inconclusive (see also Schaller, 1972; Schaller and Lowther, 1969; Poirier, 1973b). Some degree of aggressiveness among primates was and still is adaptive—selected for by natural processes to contribute to reproductive fitness. The complex trait of human aggression should not be assumed a useless, or even harmful, genetic residue from prehistoric times. It is more positively considered to have been, and in some forms and circumstances continues to be, a highly adaptive behavioral pattern.

ACKNOWLEDGMENTS

The author wishes to acknowledge the research help of Ms. Jenifer Partch.

REFERENCES

Andrew, R. (1964). The displays of primates. *In* "Evolutionary and Genetic Biology of Primates" (J. Buettner-Janusch, ed.), 227–309. Academic Press, New York.

Anthoney, T. (1968). The ontogeny of greeting, grooming, and sexual motor patterns in captive baboons (Superspecies *Papio cynocephalus*) *Behavior* **31**, 358–372.

Bates, B. (1970). Territorial behavior in primates: A review of recent field studies. *Primates* 11, 271–284.

Beach, F. (ed.) (1965). "Sex and behavior." Wiley, New York.

Beck, B., and Tuttle, R. (1972). The behavior of gray langurs at a Ceylonese waterhole. *In* "The Functional and Evolutionary Biology of Primates" (R. Tuttle, ed.), pp. 351–377. Aldine Publ., Chicago, Illinois.

Beeman, E. (1947). The effect of male hormone on aggressive behavior. *Physiol. Zool.* **20**, 373–405.

Benedict, B. (1969). Role analyses in animals and men. *Man* **4**, 203–214.

Berkowitz, L. (1965). "Aggression" McGraw-Hill, New York.

Bernstein, I. (1968). The lutong of Kuala Selangor. *Behavior* **14**, 136–163.

Bolwig, N. (1957). Some observations on the habits of the chacma baboon. *Papio ursinus. S. Afr. J. Sci.* **54**, 255–260.

Booth, A. (1957). Observations on the natural history of the olive colobus monkey, *Procolobus verus* (van Beneden). *Proc. Zool. Soc. London* **129**, 421–430.
Burt, W. (1943). Territoriality and home range concepts as applied to mammals. *J. Mammal.* **24**, 346–352.
Carpenter, C. (1934). A field study of the behavior and social relations of the howling monkeys (*Alouatta palliata*). *Comp. Psychol. Monogr.* **10**, 1–168.
Carpenter, C. (1940). A field study of the behavior and social relations of the gibbon (*Hylobates lar*). *Comp. Psychol. Monogr.* **15**, 1–212.
Carpenter, C. (1942). Societies of monkeys and apes. *Biol. Symp.* **8**, 177–204.
Carthy, J., and Ebling, F. (eds.) (1964). "The Natural History of Aggression." Academic Press, New York.
Chalmers, N. (1972). Comparative aspects of early infant development in some captive cercopithecines. *In* "Primate Socialization" (F. Poirier, ed.), pp. 63–82. Random House, New York.
Chamove, A., Harlow, H., and Mitchell, G. (1967). Sex differences in the infant-directed behavior of preadolescent rhesus monkeys. *Child Develop.* **38**, 329–335.
Chance, M., and Jolly, C. (1970). "Social Groups of Monkeys, Apes and Men." Cape, London.
DeVore, I. (1963). A comparison of the ecology and behavior of monkeys and apes. *In* "Classification and Human Evolution" (S. Washburn, ed.), pp. 301–316. Aldine, Chicago, Illinois.
DeVore, I., and Hall, K. (1965). Baboon ecology. *In* "Primate Behavior" (I. DeVore, ed.), (1965), pp. 20–53. Holt, New York.
Eimerl, S., and DeVore, I. (1965). "The Primates." Time-Life Books, New York.
Frank, L. (1957). Tactile communication. *Genet. Psych. Monogr.* **56**, 209–225.
Freeman, D. (1964). A biological view of man's social behavior. *In* "Social Behavior from Fish to Man" (W. Etkin, ed.), pp. 152–188. Phoenix, Chicago, Illinois.
Furuya, Y. (1961–1962). The social life of the silvered leaf monkeys. *Trachypithecus cristatus: Primates* **38**, 41–60.
Gartlan, J., and Brain, C. (1968). Ecology and social variability in *Cercopithecus aethiops* and *C. mitis*. *In* "Primates: Studies in Adaptation and Variability" (P. Jay, ed.), pp. 253–292. Holt, New York.
Gilula, M., and Daniels, D. (1969). Violence and man's struggle to adapt. *Science* **25**, 396–405.
Goy, R. (1968). Organizing effects of androgen on the behavior of rhesus monkeys. *In* "Endocrinology and Human Behavior" (R. Michael, ed.), pp. 12–31. Oxford Univ. Press, London and New York.
Hall, K. (1964). Aggression in monkey and ape societies. *In* "The Natural History of Aggression" (J. Carthy and F. Ebling, eds.), pp. 51–64. Academic Press, New York.
Hamburg, D. (1970). Sexual differentiation and the evolution of aggressive behavior in primates. *In* "Environmental Influences on Genetic Expression: Biological and Behavioral Aspects of Sexual Differentiation" (N. Kretchmer and D. Walcher, eds.), pp. 141–151. National Institutes of Health, Washington, D.C.
Hamburg, D. (1971). Psychobiological studies of aggressive behavior. *Nature (London)* **230**, 19–22.
Hill, D. (1970). Aggression in man and animals, *Proc. Roy. Sci. Med.* **63**, 159–162.
Imanishi, K. (1960). Social organization of subhuman primates in their natural habitat. *Curr. Anthrop.* **1**, 393–407.
Itani, J. (1959). Paternal care in the wild Japanese monkey. *Macaca fuscata fuscata. Primates* **2**, 84–98.

Itani, J. (1972). A preliminary essay on the relationship between social organization and incest avoidance in nonhuman primates. *In* "Primate Socialization" (F. Poirier, ed.), pp. 165–172. Random House, New York.

Itani, J., Tokuda, K., Furuya, Y., and Shin, Y. (1963). Social construction of natural troops of Japanese monkeys in Takasakiyama. *Primates* **4,** 2–42.

Jay, P. (1962). Aspects of maternal behavior among langurs. *In* "Relatives of Man" (J. Buettner-Janusch, ed.), pp. 468–477. New York Acad. of Sci.

Jay, P. (1963a). The Indian langur monkey (*Presbytis entellus*). *In* "Primate Social Behavior" (C. F. Southwick, ed.), pp. 114–124. Van Nostrand Reinhold, Princeton, New Jersey.

Jay, P. (1963b). Mother-infant relations in langurs. *In* "Maternal Behavior in Mammals" (H. Rheingold, ed.), pp. 282–304. Holt, New York.

Jay, P. (1965). The common langur of North India. *In* "Primate Behavior" (I. DeVore, ed.), pp. 197–250. Holt, New York.

Jensen, G., Bobbitt, R., and Gordon, B. (1967). The development of mutual independence in mother-infant pigtailed monkeys, *Macaca nemestrina*. *In* "Social Communication Among Primates" (S. Altmann, ed.), pp. 43–55. Univ. Chicago Press, Chicago, Illinois.

Johnson, R. (1972). "Aggression in Man and Animals." Saunders, Philadelphia, Pennsylvania.

Jolly, A. (1972). "The Evolution of Primate Behavior." Macmillan, New York.

Kaufman, I., and Rosenblum, L. (1969). The waning of the mother-infant bond in two species of macaques. *In* "Determinants of Infant Behavior" (B. Foss, ed.), pp. 41–59. Methuen, London.

Kawai, M. (1958a). On the system of social ranks in a natural troop of Japanese monkeys I: Basic and dependent rank. *Primates* **1,** 111–130.

Kawai, M. (1958b). On the rank system in a natural group of Japanese monkeys. *Primates* **1,** 84–98.

Knudson, M. (1971). Sex differences in dominance behavior of young human primates. Paper presented to Amer. Anthrop. Ass., New York.

Koford, C. (1963). Rank of mothers & sons in bands of rhesus monkeys. *Science* **141,** 356–357.

Kummer, H. (1967a). Dimensions of a comparative biology of primate groups. *Amer. J. Phys. Anthrop.* **27,** 357–366.

Kummer, H. (1967b). Tripartite relations in hamadryas baboons. *In* "Social Communication Among Primates" (S. Altmann, ed.), pp. 63–73. Univ. of Chicago Press, Chicago, Illinois.

Kummer, H. (1968). "Social Organization of Hamadryas Baboons." Univ. of Chicago Press, Chicago, Illinois.

Lancaster, J. (1972). Play-mothering: The relations between juvenile females and young infants among free-ranging vervet monkeys. *In* "Primate Socialization" (F. Poirier, ed.), pp. 83–104. Random House, New York.

Levy, J., and King, J. (1953). The effects of testerone propionate on fighting behavior in young male C57BL/10 mice. *Anat. Rec.* **117,** 562–653.

Marler, P. (1965). Communication in monkeys and apes. *In* "Primate Behavior" (I. DeVore, ed.), pp. 544–584. Holt, New York.

Marler, P. (1968). Aggregation and dispersal: Two functions in primate communication. *In* "Primates: Studies in Adaptation and Variability" (P. Jay, ed.), pp. 420–439, Holt, New York.

Marler, P. (1969). *Colobus guereza: Territoriality and group composition. Science* **163,** 93–95.
Marler, P. (1970). Vocalizations of East African monkeys, I. Red Colobus. *Folia Primat.* **13,** 71–81.
Mason, W. (1966). Social organization of the South American monkey *Callicebus moloch:* A preliminary report. *Tulane Stud. Zool.* **13,** 23–28.
McGrew, W. (1972). "An Ethological Study of Children's Behavior." Academic Press, New York.
Mead, M. (1935). "Sex and Temperament in Three Primitive Societies." Morrow, New York.
Michael, R., and Zimpe, D. (1970). Aggression and gonadal hormones in captive rhesus monkeys (*M. mulatta*) *Anim. Behav.* **18,** 1–10.
Mitchell, G. (1969). Paternalistic behavior in primates. *Psychol. Bull.* **71,** 399–417.
Mitchell, G., Arling, G., and Moller, G. (1967). Long-term effects of maternal punishment on the behavior of monkeys. *Psychol. Sci.* **8,** 197–198.
Mitchell, G., and Brandt, E. (1970). Behavioral differences related to experience of mother and sex of infant in the Rhesus monkey. *Develop. Psych.* **3,** 149.
Mohnot, S. M. (1971a). Some aspects of social changes & infant-killing in the hanuman langur, *Presbytis entellus* (Primates: Cercopithecidae) in western India. *Mammalia* **35,** 175–198.
Mohnot, S. C. (1971b). Ecology and behavior of the hanuman langur, *Presbytis entellus* (Primates: Cercopithecidae) invading fields, gardens and orchards around Jodhpur, western India. *Trop. Ecol.* **12,** 237–249.
Moyer, K. (1969). Internal impulses to aggression. *Trans. N.Y. Acad. Sci.* **31,** 104–115.
Moyer, K. (1971). A preliminary model of aggressive behavior. *In* "The Physiology of Aggression and Defeat" (B. Eleftheriou and J. Scott, eds.), pp. 60–81. Plenum Press, New York.
Napier, J., and Napier, P. (1967). "A Handbook of Living Primates." Academic Press, New York.
Nishida, T. (1972). A note on the ecology of the Red-Colobus monkeys (*Colobus badius tephrosceles*) living in the Mahali mountains. *Primates* **13,** 57–65.
Poirier, E. F. (1967). The Ecology and Social Behavior of the Nilgiri Langur (*Presbytis johnii*) of south India. Univ. Microfilms, Ann Arbor, Michigan.
Poirier, F. E. (1968a). Tactile communicatory patterns among Nilgiri langurs: With reflections of its role among humans. *Ann. South Anthrop. Soc.,* 3rd.
Poirier, F. E. (1968b). Analysis of a Nilgiri Langur (*P. johnii*) home range change. *Primates* **9,** 29–44.
Poirier, F. E. (1968c). The Nilgiri langur (*P. johnii*) mother–infant dyad. *Primates* **9,** 45–68.
Poirier, F. E. (1968d). Nilgiri langur (*P. johnii*) territorial behavior. *Primates* **4,** 351–365.
Poirier, F. E. (1969a). Behavioral flexibility and intertroop variability among Nilgiri langurs of south India. *Folia Primatol.* 119–133.
Poirier, F. E. (1969b). The Nilgiri langur troop: Its composition, structure, function and change. *Folia Primat.* 20–47.
Poirier, F. E. (1969c). Nilgiri langur territorial behavior. *In* "Recent Advances in Primatology: Behavior" (C. Carpenter, ed.), pp. 31–35. S. Karger, Basel.
Poirier, F. E. (1970a). Characteristics of the Nilgiri langur dominance structure. *Folia primat.* **12,** 161–187.

Poirier, F. E. (1970b). Nilgiri langur ecology & social behavior. *In* "Primate Behavior: Recent Developments in Field Laboratory Research" (L. Rosenblum, ed.), pp. 251–383. Academic Press, New York.
Poirier, F. E. (1970c). The Nilgiri langur Communication matrix. *Folia Primat.* **13,** 92–137.
Poirier, F. E. (1971). The Nilgiri langur—A threatened species. *Zoonooz,* San Diego Zoo, 10–15.
Poirier, F. E. (1972). Introduction. *In* "Primate Socialization" (F. E. Poirier, ed.), pp. 3–29. Random House, New York.
Poirier, F. E. (1973a). Primate socialization and learning. *In* "Learning and Culture" (S. Kimball, J. Burnett, eds.), pp. 3–41. Univ. Washington Press, Seattle, Washington.
Poirier, F. E. (1973b). "Fossil Man—An Evolutionary Journey." Mosby, St. Louis, Missouri.
Ransom, T., and Rowell, T. (1972). Early social development of feral baboons. *In* "Primate Socialization" (F. Poirier, ed.), pp. 105–145. Random House, New York.
Ribble, M. (1943). "The Rights of Infants." Columbia Univ. Press, New York.
Ripley, S. (1967). Intertroop encounters among Ceylon gray langurs (*Presbytis entellus*). *In* "Social Communication Among Primates" (S. Altmann, ed.), pp. 237–253. Univ. Chicago Press, Chicago, Illinois.
Sade, D. (1965). Some aspects of parent-offspring and sibling relations in a group of rhesus monkeys with a discussion of grooming. *Amer. J. Phys. Anthrop.* **23,** 1–17.
Sade, D. (1967). Determinants of dominance in a group of free-ranging rhesus monkeys. *In* "Social Communication Among Primates" (S. Altmann, ed.), pp. 99–115. Univ. of Chicago Press, Chicago, Illinois.
Sarich, V. (1970). Primate systematics with special reference to Old World monkeys: A protein perspective. *In* "Old World Monkeys: Evolution, Systematics & Behavior" (J. Napier and P. Napier, eds.), pp. 175–226. Academic Press, New York.
Scarf, M. (1972). He & She: The sex hormones and behavior. New York Times Mag. May 7, pp. 30 ff.
Schaller, G. (1972). Predators of Serengeti: Part 2. *Natur. Hist.* **81,** 60–70.
Schaller, G., and Lowther, G. (1969). The relevance of carnivore behavior to the study of the early hominids. *S. W. J. Anthrop.* **25,** 307–341.
Schenkel, R., and Schenkel-Hulliger, L. (1967). On the sociology of free-ranging colobus (*Colobus guereza caudatus* Thomas 1885). *In* "Progress in Primatology" (D. Starck, R. Schneider, and H. Kuhn, eds.), pp. 185–194. Gustar Fischer Verlag.
Scott, J. (1958). "Aggression." Univ. of Chicago Press, Chicago, Illinois.
Scott, J. (1962). Hostility and aggression in animals. *In* "Roots of Aggression" (E. Bliss, ed.), pp. 167–178. Harper, New York.
Scott, J. (1966). Agonistic behavior of mice and rats: A review. *Amer. Zool.* **6,** 683–701.
Seward, J. (1945). Aggressive behavior in the rat. I General characteristics: age and sex differences. *J. Comp. Psych.* **38,** 175–197.
Sigg, E. (1968). Relationship of aggressive behavior to adrenal & gonadal function of male mice. Paper presented to *Int. Symp. Aggression, 1st, Milano.*
Simons, E. (1970). The deployment and history of Old World monkeys (Cercopithecidae, Primates). *In* "Old World Monkeys: Evolution, Systematics and Behavior" (J. Napier and P. Napier, eds.), pp. 97–136. Academic Press, New York.
Southwick, C. (ed.). (1970). "Animal Aggression." Van Nostrand Reinhold, Princeton, New Jersey.

Southwick, C. (1972). Aggression among nonhuman primates. Addison-Wesley module in Anthropology #23, pp. 1–23.
Spitz, R. (1965). "The First Year of Life." International Univ. Press, New York.
Spitz, R. (1969). Aggression and Adaptation. *J. Nerv. Ment. Disorder* **149**, 81–90.
Suchowski, G. (1968). Sex hormones and aggressive behavior. Paper presented to *Int. Symp. Aggression, 1st, Milano.*
Sugiyama, Y. (1964). Group composition, population density and some socio-ecological observations of hanuman langurs (*Presbytis entellus*). *Primates* **5**, 7–37.
Sugiyama, Y. (1965a). Behavioral development & social structure in two troops of hanuman langurs (*Presbytis entellus*). *Primates* **6**, 73–106.
Sugiyama, Y. (1965b). On the social change of hanuman langurs (*Presbytis entellus*) in their natural condition. *Primates* **6**, 381–418.
Sugiyama, Y. (1967). Social organization of hanuman langurs. *In* "Social Communication Among Primates" (S. Altmann, ed.), pp. 221–236. Univ. Chicago Press, Chicago, Illinois.
Tanaka, J. (1965). Social structure of the Nilgiri langur. *Primates* **6**, 107–122.
Ullrich, W. (1961). Zur biologie & soziologie der Colobus affen (*Colobus guereza caudatus* Thomas, 1885) *Zool. Garten. N.F. 25* **6**, 305–368.
Vessey, S. (1971). Social behavior of free-ranging rhesus monkeys in the first year. Paper presented at Amer. Anth. Ass., New York.
Vogel, C. (1970). Behavioral differences of *Presbytis entellus* in two different habitats. *Proc. Int. Congr. Zool., 3rd, Zurich,* pp. 41–47.
Vowles, D. (1970). "The Psychology of Aggression." Edinburgh Univ. Press, Edinburgh.
Washburn, S., and Hamburg, D. (1968). Aggressive behavior in Old World monkeys and apes. *In* "Primates: Studies in Adaptation and Variability" (P. Jay, ed.), pp. 458–478. Holt, New York.
Wilson, A., and Boelkins, C. (1970). Evidence for seasonal variation in aggressive behavior by *Macaca mulatta. Anim. Behav.* **18**, 719–724.
Yoshiba, K. (1968). Local and intertroop variability in ecology and social behavior of common langurs. *In* "Primates: Studies in Adaptation and Variability" (P. Jay, ed.), pp. 217–242. Holt, New York.
Zuckerman, S. (1932). "The Social Life of Monkeys and Apes." Routledge and Kegan Paul, London.

VARIATION IN CERCOPITHECOID AGGRESSIVE BEHAVIOR

UELI NAGEL
University of Zurich

HANS KUMMER
University of Zurich

INTRODUCTION

Definitions

Aggression is often thought to be necessarily associated with violence and destruction. Carthy and Ebling (1964) express this view in stating that "an animal acts aggressively when it inflicts, attempts to inflict or threatens to *inflict damage* on another animal. The act is accompanied by recognizable behavioral symptoms and definable physiological changes [p. 1]." They exclude, however, predation from their definition. Washburn and Hamburg (1968) in their review of aggressive behavior in Old World monkeys and apes use Carthy and Ebling's definition, but argue that predation should be included to make possible evolutionary considerations on aggression in early man.

In this article we will use the term "aggression" in a more restricted sense. Aggression in animals is primarily a way of competition, not of destruction. As will be shown later, aggressive episodes rarely inflict damage in Old World monkeys under natural conditions. In the following, aggression will be understood as a massive physical impact directed at a conspecific, which, in response, will flee, withdraw, avoid the aggressor, or cease to perform an activity which habitually releases aggression from others. The term "threat" will be used to indicate noncontact behavior which is correlated in time with aggressive behavior and releases in the threatened animal the same, though less intense, responses as aggression. Aggression and threat together will be called agonistic behavior.

The frequency and the physical effect of aggression are not necessarily correlated. It is thus reasonable to differentiate between aggressive and dangerous individuals or species. An individual or species is called *aggressive* when it frequently shows aggressive behavior, no matter how *dangerous* its fighting technique is. The latter can be measured by the percentage of wounds resulting from a given number of attacks.

Motor Patterns

A comprehensive description of the agonistic behavior patterns of an Old World monkey (*Macaca speciosa*) and comparisons with the other species studied so far can be found in Bertrand (1969). Van Hooff (1967) gives a comparative analysis of the facial expressions. Bernstein (1970) compiled a preliminary catalog of social responses of the Old World monkeys. From these comparisons it becomes clear that the basic motor patterns of threat and aggression are similar throughout the Cercopithecoidea. Typical elements of threat behavior are the stare, which is often associated with raising eyebrows; the open mouth threat face; repeated jerky head movements, horizontal or vertical; slapping the ground or branches; and intention movements of charging. Typical elements of aggression are charging, grabbing, slapping, hitting, pushing, pulling, and biting. Some of these basic patterns have a specific form in a certain species; in general, they seem to be fixed action patterns (Hinde, 1966) with a small range of intraspecific variability. The patterns are either absent or present in the behavioral repertoire of an individual or group, but they are rarely modified into different forms within one species. Inhibited forms of threat and aggression are often found; for example, male hamadryas baboons normally use only their incisors but not their long canines when giving a "neck-bite" to a female.

Whereas the form of agonistic motor patterns is not modified to a great extent within a species, its frequency definitely is. We shall examine the modifying factors in the section on causes of aggressive behavior. It will be found that both individuals and groups considerably change the frequency of aggression in response to social and ecological factors. Furthermore, the sequence of behavior elements in an aggressive interaction is variable, depending on whether the attacked or threatened animals respond by aggressive, submissive, indifferent, or escape behavior. Finally, the timing and orientation of aggressive behavior (which are the bases for the establishment of a rank order) can be modified as well.

Frequency of Aggression and Threat under Natural Conditions

The available quantitative data on the spontaneous frequency of agonistic behavior in wild Old World monkeys are compiled in Table 1. Comparable data are limited to baboons, rhesus macaques, patas monkeys, and langurs. The following tentative calculations suffer from these limitations. First, in an average cercopithecoid primate group, one aggression and/or threat occurs every 4.8 hr of daytime, or an average of 2.5 aggressions and/or threats occur during a full day (12 hr daylight). Second, the average proportion between threats and contact–aggressions is about 2.5 to 1 within groups and 4.5 to 1 between groups. Thus, mere threats are much more frequent than contact–aggressions, and the latter are relatively more frequent within groups than between groups. Finally, wounds did not result from aggressions in the cases where detailed counts were made. Inflicting damage (wounds) or even killing is so rare in Old World monkeys under natural conditions that it may be regarded as an accident.

CAUSES OF AGGRESSIVE BEHAVIOR

In this section we shall not consider the causes of the form of single motor patterns of threat and aggression, but concentrate on the more flexible aspects of aggressive behavior: its frequency (the aggressiveness), its sequence, its orientation and timing. The latter three aspects will be called quality of aggression.

Genetically Determined Differences

Since every behavior is affected by both genetic and environmental factors, one can only speak of genetically caused differences in aggressiveness, not of genetically fixed levels of aggressiveness.

DIFFERENCES BETWEEN SPECIES

Although differences between species are obvious, only a few attempts have been made by field researchers to determine quantitative differences in the aggressive behavior of species living in the same habitat. Hall (1965) recorded one aggressive episode about every 2.5 hr in baboons at Murchison Falls, Uganda, whereas in the same area in patas groups (of

TABLE 1 **Frequencies of Threat and Aggression in Wild Cercopithecoid Groups**

Species and references	Hours of observation	Number of groups	Number of agonistic episodes	
		(a) Between groups		
Papio ursinus			Threatening, barking, chasing	4
(Stoltz and Saayman,			Physical contact	0
1970)	848	7	Wounds	0
(Saayman, 1971)	390	4	Total encounters	58
			Nonaggressive	44
			Fighting	14
			Wounds	?
Papio anubis			Total encounters	3
(Nagel, 1973)	ca. 100	3	Tension, barking	2
			Contact	0
			Wounds	0
Erythrocebus patas			Total encounters	2
(Hall, 1965)	627	3	Threats only	2
			Contact	0
			Wounds	0
Presbytis entellus thersites			Total aggressive encounters	30
(Ripley, 1967)	300	4	Threats only	6
			Attacks + chasing	20
			Fighting (contact)	4
			Wounds	"one serious fight"
Presbytis cristatus			Total encounters	117
(Bernstein, 1968)	1058	6	Visual contact + display	90
			Chasing or fighting	27
			Biting, wounds	0
Macaca mulatta			Total encounters	26
(Lindburg, 1971)	ca. 400	4	Threats + chasing	5
			Fights or contact	0
			Wounds	0

TABLE 1 (continued)

Species and references	Hours of observation	Number of animals	Number of agonistic episodes	
		(b) Within groups		
Papio ursinus			Total	166
(Hall, 1962)	190	35	Threats only	114
			Contact episodes	53
			Biting	20
			Wounds	0
(Saayman, 1972)	437	77	Total attacks	517
			Biting or wounds	?
Erythrocebus patas			Total	49
(Hall, 1965)	627	12	Threats only	39
			Contact	10
			Biting	2
			Wounds	0
Presbytis cristatus			Total	178
(Bernstein, 1968)	1058	ca. 30	Threats only	83
			Fighting (contact)	93
			Biting	2
			Wounds	0

about half the size of baboon groups) one episode occurred only every 12 hr. Struhsaker (1969), counting the number of "vocal agonistic encounters" in the rain forest in Cameroon, found a difference between similarly sized groups of *Mandrillus leucophaeus* (18 incidents in 3 hr 40 min) and *Cercopithecus nictitans* (2 incidents in 3 hr 15 min). Jolly (1972) summarizes the qualitative evidence from the field as follows:

> The macaques and baboons as a group, are by far the most aggressive; the guenon-patas-group, which also ranges from forest to savanna, tends to have a low level of intragroup aggression in either place . . . ; second, among forms with ritualized long-distance spacing behavior, the level of aggression is very low within the group.[1]

Among the Old World monkeys, the latter pattern is realized by the leaf-eaters of the subfamily Colobinae. Thus, a certain basic level of aggressiveness seems more or less characteristic of a genus or family; however, conspicuous differences were also found between closely related species. Bertrand (1969) compared stable captive groups of stumptails (*Macaca speciosa*) and liontails (*Macaca silenus*) in a restricted food situation;

[1] From A. Jolly, "The Evolution of Primate Behavior," Macmillan Publishing Co., Inc., New York, p. 178. © 1972 by Macmillan Publishing Co., Inc.

groups were of identical composition and lived in identical enclosures. Not only was the overall frequency in stumptails almost five times higher than in liontails, but the quality of aggression was also different. Contacts (though generally not severe) made up about three-fourths of the incidents in liontails, but only about one-fourth in stumptails.

Differences in the intensity and sequence of aggressive episodes were found by Rowell (1971) when she compared captive groups of two guenon species with a similarly sized baboon group studied earlier (Rowell 1967) in the same cages. Whereas no baboon was ever killed during the many fights in the captive group, several animals were killed in both the vervet group (*Cercopithecus aethiops*) and the sykes group [*Cercopithecus mitis* (*albogularis*)], mainly during the establishment of the groups. In agonistic situations, the two guenon species did not use the gestures of appeasement and conciliation (such as lip smacking and presenting), which are frequently used by baboons to prevent or stop dangerous situations of conflict.

SEX DIFFERENCES

Among the well-studied species of Old World monkeys, males have been found to be generally more aggressive than females. Data from the field (e.g. Gautier–Hion, 1970; Lindburg, 1971; Kummer, 1968; Saayman, 1971) are confirmed by experimental evidence from the laboratory. Thompson (1969) found significant sex differences in aggressiveness in artificially combined pairs of crab-eating macaques (*Macaca fascicularis*).

In a similar test series, fights were recorded in all of 27 male-male pairs of rhesus macaques, but only in 4 of 27 female–female pairs (Angermeier *et al.*, 1968). Harlow and Harlow (1965) found 2-month-old rhesus male infants to be several times more aggressive than female infants of the same age. The males also showed much more rough-and-tumble play. Goy (1966) repeated Harlow's experiments, using the following procedure. Immediately after birth, the infants were removed from the mother and kept for 15 days in an incubator; later, the infants lived in separate cages, with 90 min of "play-group–contact" among each other every day. They were tested at the age of 7 months. Male infants threatened four times more often than females (significant difference); they initiated plays and engaged in rough-and-tumble play twice as much as females (not significant). Asami (cited by Frisch, 1968, p. 244) separated two infant female Japanese macaques and two infant males from their groups of origin and raised the same sexed infants together; he "noticed that male infants begin to show more aggressiveness than female infants the sixth month after birth." In these studies, a differential influence of the mother can be excluded.

Males and females also differ in the quality of their aggressive behavior. In a captivity situation patas males threatened extensively before attacking, whereas females sometimes slowly approached from behind and surprisingly bit an opponent animal (Kummer, unpublished observation). The adult male in a captive group of patas monkeys studied by Hall and Mayer (1967) never inflicted a wound on a female, whereas quarrels between females sometimes resulted in bad injuries. A similar difference is observed in a captive group of Sykes monkeys [*Cercopithecus mitis* (*albogularis*)] presently studied by Nagel, where the adult male always threatens before attacking, whereas the females sometimes make unexpected lunges or slaps at a nearby sitting or passing animal. In this case, however, attacks of both males and females sometimes cause injuries. Bertrand (1969) describes the sex differences in quality of aggression in captive stumptail as follows: "Males seemed more 'status conscious,' more likely to fight in order to settle social ranks immediately. Females were more cautious [p. 125]." In a newly formed all-female group of stumptails, animals would first present to one another and even groom one another; only after some time did fights break out. In the wild and in captivity, adult female hamadryas baboons rarely fight with other adults; instead, they threaten an opponent until their leader male attacks and possibly bites it (Kummer, 1957, 1968).

To summarize, males are generally more aggressive than females, whereas aggressive behavior of females is sometimes less predictable and in this sense more dangerous than that of males. Males seem to compensate for their strength and dangerous canines by giving ample warning before they attack. Whereas it is clear that sex differences in the frequency of aggression are somehow genetically determined, it is uncertain to what degree this is true for the differences in quality of aggression.

Differences Caused by Hormones

A more immediate determinant, open to experimentation, is found in the gonadal hormones. The experiments reported in the previous section suggest that gonadal hormones influence the behavior of infants. Resko (1970) found that testosterone production in the gonads of the male rhesus starts at the 45th day of fetal life. Androgens are found in the blood of newborn male rhesus, but disappear after the second day of life. It has been shown by many workers that the presence or absence of androgens during a critical phase of fetal development determines the differentiation of external genitalia in mammals. For example, absence of androgen during the critical phase results in the differentiation of female genitalia. Young and his co-workers (summarized by Goy, 1966) showed that the development of sexual behavior and of ovulation has the same causes. Goy (1966) gave intramuscular injections of testosterone propionate to pregnant rhesus females in the 39th through 90th day of fetal life. Four pseudohermaphrodites, i.e., androgenized females, were obtained from this procedure, which showed intermediate external genitalia. When tested between the second and fifth month after birth with normal female peers of the same age, they were found to threaten and initiate play significantly more often than the normal females. They did not, however, reach the level of normal males of their age. Differential treatment by the mother could again be excluded, but modification among the peers remains a possibility. Castration of male infants at the third to fourth month, on the other hand, did not alter the subsequent behavioral development typical of males (as followed up to the fourth year); the hormonal determination had taken place in the developing male before the fourth month.

Thus, one can conclude that prenatal influence of testosterone is the most important cause for the higher aggressiveness observed in captive male rhesus monkeys from early stages on to puberty. Other influences such as differential treatment by the mother or by adult males and different experiences in the play group will be evaluated in the next section.

Not much is known as yet about the effects of the momentary level of gonadal hormones in adult monkeys. Rose *et al.* (1971) studied an all-male group of rhesus monkeys in a large enclosure. They found significant positive correlations between several measures of aggressiveness and the level of plasma testosterone. Furthermore, the 8 highest-ranking males had as a group a significantly higher level of plasma testosterone than the 26 lower ranking males. It is not yet clear, however, whether the higher plasma testosterone concentration precedes their dominant and aggressive status or is its result. Wilson and Boelkins (1970) presented evidence for a seasonal variation of aggression in free-ranging rhesus monkeys. They surveyed the Cayo Santiago population over 2 years, counting every month the number of well-defined wounds and the number of dead or missing animals. They found a significantly higher frequency of wounds and deaths among males in the mating season and ascribe this cycle to the seasonality of testosterone production in the males. Michael and Zumpe (1970) controlled the hormonal state of adult female rhesus monkeys in three heterosexual pairs in captivity and noted the concurrent changes in the female's aggressive behavior toward the male. In a pair, where the female was cycling normally, no relationship was observed between the state of the female's cycle and its aggressiveness. In a pair with a pregnant female, a significant increase was found in aggressiveness, paralleled by an increase in refusals against mounting attempts of the male. The same effects, though less strong, were observed when the hormonal state of pregnancy was artificially produced in an ovariectomized female. In free-ranging chacma baboons, Saayman (1972) observed that "females in the flat cycle state were involved in significantly more aggressive episodes than were females in other cycle states, including pregnancy. Flat females appeared to be more aggressive than other cycling females, since they were responsible for 75% of the attacks within this class [p. 81]."

It should be emphasized that in all examples the gonadal hormones influenced not only the aggressive behavior, but also sexual behavior, grooming, playing, and sometimes other behaviors. The hormone-induced differences in aggressiveness are part of an integrated behavioral syndrome.

Influences of Rearing Conditions

The descriptions of socialization in free-ranging monkeys are too few and too limited yet for a correlational analysis of the effects of different natural rearing systems on aggressive behavior. In the following, we shall review the available evidence from laboratory experiments and evaluate its importance for the understanding of socialization influences in the natural group situation.

SOCIAL DEPRIVATION

Well-known research has been done on socialization in rhesus monkeys by Harlow and his collaborators. The earlier work has been concerned with the effects of total or partial isolation of the infant from its mother and/or from peers (summarized in Harlow and Harlow, 1965, 1969). The isolated animals were all more or less socially inactive, fearful, and disturbed; social exploration and social play were depressed. The resulting changes in aggressive behavior were only a minor aspect of the whole behavioral syndrome.

In summary, deprivation of social experiences (especially tactile) in an early stage increases subsequent aggressiveness in rhesus monkeys. Social experiences from a peer group may partly compensate for the lack of a mother. Intensity of aggression is also increased in "deprived" monkeys. It appears that mammals deprived of early experiences later respond to weak stimuli generally as if they were strong stimuli (Candland, 1971; Menzel, 1963).

MATERNAL INFLUENCES IN THE LABORATORY SITUATION

As such radical changes in rearing conditions as total or partial isolation of the infant are not comparable to rearing conditions in the wild, experiments on the effects of differential maternal behavior during early stages of infant development are more valuable from this point of view. Maternal behavior can be described as varying along two scales: One scale extends from overprotective and possessive mothering to readily giving away the infant (soon after birth) to interested group members. The other scale varies between gentle and brutal responses to the infant's demands, especially during weaning. The results of a series of experiments (Mitchell *et al.*, 1966; Mitchell, 1968a,b; Møller *et al.*, 1968; Mitchell and Brandt, 1970) on the first question can be summarized as follows: When the mother is overprotective and frequently restrains the infant during the first months, the infant will later display much submissive or escape behavior, high "cooing" frequencies (a "contact-seeking" vocalization), little independence, and low levels of aggressivity and social exploration.

The effects of frequent rejecting and punishing of the infant by the mother are less clear. Infants of brutal and/or indifferent mothers (motherless mothers) were found to display high levels of peer-directed aggression in their first 6 months (Arling and Harlow, 1966). According to another report (Møller *et al.*, 1968), however, such infants showed more submissive behavior and less threat and yawns than controls raised by normal mothers. Mitchell *et al.* (1967) compared two groups of adolescent rhesus monkeys with different rearing experiences. The total frequency of threats was about equal in the two groups, but the monkeys, who had experienced much punishment in their first 6 months, threatened significantly more than the "low-punished" monkeys against two types of stimulus animals. The latter threatened, yawned, and "cage shaked" more against the experimenter instead. It appears that the quality and phasing of the mother's punishing behavior are essential for its effect upon the later behavior of the infant.

Punishing behavior, in turn, is influenced by the environment, as is exemplified by a study of Jensen *et al.* (1969). In pigtail mother–infant pairs, observed during the first 15 weeks of the infant's life, the mother hit the infant much more in a "privation environment" than in a "rich environment." The cause of this was probably that the infant in the "privation environment" had nowhere to go for protection and/or arousal reduction. The infant was found to respond to the mother's hits mainly by approaching even closer to the mother or by clinging to or climbing onto the mother. These were exactly the situations eliciting most frequently the mother's hitting behavior. Castell and Wilson (1971) compared the development of pigtail mother–infant interactions in the first 24 weeks. When the mother–infant pairs lived in a group of 20 pigtail monkeys, infant-punishing was almost absent. In pairs who were kept in individual cages, the mothers punished their infants, but did less so in large cages than in small cages.

The above-mentioned differences in maternal behavior are related to the sex of the infant. Jensen *et al.* (1968) found that pigtail mothers in the first weeks after birth restrained and cuddled male infants more than female infants and later rejected them and threatened them away more frequently. Normal rhesus mothers were found to threaten male infants more often and to punish them earlier and more often than female infants (Mitchell, 1968b); Mitchell also observed that two male infants of overprotective mothers, who were restrained and retrieved unusually often, appeared to be "feminized" in that they were less aggressive and more submissive than most males. Mitchell and Brandt (1970) write of rhesus monkeys that "one can best characterize mothers of males as 'punishers,' mothers of females as 'protectors,' male infants as 'doers' and female infants as 'watchers' [p. 149]."

This illustrates well how genetic predispositions and environmental influences act together in shaping the final behavioral characteristics. In the cases known so far in the laboratory, both "nature" and "nurture" act toward more independence and aggressiveness in males, so that it is difficult to judge the relative importance of either factor.

MATERNAL INFLUENCES IN THE GROUP SITUATION

In their experimentally controlled approach, the reported laboratory studies are necessarily artificial: The mother–infant dyad is often studied as an isolate, removed from the normal group environment. As is shown by the studies of Jensen *et al.* (1969) and of Castell and Wilson (1971), the physical environment strongly influences the mother–infant relationship; the social environment is probably of even greater importance (Hinde and Spencer-Booth, 1967).

It has been found in Japanese macaques (Kawai, 1958; Kawamura, 1958) and in the rhesus monkeys of Cayo Santiago (Koford, 1963; Sade, 1967) that infants of low-ranking females become low-ranking as adults, too. The same seems to be true in wild populations of savanna baboons (Ransom and Rowell, 1972), as well as in a zoo group of crab-eating macaques (Angst, personal communication). Possibly this mechanism is common to all large, multimale troops of the savanna baboons and most macaque species in the wild, as well as in captivity. Such dependence of the infant's behavioral development on the mother cannot be explained by early maternal influences alone.

In captive groups, the same low-ranking females who are overprotective in the first weeks after birth, become overpunishing in the second half-year of the infant's life (Bertrand, 1969; Rosenblum and Kaufman,

1967). Both maternal behaviors seem to be an expression of the almost permanent tension and anxiety which are characteristic for low-ranking animals in captivity.

Overprotectiveness of low-ranking savanna baboon females seems to be entirely a captivity artifact (Ransom and Rowell, 1972). Among rhesus monkeys on Cayo Santiago, high-ranking females retrieved and protected their infants much more in interactions with older juveniles than low-ranking females (Kaufman, 1966). In neither case can the subsequent subordination of the infants be explained by their mother's overprotectiveness.

Maternal punishing or brutal rejecting is rare in free-ranging cercopithecoid groups, even among the relatively aggressive baboons and macaques (DeVore, 1963; Kaufman, 1966; Kummer, 1968; Lindburg, 1971; Ransom and Rowell, 1972). It is fairly often observed only in the weaning period, which only begins around the ninth month of life in baboons (DeVore, 1963) and in the fourth to fifth month in rhesus macaques (Lindburg, 1971). There are great individual differences in weaning techniques of the mothers and in time of weaning onset (Ransom and Rowell, 1972). The results of the laboratory experiments on punishing, on the other hand, all describe the effects of punishment experiences *in the first 6 months.* They give no explanation for the subordinate behavior which infants of low-ranking females display as adults, nor can they be taken as evidence for the hypothesis that rough weaning techniques are a major cause of high aggressiveness in free-ranging cercopithecoids.

This raises the question whether, in the natural group situation, the mother's position in the group and her attitude toward other group members (as manifested in her behavior toward them) are more important for the infant's socialization (and especially the development of its aggressive behavior) than the mother's actual behavior toward the infant. Possibly, learning through association with the mother is more important than the conditioning effects of maternal behavior.

INFLUENCES OF ADULT MALES

No experimental research has been done yet on the influence of adult males on the development of the infant's or juvenile's aggressive behavior. From recent research (review by Mitchell and Brandt, 1972; field reports by Ransom and Rowell, 1972; Deag and Crook, 1971; Burton, 1972), it becomes more and more evident that there are great differences in the amount of paternal behavior in Old World monkeys and that males play an important part in socialization among at least some species. A comparison of the species of Old World monkeys studied in the field up to now reveals a rough positive correlation between overall aggressiveness of a species and the males' involvement in socialization. Overall aggressiveness is relatively low in species where the single adult male or the dominant adult male has a relatively peripheral position and/or is not much involved with the infants and juveniles, as in *Erythrocebus patas,* most *Cercopithecus* species, *Presbytis entellus, Presbytis johnii, Presbytis cristatus, Colobus abyssinicus.* On the other hand, overall aggressiveness is relatively high in baboons and macaques, where the dominant males play a central role in the group's life and are (in some species) regularly involved in socialization. Since aggression in monkeys is mainly a male affair, this may be explained by means of the identification theory (Imanishi, 1957). The more central the figure of the adult and dominant male in the group, the more the male infant learns and begins to develop aggressive patterns through identification with the central male. The role of the adult male—direct or indirect—in the development of the infant's aggressive behavior seems to have received too little attention in the past.

Results on the socializing effects of the peer group are as yet too scanty to evaluate its influence on aggressiveness.

Immediate Causes of Aggressive Behavior

SOCIAL FACTORS

Competition over Females. In many species of Old World monkeys, conflicts and aggressive interactions are especially intense during mating periods. The age–sex class involved mainly in aggressive interactions and the sequence and direction of the aggression may differ considerably between species, e.g., between patas monkeys (Hall, 1965) and baboons (DeVore and Hall, 1965).

Differences within species have been observed in baboons, langurs, and mangabeys. Hall (1962) found a significant correlation between copulation frequency of the alpha male chacma baboon and the frequency of aggression between adult females. Flat females would attack the consorted estrous female and chase it away from the male. In the anubis baboons of the Nairobi Park, on the other hand, adult males "harass" the copulating male (DeVore and Hall, 1965). Male competition over females was not observed in the forest-living anubis baboons studied by Rowell (1966). Paterson (1973) found pronounced differences in male competition between two groups of anubis baboons living in different habitats. When there was only a single estrous female in the one group, the day would begin with long fights among the adult males to decide the "consort of the day." In the other group, no fights over estrous females were observed. A similar difference was described by Vogel (1971) for two populations of hanuman langurs from different habitats, which might be genetically different at the subspecific level. In the groups of the Sariska population (Rajastan) the adult males were tolerant of each other's sexual relations, whereas in the groups of the Kumaon-Hills (Himalaya) the dominant male would aggressively interrupt copulations of subordinate males. Chalmers (1968) studied two groups of black mangabeys living about 80 km apart in similar habitats. In one group, conflict or competition over estrous females was never observed, whereas in the other group such females were the source of conflict in 6 of 8 cases where this source could be determined. At present, there is not enough evidence for identifying the main cause for these differences. It seems to us that one group could be more competitive than another because of different social traditions or because it includes particularly incompatible individuals.

Conflict over Infants. In wild and captive patas monkeys, many (in the laboratory two-thirds of all) aggressive episodes were precipitated by the crying of a distressed infant, with the mother immediately attacking any animal who really or presumedly had treated the infant roughly (Hall, 1965; Hall and Mayer, 1967). Attacks in defense of their infants are a major source of aggressiveness of rhesus monkey mothers in India (Lindburg, 1971).

In baboons and macaques, dominant males have the role of "control animals"; they intervene in conflicts, especially when an infant or young juvenile is roughly handled. The control animal's aggressive interventions are important for keeping intragroup aggression at a low level, as has been experimentally demonstrated in pigtail macaques by Tokuda and Jensen (1968). In one group of black mangabeys, adult males were attracted by infants; conflicts between adult males over infants were the second most frequent cause of aggression in this group (Chalmers, 1968).

Disturbance of Social Order. In the wild, the social order of a group is disturbed only under exceptional instances, such as mass disease or when important members of a group are killed by predators or humans. In captivity and in controlled populations, the influence of introducing new animals or taking away important group members has been experimentally investigated several times (Southwick, 1967; Marsden, 1968; Tokuda and Jensen, 1968; Bertrand, 1969; Bernstein, 1969; Vessey, 1971). Aggressiveness increased dramatically upon introduction of strangers; when group members were removed it often also increased, depending on the removed animal's former position. In Southwick's (1967) experiments on a captive group of 17 rhesus monkeys in the Calcutta Zoo, aggressive interactions (also among group members) increased 4 times in response to the introduction of 2 strange juveniles, 10 times upon a first introduction of 2 strange adult females, 8 times upon introduction of 2 strange adult males and, finally, 4 times upon a second introduction of 2 strange females. Between the experiments, aggressiveness always fell to a base-line level. The effects were much stronger than those of reducing food quantity or space (see the following). In the many introductions of new animals, which are known from zoo groups or other captive groups, not only the frequency but also the intensity of aggressive behavior was dramatically increased. Often the fight ended in severe wounding or even in killing. Indeed, in most cases known so far, killing between monkeys has occurred only after disturbance of the social order and mainly after strangers were introduced.

ECOLOGICAL FACTORS

Food. Wild monkeys find their food in small and spatially dispersed items; normally, conflict over food does not occur. In some cases, where preferred or necessary food was naturally restricted to a few places, conflict and high levels of intragroup aggressive interactions were observed in wild monkeys. Descriptions of such a situation are given for mangabeys by Chalmers (1968), for patas by Hall (1965), and for talapoins by Gautier-Hion (1970). In baboons and macaques, it was found repeatedly that fights within or between wild groups could be provoked by presenting the monkeys a preferred food in one or a few heaps (e.g., Kummer, 1968). This method of feeding is generally used in captivity; most species of Old World monkeys adapt to this unnatural feeding situation by establishing a rigid feeding rank order.

The effects of a change in food quantity—while the spatial distribution of food sources remains unchanged—are of a different kind. Loy (1970) observed the free-ranging rhesus population of Cayo Santiago, where the staple food is monkey chow permanently provided at six feeding stations around the island. Due to a delayed shipment of the chow, the monkeys obtained only one-tenth of the normal quantity of staple food during 10 days. The frequency of fights at feeding stations as well as away from them decreased significantly to about half the normal value. The frequency of grooming, play, and mating also decreased significantly, whereas the time spent foraging from the vegetation cover increased greatly. Southwick (1967) made similar observations on a rhesus group of 17 animals in an enclosure of 100 m^2. Providing 25% more food or 25% less food did not significantly change aggressiveness. When only half the normal amount of food was given, aggressiveness decreased significantly and a similar decrease was found in the other social behaviors. The animals became socially lethargic and spent most of the time slowly searching for food.

Space. It is well documented that crowding produces an increase in aggressiveness in many mammals (summarized by Archer, 1970). In Southwick's experiments, aggressiveness in the group increased significantly to about the double of the base-line value when it was limited to half the previous space. Alexander and Roth (1971) produced three subsequent short-term crowding situations in a captive group of 84 Japanese macaques. Instead of both a large corral and a small connected pen, only the 187-m^2 pen (providing only 2.3% of the previous total area) was allowed as living space for periods of 4, 5, and 6 days. Both group attacks ("mobbing") and dyadic aggressive interactions were significantly more frequent during crowding periods than during control periods. The strongest increase was in adult male–male dyads, whereas in adult female–female dyads there was a significant *de*crease under crowding conditions. The rank order became more rigid in the crowded group. In these two cases, population density was changed by diminishing the available space, while the number of animals was not changed. These are changes in "spatial density" which should be differentiated from changes in "social density," i.e., changes in the number of animals (group size) while the available space remains unchanged. This second aspect of crowding has not yet been experimentally investigated in Old World monkeys.[2] From field observations on growing populations of monkeys (e.g., Furuya, 1969), it appears that the tension which results from this increase in social density is normally resolved by group fission.

Captivity and Zoo Groups. Lack of space seems to be a main cause of the high scores on all social behaviors which are found in captive and zoo groups when compared with wild groups of the same species (Kummer and Kurt, 1965; Rowell, 1967). Aggressive behavior was 9 to 15 times more frequent in adult hamadryas baboons in the Zurich Zoo than in the wild in Ethiopia. In the anubis baboons compared by Rowell (1967), approach–retreat interactions ("agonistic interactions") were significantly more frequent in the captivity group, but no comparison was made for threats and aggressive patterns alone. The captivity group was organized in a rigid rank order, whereas no rank order was observed in the wild groups. Masure and Bourlière (1971) found aggressive interactions with vocalization to be three to ten times more frequent in a

[2] McGrew (1971) has compared the effects of social density changes with spatial density changes in nursery school children. He found that at higher social density the children tended to avoid each other, whereas at higher spatial density proximity and peer contacts increased proportionally.

group of savanna baboons (*Papio papio*) in the Vincennes Zoo than in the wild populations of savanna baboons (*Papio ursinus*) studied by Hall (1962) and Saayman (1971a,b).

Yet the limitation of space is not the only cause of the increased frequency and intensity of aggression in captive cercopithecoid groups. Other possible causes are: (*1*) Vegetational or artificial "cover" is usually absent or rare in a captivity situation. This may account for the fact that individual distances were higher in the hamadryas baboons of the Zurich Zoo than in wild hamadryas groups. (*2*) Food is often given during a few short feeding periods and spatially not sufficiently dispersed. Prolonged and intensive searching for food is unnecessary in the zoo, and therefore more time is free for social behavior. (*3*) In captivity, the alpha male has none of the ecological functions and preoccupations observed in the wild. It is possible that therefore more aggressive and group-directed animals, whose qualifications come only from social functions, can take the alpha position. (*4*) Group composition is not determined by the animals. In most cases, previously alien animals have been forced together. Incompatibilities among its members cannot be solved by emigration. Later introduction of new members, as was described earlier, strongly increases aggression within the whole group.

FUNCTIONS OF AGGRESSION

The main functions of aggressive behavior in cercopithecoid monkeys are to establish and maintain social structure and to relate it to ecological resources. It rarely exerts such functions alone, but by interacting with such factors and forces as social attraction, escape, submission, or the familiarity of a place or area. An attempt at grasping these functional networks best begins with the role of aggression among two animals only.

The primary effect of aggression among two monkeys is an increased distance between them. Differential aggressiveness plus differential social attraction result in a median pair distance which can be typical of species and sex–age classes. Thus, Table 2 shows that pair distances of patas monkeys in a large enclosure are greater than those of gelada baboons, and that the male–male pairs keep longer distances than male–female and female–female pairs in either species.

In this case, the differences were not caused by a higher level of aggressiveness in patas, but by its persistence over time. After initial fights, geladas gradually proceeded to a friendly relationship by using submissive gestures which are not found among patas monkeys. In

TABLE 2 **Median Intrapair Distances (Meters) of *Erythrocebus patas* and *Theropithecus gelada* in an Open 30 by 100 m Enclosure**[a]

	Male–male	Male–female	Female–female
E. patas	62 m	12 m	13 m
T. gelada	8 m	1 m	1 m

[a] Source: Kummer, unpublished observation.

addition, social attraction was greater in geladas than in patas. The frequencies of social grooming were negatively correlated with distance medians throughout the sample of Table 2; nevertheless, grooming alone did not a count for the differences in distance. The combination of these dyadic factors probably lays the groundwork for the composition and the cohesiveness of groups. Aggressiveness and the inability of resolving it by submission may be responsible for the fact that the adult males of wild *E. patas* do not tolerate a second male in their heterosexual groups (Hall, 1965), in contrast to macaques and baboons.

In some cercopithecoids aggression can have a reverse effect in that it reduces dyadic distances. Median distances between a male and its nearest female in free-ranging groups are significantly shorter in hamadryas baboons than in anubis baboons (Nagel, 1971, see Fig. 1). The hamadryas female keeps close to the male, i.e., at a distance where her submissive behavior is effective, in response to his aggressive herding technique (Kummer, 1968). Chance (1956) has observed such "reflected escape" toward the aggressor in captive rhesus monkeys, and Gartlan (1970) saw similar sequences in wild drills (*Mandrillus leucophaeus*).

Aggression, then, affects distances, and these, in turn, must be investigated as to their function. One such function is an efficient attribution of monkeys to food sources, as Ripley (1970) has shown for the tree-feeding of langurs (*Presbytis entellus*). The patas monkeys observed by Kummer in a large enclosure attributed sleeping trees to individuals by a bout of threat behavior at dusk. They thus prevented more than one monkey from sleeping in the same tree. In this ecological function, aggression may again interact with other factors. Social attraction may decrease during foraging, and the attraction to individual food sources further increases and shapes the scattering of group members. For example, hamadryas females keep an average distance of 0.65 m from their males during rest periods. This distance increases to 3 m when the

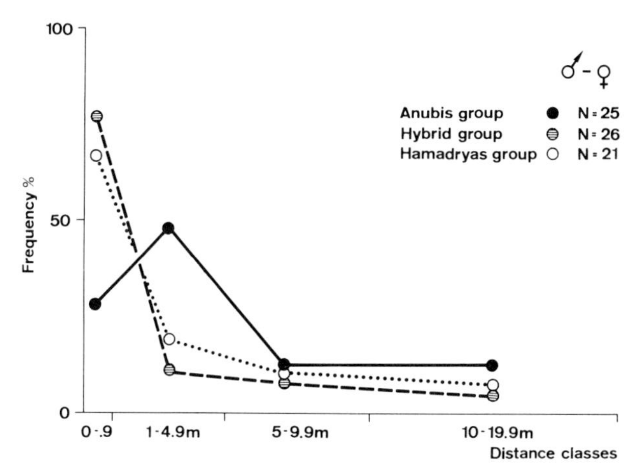

Fig. 1 Frequency distributions of the estimated distances from a randomly selected male to the nearest female in free-ranging anubis baboons, hamadryas baboons, and their hybrids. Median distances are significantly shorter in hamadryas and hybrids than in anubis, reflecting the fact that hamadryas males (and partly hybrid males also) aggressively herd their females. (From Nagel, U.: Social Organization in a Baboon Hybrid Zone. *Proc. 3rd Int. Congr. Primat.* Zurich 3: 48–57 (Karger, Basel 1971).)

group is foraging; this increase is not caused by any apparent aggressive behavior.

Aggression is effective only at distances of a few meters. At larger distances, e.g., between groups, it is replaced by other forms of communication, such as the tree shaking of macaques, the tree jumping of vervets (Gartlan and Brain, 1968), the jumping displays of several langur species (Jay, 1965; Poirier, 1969; Bernstein, 1968) and of the conspicuously maned guereza (*Colobus abyssinicus,* Ullrich, 1961), or the male loud calls of forest *Cercopithecus* species (Gautier, 1969; Struhsaker, 1969).

In several important contexts of group life, aggression functions in a triadic setting, for example, among two groups and their home range. Qualitative field evidence strongly suggests that a group (or an individual) is more likely to attack on familiar ground, whereas it tends to flee in an unfamiliar area. The impression is supported by Mason's (1965) important laboratory findings on young chimpanzees; these subjects compensate for high arousal caused by an unfamiliar room in that they reduce the rate of play–fighting.

The flight distance of hamadryas baboons toward observers is much greater in unfamiliar parts of their home range than in familiar ones (Kummer, 1968). Territorial behavior in vertebrates generally reveals a negative correlation between aggressiveness and familiarity of the surroundings.

This relationship may explain why open country cercopithecoids with their large home ranges are not territorial, whereas the groups of several forest cercopithecoids approach some form of territorial behavior in their smaller home ranges. Table 3 demonstrates a negative correlation between territorial behavior and home range size in Indian hanuman langurs. A similar relationship has been reported for two populations of *Cercopithecus aethiops* by Gartlan and Brain (1968).

TABLE 3 **Home Range Size and Frequency of Intergroup Fighting in *Presbytis entellus*[a]**

Area	Mean home range size (mi^2)	Intergroup fighting
Orcha	1.5	Rare
Kaukori	3.0	Rare
Dharawar	0.07	Frequent
Polonnaruwa, Ceylon	0.1	Frequent

[a] After Ripley (1967) and Yoshiba (1968).

The organizing functions of aggression are particularly interesting among three monkeys of the same species. Monkeys learn to initiate or suppress their own social interactions depending on which other animals are nearby. Instead of keeping their distance from a group member who has frequently attacked them, they merely refrain from grooming or sitting close to a particular partner in the presence of a third animal who, according to their experience, is intolerant of that particular relationship. Coupled with such learning abilities, cercopithecoid aggression can structure the society instead of scattering it. An example was reported for gelada baboons by Kummer (1971). In this case, the final social structure was a direct effect of the success with which pair relationships were aggressively suppressed by other group members.

Which social interactions are suppressed by others varies among species. Baboon and patas mothers are generally intolerant of interactions between their infants and adults. In barbary macaques (Deag and Crook, 1971) and langurs (Jay, 1965; Poirier, 1968), infants are freely passed among many group members. Hamadryas males allow their females to groom but not to mate with subadult followers, and they tolerate no female interaction whatsoever with an adult male. In anubis baboons, the latter is true when the female is in estrus but not at other times. Exclusive claim of a social partner has been found most often in cercopithecoid monkeys, and apparently more so among the Cercopithecinae than among the Colobinae. The control of aggressive interactions among group members can be a particular role of the highest-ranking male, as Tokuda and Jensen (1968) experimentally demonstrated on a group of *Macaca nemestrina.*

A typically cercopithecoid pattern of aggression in triads is the use of an ally. A patas female intending to attack the group male can gain the support of his female consort by choosing the moment when the consort happens to be frightened by a sudden movement of the male. The hamadryas female uses its male to attack another female by presenting to the male while threatening the opponent (Kummer, 1957). Thus, the general definition of aggression given in the introduction to this chapter must be modified for cercopithecoids: Aggression is a massive impact of one's own *or another's* strength against a conspecific, with the function of removing him from a place *or of detaining him from an object, from a social partner, or from an activity.*

That groups of the same species rarely intermingle in cercopithecoids has obvious ecological advantages. It is adaptive for each group to be tied to an adequate home range which is small enough to be remembered in detail. It is less clear how group separation is maintained. The attractions of the familiar area and of familiar group members are important factors. A third mechanism, named "triadic differentiation," seems to support the xenophobic response. Experiments by Kummer *et al.* (in press) show that two newly convened adult baboons often go through an aggressive phase before they groom one another. When three baboons are convened, the dyad which was already most advanced further gains in cohesiveness, whereas the other two dyads regress to an earlier stage which is characterized by greater distance and stronger aggressive motivation. An individual can thus become permanently segregated from a pair (Fig. 2). Among wild populations, triadic differentiation should almost invariably strengthen the older bonds among group members while keeping intergroup relationships at a less friendly or agonistic level.

The cercopithecoids are probably the most aggressive among the primates. To date, they also seem to have the most clearly and thoroughly organized societies. Aggression is most probably one of the organizing factors, but its effects cannot be assessed, and even less be judged as adaptive or problematic, without investigating the still poorly understood network of social processes within which it operates.

Fig. 2 Triadic differentiation in baboons. The male in the foreground is excluded from the male–female pair in the rear. Both his aggression and his associative responses are strongly inhibited. (Photograph by Hans Kummer.)

REFERENCES

Alexander, B. K., and Roth, E. M. (1971). The effects of acute crowding on aggressive behavior of Japanese monkeys. *Behaviour* **39,** 73–90.

Angermeier, W. F., Phelps, J. B., Murray, S., and Howanstine, J. (1968). Dominance in monkeys: Sex difference. *Psychon. Sci.* **12**(7), 344.

Archer, J. (1970). Effects of population density on behaviour in rodents. *In* "Social Behaviour in Birds and Mammals" (J. H. Crook, ed.), pp. 169–210. Academic Press, New York.

Arling, G. L., and Harlow, H. R. (1967). Effects of social deprivation on maternal behavior of rhesus monkeys. *J. Comp. Physiol. Psychol.* **64,** 371–377.

Bernstein, I. S. (1968). The lutong of Kuala Selangor. *Behaviour* **32,** 1–16.

Bernstein, I. S. (1969). Introductory techniques in the formation of pigtail monkey troops. *Folia Primat.* **10,** 1–19.

Bernstein, I. S. (1970). Some behavioral elements of the Cercopithecoidea. *In* "Old World Monkeys: Evolution, Systematics and Behavior" (J. R. Napier and P. H. Napier, eds.), pp. 263–295. Academic Press, New York.

Bertrand, M. (1969). The behavioral repertoire of the stumptail macaque. *Bibl. Primat.* No. 11.

Burton, F. D. (1972). The integration of biology and behavior in the socialization of *Macaca sylvana* of Gibraltar. *In* "Primate Socialization" (F. E. Poirier, ed.), pp. 29–62. Random House, New York.

Candland, D. K. (1971). The ontogeny of emotional behavior. *In* "Ontogeny of Vertebrate Behavior" (H. Moltz, ed.), pp. 95–169. Academic Press, New York.

Carthy, J. D., and Ebling, F. J. (eds.) (1964). "The Natural History of Aggression." Academic Press, New York.

Castell, R., and Wilson, C. (1971). Influence of spatial environment on development of mother-infant interaction in pigtail monkeys. *Behaviour* **39**, 202–211.

Chalmers, N. R. (1968). The social behavior of free-living mangabeys in Uganda. *Folia Primat.* **8**, 263–281.

Chance, M. R. A. (1956). Social structure of a colony of *Macaca mulatta. Brit. J. Anim. Behav.* **4**, 1–13.

Deag, J. M., and Crook, J. H. (1971). Social behavior and 'agonistic buffering' in the wild barbary macaque, *Macaca sylvana* L. *Folia Primat.* **15**, 183–200.

DeVore, I. (1963). Mother-infant relations in free-ranging baboons. *In* "Maternal Behavior in Mammals" (H. L. Reingold, ed.), pp. 305–335. Wiley, New York.

Frisch, J. E. (1968). Individual behavior and intertroop variability in Japanese macaques. *In* "Primates: Studies in Adaptation and Variability" (P. C. Jay, ed.), pp. 243–252. Holt, New York.

Furuya, Y. (1969). On the fission of troops of Japanese monkeys, II: General view of the troop fission of Japanese monkeys. *Primates* **10**, 47–70.

Gartlan, J. S. (1970). Preliminary notes on the ecology and behavior of the drill, *Mandrillus leucophaeus* Ritgen, 1824. *In* "Old World Monkeys: Evolution, Systematics and Behavior" (J. R. Napier and P. H. Napier, eds.), pp. 445–480. Academic Press, New York.

Gartlan, J. S., and Brain, C. K. (1968). Ecology and social variability in *Cercopithecus aethiops* and *C. mitis*. *In* "Primates: Studies in Adaptation and Variability" (P. C. Jay, ed.), pp. 253–292. Holt, New York.

Gautier, J. P. (1969). Emissions sonores d'espacement et de ralliement par deux cercopithèques arboricoles. *Biol. Gabon.* **5**, 118–145.

Gautier-Hion, A. (1970). L'organisation sociale d'une bande de talapoins (*Miopithecus talapoin*) dans le Nord-Est du Gabon. *Folia Primat.* **12**, 116–141.

Goy, R. W. (1966). Role of androgens in the establishment and regulation of behavioral sex differences in mammals. *J. Anim. Sci.* **25**, 21–31.

Hall, K. R. L. (1962). The sexual, agonistic and derived social behaviour patterns of the wild chacma baboons *Papio ursinus*. *Proc. Zool. Soc. London* **139**, 283–327.

Hall, K. R. L. (1965). Behaviour and ecology of the wild patas monkeys, *Erythrocebus patas,* in Uganda. *J. Zool.* **148**, 15–87.

Hall, K. R. L., and DeVore, I. (1965). Baboon social behavior. *In* "Primate Behavior" (I. DeVore, ed.), pp. 53–110. Holt, New York.

Hall, K. R. L., and Mayer, B. (1967). Social interaction in a group of captive patas monkeys, *Erythrocebus patas*. *Folia Primat.* **5**, 213–236.

Harlow, H. R., and Harlow, M. K. (1965). The affectional systems. *In* "Behavior of Nonhuman Primates" (A. M. Schrier, H. R. Harlow, and F. Stollnitz, eds.), Vol. 2, pp. 287–334. Academic Press, New York.

Harlow, H. R., and Harlow, M. K. (1969). Effects of early rearing experience on the behavior of rhesus monkeys. *In* "Determinants of Infant Behavior" (B. M. Foss, ed.), Vol. 4, pp. 15–36. Wiley, New York.

Hinde, R. A. (1966). "Animal Behaviour." McGraw-Hill, New York.

Hinde, R. A., and Spencer-Booth, Y. (1967). The effect of social companions on mother-infant relations in rhesus monkeys. *In* "Primate Ethology" (D. Morris, ed.), pp. 267–286. Weidenfield and Nicolson, London.

Imanishi, K. (1957). Identification—a process of socialization in the subhuman society of *Macaca fuscata, Primates* **1**(1), 1–29 [English Transl.: "Japanese Monkeys" (K. Imanishi and S. A. Altmann, ed.), pp. 30–51. Altmann, Alberta, Canada, 1965].

Jay, P. C. (1965). The common langur of north India. *In* "Primate Behavior" (I. DeVore, ed.), pp. 197–249. Holt, New York.

Jensen, G. D., Bobbitt, R. A., and Gordon, B. N. (1968). Sex differences in the development of independence of infant monkeys. *Behaviour* **30**, 1–14.

Jensen, G. D., Bobbitt, R. A., and Gordon, B. N. (1969). Mother-infant interactions in *Macaca nemestrina:* Hitting behavior. *Proc. Int. Congr. Primat., 2nd,* **1**, 186–193.

Jolly, A. (1972). "The Evolution of Primate Behavior." Macmillan, New York.

Kaufman, J. H. (1966). Behavior of infant rhesus monkeys and their mothers in a free-ranging band. *Zoologica* **51**, 17–28.

Kawai, M. (1958). On the system of social ranks in a natural troop of Japanese monkeys: (I) Basic and dependent rank. *Primates* **1** (2), 111–130 [English transl.: "Japanese Monkeys" (K. Imanishi and S. A. Altmann, eds.), pp. 66–86. Altmann, Alberta, Canada, 1965].

Kawamura, S. (1958). Matriarchal social ranks in the Minoo-B troop: A study of the rank system of Japanese monkeys. *Primates* **1** (2), 149–156 [English transl.: "Japanese Monkeys" (K. Imanishi and S. A. Altmann, eds), pp. 105–112. Altmann, Alberta, Canada, 1965].

Koford, C. (1963). Ranks of mothers and sons in bands of rhesus monkeys. *Science* **141**, 356–357.

Kummer, H. (1957). "Soziales Verhalten einer Mantelpavian-Gruppe." Huber, Bern.

Kummer, H. (1968). Social organization of Hamadryas baboons. *Bibl. Primat.* No. 6.

Kummer, H. (1971). Immediate causes of primate social structures. *Proc. Int. Congr. Primat. 3rd, Zurich 1970* **3**, 1–11.

Kummer, H., and Kurt, F. (1965). A comparison of social behavior in captive and wild Hamadryas baboons. *In* "The Baboon in Medical Research" (H. Vagtborg, ed.), Vol. 2, pp. 65–80. Univ. Texas Press, Austin, Texas.

Kummer, H., Goetz, W., and Angst, W. (1970). Cross-species modifications of social behavior in baboons. *In* "Old World Monkeys: Evolution, Systematics and Behavior" (J. R. Napier and P. H. Napier, eds.), pp. 351–364. Academic Press, New York.

Kummer, H., Goetz, W., and Angst, W. (in press). Triadic differentiation: An inhibitory process protecting pair bonds in baboons. *Behaviour.*

Lindburg, D. G. (1971). The rhesus monkey in North India. *In* "Primate Behavior" (L. A. Rosenblum, ed.), Vol. 2, pp. 1–106. Academic Press, New York.

Loy, J. D. (1970). Behavioral responses of free-ranging rhesus monkeys to food shortage. *Amer. J. Phys. Anthrop.* **33**, 263–271.

Marsden, H. M. (1968). Agonistic behavior of young rhesus monkeys after changes induced in the social rank of their mothers. *Anim. Behav.* **16**, 38–44.

Mason, W. A. (1965). Determinants of social behavior in young chimpanzees. *In* "Behavior of Nonhuman Primates" (A. M. Schrier, H. F. Harlow, and F. Stollnitz, eds.), Vol. 2, pp. 335–364. Academic Press, New York.

Masure, A. M., and Bourlière, F. (1971). Surpeuplement, fécondité, mortalité et aggressivité dans une population captive de *Papio papio*. *Terre et Vie* **25**, 491–505.

McGrew, W. C. (1972). "An Ethological Study of Children's Behavior." Academic Press, New York.

Menzel, E. W., Jr. (1963). The effects of cumulative experience on responses to novel objects in young isolation-reared chimps. *Behaviour* **26**, 130–150.

Michael, R. P., and Zumpe, D. (1970). Aggression and gonadal hormones in captive rhesus (*Macaca mulatta*). *Anim. Behav.* **18**, 1–10.

Mitchell, G. D. (1968a). Persistent behavior pathology in rhesus monkeys following early social isolation. *Folia Primat.* **8**, 132–147.

Mitchell, G. D. (1968b). Attachment differences in male and female infant monkeys. *Child Develop.* **39**, 612–620.

Mitchell, G. D., and Brandt, E. M. (1970). Behavioral differences related to experience of mother and sex of infant in the rhesus monkey. *Develop. Psychol.* **3**, 149.

Mitchell, G. D., and Brandt, E. M. (1972). Paternal behavior in primates. *In* "Primate Socialization" (F. E. Poirier, ed.), pp. 173–206. Random House, New York.

Mitchell, G. D., Ruppenthal, G. C., Raymond, E. J., and Harlow, H. F. (1966). Long-term effects of multiparous and primiparous monkey mother rearing. *Child Develop.* **37**, 781–791.

Mitchell, G. D., Arling, G. L., and Møller, G. W. (1967). Long-term effects of maternal punishment on the behavior of monkeys. *Psychon. Sci.* **8**, 209–210.

Møller, G. W., Harlow, H. F., and Mitchell, G. D. (1968). Factors affecting agonistic communication in rhesus monkeys (*M. mulatta*). *Behavior* **31**, 339–357.

Nagel, U. (1971). Social organization in a baboon hybrid zone. *Proc. Int. Congr. Primat., 3rd, Zurich, 1970* **3**, 48–57.

Nagel, U. (1973). A comparison of anubis baboons, hamadryas baboons and their hybrids at a species border in Ethiopia. *Folia Primat.* **19**, 104–165.

Paterson, J. D. (1973). Ecologically differentiated patterns of aggressive and sexual behavior in two troops of Ugandan baboons, *Papio anubis*. *Amer. J. Phys. Anthropol.* **38**, 641–648.

Poirier, F. E. (1968). The nilgiri langur (*Presbytis johnii*) mother-infant dyad. *Primates* **9**, 45–68.

Poirier, F. E. (1969). Nilgiri langur (*Presbytis johnii*) territorial behavior. *Proc. Int. Congr. Primat., 2nd, Atlanta, 1968* **1**, 31–35.

Ransom, T. W., and Rowell, T. E. (1972). Early social development of feral baboons. *In* "Primate Socialization" (F. E. Poirier, ed.), pp. 105–144. Random House, New York.

Resko, J. A. (1970). Androgen secretion by the fetal and neonatal rhesus monkey. *Endocrinology* **87**, 680–687.

Ripley, S. (1967). Intertroop encounters among Ceylon gray langurs (*Presbytis entellus*). *In* "Social Communication among Primates" (S. A. Altmann, ed.), pp. 237–254. Univ. Chicago Press, Chicago, Illinois.

Ripley, S. (1970). Leaves and leaf-monkeys: The social organization of foraging in gray langurs *Presbytis entellus thersites*. *In* "Old World Monkeys: Evolution, Systematics and Behavior" (J. R. Napier and P. H. Napier, eds.), pp. 481–509. Academic Press, New York.

Rose, R. M., Holaday, J. W., and Bernstein, I. S. (1971). Plasma testosterone, dominance rank and aggressive behavior in male rhesus monkeys. *Nature (London)* **231**, 366–368.

Rosenblum, L. A., and Kaufman, I. C. (1967). Laboratory observations of early mother-infant relations in pigtail and bonnet macaques. *In* "Social Communication among Primates" (S. A. Altmann, ed.), pp. 33–41. Univ. Chicago Press, Chicago, Illinois.

Rowell, T. E. (1966). Forest living baboons in Uganda. *J. Zool.* **149**, 344–364.

Rowell, T. E. (1967). A quantitative comparison of the behaviour of a wild and a caged baboon group. *Anim. Behav.* **15**, 499–509.

Rowell, T. E. (1971). Organization of caged groups of *Cercopithecus* monkeys. *Anim. Behav.* **19**, 625–645.

Saayman, G. S. (1971a). Behaviour of chacma baboons. *Afr. Wild Life* **25**, 25–29.

Saayman, G. S. (1971b). Behavior of adult males in a troop of free-ranging chacma baboons (*Papio ursinus*). *Folia Primat.* **15**, 36–57.

Saayman, G. S. (1972). Aggressive behavior in free-ranging chacma baboons (*Papio ursinus*). *J. Behav. Sci.* **1**, 77–83.

Sade, D. S. (1967). Determinants of dominance in a group of free-ranging rhesus monkeys. *In* "Social Communication among Primates" (S. A. Altmann, ed), pp. 99–115. Univ. Chicago Press, Chicago, Illinois.

Southwick, C. H. (1967). An experimental study of intragroup agonistic behavior in rhesus monkeys (*Macaca mulatta*). *Behaviour* **28**, 182–209.

Stoltz, L. P., and Saayman, G. S. (1970). Ecology and behaviour of baboons in the Northern Transvaal. *Ann. Transv. Mus.* **26**, 99–143.

Struhsaker, T. T. (1969). Correlates of ecology and social organization among African cercopithecines. *Folia Primat.* **11**, 80–118.

Thompson, N. S. (1969). The motivations underlying social structure in *Macaca irus*. *Anim. Behav.* **17**, 459–467.

Tokuda, K., and Jensen, G. D. (1968). The leader's role in controlling aggressive behavior in a monkey group. *Primates* **9**, 319–322.

Ullrich, W. (1961). Zur Biologie und Soziologie der Colobusaffen (*Colobus guereza candatus* Thomas 1885). *Der Zool. Garten* (*NF*) **25**, 305–368.

Van Hooff, J. A. R. A. M. (1967). The facial displays of the catarrhine monkeys and apes. *In* "Primate Ethology" (D. Morris, ed.), pp. 7–68. Weidenfield and Nicolson, London.

Vessey, S. H. (1971). Free-ranging rhesus monkeys: Behavioural effects of removal, separation and reintroduction of group members. *Behaviour* **40**, 216–227.

Vogel, C. (1971). Behavioral differences of *Presbytis entellus* in two different habitats. *Proc. Int. Congr. Primat., 3rd, Zurich, 1970* **3**, 41–47.

Washburn, S. L., and Hamburg, D. A. (1968). Aggressive behavior in Old World monkeys and apes. *In* "Primates: Studies in Adaptation and Variability" (P. C. Jay, ed.), pp. 458–478. Holt, New York.

Wilson, A. P., and Boelkins, R. C. (1970). Evidence for seasonal variation in aggressive behavior by *Macaca mulatta*. *Anim. Behav.* **18**, 719–724.

Yoshiba, K. (1968). Local and intertroop variability in ecology and social behavior of common Indian langurs. *In* "Primates: Studies in Adaptation and Variability," pp. 217–242. Holt, New York.

XENOPHOBIA AMONG FREE-RANGING RHESUS GROUPS IN INDIA[1]

CHARLES H. SOUTHWICK

The Johns Hopkins University

M. Y. FAROOQUI

Aligarh Muslim University

M. F. SIDDIQI

Aligarh Muslim University

B. C. PAL

University of Kalyani

INTRODUCTION

Several studies on captive primates have shown aggressive intolerance of established groups toward strangers of the same species. When social strangers have been introduced into groups of confined monkeys or baboons, they have often been severely attacked. In a group of 17 baboons in the Bloemfontein zoo of South Africa, violent fighting erupted when an alien adult male and adult female were introduced into the group, and many of the animals were killed or died of injuries (Hall, 1964). In captive groups of rhesus monkeys, the introduction of new animals has also sparked aggression and violence in otherwise well-adjusted groups (Bernstein, 1964; Southwick, 1967, 1969, 1970).

In field studies, similar xenophobic responses[2] have been observed in

[1] Financial support for this work was provided by U.S. Public Health Service grants RO7 AI-10048-10 and NIMH grant 18440-01 to The Johns Hopkins University.

[2] Xenophobia, defined as the "fear or hatred of strangers. . . ." (Websters New 20th Century Dictionary, 2nd ed., 1968) is used in this paper primarily in the sense of hatred or intolerance and aggression toward social strangers of the same species.

Japanese macaques by Kawai (1960), in groups which have been provisioned and attracted to particular places for feeding. Field studies have also noted that the most conspicuous aggression in nonhuman primates often occurs in intergroup conflicts, when members of different social groups come into contact or when a member of one group attempts to enter another group.

All of these observations have suggested that xenophobia is a common social and behavioral trait of primates. Since many of the reports of xenophobic violence have been on captive or artificially provisioned animals, however, the question has persisted as to how much of the violence observed has been a function of confinement or artificial feeding.

The present study was undertaken as a series of field experiments on natural rhesus groups in India to test their responses to introduced strangers. The main questions were, "How would natural, free-ranging groups of rhesus react to the introduction of social strangers?" "Would alien individuals be accepted or rejected by the group?" "Would the xenophobic aggression of captive rhesus be exhibited by natural groups which are not artificially constrained in any way?" This study was a specific part of a larger investigation on the aggressive behavior of rhesus monkeys in relation to habitat and environment in northern and eastern India.

METHODS

This study involved four free-living groups of rhesus monkeys, selected from a larger sample of 10 groups under observational study for aggressive behavior. Three of the groups were in rural settings in northern Uttar Pradesh, and one was in an urban environment in Calcutta. Three of the groups were large, around 60 to 70 individuals, and one group was of moderate size with 22 individuals. Basic information on the size and loca-

TABLE 1 **Location, Habitat, and Total Sizes of Rhesus Groups in This Study**

Place name	Group	General location	Type of habitat	Total group size
Tughlaqabad	Central	8 miles south of New Delhi	Rural	71
Tughlaqabad	East	8 miles south of New Delhi	Rural	22
Chhatari-da-Raho	Main	12 miles north of Aligarh	Rural	65
Hastings Road		Calcutta, on southwest side of maidan	Urban	62

TABLE 2 **Group Compositions, February–March 1971**

Group	Adult males	Adult females	Infants	Juveniles	Total size
Tughlaqabad Central	13	27	16	15	71
Tughlaqabad East	4	9	4	5	22
Chhatari-da-Raho	6	22	20	17	65
Calcutta Hastings	8	21	15	18	62

tion of the groups is provided in Table 1. Group compositions are given in Table 2, and the habitats of these groups are shown in Figs. 1, 2, 3, and 4.

Observational study on the aggressive behavior of these groups began in December 1970, with the purpose of defining the base-line levels of aggressive behavior within each group. These observations provided data

Fig. 1 Tughlaqabad central group, 8 miles south of New Delhi. The ruins of the ancient city of Tughlaqabad (fourteenth century AD) which form the main habitat of the group are in the background. The monkeys also inhabited roadside trees and agricultural fields adjacent to the ruins, and had a total home range of more than 100 acres.

Fig. 2 Tughlaqabad east group, which occupied a range approximately one-quarter mile east of the ruins of Tughlaqabad. Roadside trees and adjacent fields comprised its main habitat, and its home range consisted of approximately 50 acres.

on the frequency and pattern of aggressive behavior prior to experimental introductions. Both the base-line observations and experimental releases were made during the nonbreeding season, since it is known that rhesus groups are less aggressive and more stable at this time than during the breeding season (Wilson and Boelkins, 1970; Lindburg, 1971).

In the experimental phase of the study, begun in March 1971, introductions of social strangers of different sex and age classes were made to test the responses of the group to these strangers. The monkeys used for introduction were purchased from commercial primate dealers in Calcutta and New Delhi. A total of 23 introductions were made into the four groups from March 2, 1971, to June 15, 1971. In most cases, two animals were released at a time, though single releases were made on three occasions. All animals released were wild-caught individuals which had been in captivity 2 or 3 weeks.

The animals to be released were transported to the home range of the group in a large wood and wire holding cage. The releases were made at various times of the day when the group was resting or grooming; that is,

Fig. 3 Chhatari-da-Raho main group, 12 miles north of Aligarh. Its habitat consisted of this rural schoolyard, an adjacent road and irrigation canal, and surrounding agricultural fields. The group occasionally visited a village about one-quarter mile away. Its total home range was 50 to 75 acres.

in a relatively quiet state. We attempted to make the introduction unobtrusively, without attracting great attention, but this was usually difficult, because the monkeys could quickly spot that something unusual was happening. The groups were, of course, all habituated to our presence and that of our station wagon, for we had driven up to these groups many times in the 3–4-month observational studies prior to release.

In making the releases, the holding boxes were taken from the station wagon as quietly and casually as possible, sliding doors were lifted, and the monkeys allowed to run out. The releases were done near some cover, such as bushes, trees, or rocks so the released animal was not totally exposed for a long time. Two releases were made from a burlap bag, and two releases were made by simply withdrawing the cage doors while the boxes were still within the wagon, in hopes that these methods might attract less attention than actually removing boxes from the wagon. In all cases, however, monkeys of the resident group were aware of something novel occurring—we found it impossible to make releases without arous-

Fig. 4 Calcutta Hastings Road group, which occupied a home range of less than 10 acres near the southeast corner of the Calcutta maidan. The area had pedestrian traffic, automobiles, and herds of domestic animals passing throughout the day.

ing the interest of at least some of the resident monkeys. This is an inherent difficulty with this type of field experiment.

The introduced animals were marked with nyanzol dye on one foreleg and one hind leg to permit individual identification in a way that would not be too obtrusive for monkeys.

Releases on the same group were made at intervals of at least 2 days, and usually much longer, to ensure that the group's activity and social behavior would be back to normal base-line levels. In most cases this occurred within a day.

Aggressive behavior patterns were tallied throughout the study in four major categories: threat, chase, attack, and fight. These have been defined in a previous paper as follows (Southwick, 1967, 1969):

Threat—any conspicuous aggressive gesture of one animal toward another involving an open-jawed stare gesture, head bobbing, or ear flattening; usually accompanied by aggressive vocalization sounding like a hoarse "ho" or "hough" (Fig. 5).

Chase—an active chase or lunge, usually involving one or several threat gestures; distinguished from a playful chase by its vigor and threat context (Fig. 6).

Fig. 5 Typical aggressive threat posture of adult male rhesus. Ears are flattened, eyebrows lifted, and a coarse vocalization is usually given.

Fig. 6 An aggressive chase, in which an adult female (right center), and an adult male and juvenile (center), vigorously chase another juvenile (shown in an extended leap in the left center).

Attack—an aggressive chase or lunge terminating in a rough slap, hit, or bite of one animal upon another (Fig. 7).

Fight—mutual rough slaps, hits, and/or bites between two animals, usually with vigorous wrestling.

These basic categories were scored so that each discrete incident or bout by each animal was scored as one. Thus a single threat gesture by one animal was scored as one; if each of three animals threatened another, it was scored as three threats. The data were collected continuously during observation periods, tallied into 5-min blocks, and then summarized and expressed as the average number of acts per hour of observation.

Records were also taken of the number of individuals under surveillance during each 5-min block, and whether or not the majority of these individuals were feeding or not feeding. The base-line data presented in this paper were all obtained during nonfeeding times, since the rate of aggressive interaction increased greatly (usually three- to five-fold) during feeding times. All of the groups in this study were fed by villagers

Fig. 7 Attack behavior, in which an adult male catches, bites, and shakes a juvenile. Other juveniles and females watch, and one juvenile (right center with teeth showing) gives an alarm "screech."

and local passersby, usually in the early morning and evening, and during these feeding periods, considerable aggressive and competitive activity centered around getting the food which people threw out for the monkeys. The food given to the monkeys varied considerably and included sugar cane, gram nuts or pulse (a hard leguminous bean), bananas, vegetables, rice, and chapatis or roti (a pan-fried flat cake). Large foods, such as bananas and chapatis, usually elicited more aggression than rice and gram nuts. For these reasons, all of the base-line data presented in this paper, as well as all the response data to introduced strangers, were taken at nonfeeding times. Most of the observations were made from 8:00 to 11:00 in the morning and 3:00 to 6:00 in the afternoon.

RESULTS

Base-Line Levels of Aggression

The base-line levels of aggression for the four groups studied varied from an average of 7.6 aggressive interactions per hour to 25.8 (Table 3). When reduced to the basis of total aggressive interactions per hour per monkey, the average varied from 0.231 to 0.397. At the outset of the study we anticipated that the Calcutta group would have a higher level of aggression than the rural groups, since it was crowded into a smaller home range in an urban environment, but its level of aggression was not higher. In fact, the Calcutta group of 62 monkeys had a significantly lower intensity of aggression than two of the rural groups, the Tughlaqabad east group and the Chhatari-da-Raho main group (Table 3).

TABLE 3 **Base-Line Patterns of Aggressive Behavior: Average Numbers of Aggressive Interactions per Hour in Nonfeeding Situations**[a]

Group	Aggressive interactions					Total AI/hr/monkey
	Threat	Chase	Attack	Fight	Total	
Tughlaqabad central	10.1	6.1	0.6	0.1	16.9	0.2383
(120 hr of obs.)	±0.85	±0.49	±0.09	±0.03	±1.34	±0.0189
Tughlaqabad east	4.2	2.9	0.4	0.03	7.6	0.3461
(60 hr of obs.)	±0.77	±0.51	±0.12	±0.03	±1.35	±0.0616
Chhatari-da-Raho	18.2	6.5	0.9	0.1	25.8	0.3968
(100 hr of obs.)	±1.31	±0.58	±0.16	±0.05	±1.75	±0.0269
Calcutta Hastings	10.3	2.6	1.3	0.1	14.3	0.2306
(60 hr of obs.)	±0.74	±0.27	±0.24	±0.03	±1.17	±0.0189

[a] (± standard errors)

Over 90% of the total aggressive acts in all groups were threats and chases which involved no physical contact. Attacks and fights constituted only 6.9% of all aggressive interactions in the Tughlaqabad group to a high of 9.8% in the Calcutta group. Although the urban group in Calcutta had a lower total prevalence of aggression, it did have the highest ratio of attacks involving physical violence in relation to total aggression.

Responses to Introduced Strangers

All of the introduced monkeys were met with aggressive responses except three infants released into the Calcutta group. The infants (all three of whom were 6 to 9 months of age) aroused immediate interest from the group, but the initial response of the resident monkeys was one of curiosity and investigation, whereas for all other releases of juveniles and adults the initial response was aggression.

The aggressive response occurred within a matter of 5–15 sec in some cases, whereas in others it was delayed as long as 5 min. The variation depended on several factors including the behavior of the released animal, the arousal state of the resident group, the personality characteristics of the resident individuals who first spotted the new animal, etc.—these various factors will be illustrated with some specific examples later in the paper.

In most cases, the initial aggressive response from the residents was a series of threats toward the alien. In a few cases the alien returned the threat, but in most the alien fled or cowered and this led to increased threatening, chasing, and in some cases, direct attacks.

These events usually produced a great outburst of aggressive activity from the resident group, and within the first hour after release, aggression levels increased from 42% to 822% above the normal base-line levels of the group (Tables 4, 5, and 6). Figure 8 shows the aggressive response of

TABLE 4 **Aggressive Interactions in First Hour Following Introduction: Tughlaqabad Central**

Circumstance	Aggressive interactions/hr					Percent change in total
	Threat	Chase	Attack	Fight	Total	
Base-line averages	10.1 ±0.85	6.1 ±0.49	0.6 ±0.09	0.1 ±0.03	16.9 ±1.34	
Add 2 new JJ	70	49	7	1	127	+651%
Add 2 adult ♀♀	16	8	0	0	24	+ 42%
Add 2 infants	22	10	0	0	32	+ 89%
Add 2 adult ♂♂	56	22	0	0	78	+361%

TABLE 5 **Aggressive Interactions in First Hour Following Introduction: Tughlaqabad East and Chhatari Main**

Circumstance	Aggressive interactions/hr					Percent change in total
	Threat	Chase	Attack	Fight	Total	
Base line[a]	4.2 ±0.77	2.9 ±0.51	0.4 ±0.12	0.03 ±0.03	7.6 ±1.35	
Add 2 new JJ	18	14		1	33.0	+334.2%
Base line[b]	18.2 ±1.31	6.5 ±0.58	0.9 ±0.16	0.1 ±0.05	25.8 ±1.75	
Add 2 new JJ	110	25	5	1	141.0	+446.5%
Add 2 adult ♀♀	142	49	45	2	238.0	+822.5%
Add 2 adult ♂♂	129	71	8	0	208.0	+706.2%

[a] Tughlaqabad east.
[b] Chhatari main.

TABLE 6 **Aggressive Interactions in First Hour Following Introduction: Calcutta Hastings**

Circumstance	Aggressive interactions/hr					Percent change in total
	Threat	Chase	Attack	Fight	Total	
Base line	10.3 ±0.74	2.65 ±0.27	1.3 ±0.24	0.1 ±0.03	14.3 ±1.17	
Add 1 J and 1 infant	47	32	5	0	84.0	+487.4%
Add 2 infants	8	0	0	0	8.0	− 44.0%
Add 1 adult ♀	62	18	2	0	82.0	+473.4%
Add 1 adult ♂	63	24	8	2	97.0	+573.3%
Add 1 adult ♂	89	19	4	0	112.0	+683.2%

part of the Tughlaqabad central group to an introduced male. Descriptions of the main events following each release are given in the following.

Tughlaqabad Groups

Sometimes the alien animal fled immediately, and the entire interaction was over within 5 min. This occurred in the releases of two adult females in the Tughlaqabad central group on April 2, 1971 (Table 4), in which the two females immediately fled upon release over the high rock ramparts of the Tughlaqabad ruins (Fig. 1). They were followed about one-quarter of a mile by some of the resident monkeys, but they never stopped until the resident monkeys retired from the chase. The entire incident was

Fig. 8 Aggressive response of Tughlaqabad central group to alien male. The introduced male is shown in the center with tail held upward at the "one o'clock" position, while resident monkeys are grouped around it in threatening behavior.

over in 5 min, and the group returned to normal activities. As a result, there was only a slight elevation (42% increase) in aggression during the first hour postrelease. A similar pattern occurred following the release of two infants to this group on April 10, 1971. The initial response of the group was curiosity, and they clustered around the new infants (Fig. 9). Only mild threats were given, but within 8 min the infants fled, going over the rock ramparts and striking out across the countryside. Of these four individuals which fled immediately in this manner, only one was seen subsequently—one of the adult females was seen at intervals of 2, 5, and 6 weeks in the vicinity of the Tughlaqabad east group, but it was also aggressively repelled from this group, and never became an established member of the group.

The juveniles and adult males which were released into the Tughlaqabad central group did not flee immediately from the entire vicinity and they circled repeatedly around the fringes of the resident group's home range. When they returned to a certain point, they would again receive a volley of threats and chases. This occurred for 2 to 3 hr with the juveniles (released on March 23, 1971), and 1 to $1\frac{1}{2}$ hr in the case of the adult

Fig. 9 Curiosity response of Tughlaqabad central group to an introduced infant. The infant cannot be seen, but is in the center of the cluster and is being closely examined by several adults and juveniles of the group.

males (released on May 19, 1971), so in both cases there was a substantial elevation of aggressive interaction above base-line levels (Table 4). In all cases, the aliens finally fled the area within 3 hr, and were not seen again. It was particularly interesting that the adult males were not actually attacked, and though large and robust, they did not attempt to attack or fight their way into the group.

Two juveniles released into the smaller Tughlaqabad east group on April 24, 1971 behaved in a similar way, and were met by the same type of aggressive response as in the Tughlaqabad central group. They attempted to stay on the periphery of the resident group for 2 or 3 hr, but finally left the area entirely.

One of the adult females introduced into the Tughlaqabad central group on April 2, 1971 was seen 6 weeks later on May 18 attempting to enter the Tughlaqabad east group. She approached the edge of the group and entered a tall tree near an adult male and juvenile who did not threaten her. Within a few minutes the alpha male and another male in the group came toward her, threatening vigorously. She descended from the tree and fled across a field with the two adult males in aggressive pursuit. She was seen on two subsequent occasions in the next 2 days trying to approach the group, but on both occasions she was aggressively repelled. Each time she fled south toward a village one-quarter mile away and disappeared around the village. Like most of the retreating introductions, she was not seen again.

Chhatari Group

Two juveniles released into the Chhatari main group on April 13, 1971 behaved much like those in the Tughlaqabad groups, and they received very much the same response. One additional factor was the presence of an irrigation canal along the eastern boundary of this group's home range. About 10 min after release, during which the juveniles were threatened often and chased from tree to tree and around several buildings, the two juveniles were chased eastward to the edge of the canal. Approximately 10 to 12 juveniles, females, and young adult males from the resident group approached them threateningly, whereupon the two juveniles jumped into the deep water of the canal and swam to the opposite bank, about 15 ft away. One juvenile (a female) continued going east to a mango grove about 300 yd away and was not seen again; the other juvenile (a male) continued to run along the canal bank and made several attempts to swim back to the group. On each attempt, he was met by a volley of threats

Fig. 10 Chhatari-da-Raho group at edge of canal after chasing an alien juvenile across the canal and threatening it from the bank.

from the group (Fig. 10), so he turned around in midstream and went back to the safe bank. After 2 hr, he finally gave up, and he also retreated further east to the mango grove where his companion had gone earlier.

The two adult females released into the Chhatari main group on April 15, 1971 had an even more unusual history. Upon release both were threatened and chased away by the group, but both persisted in trying to return to the group. The younger adult female came directly back into the group despite a tremendous barrage of threats and some direct attacks. She ran back into the center of the group's home range and cowered on the ground in a state of total submission. In this condition she was attacked and bullied by many members of the group, especially juveniles and other adult females. They bit her rump and tail and tugged at her ears and scalp. We fended off some of the most violent attacks to prevent serious wounding. Although there were no bleeding wounds, the female went into a state of shock—her face paled, breathing became shallow, and she became very weak and lost motor coordination. We chased away the other monkeys and took her to a refuge under a bush 200 yd away where she recovered some color and strength. She was still there at nightfall 3 hr after release.

The other adult female which was released fared somewhat better, but she was also under attack. She remained on the edge of the group's home range and did not attempt to come directly into the center of the group. She was last seen about 300 yd away on the edge of a small irrigation ditch. Due to the intensive threats, chases, and attacks directed to these two females, the aggression score increased 822% above base-line levels (Table 5).

We returned to this group at sunrise the next morning and found group activity in a normal condition. After considerable searching, we found one of the adult females dead about 100 yd from the center of the group's range. The skeleton had been thoroughly picked by vultures and hyenas or jackals, and we could not be sure which of our females it was. There was sufficient nyanzol dye on some of the remaining hair to confirm it as one of our releases. We were unable to find any trace of the other female or the two juveniles.

Two adult males were introduced into the Chhatari main group on June 15, and their introduction sparked an immediate aggressive response from the group. They fled across fields, and made two attempts to return to the group before retreating entirely from the home range of the Chhatari group. Their introduction resulted in a 706% increase in aggressive activity in the first hour.

Calcutta Group

The first releases into the Calcutta group were one juvenile female (4.5 kg, estimated 2 years of age) and one infant male (2 kg, estimated 8–9 months of age) released on the morning of March 2, 1971. Both animals were recognized as strangers within 10 sec—the juvenile female was threatened at once and aggressively chased over a wall and into an adjacent compound. The initial threatening and chasing was done by both juveniles and adult females within the group. The infant retired in plain view near the wall, and was approached cautiously for investigation by several juveniles, but was not threatened or chased. The infant was sniffed and investigated, but no aggressive responses were directed to it by any member of the resident group. When three adult males arrived on the scene, however, and saw the infant, they attacked us (CHS and our field assistant, B. S. Chouhan). These males had known us for over 3 months, and had never so much as threatened us previously, let alone attacked us. The attacks toward us this morning were severe and bold and caught us by surprise. Fortunately, we were just slapped and hit hard on the legs and were not bitten.

Within the next hour, threats and aggressive chases continued against the introduced juvenile, but not against the infant. The infant continued to be the object of curious investigation. The attacks of the adult males toward us stopped after several rushes and slaps, when we threatened them with a stick. The group remained agitated and aggressive for about 90 min, with all of the aggression directed toward the introduced juvenile and toward us, none toward the infant. Although we were not attacked again by the adult males, we were threatened by them and even by females and juveniles over the course of the next 90 min.

The next morning, March 3, there was no sign of the introduced juvenile, but the infant was getting along fine. The group was relaxed, and the infant was showing shy, quiet behavior on the edge of the group, but it was definitely acting like a member of the group. In the next few days, it became better integrated into the group, and on March 6, it was first seen to be adopted by an adult female. The female carried it both ventrally and dorsally; it played near the female, and the pair showed all the general traits of a mother–infant bond except nursing. This was the first of two infant adoptions which occurred in this group.

Two additional infants were introduced into the Calcutta group on March 4. They were 2.1 kg (female) and 2.4 kg (male), estimated to be 9 to 10 months of ago. Upon release, the initial group response was curiosity and investigation. There were no threats, aggressive chases, or attacks. Juveniles sniffed the two infants who ran along a wall about 100 ft and then climbed a banyan tree. B. S. Chouhan and I (CHS) were threatened by one adult female, but we were not attacked by any adults in the group. The newly introduced infants remained in the banyan tree for over an hour. Juveniles continued to approach them, but there was a notable absence of threats. The group remained quiet and relaxed, and group aggressive activity in the first hour after release was actually 44% less than normal (Table 6).

The next day, one of the infants introduced on March 4 was adopted by an adult female without an infant of her own. The infant was seen to play around the female, follow her closely, and even ride dorsally for 20 ft. When I (CHS) approached the pair, the adult female put her arm around the infant and threatened me. The other infant remained in the group, acting very shy and nervous on the fringe of the group.

Thus, of three infants introduced into the Calcutta group, all three became members of the group, and two were socially adopted by resident adult females. These adoptions persisted successfully as long as the group was studied until the end of May. We assume that these infants became permanent members of the group.

On March 8, an adult female was released, and was initially ignored by most of the group. Several females approached with cautious curiosity, and the introduced female, a fairly old and aggressive adult, began to threaten them. They returned the threat and were joined by several juveniles. Within a few minutes, the intensity of threats increased, the introduced female retreated, and several members of the group chased her aggressively. They did not attack, but a volley of threats were exchanged for 15 minutes, resulting in a fourfold increase in aggression. Finally the old female retreated over a wall into a compound not normally used by the group, and they did not follow her. The female disappeared into another area and was not seen around the group again.

An adult male was released into the Calcutta group on March 14, and serious threats began in 2 min. The initial threats were from resident females and juveniles, but quickly several adult males came and increased the intensity of the threats. The introduced male returned the threats, but was soon attacked and chased by subordinate adult males in the group. The introduced male took refuge on a wood pile, where some nearby woodcutters jokingly entered the "fight" and began to join the side of the introduced male and threaten the resident group males. They effectively kept the group males at bay, and there was a remarkable instance of "cooperation" between the harassed introduced rhesus male and the woodcutters. Without this protection from the woodcutters, this introduced male may have been killed because it was being severely attacked. The woodcutters obviously enjoyed coming to the aid of the disadvantaged monkey, and they kept up the game until nightfall. When they departed, the monkey also left and it was not seen again. It moved in the same direction as the departing men and disappeared into the city.

The final introduction into the Calcutta group was that of another adult male on May 4. The introduced male was instantly recognized as a stranger and initial threats came from two resident males. Adult females also joined the threats. The initial response of the introduced male was not apparent fear or retreat, but when threats and aggressive lunges toward him increased, he finally turned and fled over a wall. Approximately 15 to 20 group members followed him along the wall and threatened him from the wall or nearby rooftops. These threats and chases resulted in more than a sixfold increase in aggressive activity in the first hour (Table 6), but this aggression quieted down in an hour and a half and the group returned to normal activity. This male also disappeared from the area and was not seen on the following day.

DISCUSSION

The behavioral responses of the free-ranging rhesus monkey groups in this study were very similar to those of captive groups in previous studies (Bernstein, 1964; Southwick, 1967, 1969, 1970); that is, established groups were aggressively intolerant of introduced social strangers. Hence, this study demonstrated that this type of xenophobic aggression is not an artifact of confinement.

The only exceptions to this aggressive response occurred with introduced infants. Infant rhesus released into a natural group in Calcutta were not threatened or attacked—they became members of the group—and two out of three were adopted by adult females without infants of their own.

All introduced juveniles and adults, however, were met by severe aggression within a few minutes after introduction. In previous studies of small captive rhesus groups in Calcutta, aggression increased 4 to 11 times above base-line levels in the first hour after introduction (Southwick, 1967, 1969, 1970). In the free-ranging groups of this study, aggression usually increased three to eight times above base-line levels, but in a few cases it was less than this because the introduced animal fled immediately. Most of the introduced animals fled within a half hour and left the area entirely, but a few individuals remained in the area and made repeated attempts to join the group. One adult female which did so was killed by the resident group during the night when we were unable to prevent her death.

Considerable variation occurred in the details of how quickly the strangers were perceived after introduction and how quickly or severely they were attacked and driven away. In some cases, aggression toward the introduced animal occurred within a few seconds after release, and in other cases it was preceded by considerable investigation and did not occur for 5 or 10 min. Some of the factors involved in this variation were: (*1*) which members of the group first recognized the stranger; (*2*) the personality characteristics and social rank of these individuals; that is, whether they were mild or aggressive, subordinate or dominant; (*3*) the activity state of the individuals, whether they were resting or actively investigating their surroundings; (*4*) the behavior of the introduced animal, whether he or she cowered and hid, showed fright responses, fled, or approached the group shyly or boldly.

The aggressive response of the group was very rapid, almost instantaneous, if the stranger cowered with alarm or distress reactions upon release and was then spotted by an aggressive individual within the resident group. In such a case, the aggressive group member would instantly threaten the alien and arouse other group members to do so. A very rapid aggressive response also occurred if the introduced animal approached the group issuing his or her own threats. This would then be met by aggressive threats and would quickly create a tense situation, arousing other members of the group.

On the other hand, if the introduced animal simply ran into neighboring cover, and was not seen by aggressive individuals within the group, it could often sit quietly for 5 or 10 min before some group member found it and recognized it as a stranger. No matter how casual or quiet the introduction, however, the new animal was identified as a stranger within at least 15 min in all cases and then the process of aggression began.

Considering the variability in age, sex, temperament, social rank, activity state, and so forth, of both the resident group members and the introduced animal, it was striking that the general response, namely xenophobic aggression, was so consistent.

The question arose, of course: "How did the group recognize the introduced animal as a stranger?" In previous studies on small captive groups of only 15 to 17 individuals in colony cages, all animals could see exactly what was happening, and the introduction of a stranger was a very obvious event. In these natural groups, however, with 60 to 70 individuals, with a conscious attempt on our part to make the introduction as quietly and unobtrusively as possible, there were some circumstances when the introduced animal could quickly disappear into cover without attracting a great deal of attention. Even under these circumstances, the stranger was recognized as such in less than 15 min. Sometimes this recognition was evident by the awkward behavior and fright responses of the introduced animal. This was not the case with all individuals, however, and we were led to the hypothesis that group membership, even in a large group, is ultimately based on individual recognition. We cannot prove this contention, but feel at this time that it is the most reasonable explanation of our observations. We cannot begin to identify the sensory modalities involved in this recognition. Vision, olfaction, and audition may all play a role.

In captive studies, the initial aggressive response frequently came from individuals of the sex and age class approximating that of the introduced animal (Southwick, 1967, 1969, 1970). In other words, the initial aggression against introduced juveniles often came from the resident juveniles, whereas initial aggression against adult females came from the resident adult females, etc. It was postulated, therefore, that a "social template" existed, so that the sex and age class most affected by the introduced stranger would be the one showing the initial aggressive response.

Evidence for this did not occur in this field study, probably because of the much wider range of variables affecting the results. In all introductions, the initial aggressive response was often made by juveniles or adult females. Table 7 shows the number of times in which juveniles, females,

TABLE 7 **Initiator of Aggressive Response**

Introduced animal	Animals which initiated aggressive response			
	Juveniles	Adult females	Subadult males	Adult males
Infants	0	0	0	0
Juveniles	3	2		1
Adult females	3	3		
Adult males	3	3	1	2
Totals	9	8	1	3
Percents	43%	38%	5%	14%

and males initiated the aggressive response. In 21 cases where the initiators were identified, 17 (81%) were juveniles and/or females. High-ranking adult males often ignored the entire proceedings, except in one case in the Tughlaqabad central group where the alpha male aggressively chased an introduced juvenile. Evidence of a social template was slight, and consisted of the fact that subadult and adult males did not respond aggressively to introduced females, whereas they did to introduced males.

Xenophobia is a widespread trait throughout the animal kingdom, but it is by no means universal in all animals. Among invertebrates, it has been shown experimentally in various species of social insects, especially bees and ants (Chauvin, 1968; Wallis, 1970; Wilson, 1971), but it obviously does not occur among other aggregational insects such as locusts, butterflies, gnats, mosquitos, and many beetles. Among vertebrates, it has been demonstrated experimentally in the Syrian hamster (Lerwill and Makings, 1971), white-footed mouse (Southwick, 1964), Chacma baboon (Hall, 1964), and herring gull (Tinbergen, 1961). It may be presumed to occur in a number of other vertebrates, especially those with prominent territorial behavior and/or relatively closed social groups. Thus, xenophobia is a reasonably predictable trait in howler monkeys (Carpenter, 1934; Altmann, 1959), gibbons (Carpenter, 1940; Ellefson, 1968), timber wolves (Mech, 1970), and African lions (Schaller, 1972). It is also a characteristic of some human social groups in certain circumstances—for example, in urban gangs of Chicago (Hall, 1966; Suttles, 1968), and in many tribes of hunter-gatherers (Livingstone, 1968).

On the other hand, xenophobic behavior has not been observed, nor would it be expected to occur in most typical encounters of vertebrates with relatively open societies, such as bandicoot rats (Frantz, 1972), red deer (Darling, 1937), chimpanzees (Reynolds, 1966; Van Lawick-Goodall, 1971), jackdaws (Collias, 1944), and many species of birds during migratory and winter flock formation. Collias (1970) pointed out that the latter are examples of large compound groups which are more or less open to newcomers.

In the long-tailed macaque, *Macaca fascicularis,* strangers introduced into a captive group were met by strong aggression and often killed (Dollinger, 1971), but field studies of the same species showed a milder and different response. At the western tip of Java, field work on *M. fascicularis* showed that neighboring groups had considerable home-range overlap and were mutually tolerant within these areas (Angst, 1973). When individuals were trapped and introduced into an adjacent group or one several home-ranges away, they made their way back unharmed to their own group. They were not attacked enroute. When three individuals from an offshore island were transported 18 km and released near a new group, they remained distinct most of the time, but did associate with the peripheral males and were not attacked by the main group. Although the field transplants were not precisely comparable with those in our study, we can tentatively conclude that *M. fascicularis* does not show such pronounced xenophobia in natural populations as *M. mulatta.*

Field experiments on bandicoot rats (*Bandicota bengalensis*) in Calcutta were particularly instructive and clear in illustrating a lack of xenophobic aggression in this species, and showing social groups which were completely open to strangers. Frantz (1972) released alien rats into crowded natural populations in grain storage warehouses, and found that they were either ignored and/or completely accepted. In 42 releases of alien rats, none was threatened or attacked. This demonstrated a completely open group with no social boundaries. In captive colonies, however, the same species of bandicoot rat became intolerant and xenophobically aggressive toward social aliens (Frantz, 1972).

It is important to point out that xenophobic aggression is not the same as territorial aggression or aggression related to dominance hierarchies. In both territorial and hierarchical behavior, aggression is often directed toward socially familiar animals. The territorial animal may interact aggressively most often with his nearest neighbor, and the social-ranked animal may interact most often with his nearest-ranked peer. The essence of xenophobia is an aggressive response toward a complete social stranger. This may occur, of course, in either territorial or socially ranked animals, so there is considerable overlap in these behaviors, but there are also significant differences. Territorialism and hierarchical behavior are very often maintained by display, whereas xenophobic aggression frequently involves violence.

In human terms, xenophobic aggression may have some analogous relevance to certain patterns of war, where human social groups make a sharp distinction between friend and foe, and all members of the foe are endowed with certain alien traits (Frank, 1967). On the other hand, xenophobia probably does not have much relevance to most cases of homicidal violence which are usually expressed between social acquaintances. We do not wish to carry the animal–human parallels too far, except to point out that certain instructive analogous comparisons can be cautiously drawn in some cases, but not in all.

It can be concluded from this brief survey of the literature that xenophobia is a variable social trait, evident in some species but not in others which may be closely related. Even the same species may display xenophobic aggression in some circumstances, or some seasons, but not in others. When it occurs in natural settings, xenophobia is a functional and adaptive trait in that it maintains the integrity of the social group. As Ardrey (1970) pointed out, it ensures that group members will be socially familiar. It limits the flow of individuals between groups, and can therefore affect patterns of both social and genetic evolution. Xenophobia has apparently arisen in the course of natural selection and social evolution in those species and populations (such as rhesus and howler monkeys) where discrete social groups are adaptively favored. It has not arisen in those species and populations (such as the Calcutta bandicoot rat) where open populations and social flexibility are more adaptable and are selectively favored. We therefore consider xenophobia an ecologic and social adaptation and not necessarily a fixed behavior pattern. It is, however, a trait of great ecologic and evolutionary significance in both animal and human populations, and one that deserves substantially more study.

SUMMARY

Field studies on four groups of free-ranging monkeys in northern and eastern India in 1970–1971 showed them to be aggressively intolerant of social strangers. Introduced juvenile and adult rhesus were consistently repelled by threats, aggressive chases, and direct attacks from both rural and urban groups. In the first hour after introduction, aggressive interactions usually increased three to eight times above base-line levels. Most of the introduced animals fled, but some made repeated attempts to join the group. Introduced infants were not threatened, except very mildly, nor were they attacked. Two out of three infants introduced into one group were adopted by adult females who did not have infants of their own. These infants became accepted group members.

Xenophobia, defined as intolerance and aggression toward social strangers of the same species, is a characteristic of both captive and free-ranging groups of rhesus monkeys; the only exception to its occurrence in this study occurred with the introduction of infants. Xenophobia is a widespread behavioral trait throughout the animal kingdom, most commonly seen in animals with discrete and relatively closed social groups. It is related to territorial and hierarchical behavior, but is distinct from both in that it is a response to social strangers, whereas territorialism and hierarchical behavior are often directed toward social acquaintances.

ACKNOWLEDGMENTS

This work was done under the auspices of the Johns Hopkins Center for Medical Research and Training in Calcutta, India. We are indebted to Frederik Bang and Thomas Simpson for encouragement and administrative support. K. K. Tiwari, Zoological Survey of India; Jamil Khan, Aligarh University; and John and Lisa Oppenheimer of the Johns Hopkins C.M.R.T. all provided field assistance and important advice. B. S. Chouhan and Shymal Roy contributed greatly to the field work.

REFERENCES

Angst, Walter. (1973). Pilot experiments to test group tolerance to a stranger in wild *Macaca fascicularis*. *Amer. J. Phys. Anthro.* **38,** 625–630.

Altmann, S. A. (1959). Field observations on a howling monkey society. *J. Mammal.* **40,** 317–330.

Ardrey, R. (1970). "The Social Contract." Atheneum, New York.

Bernstein, I. S. (1964). The integration of rhesus monkeys introduced to a group. *Folia Primat.* **2,** 50–63.

Carpenter, C. R. (1934). A field study of the behavior and social relations of howling monkeys (*Alouatta palliata*). *Comp. Psych. Monogr.* **10**(2), 1–168.

Carpenter, C. R. (1940). A field study in Siam of the behavior and social relations of the gibbon (*Hylobates lar*). *Comp. Psych. Monogr.* **16**(5), 1–212.

Chauvin, R. (1968). "Animal Societies" (translated from the French by George Ordish). Hill and Wang, New York.

Collias, N. E. (1944). Aggressive behavior among vertebrate animals. *Physiol. Zool.* **17**(1), 83–123.

Collias, N. E. (1970). Aggressive behavior and evolution. *In* "Animal Aggression" (C. H. Southwick, ed.), Chapter 3, pp. 40–59. Van Nostrand Reinhold, Princeton, New Jersey.

Darling, F. F. (1937). "A Herd of Red Deer." Oxford Univ. Press, Oxford.

Dollinger, P. (1971). "Tod durch verhalten bei zooterian." Juris Druck und Verlag, Zürich.

Ellefson, J. (1968). Territorial behavior in the common white-handed gibbon, *Hylobates lar*. *In* "Primates: Studies in Adaptation and Variability" (P. C. Jay, ed.), pp. 180–199. Holt, New York.

Frank, J. D. (1967). "Sanity and Survival." Random House, New York.

Frantz, S. C. (1972). Behavioral ecology of the lesser bandicoot rat (*Bandicota bengalensis*) in Calcutta. Ph.D. Thesis, Johns Hopkins Univ., Baltimore, Maryland.

Hall, E. T. (1966). "The Hidden Dimension." Doubleday, Garden City, New York.

Hall, K. R. L. (1964). Aggression in monkey and ape societies. *In* "The Natural History of Aggression" (J. D. Carthy and F. J. Ebling, eds.), pp. 51–64. Inst. of Biology, London.

Kawai, M. (1960). A field experiment on the process of group formation in the Japanese monkey (*Macaca fuscata*), and the releasing of the group at Chirayama. *Primates* **2**(2), 181–253.

Lerwill, C. J., and Makings, P. (1971). The agonistic behavior of the golden hamster, *Mesocricetus auratus* (Waterhouse). *Anim. Behav.* **19**, 714–721.

Lindburg, D. (1971). The rhesus monkey in north India: an ecological and behavioral study. *In* "Primate Behavior" (L. A. Rosenblum, ed.). Academic Press, New York.

Livingstone, F. B. (1968). The effects of warfare on the biology of the human species. *In* "War: The Anthropology of Armed Conflict and Aggression" (M. Fried, M. Harris, and R. Murphy, eds.), pp. 3–15. Natur. Hist. Press, Garden City, New York.

Mech, L. D. (1970). "The Wolf." Natur. Hist. Press, Garden City, New York. 384 pp.

Reynolds, V. (1966). Open groups in hominid evolution. *Man* **1**, 441–455.

Schaller, G. B. (1972). "The Serengeti Lion. A Study of Predator-Prey Relations." Univ. of Chicago Press, Chicago, Illinois.

Southwick, C. H. (1964). *Peromyscus leucopus:* an interesting subject for studies of socially induced stress responses. *Science* **143**, 55–56.

Southwick, C. H. (1967). An experimental study of intragroup agonistic behavior in rhesus monkeys (*Macaca mulatta*). *Behaviour* **28**, 182–209.

Southwick, C. H. (1969). Aggressive behaviour of rhesus monkeys in natural and captive groups. *In* "Aggressive Behaviour" (S. Garattini and E. B. Sigg, eds.), pp. 32–43. Excerpta Medica, Amsterdam.

Southwick, C. H. (ed.) (1970). "Animal Aggression." Van Nostrand Reinhold, Princeton, New Jersey.

Suttles, G. B. (1968). "The Social Order of the Slum." Univ. of Chicago Press, Chicago, Illinois.

Tinbergen, N. (1961). "The Herring Gull's World. A Study of the Social Behaviour of Birds." Basic Books, New York.

Van Lawick-Goodall, J. (1971). "In the Shadow of Man." Houghton-Mifflin, Boston, Massachusetts.

Wallis, D. I. (1970). Aggression in social insects. *In* "Animal Aggression" (C. H. Southwick, ed.), Chapter 6, pp. 94–102. Van Nostrand Reinhold, Princeton, New Jersey.

Wilson, E. O. (1971). "The Insect Societies." Belknap Press, Cambridge, Massachusetts.

Wilson, A. P., and Boelkins, R. C. (1970). Evidence for seasonal variations in aggressive behaviour by *Macaca mulatta*. *Anim. Behav.* **18**, 719–724.

FACTORS INFLUENCING THE EXPRESSION OF AGGRESSION DURING INTRODUCTIONS TO RHESUS MONKEY GROUPS [1]

IRWIN S. BERNSTEIN

University of Georgia and Yerkes Regional Primate Research Center of Emory University

THOMAS P. GORDON

Yerkes Regional Primate Research Center, Field Station and University of Georgia

ROBERT M. ROSE

Boston University School of Medicine

INTRODUCTION

Rhesus monkeys (*Macaca mulatta*) normally live as members of well-organized heterosexual social groups. Although individual adult males may leave a troop and live as solitaries for brief periods, they ordinarily quickly join another heterosexual troop (Carpenter, 1942; Koford, 1963). Most such transfers occur during the breeding season and presumably serve as the mechanism for genetic exchange among troops (Lindburg, 1969; Neville, 1969). The entry of a new male into a troop is, however, frequently the occasion for fierce fighting and, in general, aggression reaches highest levels during the breeding season (Wilson and Boelkins,

[1] This research was supported by USAMRDC contract DADA 17-69-C-9014, NIMH grant MH-13864, and in part by PHS grant RR 00165 from NIH. In conducting this research, the investigators adhered to the "Guide for Laboratory Animal Facilities and Care" prepared by the Committee on the Guide for Animal Resources, National Academy of Sciences, National Research Council.

1970). Social mechanisms facilitating the integration of new members into a troop do exist and Sade (1968) has commented that the presence of genealogically related animals already in a troop (such as mother or an older brother) may be one such mechanism.

Rhesus monkeys in the laboratory have well-earned reputations for their aggressive responses and near-intractable dispositions. Their initial response to new stimuli is most often some agonistic display (for example, see Bernstein *et al.*, 1963 in their comparisons of gibbon and rhesus monkey responses to novel stimuli). The ready elicitation of aggressive or submissive responses and their rapid organization into dominance relationships has focused much research on the importance of the status hierarchy to social organization (Bernstein, 1970). Nonetheless, rhesus monkey societies may be examined for a variety of role relationships (Bernstein and Sharpe, 1965) and the time in which group members are occupied in agonistic encounters rapidly falls to less than 5% as a group stabilizes, whereas during group formation more than 80% of social interactions involve one form of agonistic response or another (Bernstein and Mason, 1963a,b; Bernstein, 1964a).

If group formation and the integration of newly introduced animals into a group are considered as aspects of the same social processes, then these processes may be considered as among the most potent stimuli conditions for eliciting agonistic interactions in rhesus monkeys. Southwick (1967), in a series of carefully controlled and documented studies of a captive group maintained in the Calcutta zoo, clearly demonstrated that the introduction of new animals to this group was a far more potent stimulus for eliciting aggression than was severe food deprivation or crowding as produced by successive halving of the available living space. A similar study performed by Marsden (1971) even indicates that food stress decreases agonistic behavior whereas troop-to-troop encounters produce dramatic agonistic interactions.

The reason for the potency of these introductions as stimuli for eliciting aggression can perhaps be explained in the following theoretical formulation: One of the primary social functions of a troop is to maintain its integrity. Any threat to the integrity of a troop is met with by all the resources at the command of the troop. This may mean avoidance and coordinated flight from the source of disturbance, aggressive response and even attack upon the source of disturbance, or some combination thereof. Disturbances may be seen as emanating either from within the troop or from outside of the troop. Any individual violating the social code which serves to maintain social order within the troop would then be regarded as a source of disturbance to the organization and as such would become the target of group aggression (Hall, 1964). It is perhaps for this reason that the status hierarchy of macaque troops tend to be so stable (Bernstein, 1969a, 1970). Furthermore, the alpha male, or the animal serving in the control role capacity, is a key animal in most groups who functions both to break up episodes of prolonged or intensive intragroup fighting and is also active in meeting threats to the troop from external sources (Bernstein, 1964c, 1966; Hall, 1964, 1965a; Kaufmann, 1967).

Such extragroup sources of disturbance may arise from contact with extraspecific individuals or groups, or conspecific individuals or groups. In groups such as patas monkeys (*Erythrocebus patas*), the single adult male appears to spend much of his time visually searching for other patas (Hall, 1965b, 1967), and in the territorial gibbons (*Hylobates lar*), the adult male appears to seek opportunities to defend the territorial boundaries against neighboring males (Ellefson, 1968). In fact, much of the organization of space within a primate species involves spacing mechanisms which separate different social units (Kummer, 1970). Dramatic examples may be seen in a wide variety of primate taxa such as hamadryas (*Papio hamadryas,* Kummer, 1968), *Cercopithecus aethiops* (Gartlan and Brain, 1968; Struhsaker, 1967), *Presbytis cristatus* (Bernstein, 1968), *Callicebus moloch* (Mason, 1966), and others.

When wild troops meet they usually employ some social mechanism to maintain the integrity of social units by appropriate spacing. When troops come together suddenly and dramatically, severe fighting may erupt. Southwick *et al.* (1965) describes such occurrences in rhesus monkey troops living in a temple area whenever troops suddenly met one another, as when coming around a corner wall of the temple. The leading young males were almost immediately engaged in fighting and the fully adult and alpha males usually came forward to decisively settle the fight, after which the two troops would separate into their original units.

In the case of geladas (*Theropithecus gelada,* Crook, 1966) and hamadryas (*Papio hamadryas,* Kummer, 1968), and perhaps the drill (*Mandrillus leucophaeus,* Gartlan, 1970) the ability of one-male social units to come together peaceably is explained in terms of a higher order social organization which still permits the basic one-male units to preserve social integrity, primarily by some sort of spatial ordering either within or between the one-male units (*P. hamadryas* and *T. gelada* respectively).

In the case of the chimpanzee (*Pan troglodytes,* Goodall, 1965; Itani and Suzuki, 1967; Reynolds, 1963), the usual association pattern seems to be a band with fluid membership. Association within a band is postulated to be open to any member of a particular community, but presumably members of another community are not tolerated. Here the community would be the basic permanent social unit which maintains its social integrity.

The gorilla (*Gorilla gorilla,* Emlen and Schaller, 1960) on the other hand, appears to live in troops which may meet and mingle and then go their separate ways without exchange of membership. This does not seem to parallel the chimpanzee situation nor is it readily explained using the gelada or hamadryas models. The exact social mechanisms remain undescribed as yet and rather than pursue speculative explanations, we will hope that current field research will produce empirical data describing the pertinent social mechanism, or perhaps demonstrating that the gorilla is truly exceptional to our notions of the means by which primate societies resist intruders and preserve their social integrity.

Primate social units thus appear to be intolerant of close proximity to extra group conspecifics. For those taxa showing multiple levels of social organization, toleration of other social units is limited to those units within a common higher order grouping system. The meeting of two unfamiliar conspecific social units is thus the occasion for considerable tension, if not outright aggression; Altmann (1962), Hall (1960), and Stoltz and Saayman (1970) have all described aggressive intertroop reactions when conspecific troops of Cercopithecinae meet. Stoltz and Saayman (1970), moreover, indicate that not only do most of the adult males in a baboon troop move forward and aggress against another troop, but that one male typically remains behind and herds the females and young from the scene of conflict; a mechanism which certainly serves to maintain the integrity of the troop regardless of the outcome of intertroop conflict. This mechanism effectively prevents the absorption of the defeated group's members by the victorious group, an outcome common when two captive groups are introduced to each other in spatially restricted areas. For example, read the accounts of Bernstein for *Macaca nemestrina* (1969b) and *Cercocebus atys* (1971) and Castell (1969) for *Saimiri sciureus.*

The primary threat to one group by another seems to lie in the males, as described clearly by Sugiyama (1964) for *Presbytis entellus* when other troops, all-male bands or solitary males were in the vicinity of a troop. In the encounters described, group males not only repelled one another but showed extremely vigorous response to all-male bands moving through their ranges.

Based on these reports, it is suggested that those individuals posing the most significant threat to the existing social order would be most vigorously resisted and rejected by the group. In a rhesus monkey group with a strongly organized and apparently fundamental hierarchy, we might expect that individuals threatening to disrupt the existing hierarchy would be most strongly resisted. Accordingly, infants and small immatures would pose the least threat as they can be readily subdued by the adult members of the society and the immature members have ready recourse to adult aid through genealogical social bonds.

In a strongly sexually dimorphic species such as the rhesus, an adult female would constitute little threat to the status of an adult male, although by alliance with one or more adult males a new female may readily assume high rank in a group. Resistance to such restructuring of the social order, however, would be expressed in terms of aggressive interactions involving the adult male enforcing the female's position within the group. The female may also be expected to show aggressive behavior to lower ranking males in assertion of her rank position.

Adult female group members, on the other hand, may find that a new female represents a more direct threat to their social position and alliances, and resident females are therefore considered more likely to attack an adult female intruder, perhaps also successfully enlisting males in support of their attack. Such female resistance to female additions may be thus contingent upon either the number of adult females already in the group or their numerical density relevant to the adult male population. When relatively large numbers of adult males are present, one of these may ally himself with the new females on introduction, thus supporting her entry into the group. Such alliances need not be based on immediate sexual receptivity and are therefore not considered as consort relationships. In general, alliance patterns and individual male–female preferences do not seem dependent on immediate female receptivity and it must be assumed that males recognize the status of females as such, even when they are not in estrus, a finding which comes as no surprise to dog owners and most students of mammalian behavior, but which has frequently been overlooked by students of primate behavior.

The greatest resistance to newcomers introduced to a rhesus monkey society would be expected in the case of adult male intruders, despite the fact that this is the most common form of membership exchange under natural conditions. An adult male represents a potential alpha male and, as such, a competitor for the key role in the group. Occupation of such a role would cause dramatic restructuring of the social order as new alliances are formed and the relative standings of different matriarchal units would be threatened; as has, in fact, been reported during spontaneous reorganization of pigtail monkey (*Macaca nemestrina*) groups (Bernstein, 1969a). An animal challenging every animal for dominance supremacy would thus meet the greatest resistance, whereas one accepting immediate low rank and then gradually working his way through the status hierarchy would meet less immediate combined resistance. The latter course of action is apparently the usual mechanism under natural conditions when males join other social units, first by following, then by peripheral association, and finally by challenge for rank position.

When considering the reintroduction of animals removed from a group, one must consider whether the animal being returned can easily claim his old social position. This will depend on both individual recognition and on whether any significant reordering of the group social structure has occurred during his absence. If such reordering has taken place and the individual is recognized, then his reception will be determined by his reactions; that is, whether he accepts his new position at the bottom of the group or strives to regain his old position. Thus, the activities of the animal introduced, or reintroduced, to the group, can influence his reception. Individuals accepting immediate low status would thus be expected to receive less aggression than those challenging the group or resisting efforts to subordinate them. Vessey (1971) reported exactly such a finding in a series of 26 reintroductions of rhesus monkeys. Obviously, rhesus monkeys have considerable latitude in the types of aggressive or submissive responses they can display and we should expect differential utilization of such responses to influence the course of attacks against newly introduced animals. Some support for this model may be found in both Vessey (1971) and in Kawai's (1960) description of the formation of a group of Japanese macaques. Some individuals were already familiar to other animals and were readily accepted; others resisted the nucleus group and met severe aggression; while the reception received by most seemed to vary as a function of age and sex. Vessey's rhesus studies indicated that almost all reintroduced females rejoined their groups, whereas nearly half of the males failed to regain entry. Furthermore, low-ranking females had the least difficulty in reclaiming their former social positions.

In summary then, it is hypothesized that the age and sex of the animal being introduced, the sex composition of the group the animal is being introduced to, the previous social history of the introduced animal with that group, and the individual response patterns of animals introduced to a group will all influence the amount and type of aggressive responses directed to newcomers by a host group.

THE EXPERIMENTS

In testing the model outlined, a series of over 75 introductions of rhesus monkeys into established groups was conducted over a 2-year period at the Yerkes Field Station. Systematic data collection allowing for quantitative comparison was available for 56 of these introductions.

The basic facilities used consisted of four outdoor compounds each 125 ft (38.4 m) on a side with attached indoor quarters which were ordinarily not available to the animals during these introductions. The photograph in Fig. 1 demonstrates the basic layout of the facilities.

Fig. 1 Aerial view of the compounds at the Yerkes Field Station where the experiments took place. Each of the four outdoor areas in the foreground complex is 38.1 m on each side.

Well over 100 animals were involved in these studies, either as subjects or members of the host troops. Some individuals served as subjects in more than one experiment, either as subjects introduced to groups of different composition, or as subjects reintroduced to groups after specified periods of absence. The series of conditions and subject codes is summarized in Table 1.

In the 56 introductions in which systematic data were collected, 39 involved males being introduced and 17 involved females; 25 were to an all-male group, 8 to an all-female group, and 23 to a heterosexual group; 38 were single animal introductions and 18 were multiple animal introductions. A total of 37 introductions involved the first known contact

TABLE 1 **Conditions and Subject Codes**[a]

Condition	Subject
Single male introduced to male group 1	Pe, Qe, Re, Se, Nd
Single male introduced to male group 2	Ma, Eb, Ka
Single male introduced to heterosexual group 3	Zb, Eb, Fc, Ec, Wb, Kb, Ib
Single male introduced to heterosexual group 4	Kd, Gd
Single male introduced to female group 5	Pe, Qe, Re, Se
Single male introduced to female group 6	Vb
Isolation reared male introduced to female group 5	Ze
Single male introduced to isolation reared group 7	Ta
Multiple males introduced to heterosexual group 3	Four males
Single male reintroduced to female group 5	Qe, Re
Single male reintroduced to heterosexual group 4	Ga
Multiple males reintroduced to male group 1	Kb, Kc, Rb, Ta, Ab, Dc, Nc, Kb, Zb
Multiple males reintroduced to male group 8	Two males
Multiple animals reintroduced to heterosexual group 4	Kd, Qc
Single female introduced to male group 1	Ed, Od, Ua, Ya
Single female introduced to heterosexual group 3	Ed, Fa
Isolation reared female introduced to male group 1	Ff
Multiple females introduced to male group 8	8 females
Multiple females introduced to heterosexual group 4	4 females, 11 females
Single female reintroduced to male group 1	Ya
Single female reintroduced to heterosexual group 9	Ed, Oa, Fa
Multiple females reintroduced to heterosexual group 3	8 females, 4 females

[a] Group 2 consists of males alone from group 3. Group 3 consists of groups 4 and 5 plus multiple additions several months after formation. Group 6 consists of the females alone from group 3. Group 9 consists of group 1 plus female additions. Group 8 was a subset of group 1 after several months of separation.

between subject and hosts and 19 consisted of subjects being reintroduced after periods varying from a few weeks to almost a full year.

Owing to the large number of variables under consideration, 56 introductions is actually too few to permit a factor analysis demonstrating the relative contributions of each of the variables and their interactions. In addition, several cells (such as females being introduced to an all-female group) are vacant or represented by only one or two cases. Other variables which might reasonably be expected to influence the course of individual receptions, such as the absolute number of animals in a group, the ratio of males to females in heterosexual groups, the presence in variable numbers of immature animals in some heterosexual groups, and the season of the year, were not rigorously controlled.

Inasmuch as a definitive statistical analysis was thereby precluded, the data were subjected to multiple independent analyses examining relationships to each of the variables hypothesized as producing specific effects in the model. Wilcoxon rank tests were performed as well as median sign tests where appropriate. It is therefore acknowledged that one or more significant findings (as well as failures to achieve significance) may be spurious due to the number of statistical tests conducted. In evaluating such findings, one should consider the total number of tests which proved significant and the confidence limits achieved. The probability that the tests relating to the hypotheses stated, and that only these tests would have proven significant, is exceedingly small.

Data collection consisted of recording the responses performed by the subjects introduced and the responses directed toward the subjects, according to a predetermined catalog of defined items independently demonstrated to be capable of reliable measure under the prevailing conditions. Data were collected continually during the first hour following introduction. When multiple animals were released, multiple observers were used. Still and motion picture photography supplemented most observation sessions and a videotape recording was used as an experimental adjunct to data collection. Outside of the formal data collection routine, notes were made of other observations, but these could not be used in quantitative analyses, for obvious reasons. The data were analyzed into three successive 20-min blocks to permit analysis for effects over time, but all entries in this chapter are expressed in terms of hourly frequency rate or in terms of the percentage of total responses scored, accounted for by any one class of responses under discussion. The responses scored were divided into the following classes for more ready analysis:

Contact aggression: all responses involving physical contact with the potential to produce bodily damage; in practice: bite, slap, pull.

Noncontact aggression: responses relating to the threat of bodily injury; in practice: open mouth expression, charge.

Submission: other than aggressive response to aggression; in practice: grimace, flee, avoid, squeal, crouch passively.

Grooming: manual and/or oral manipulation of the hair

or skin of another, includes picking at the genitalia.

Sex: responses included in copulatory sequences even when divorced from such context; in practice: sexual presentation, hip touch, mounting.

Other: all other social contacts or interactions not otherwise scored; in practice includes: non-specific contact, lying on, sniffing and similar directed responses.

The terms "aggression" and "agonistic responses" are used to mean the combination of contact and noncontact aggression, and the combination of these with submission, respectively. All specific responses were defined by their motor elements rather than interpreted function.

Most introductions were conducted in the morning following venipuncture of either the subjects or hosts as part of a routine survey of cortisol and testosterone responses in the males as a function of the stress produced by such social introductions. Cortisol responses to venipuncture were low and stable, suggesting that this procedure did not constitute an appreciable influence on the subsequent behavior exhibited during introductions. There were no apparent differences between introductions preceded by such procedures and those which were not.

THE DATA

Changes within the First Hour

As had been found previously (Bernstein and Mason, 1963b; Bernstein, 1964c), the 20-min period immediately following an introduction was the period of greatest activity. In general, the most intensive and violent responses occurred within the first few minutes during which most newly introduced animals attempted to flee the area and avoid the group, and the group tried to chase the newly introduced animal from the area. Since the boundaries of the enclosure made escape impossible, it was only a few minutes before the introduced animal either retreated into a corner or was caught by the host group. This general pattern of high activity was consistent although details of responses varied as a function of the differential sex of host and newcomer and past history as a group member.

Three introductions were terminated during the first 20 min to rescue the newcomer, and three other introductions were not allowed to run a full hour for similar reasons. In another three introductions, the host group was temporarily distracted to determine the extent of injuries suffered by the introduced animal. If we eliminate only the first three such cases, that leaves 53 introductions during which we can examine the course of events during the first hour.

In 45 of the 53 introductions, maximum response rates for subjects were obtained during the first 20 min, and in 42 cases the group's response rate to the subject peaked during this same period. The rate of decline was rapid and continuous over the three 20-min blocks with regard to total responses done by 29 of the subjects and with regard to total responses received by 32 of the subjects. In the remaining cases there was a tendency for a slight increase in the final 20-min period as compared to the middle 20-min period. One might speculate that the sheer vigor of initial interactions gave subjects some respite during the middle period because of group fatigue, but an examination of median response rates for each type of interaction listed in Table 2 suggests that there

TABLE 2 **Response Medians, Hourly Rates for Introduced Subjects**[a]

Blocks of 20 min after entry		All social		AGO		AGG		CAG		NCA		SUB		GRM		SEX		OTH	
		D	R	D	R	D	R	D	R	D	R	D	R	D	R	D	R	D	R
Males	1	89	120	74	96	24	83	14	45	6	32	45	3	0	0	0	3	0	12
	2	39	45	27	39	9	18	0	6	0	15	18	0	0	0	0	0	0	0
	3	42	29	32	23	9	8	0	0	0	20	14	0	0	0	0	0	0	3
Females	1	54	126	48	72	3	33	0	9	0	18	45	0	0	0	0	3	0	12
	2	18	63	18	27	0	6	0	3	0	3	15	0	0	0	0	0	0	3
	3	18	18	15	12	0	6	0	3	0	3	12	0	0	0	0	0	0	3
All animals	1	72	126	65	87	18	60	6	29	3	30	45	3	0	0	0	3	0	12
	2	35	42	27	38	3	18	0	3	0	12	15	0	0	0	0	0	0	0
	3	33	33	27	18	3	6	0	3	0	3	12	0	0	0	0	0	0	3

[a] CAG is contact aggression, NCA is noncontact aggression, SUB is submission, GRM is groom, SEX is sexual, OTH is other social, AGG is aggression, AGO is agonistic. D and R stands for "does" and "receives."

was a transition taking place in the types of responses being exchanged, even within the first minutes of introduction.

Table 2 is expressed in terms of medians because of the enormous range of variation seen in the many diverse types of introductions. For example, the range of total rate of social response included two subjects with rates of less than 30 per hour and two subjects with rates of more than 250 per hour. The host groups likewise responded to two subjects with rates of less than 20 per hour and to two subjects with rates in excess of 350 per hour. It should be noted, however, that after the first 20 min only one subject continued to respond at a rate even as high as 150 per hour and two animals ceased interacting entirely. The host group showed similar sharp declines in all cases save one.

The total pattern of response peaks in the first 20 min was not reflected in each of the response classifications used for analysis. Table 3 presents

TABLE 3 **Subjects with Initial 20-min Peak Response Rates and Subjects with Continuously Decreasing Response Rates during the First Hour**[a]

Response category		Initial peaks: All subjects $n = 53$	Initial peaks: Males $n = 36$	Initial peaks: Females $n = 17$	Steady declines: All subjects $n = 53$	Steady declines: Males $n = 36$	Steady declines: Females $n = 17$
All social	D	45	31	14	29	20	9
	R	42	27	15	32	20	12
AGO	D	44 (1)	29 (1)	15	30 (1)	21 (1)	9
	R	38 (1)	24 (1)	14	30	19	11
AGG	D	36 (8)	27 (3)	9 (5)	29 (7)	21 (2)	8 (5)
	R	34 (4)	22 (3)	12 (1)	25 (4)	16 (3)	9 (1)
CAG	D	28 (12)	24 (2)	4 (10)	23 (14)	19 (4)	4 (10)
	R	34 (8)	32 (6)	12 (2)	30 (8)	19 (6)	11 (2)
NCA	D	26 (18)	19 (10)	7 (8)	21 (4)	15 (10)	6 (8)
	R	34 (4)	23 (3)	11 (1)	22 (3)	15 (3)	7 (1)
SUB	D	34 (4)	21 (4)	13	24 (3)	16 (3)	8
	R	22 (15)	12 (11)	10 (4)	20 (20)	10 (16)	10 (4)
GRM	D	0 (50)	0 (34)	0 (16)	0 (49)	0 (33)	0 (16)
	R	9 (26)	6 (17)	3 (9)	10 (23)	7 (14)	3 (9)
SEX	D	9 (35)	6 (23)	3 (12)	8 (34)	4 (23)	4 (11)
	R	21 (22)	15 (13)	6 (9)	20 (20)	12 (13)	8 (7)
Other social	D	18 (26)	12 (19)	6 (7)	17 (25)	12 (18)	5 (7)
	R	27 (16)	16 (13)	11 (3)	24 (12)	17 (9)	7 (3)

[a] D and R rows indicate "does" and "receives." Numbers in parentheses indicate subjects with zero or equal scores.

the data for the number of animals showing peak frequencies in the first 20-min period for each of the nine response classifications used. "All social," "agonistic" and "aggressive" classifications subsume one another as well as the other categories and therefore relate to one another. Since many of the subjects could not be ranked for each of the three first 20-min periods for all categories due to zero scores, the number of subjects for which no determination was possible is shown in parentheses next to the peak numbers.

In a similar manner, the number of subjects showing steady declines in frequency rates for each response classification is shown in the adjacent columns with numbers in parentheses once again indicating the number of subjects with either tied scores or no data points from which to make a judgment.

Table 3 thus indicates that the peak in social interaction rate can be attributed primarily to the peak in agonistic responses during the first 20 min (44 subjects did the most and 38 received the most social interactions during this interval). Furthermore, the most consistent steady declines were seen in total agonistic data (30 subjects for both "does" and "receives"). The general tendency for agonistic responses to peak in the first 20 min can be seen in both aggressive and submissive response classifications (36 subjects did most of their aggressive responses in the first 20 min and 34 subjects did most of their submissive responses in the same interval, but whereas 34 subjects received most of the group's aggression during this time period, only 22 of the subjects received the majority of group submissive responses directed towards them in this time interval and 19 received greater scores during some other interval.) The same situation is reflected in the decline in aggressive and submissive scores. An analysis of aggressive responses shows an essentially equal tendency for contact and noncontact forms to peak in the first 20 min, but the decline in contact aggression received was steady for 30 subjects whereas 30 subjects did not show a steady decline in noncontact aggression received. Sexual, grooming, and other social responses did not show a clear tendency to peak or decline and, if anything, subjects tended to perform sexual and grooming responses more often after the first 20 min, but the number of data points makes such a determination impossible since more than half of the subjects (34) performed no sexual responses or distributed them equitably during the hour. A similar situation for grooming responses exists for 49 of the subjects.

Agonistic responses accounted for more than 50% of all responses performed by subjects during the first 20-min period in 52 of the 56 introductions. Similarly, in 49 of 56 cases agonistic responses accounted for more than 50% of the responses directed toward subjects by group members. Submissive responses accounted for more than 60% of the agonistic responses performed by 36 of the subjects, including 11 subjects who showed only submissive responses. In contrast, 44 subjects were the target of aggression in 60% or more of agonistic interactions. The remaining 12 subjects were sharply distinct in that they received aggression in less than

20% of their agonistic interactions and 6 of these subjects never received any aggression from the host group. As might be inferred, such introductions usually involved animals who either assumed control of the host group or achieved very high status as the result of rapid alliance with the resident control animals.

Although the rates for contact and noncontact forms of aggression were similar and declined rapidly, the sex of the introduced animal did account for some of the wide individual variation with regard to exactly which of these predominated in any one introduction. We shall deal with these influences later, but for now it is sufficient to note that the relative uniformity in group scores masks consistent individual variation.

Grooming, which was rarely observed during introductions, was scored for only three of the subjects whereas 31 subjects received grooming from group members. Such grooming, however, was often mixed with aggressive responses and in some cases wounds inflicted by an animal received grooming attention shortly thereafter, but such attention did not preclude the infliction of additional wounds by the same animal. The relationship of grooming and mounting relative to rank as reported by Vessey (1971) during reintroductions was not confirmed in these introductions.

Sexual responses were also rarely observed as initial interactions, but host animals were seen to mount one another prior to and during attacks upon the subjects. These mountings were often between two males who combined in attacking the newly introduced subject. Only 18 subjects showed any form of sexual response, but 35 received sexual responses of one form or another, and most of these (21 of the 35 subjects) received the majority of such responses sometime during the first 20 min following introduction.

The remaining social responses were classified under "other" and this category was scored for only half of the subjects, and in no case accounted for as much as half of the responses by any animal introduced. Such responses were received more often than given and when they consisted of sniffing or touching might easily have been considered as exploration of the newly introduced subject. Such an explanation cannot be invoked in cases where host members literally sat or sprawled on a passive animal after it had ceased to resist the group.

In general, subjects tended to receive more responses from the host group than they performed. The only notable exception is that 43 of the 56 subjects performed more submissive responses than were directed toward them. Thus, whereas 36 subjects received more agonistic responses than they performed, 45 received more aggression than they performed.

Differences in Reception as a Function of the Sex of the Subject and Host Group Members

Not only did the sex of the subject influence the course of introductions, but host groups of different sex composition reacted differentially to the subjects. Three basic comparisons can be made: (*1*) males versus females introduced to male groups, (*2*) males introduced to male versus female groups, and (*3*) males versus females introduced to heterosexual groups. The last case includes groups of disparate sex ratios and certain extreme cases acted more like unisexual than heterosexual groups, a finding consistent with the original model proposed.

Despite the similarity of initial group responses to all animals introduced—an excited rapid approach by at least some group members—sexual recognition was apparent in a very short time and the nature of responses directed to males and females differed from the very start. Although females in an undisturbed group ordinarily show greater frequencies of noncontact and manual forms of aggression than do males (Bernstein, 1970), the stimulus of an intruder or a strange group appears to be more powerful for males than for females, and, if anything, under the circumstances males are more aggressive than females, thus supporting the hypothesis that male aggressive potential, although ordinarily controlled, is specifically released under these disruptive circumstances. An introduced male, therefore, is not only more of a challenge to a host group by virtue of his size and sex, but is also more likely to respond to the host group with vigorous aggressive responses, thus compounding his challenge to the host group.

Due to the interactions between sex of the subject and sex of the host group, male and female subject introductions could not be readily distinguished over all host groups. Male subjects, however, did perform significantly more contact aggression responses and both carried out and received more aggression than did females. Male subjects generally resisted group aggression and fought back when attacked and although most defensive fighting was manual, male subjects also bit and slashed males from the host group. Male host group members sometimes suffered significant wounds in these encounters whereas female host group members were never seen to suffer slash wounds, although male subjects did sometimes bite females. After witnessing several introductions, it was subjectively clear that males exhibited considerable restraint in fighting with females, even in those cases where females in a host group succeeded in defeating a subject male and beat him into passive submission.

Males fought with females in a somewhat stylized fashion, pulling at their heads and necks manually and then biting the female on the neck using incisor teeth only. Acting in such a restrained manner, as compared to the fighting patterns exhibited in fighting males, a male subject often lost to the combined attack of several females despite his superior size and abilities.

Females with significantly lesser use of contact aggression responses also had significantly lower median aggression response scores as compared to male subjects. Furthermore, when scores obtained during introductions of multiple females were removed from consideration, it was found that not only did single females show significantly lower scores for contact and total aggression responses performed, but that they also received significantly fewer contact and noncontact aggression responses.

INTRODUCTION TO ALL-MALE GROUPS

Lower female scores for participation in aggressive interactions were due primarily to the treatment females received when introduced to all-male groups. As noted, the females performed significantly less contact aggression in all introductions, including these, and in these introductions they received significantly less aggression of any type from the males.

In contrast, when male subjects were introduced to an all-male host group, significantly higher interaction rates were observed. Much of this was due to extremely high agonistic response rates both received and performed by introduced males. Contact aggression rates were particularly high and introduced males also performed submission responses at rates significantly in excess of median scores. Despite the high levels of contact aggression performed by males introduced to an all-male group, they received significantly fewer submission responses, thus indicating that their intense and concentrated agonistic responding was not effective in subduing any number of host group males and was primarily in response to attacks upon themselves. The eight males individually introduced to an all-male group for the first time received much more aggression from host animals than was recorded in any other introduction. In fact, one male was killed, and only by virtue of active experimenter interference were four other animals able to survive.

The sharp contrast between male and female introductions to an all-male group was not due to females passively submitting to males, as indeed male subjects tended to submit more frequently than did females in these circumstances. Male subjects resisted attack, but even after submitting frequently received further attack. One notable finding was that submissive males would tolerate such attack passively, emitting only submissive signals, up to a point. Passive submissive males receiving additional attacks would appear to suddenly respond vigorously to the animal attacking them, even after having suffered repeated prolonged attack without resistance. In such cases the subject would fight back, but continue to emit submissive signals. After several repetitions and upon close examination of movie films taken at the time, it was strongly suggested that a male showing submission to physical attack would tolerate bites inflicted with the incisor teeth even when such bites were vigorous and prolonged, but that when an attacking male sank a canine tooth into the subject, the subject would turn and resist vigorously.

Male receptions in all-male groups were not uniformly disastrous for the subjects. Many of the reintroductions in particular passed without serious incident and some of the initial introductions did not produce serious wounding despite the vigor and frequency of aggressive interactions. Furthermore, agonistic interactions were not the only response categories showing significantly higher response rates than seen in other types of introductions. Other nonspecified social contact scores also reached significantly higher-than-average levels during these introductions.

MALES INTRODUCED TO FEMALE HOST GROUPS

In contrast to the intensity and frequency of aggression seen when males were introduced to an all-male group, when males were introduced to a group consisting exclusively of females, the males received significantly fewer contact aggression responses and significantly more submissive responses than in the average introduction. As a consequence perhaps, male subjects showed significantly fewer than average submissive responses while performing significantly more noncontact aggression responses. In general, the males seemed capable of taking over an all-female group resorting more to threat than to active physical attack. This is not to suggest that they were completely unopposed. The all-female host group initially did respond to the subject male by attacking him, but such attacks were seldom prolonged or successful for any number of repetitions. Even when groups consisting of females with a single immature or adult male included were considered as if they were all-female host groups, male subjects continued to show less submission and to receive significantly less aggression, especially contact aggression, than in other introductions. Total agonistic response rates for male subjects were significantly below average despite increased frequencies of noncontact forms of aggression and increased total aggression scores, a finding which underscores the typically large contribution to agonistic response rates contributed by submissive responses. The frequency with which submissive responses were received by the males was inversely related to the effort they expended in aggression. In these introductions the males received significantly higher frequencies of submissive responses although they themselves devoted the smallest percentage of their own interactions to agonistic responses. Furthermore, it was only when introduced to predominately female groups that the males devoted less than half of their aggression to contact forms. In roughly half of these introductions, males received more submission from group members as the hour progressed. In contrast, when males had been introduced to all-male groups, only 2 of 11 ever received submission from host group members after the first 20 min. Females introduced to all-male groups did receive increasing frequencies of submissive responses from male group members. In the latter case, females rapidly allied themselves with top-ranking males and proceeded to assert themselves over all lesser males by judicious use of these alliances.

Males introduced to all-female groups and females introduced to all-male groups did experience the highest sexual interaction rates. Only in the case of one male, however, did participation in sexual activity account for as much as half of the subject's social responses. Subjects generally received sexual responses more often than they initiated sexual interactions, but, in general, response rates were low and even zero in some cases.

INTRODUCTIONS TO HETEROSEXUAL GROUPS

When female subjects were introduced to heterosexual groups, they performed less aggression, particularly contact aggression, than the average subject, a finding consistent with general female response patterns. In general, what aggression females did receive on introductions was more likely to be in the form of noncontact rather than contact responses. Female subjects, however, had received increasing frequencies of submissive responses as a function of time in all-male groups, but in heterosexual groups only 2 of 10 female subjects ever received submissive responses at all after the first 20 min.

Although little grooming was seen, female subjects did receive more with time in five of eight cases. Only one female subject received more than 30% of social responses in the form of grooming and the next highest subject score was less than 12%. Another female devoted one third of her initial interactions with a heterosexual group to grooming, but this was clearly an exceptional case as no other subject devoted even as much as 2% of its interaction to grooming. There was no apparent explanation for the single anomalous case.

Males, on the other hand, when introduced to heterosexual groups were subjected to repeated attacks. All forms of social response to male subjects were higher than in most introductions and male subjects received significantly more agonistic responses, primarily aggression, and that mostly contact aggression, than in other introductions. They also received more general social responses, and performed high rates of submissive responses. Heterosexual groups thus responded to male subjects with as much vigor as had all-male groups; thus the differential response to all-female groups could be attributed primarily to the absence of males in the host group. Females in heterosexual groups contributed vigorously to the attacks upon subject males and a group of several females often totally overpowered a male. These females had the support of resident males, but perhaps by the very vigor and persistence of female attacks, male subjects suffered less damage than when introduced to an all-male group. The females, with their smaller canine teeth and lessened potential to inflict serious injury, monopolized much of the attack upon intruder males and in so doing precluded many attacks upon the submissive introduced males on the part of resident males. Thus, although females were successful in attacking introduced males, these vigorous and prolonged attacks on the part of females were only possible in the presence of supporting resident males, and also served to protect the introduced males from the resident males.

If we contrast male subject reception by groups consisting only of females with their reception by groups consisting of at least some males, the two types of situations are significantly different. When males are present in the host group, male subjects do significantly more social responding and this is primarily agonistic. The subject males show higher submission as well as contact aggression response rates, but also receive higher agonistic response rates, consisting almost entirely of aggressive responses, in which contact aggression is significantly higher than in other introductions. In addition, other social responses are directed at male subjects at higher-than-average rates, thus contributing to a general picture of maximum interaction. Host groups containing males respond more intensely to intruders and these responses are specific to male intruders. The percentage of group effort devoted to agonistic responses, is, however, less than the percentage of effort devoted to agonistic responses on the part of the subjects. The percentage of agonistic responses devoted to aggression is, however, generally less for subjects than for host groups. Expressed another way, whereas only 4 of the 56 introductions did not reveal submissive responses on the part of the introduced animals, many of the animals introduced never received submissive responses from any group member. Male subjects persisted in exhibiting higher levels of submission than did females over time, but they also received much more intensive and prolonged attack.

Multiple Animal Introductions and the Influence of Past History on the Reception of Subjects

Eighteen of the introductions consisted of multiple animals introduced simultaneously to a host group. The majority of these, however, consisted of reintroductions of former group members. It was originally thought that the distraction of multiple targets for group response on introductions would not only reduce interactions with any one subject, but that this might confuse matters sufficiently so that total response rates would decline. This did not prove to be the case.

When multiple females were introduced or reintroduced to a group, the total social interaction rate on the part of the host group exceeded typical response rates. On the other hand, when multiple males were reintroduced to an all-male group, social interaction rates were always less than when single males were initially introduced. Multiple female introductions were notable because only in these cases did females receive high contact aggression scores as well as noncontact aggression scores from host groups. No ready explanation for the increased vigor of aggressive responses to multiple females can be found. Reintroductions to their own group after a period of absence produced remarkable resistance to reentry, and even when introduced to an all-male group, multiple females fared less well than had most single females.

Reintroductions were otherwise generally more pacific than were original introductions. Males reintroduced to all-male groups generally suffered few contact aggression attacks and used significantly less aggression in self-defense. In the two cases where an alpha male had been removed for a short period of time (a few weeks maximum), the alpha male reassumed his old status with no challenge. These reintroductions stand out sharply from all other introductions in that the subject showed high aggression response rates but received none in return. Group members submitted immediately.

Other reintroductions can be divided into two classes—those involving short-term removals of only a few days and which produced no change in the original group social organization, and those reintroductions which occurred after periods of from several weeks to up to 1 year, during which time the group social organization had apparently restructured. Males returned under such conditions could not readily claim their old rank positions and in most cases accepted new status positions at the bottom of the hierarchy without challenge. Aggressive responses were directed to these animals by former subordinates and were generally accepted with submissive responses. Some individual subjects were entirely submissive in their responses to group members whereas others submitted but defended themselves against attack. These individuals suffered more attacks and injury than those passively submitting.

In two cases, reintroduced males tried to reassert themselves in their former status positions. One male, formerly third ranking in the group, was killed in the consequent fighting. The second case was of a relatively low-ranking male who fought and defeated the few animals whom he had formerly outranked, before the rest of the group attacked him. He submitted to the attack of his former superiors although he did attempt to defend himself. His former inferiors, however, failed to rejoin the attack against him and he apparently succeeded in reclaiming his former rank position.

Such an account is, of course, anecdotal, but does illustrate a repeated observation. Male subjects in their own self-defense did seem to provoke more vigorous group attack upon themselves, but in each of the cases where a group member was injured by the subject, the group member detached himself from the attacking party and left the area. In a small group with only a few adult males, a newly arrived male may thus effectively join a group with high initial status if he can inflict a few quick injuries to the resident males. It is only when the sheer number of resident adult males being challenged is so great that the subject cannot hope to injure any significant number of opponents that vigorous aggressive defense on the part of a new male is clearly ill adapted to his ready integration into a group. Such an explanation may at least in part account for the increased evidence of severe wounding reported in wild troops of rhesus monkeys during the breeding season when most male exchanges among troops take place (Lindburg, 1969; Neville, 1969).

One other aspect of past history influencing reception of newcomers concerns the apparent social failings of animals reared under various conditions of social deprivation. A normal male was introduced to a heterosexual group of deprivation-reared animals and although his initial responses were similar to that of any male introduced to a heterosexual group, the group failed to respond to him as a coherent unit and after 20 min the introduced male succeeded in breaking off contact with the residents and hid in some low vegetation where he apparently was unseen for the next 40 min. In none of the normal groups was any individual ever successful in eluding the group by hiding for more than a few minutes.

In two cases, socially deprived subjects were introduced to host groups. A socially deprived female introduced to an all-male group was clearly treated as a female and not a male intruder, but nonetheless received only aggressive responses from the males, and received more contact aggression than did any other female. She also received high levels of noncontact aggression and showed only submissive responses to the group. All other females included a rich variety of social responses in their interactions with the males.

A socially deprived male introduced to a group of females also received extraordinarily high levels of aggression from the females. Ninety-four percent of all responses directed to him were aggressive and whereas females seldom showed contact aggression responses to males, they directed high levels to this male. This male also both received and performed high levels of noncontact aggression responses and after a period of more than a week it was clear that he had not only failed to take charge of the group, but had also failed to join the group. The females, nonetheless, had not defeated this male and their persistent attacks had been successfully defended against, despite the male's history of social deprivation. Perhaps it is also worth noting that even this male refrained from using his canines in fighting females and showed the usual male manual fighting pattern when engaged by females. The significance of this statement is belied by the fact that he also used the same pattern in fighting with other socially deprived animals, regardless of sex. As reported elsewhere (Rose *et al.*, 1972), this male failed to show a testosterone rise in response to his introduction to females whereas all normal males had shown dramatic testosterone rises in the same circumstances. Thus, it is quite possible that this deprivation-reared male was not responding to the sexual stimuli of the females in the host group in any way.

One other point is worth making and that concerns the remarkable consistency of individual animals in these introductions. Individual differences are well known to investigators working with primate subjects, but it is perhaps comforting to know that such individual differences are at least consistent within the individuals. Individual subjects introduced to different groups and even to the same group after only brief previous experience with the group (two male and two female subjects) were remarkably consistent in the pattern of their responses in these introductions. A female twice introduced to an all-male group both times was extremely vigorous in claiming beta animal position and showed consistent high levels of aggression in her assertion of rank which far exceeded those seen in any other female subjects. In contrast, a second female showed consistent low aggressive profiles on her initial and second introduction. The males participating in more than one introduction showed consistent levels of aggressive, sexual, and other social responses both absolutely and as a function of relative distribution of total social interactions. Other animals serving as subjects in introductions to more than one group likewise showed response consistencies, thus lending credence to notions that one may be better able to predict a particular individual's response to a given situation than one can predict the outcome of any specified experimental manipulation.

THE DATA AND THE THEORY

In the introduction to this chapter we proposed a theory which states that one of the most potent causes of aggression is the threat of disruption of an established social organization. Implied in this was that a social order, once established, serves to maintain itself and any threat to the integrity of that organization would be met with by all the resources at the command of the social unit. In some situations, aggression might be the primary response of some, if not all, group members to any challenge of the social system.

From this theory a series of hypotheses were derived relative to the elicitation of aggression in rhesus monkey groups. It was proposed that the introduction of animals into an established group was one significant challenge to the group, and after examining some of the available literature, we suggested that this might be one of the most potent stimuli for eliciting aggression in rhesus monkeys. A series of experiments were performed to test the hypotheses related to various refinements of the initial statement related to contributory variables. Of these experiments, 56 introductions were followed with sufficiently rigorous data collection sequences to permit quantitative analyses.

Although much of the data collected are in support of particular hypotheses under test, acceptance of these hypotheses does not constitute confirmation of the explanatory theory proposed. Moreover, many variables hypothesized to be significantly related to responses following introductions could not be adequately tested in this sequence of experiments simply because the large number of variables postulated, and their interactions, requires much more exhaustive testing than that which exhausted the testers.

The theory regards all of the following variables as included among those which must be considered in predicting the extent and quality of aggression resulting upon the introduction of rhesus monkey subjects to groups of rhesus monkeys.

1. *Sex of the Subject.* Both male and female subjects were used and qualitative and quantitative differences in their responses were demonstrated.

2. *Age of the Intruder.* Only fully adult subjects were used and no data relevant to this variable are thus available. Adult subjects in good health are considered as one extreme in stimulus potency.

3. *Sex Composition of the Group.* Groups composed entirely of fe-

males and groups composed entirely of males were used as hosts as well as heterosexual groups. No systematic manipulation of sexual composition was otherwise attempted although there is some evidence suggesting it would produce differential effects. Certainly the two unisexual groups responded differentially to subjects as compared to each other and to heterosexual groups. The behavior of animals in a heterosexual group was not simply a reflection of the combination of unisexual group patterns. Both sexes in the presence of the other sex responded differently to newcomers than when alone. This was especially clear with regard to oppositely sexed newcomers and emphasizes the importance of another interaction: that between the sex of the host and the sex of the intruder.

4. Number of Animals Introduced Simultaneously. Only two dichotomous classes were considered: single animals versus multiple animals. Interaction effects with sex and past histories prevented a clear demonstration of the influence of this variable. Worse still, the few introductions of multiple females into a heterosexual group appeared to produce more aggressive receptions than had single female introductions. If multiple animals serve as a diffuse target-fragmenting group attack, and if in reintroductions at least some individuals respond submissively to the group and are thus readily reaccepted, then we should have seen lowered frequencies of aggressive response. Although the number of cases is too small for significant statistical treatment, and since many of the relevant comparison cells are vacant, we cannot at this point demonstrate whether it is the interaction effect, the confounding variables, or a misconstrued hypothesis that is at fault. Hopefully it is not the theory.

5. Social Relationship of the Animals Being Introduced. This variable, of course, is limited in its application to multiple animal introductions. In its most extreme forms, intruders could either be strangers to each other, or representatives of another intact social unit. Although both extremes were represented, too few cases were tested for suggesting the effects of this variable. One extreme, the introduction of intact groups to one another, was in fact tested on multiple occasions. Due to the complexities involved when two social orders come into conflict and each strives for supremacy, these findings have been set aside for separate analysis and imminent future publication.

6. Past History of the Intruder with the Host Group. The particular facet of this variable tested was the reintroduction of animals after specified periods of separation from their social units. Although no systematic study of separation interval was conducted, periods of up to 1 year of separation did not seem to preclude recognition of individuals. The reception on such reintroductions clearly differed from those witnessed during original introductions. Furthermore, animals removed for short periods of time reclaimed their former positions in the group without incident beyond a brief period of active locomotion and unspecified social exchange. After more prolonged separations, periods in excess of a few weeks, individuals were reaccepted into the group in new social positions and aggression related to establishing and enforcing this new position was seen. In only a few cases did individuals readily reclaim their original social positions rather than accept attempts to force them into omega animal slots in the status hierarchy as an initial position.

7. *Degree of Organization of the Host Group.* This variable remained untested. Its extreme forms would consist of a group constituted of mutually unfamiliar animals into which another was inserted as opposed to a multigenerational group, with the degree of social cohesion typical of natural troops of the species. The first extreme is represented by group formation experiments previously reported for rhesus and other species (Bernstein and Mason, 1963b; Bernstein, 1964a, 1969b, 1971). The other extreme is only approximated by the study groups used which, although of long-standing duration (periods of up to 3 years and including multiple offspring in heterosexual groups), might easily be defined more by confinement than by social bonds. It is considered more likely, however, that the degree of social organization present would have bound most of the group members to each other much as they united when threatened by human intruders or when alarmed by the sight of indigenous predatory animals. Similarities to the reintroductions reported by Vessey (1971) further support this notion.

8. *Social Skills of the Intruder and/or Host Group.* The only available objective evidence concerns the use of animals reared under conditions of social deprivation. The single male and the single female subject used as well as the single introduction involving a group of such animals all suggest that this prior social history variable exercised a profound effect upon the interactions observed, although it did not obscure the operation of other variables, such as sexual class membership, upon the reception received by those subjects.

9. *"Personalities" of the Animals Involved.* This rubric is used as an admission of the existence of many variables which influence the use of available responses by an individual. Even in general agonistic episodes, differential tendencies to flee, submit, defend, redirect aggression or to aggress in response to group aggression could all be seen. Furthermore, examining the records of individual animals used as subjects in multiple host group situations revealed consistent individual response profiles. These profiles were sufficiently characteristic of individuals that after familiarity with the subject was obtained, considerable success could be achieved in predicting individual specific response patterns in the general

context of a specified introductory procedure. Fortunately, however, this was one primate study in which individual variability did not obscure the operation of the other variables under consideration.

10. Physical Properties of the Test Environment. This was not examined. It was assumed that introductions in unlimited spatial areas would not have resulted in the failures to escape by flight as attempted by most intruders. Further, specific features of cover are conceived of as influential in such introductions, although in the present study, the meager vegetational cover available proved significant in only one introduction when a group of animals raised under conditions of social deprivation lost all contact with an intruder who broke off contact with the group and hid in the vegetation.

11. Physical Conditions of the Animals Involved. Only animals in apparent good physical condition were used in these experiments. It was assumed that physical debilitation would influence such introductions and, to be consistent with the theory, we would predict that unconscious or badly injured or severely ill animals would constitute less of a threat to a group than robust, active individuals. No test of this hypothesis has been made. Clearly, this and all other variables will be influenced by the next consideration.

12. Interaction Effects among All of the Above Plus Other Unspecified Variables Such as Weather, Season of the Year, etc. The interaction effects were most marked between sex of the subject and sex of the host group. In this case, at least, the results obtained were more than simply those obtained by algebraic summation of the effects of the contributing variables.

Several of the above variables are sometimes assumed or ignored and many are beyond ordinary experimental control. The constraints of the present study thus limit the application of the data relative to the variables postulated as significantly related to the theory. At this point we can only say that some of the variables have been subjected to test but that our knowledge is fragmentary and limited to anecdotes, or secondary analyses of other experiments, with regard to the effects of many of the variables and their interactions.

Thus, we once more find ourselves examining a theory for which we have performed no crucial test. The series of experiments reported provides data relevant to some of the hypotheses proposed, but cannot be considered as an adequate test of the theory. What we have demonstrated is that the introduction of rhesus monkeys to an established rhesus monkey social unit can be a potent stimulus for eliciting aggression. Furthermore, the most extreme forms of aggression and near maximum frequency rates were obtained in several types of encounters, thus qualifying this situation as equal to the most powerful determinants of rhesus monkey aggressive behavior. The variability of the results, however, suggests the multivariate contributions and should warn against any simplistic modeling.

The effect of such introductions was powerfully influenced by the sex of the animals involved, both absolutely and relatively. Male subjects clearly utilized certain responses in the species repertoire differently than did females. Furthermore, male subjects used these responses differentially when introduced to all-male and all-female host groups. Yet a consideration of sex alone would not account for the differences between initial introductions and reintroduction, nor between the receptions of individuals of different "personalities." By this last we are referring to such factors as those responsible for producing individuals who accepted and individuals who resisted rank reductions, relative to their former status, on reintroductions. Such differences could not be predicted on the basis of former rank alone, nor duration of separation, or on the basis of any other objective data available. Our predictions based on subjective impressions were not always borne out either, although we could frequently predict what form of response an individual would use in the event of contact aggressive interactions or group attack.

Prior histories of social deprivation proved to influence individual response patterns in introductions and this accounted for certain individual variation. One might therefore posit that all such individual variability is a function of past social histories and their inadequacies. Such an attempt would, of course, deny any more biological determinants of individual behavioral variability, and we are convinced that just as both genetics and individual history contribute to the size, conformation, and physical condition of the individual, so must both be considered in discussions of behavioral variability. The consideration of either alone is incomplete. It does not matter whether one is primary or secondary in its degree of contribution, both surely contribute and the interaction effect may outweigh either contribution alone, and in extreme cases, either may become the primary determinant.

At this point, with a little supporting data, it is all too easy to generalize widely and go much beyond the constraints of the supporting evidence. Our experimental data have dealt only with a single species of monkey and although other observations suggest similar results might be found in related macaques, and anecdotal evidence can be found to support extrapolation to humans, the paucity of data should be sufficient warning for us to refrain from so doing. Rhesus monkeys are representative of primates in laboratories, but perhaps they are not representative of the entire biological order. Considerable additional sampling would thus be required in addition to the completion of testing within even the initial taxon examined.

What then is the value of proposing the theoretical explanation? Surely we could content ourselves with sticking strictly to the data and seek the safety of an empirical approach. We have instead proposed a theory and made ourselves vulnerable to criticism while disclaiming any ability to solve the relevant problems of human society. We do so solely because we believe that our theory is predictive and will guide our future research along such lines that the data we collect will allow cautious expansion and generalization until such time as perhaps we might dare to suggest relevance to such desperate problems as the control of aggression in our most dangerous species.[2]

REFERENCES

Altmann, S. A. (1962). The social behavior of anthropoid primates: An analysis of some recent concepts. *In* "Roots of Behavior" (E. L. Bliss, ed.), pp. 277–285. Harper (Hoeber), New York.

Bernstein, I. S. (1964a). The integration of rhesus monkeys introduced to a group. *Folia Primatol.* **2,** 50–63.

Bernstein, I. S. (1964b). Group social patterns as influenced by removal and later reintroduction of the dominant male rhesus. *Psychol. Rep.* **14,** 3–10.

Bernstein, I. S. (1964c). The role of dominant male rhesus in response to external challenges to the group. *J. Comp. Physiol. Psychol.* **57,** 404–406.

Bernstein, I. S. (1966). Analysis of a key role in a capuchin (*Cebus albifrons*) group. *Tulane Stud. Zool.* **13,** No. 2, 49–54.

Bernstein, I. S. (1968). The lutong of Kuala Selangor. *Behaviour* **32,** 1–16.

Bernstein, I. S. (1969a). Stability of the status hierarchy in a pigtail monkey group (*Macaca nemestrina*). *Anim. Behav.* **17,** 452–458.

Bernstein, I. S. (1969b). Introductory techniques in the formation of pigtail monkey troops. *Folia Primat.* **10,** 1–19.

Bernstein, I. S. (1970). Primate Status Hierarchies. *In* "Primate Behavior" (L. A. Rosenblum, ed.), Vol. 1, pp. 71–109. Academic Press, New York.

Bernstein, I. S. (1971). The influence of introductory techniques on the formation of captive mangabey groups. *Primates* **12,** 33–44.

Bernstein, I. S., and Mason, W. A. (1963a). Activity patterns of rhesus monkeys in a social group. *Anim. Behav.* **11,** 455–460.

Bernstein, I. S., and Mason, W. A. (1963b). Group formation by rhesus monkeys. *Anim. Behav.* **11**(1), 28–31.

Bernstein, I. S., and Sharpe, L. G. (1965). Social roles in a rhesus monkey group. *Behaviour* **XXVI,** 1–2.

[2] Our apologies for the many potshots taken at colleagues along the way, some of whom are most esteemed. Although their names are concealed, their identities are revealed in criticism of their foibles. They shall, no doubt, exact fitting retribution by similar oblique attack upon the fallacies contained in this manuscript.

Bernstein, I. S., Schusterman, R. J., and Sharpe, L. G. (1963). A comparison of rhesus and gibbon responses to unfamiliar situations. *J. Comp. Physiol. Psychol.* **56**, 914–916.

Carpenter, C. R. (1942). Societies of monkeys and apes. *Biol. Symp.* **8**, 117–204.

Castell, R. (1969). Communication during initial contact: A comparison of squirrel and rhesus monkeys. *Folia Primat.* **11**, 206–214.

Crook, J. H. (1966). Gelada baboon herd structure and movement—A comparative report. *Symp. Zool. Soc. London* No. 18, 237–258.

Ellefson, J. O. (1968). Territorial behavior in the common white-handed gibbon, *Hylobates lar* Linn. *In* "Primates" (P. C. Jay, ed.), pp. 180–199. Holt, New York.

Emlen, J. T., and Schaller, G. (1960). In the home of the mountain gorilla. *Bull. N. Y. Zool. Soc.* **63**, 98–108.

Gartlan, J. S. (1970). Preliminary notes on the ecology and behavior of the drill, (*Mandrillus leucophaeus*) Ritgen, 1824. *In* "Old World Monkeys: Evolution, Systematics, and Behavior" (J. R. Napier and P. H. Napier, eds.), pp. 445–480. Academic Press, New York.

Gartlan, J. S., and Brain, C. K. (1968). Ecology and social variability in *Cercopithecus aethiops* and *C. mitis*. *In* "Primates" (P. C. Jay, ed.), pp. 253–292. Holt, New York.

Goodall, J. (1965). Chimpanzees of the Gombe Stream Reserve. *In* "Primate Behavior, Field Studies of Monkeys and Apes" (I. DeVore, ed.) Chapter 12, pp. 425–473. Holt, New York.

Hall, K. R. L. (1960). Social vigilance behavior of the Chacma baboon (*Papio ursinus*). *Behaviour* **XVI**, Part 3–4, 261–294.

Hall, K. R. L. (1964). Aggression in monkey and ape societies. *In* "The Natural History of Aggression" (J. D. Carthy and F. J. Ebling, eds.), pp. 51–64. Academic Press, New York.

Hall, K. R. L. (1965a). Social organization of the old-world monkeys and apes. *Symp. Zool. Soc. London* **14**, 265–289.

Hall, K. R. L. (1965b). Experiment and quantification in the study of baboon behavior in its natural habitat. *In* "The Baboon in Medical Research" (H. Vagtborg, ed.), pp. 29–42. Univ. of Texas Press, Austin, Texas.

Hall, K. R. L. (1967). Social interactions of the adult male and adult females of a Patas monkey group. *In* "Social Communication among Primates" (S. A. Altmann, ed.), pp. 261–280. Univ. of Chicago Press, Chicago, Illinois.

Itani, J., and Suzuki, A. (1967). The Social Unit of Chimpanzees. *Primates* **8**, 335–381.

Kaufmann, J. H. (1967). Social relations of adult males in a free-ranging band of rhesus monkeys. *In* "Social Communication among Primates" (S. A. Altmann, ed.), pp. 73–98. Univ. of Chicago Press, Chicago, Illinois.

Kawai, M. (1960). A field experiment on the process of group formation in the Japanese monkey (*Macaca fuscata*) and the releasing of the group at O'Hirayama. *Primates J. Primatol.* **2**, No. 2, 181–253.

Koford, C. B. (1963). Group relations in an island colony of rhesus monkeys. *In* "Primate Social Behavior" (C. H. Southwick, ed.), Chapter 12, pp. 136–152. Van Nostrand Reinhold, New York.

Kummer, H. (1968). Social organization of hamadryas baboons. Univ. of Chicago Press, Chicago & London, pp. 189.

Kummer, H. (1970). Spacing mechanisms in social behavior. "Man and Beast" (J. Eisenberg and W. Dillon, eds.), Chapter 6. Comparative Social Behavior, Smithsonian Inst. Pr., Washington, D. C.; *Soc. Sci. Informat.* **9**(6), 109–122.

Lindburg, D. G. (1969). Rhesus monkeys: mating season mobility of adult males. *Science* **166**, 1176–1178.

Marsden, H. M. (1971). Intergroup relations in rhesus monkeys (*Macaca mulatta*). *In* "Behavior and environment: The use of space by animals and men" (A. H. Esser, ed.), pp. 112–113. Plenum Press, New York.

Mason, W. A. (1966). Social organization of the South American monkey (*Callicebus moloch*) a preliminary report. *Tulane Stud. Zool.* **13**, 23–28.

Neville, M. K. (1969). Male leadership change in a free-ranging troop of Indian rhesus monkeys (*Macaca mulatta*). *Primates* **9**, 13–27.

Reynolds, V. (1963). An outline of the behavior and social organization of forest living chimpanzees. *Folia Primat.* **1**, 95–102.

Rose, R. M., Gordon, T. P., and Bernstein, I. S. (1972). Plasma testosterone levels in the male rhesus: Influences of sexual and social stimuli. *Science* **178**, 643–645.

Sade, D. S. (1968). Inhibition of son-mother mating among free-ranging rhesus monkeys. *Sci. Psychoänal.* **12**, 18–37.

Southwick, C. H. (1967). An experimental study of intragroup agonistic behavior in rhesus monkeys (*Macaca mulatta*). *Behaviour* **28**, 182–209.

Southwick, C. H., Beg, M. A., and Siddiqi, M. R. (1965). Rhesus monkeys in North India. *In* "Primate Behavior. Field Studies of Monkeys and Apes" (I. DeVore, ed.), pp. 111–159. Holt, New York.

Stoltz, L. P., and Saayman, G. S. (1970). Ecology and behavior of baboons in the Northern Transvaal. *Ann Transv. Mus.* **26**, 99–143.

Struhsaker, T. T. (1967). Behavior of vervet monkeys (*Cercopithecus aethiops*). *Calif. Publ. Zool.* **84**, 1–74.

Sugiyama, Y., Yoshiba, K., and Pathasarathy, M. D. (1965). Home range, mating season, male group and inter-group relations in hanaman langurs (*Presbytis entellus*). *Primates* **6**, 73–106.

Vessey, S. H. (1971). Free-ranging rhesus monkeys: Behavioral effects of removal, separation and reintroduction of group members. *Behavior* **40**, 216–227.

Wilson, A. P., and Boelkins, R. C. (1970). Evidence for seasonal variation in aggressive behaviour in *Macaca mulatta*. *Anim. Behav.* **18**, 719–724.

AGGRESSION IN NATURAL GROUPS OF PONGIDS

THOMAS K. PITCAIRN

Max-Planck-Institut für Verhaltensphysiologie

INTRODUCTION

There are many definitions of aggression in the animal behavior literature; it has been defined as the initiation of attack (Scott, 1958) and as behavior involved when an animal "inflicts, attempts to inflict or threatens to inflict damage on another animal [Carthy and Ebling, 1964, p. 1]." However it is defined, the behavior commonly known as aggression has two important concomitants: First, one animal may inflict damage on another, which under certain circumstances may be maladaptive, and second, it eventually increases the physical distance between two animals involved in agonistic conflict. These parameters are important for group-living animals such as most primates, for the possible maladaptive features become exacerbated within the group. Many workers have sought to utilize aggressive interactions as the basis for measurement of dominance orders. Their success has varied, especially when concerned with the pongids, chimpanzee, gorilla, and orangutan. The field data for these animals are rather sparse, especially for the latter two, but it does seem that they show comparatively little direct aggression between members of the same group. For example, Schaller (1963) used displacements as his measure of dominance in the mountain gorilla (*Pan gorilla berengei*) but the rate of this activity was only 0.23 episode per hour of troop observation. A similar situation occurs in the gibbon and howler monkey troops studied by Carpenter (1940, 1965). However, he concludes that dominance and aggressive relations influence the probability of all other behaviors (Carpenter, 1954) and thus have an extensive if subtle effect on the social structure.

THE PONGIDAE

The family Pongidae, or great apes, consists of three living types: the chimpanzee with one or two species, *Pan troglodytes,* and perhaps as a separate species, the pygmy chimpanzee *Pan panicus;* the gorilla, *Pan* (= *Gorilla*) *gorilla* with two important subspecies, *P.g. berengei,* the mountain gorilla, and *P.g. gorilla,* the lowland gorilla; and the orangutan, *Pongo pygmaeus.* All pongids are to some extent brachiators in structure (Napier, 1963), even though the gorillas seen in the wild rarely climb trees and never brachiate, presumably because of their large bulk, and all build tree-type nests at night. The African pongids (gorilla and chimpanzee) are black in coloration, contrasting with the bright reddish orange of orangutans, which come only from the islands of Sumatra and Borneo. These are relics of a wider Pleistocene distribution, and even now are greatly threatened by programs of forest clearance.

Both the orangutan and chimpanzee are essentially fructiverous, but the latter has a much more varied diet that includes some animal protein. Gorillas exist on the much more bulky vegetable foods that have a high proportion of roughage, such as bamboos, vines, and foliage.

REPERTOIRE OF AGGRESSIVE BEHAVIOR

The great apes, in common with other animals including man, show aggressiveness by means of facial expressions, bodily posture, and vocalizations. These patterns of behavior are species specific, and have been observed and recorded to some extent in all three species in the wild. The observations, however, are far from complete, especially for the orangutan and to a lesser extent the gorilla, and some reliance will have to be placed on observations made in captivity.

One difficulty that arises in discussing the repertoire of behavior is in the distinction made between aggressive acts and threat displays. This distinction is a difficult one to achieve theoretically, though it is usually comparatively easy in the observation situation. Eibl-Eibesfeldt (1970, p. 79) states that in fish, display constitutes "the lowest intensity of fighting behavior" and Walther (1968) distinguishes display from threat on the basis of the absence of a direct willingness to fight. Certainly many displays occur in conflict situations and can be derived from many different behavior patterns. Display components are of great importance in the pongids; Sugiyama (1969) distinguishes five different displays commonly seen in the chimpanzee, and Schaller (1963) two in the mountain gorilla. Van Lawick-Goodall (1965b) also suggests that a modification to Mike's (a low-ranking adult male chimpanzee) normal display repertoire enabled him to become dominant in rank. This chimpanzee was seen to use empty kerosene cans in a hooting display, throwing or dragging these along the ground, making a very loud noise. The correlation between his rise in rank and the use of these objects is suppositional as the first use was not seen. Thus the modification of a display behavior affected the essentially agonistic interactions of the dominance relationships.

It is always difficult to separate behaviors into discrete groups, and this has been shown quite clearly by van Hooff (1971), who, working with the captive chimpanzee group at Holloman Air Force Base in New Mexico, used component and cluster analysis of the transition frequency between the various elements in the chimpanzee's repertoire to reveal the integrated structure lying beneath. He included some display and aggressive elements in his "aggressive and bluff" system. These two subsystems had a level of concordance of 0.6 (McQuitty, 1966), which means they were quite closely related, each forming a discrete subsystem on further component analysis. Interestingly, the display elements were also related to the play system. A distinction was drawn, however, between elements of display in the aggressive series, and a group of behaviors he calls "excitatory." These further display elements were much less closely associated with aggression on the cluster analysis, but did relate to the aggression and bluff system in the component analysis.

Reynolds and Luscombe (1969), working with the same captive colony, showed that display elements were correlated with high dominance status and access to preferred foods, while the initiation of aggressive acts was not. This does not mean, of course, that there is a direct causal relationship between one and the other, but provides evidence for the close relationship between display aggression and status within the group. Chance (1967) suggests that the relationship may lie in the amount of attention paid by members of the colony to each other. He states "discrete coherence of the cohort males may depend on their maintaining a predominant degree of attention directed towards themselves [p. 508]." Thus the displays serve to focus the attention, and complete displays are most often followed by associative or submissive acts performed toward the displayer by other members of the troop (Reynolds and Reynolds, 1965; Reynolds and Luscombe, 1969; van Lawick-Goodall, 1967a).

Display components are also found in the repertoire of gorilla and orangutan. They seem to perform a function similar to those of the chimpanzee, but less information is available to verify the hypothesis. Riess *et al.* (1949) note an increase in chest-beating displays in captive lowland gorillas when animals are together, and greater occurrence in the male than the female in their two animal groups. Schaller (1963) in his extensive study of the wild mountain gorilla states that the display is a ritualized behavior seen in conflict situations, where the conflict is between aggression and flight. The situations commonly giving rise to displays were the approach of human beings, the presence of another group or lone male, a response to general disturbance, a response to other displays in the group, and in play. The response to others in the group is particularly predominant when the displayer is the dominant silver-backed male. He also states that "group members become alert if the dominant male beats his chest without obvious reason [Schaller, 1963, p. 233]."

The orangutan similarly displays in response to disturbance. Various features of the display, such as inflation of the laryngeal pouch, hair erection, branch shaking, and rocking indicate that during a vocal display (the long-call) male orangutans are "angry" (Mackinnon, 1971), and they also display during disturbance by man (Mackinnon, 1971; Harrisson, 1960; Schaller, 1961). In his study of captive orang groups, Jantschke (1972) distinguishes threat and display behavior in the direct intention to hit and bite, while many aggressive elements are included in the display.

The Chimpanzee

The chimpanzee is the most fully observed member of the pongids both in captivity and in the wild. Observations have been made by Nissen (1931), Yerkes (1943), and many earlier workers such as Garner (1896). More recent work has been published by Kortlandt (1962, 1967), Kortlandt and Kooij (1963), Reynolds and Reynolds (1965), van Lawick-Goodall (1965a, 1967, 1968a,b), and Sugiyama (1969). The behavioral repertoire has been listed in detail by some five authors, and Table 1 represents an attempt to correlate these listings. All the studies named have been on wild groups, with the exception of that of van Hooff (1967, 1971).

A cautionary note must be sounded here, for it has proved difficult to compare all the behavior categories listed by the different authors. A repertoire list depends on many factors such as length of time spent analyzing the behaviors and the primary desires of the investigator, who will "lump" and "split" the various elements accordingly. Thus, while van Hooff (1967, 1971) lists a total of 18 aggressive and display elements, Sugiyama (1969) has 4 aggressive and 5 display, and Nishida (1970) 10 agonistic interactions which include threat displays. Some authors also list behaviors in different categories, such as display or agonism and do not necessarily give a full description. The table, therefore, is only an approximation to be considered within the levels of operation of each author.

All authors who have observed chimpanzees in the wild state that the level of actual aggressive interactions involving body contact is low, and that concomitantly the level of display (threat display, agonistic display) is comparatively high. Most group-living animals, of course, settle their disputes with some form of ritualized aggression. Fear is quickly learned and difficult to extinguish (Scott, 1958) and this promotes stability in the conspecific intragroup interactions. In the chimpanzee (and in all great apes for that matter) this control mechanism seems to be highly developed, and the amount of display correspondingly high. Such display elements can be seen in locomotory behaviors, such as van Hooff's smooth approach (1) and the various bipedal gaits recorded by van Lawick-Goodall and van Hooff (2, 3, 4). Bipedal arm waving and running is a high-intensity threat behavior, often found with the waa-bark (26). Van Lawick-Goodall (1968a) states that this behavior also may be directed (in fact, most frequently) against baboons who approach too closely in the feeding area. The bipedal swagger, sway walk (2) is seen much more frequently in males than in females (van Lawick-Goodall 1968a). Head tipping (5), a sharp backward movement of the head with a "soft bark," has been recorded only by van Lawick-Goodall, and quadripedal hunch (6) by that author and Nishida (1970). Van Hooff considers it the lowest intensity level of the stamp trot (14). Various other displays (8–16) involve stamping on the ground, shaking the body or tree branches, waving sticks, and beating on the buttresses or trunks of trees. These latter displays may well have a large frustration or fear connotation also, as they are often seen in situations where one individual cannot attack another, such as adolescent males unable to obtain food because of mature males nearby (van Lawick-Goodall, 1968a), or on the approach or observation of a human being (Sugiyama, 1969). They may be accompanied by waa-barks (26) or screams. The highest intensity form of these displays is the stamp trot (van Hooff) (14), described by many authors from Garner (1896) and Köhler (1931) onward.

TABLE 1 Elements of Aggressive Threat and Display Behavior in Chimpanzees[a]

Reynolds and Reynolds (1965)	van Lawick-Goodall (1965, 1968a)	van Hooff (1965, 1971)	Sugiyama (1969)	Nishida (1970)	Others
(1) Relaxed and confident bearing and unhurried gait		Smooth approach	Confident gesture		
(2)	Bipedal swagger	Sway walk			
(3)	Bipedal arm waving and running	Sway walk			
(4)	Bipedal arm waving and running	Arm sway		Swinging one arm sideways up or down	
(5)	Head tipping (and soft bark)				
(6)	Quadripedal hunch			Standing on all fours with head down, often one leg raised	
(7)	Hitting away	Upsway		Swinging one arm sideways up or down	
(8) Stamping on ground	Stamping	Stamp	Light attack	Stamp	
(9) Slap ground			Light attack	Slap ground	
(10) Shaking of body			Light attack	Shaking the body	
(11) Hitting tree trunks with hand			Light attack	Drumming on tree trunks	
(12) Branch shaking	Shaking		Branch shake		Kortlandt (1962)

(13)	Buttress beating	Drumming		Buttress beating	Drumming	Garner (1896) drumming
(14)	Display	Charging display	Stamp trot	Chase and whooping display	Explosive display	Köhler (1921) description Hamburg (1971) aggressive display
(15)	Branch break and throw	Aimed throwing		Branch throw	Swinging or throwing sticks	Kortlandt (1962) stick throwing
(16)				Exaggerated eating		
(17)	Attacks	Attacking charge	Brusque rush	Grab and bite	Chase	Hamburg (1971) attack
(18)	Hitting	Slapping	Hitting	Grab and bite	Hit	
(19)		Hit with object	Hitting			
(20)		Biting	Biting	Grab and bite		
(21)	Scuffles	Lift	Tugging	Grab and bite	Seize	
(22)	Scuffles	Hair pull	Tugging	Grab and bite	Seize	
(23)		Stamping on the back	Trample		Kicking or stamping on object	Wilson and Wilson (1968) kick and stamp
(24)	Scowl	Glaring	Tense-mouthed face		Glaring	
(25)	Open mouth	Glaring	Open-mouthed stare		Glaring (with open mouth)	
(26)	Harsh and shrill barks, waa-bark	Waa-bark	Shrill bark	Bark	Aggressive vocalization	Marler (1969) waa-bark
(27)	Grunt, gruff-bark, panting bark	Grunt, groan, huu-calls	Grunt-bark		Aggressive vocalization	
(28)	Panting, hoots	Hooting	Rising hoot			Marler (1969) hooting

[a] Recorded by various authors.

This spectacular display takes many forms and "it sometimes occurs that an adult male chimpanzee goes wild without any apparent reason [Nishida, 1970, p. 61]." This involves stamping and running, arm waving, stick throwing and beating with arms and legs on tree trunks and buttresses. Various vocalizations can be heard, hooting (28) and screaming being predominant. This behavior may be infectious and Reynolds and Reynolds (1965) describe "carnivals" (from Garner, 1896) in which many animals take part, with drumming and calling, carrying on for many hours. This has been described by Nissen (1931) and by Sugiyama (1969) as "booming." Some confusion arises over these displays, for Sugiyama lists both chase, involving stamping, branch swinging, and bark, and whooping display, involving excited running and brachiating with violent high-pitched cries. Van Lawick-Goodall lists a general description of the display under "frustration" and points out that the highly aggressive charging displays always involve branch shaking. It seems that this display can be found in contexts of high excitement and of high levels of aggression; while the motor patterns may be slightly different, the main differentiation lies in the context of the behavior. In describing the carnivals, Reynolds and Reynolds state that these often seem to occur on the gathering of males at a fruit tree, and at the conflux of two different groups, but they were not able to observe these closely enough (because of the extreme rapidity of movement) to see whether patterns of attack are also seen.

Two other displays should also be mentioned. Aimed throwing (15), in which a stick or clod of earth (Kortlandt and Kooij, 1963) is taken up and thrown directly at another chimpanzee, has been recorded by many authors in the contexts of aggressive throwing at conspecifics, at baboons, leopards, and at man. This use of tools will be more fully discussed later. Exaggerated eating (16), as a conflict display, has also been reported by Sugiyama (1969).

Patterns of attack involving body contact (17–23) are also seen in which an animal rushes quickly at another (17) and slaps, hits, bites, pulls hair, and tramples it. This is accompanied often by an open-mouthed stare (25, van Hooff, 1967) and often preceded by a tense-mouthed face (24, van Hooff, 1967). Both the waa-bark (26) and the various grunts (27) are heard. These violent attacks constituted only 10% of the 284 attacks recorded by van Lawick-Goodall (1968a), and Sugiyama (1969) saw only 21 cases of "severe attack" in 6 months' observation and no cases of biting, even though he did see wounded animals. Similarly, Reynolds and Reynolds (loc. cit.) saw only 17 "quarrels" in 300 hr of observation and these are described as being short scuffles lasting only a few seconds.

The Gorilla

The major work done on the gorilla to date is that of Schaller (1963) and Schaller and Emlen (1962) who studied the mountain gorilla in the Virunga Volcano region of Albert National Park in the Congo. The gorillas in this region have more recently been studied by Fossey (1970, 1971), but no full report has yet been published. The ecology of the mountain gorilla has also been studied by Baumgartel (1958), Imanishi (1958), Kawai and Mizuhara (1959), and Itani (1961).

Yerkes and Yerkes (1929) state that "dispositional contrast between chimpanzee and gorilla diminishes with age [p. 456]" and many facial expressions resemble those of chimpanzees (Schaller, 1963). Schaller states that the most useful features of the gorilla's face for indicating emotion were the eyes, lips, and mouth. He found the expression and movements of the eyes particularly useful in the prediction of the animal's actions. Arousal of interest was shown by fixation of the eyes, while the "soft hue" remained. With this the lips were often parted and the head cocked to one side (cf. also in humans—Pitcairn and Grant, 1972). With the onset of mild disturbance (arrogance), the eyes would become hard and fixed, the lips pursed and slightly parted, and the head often tipped downward.

With an increase of "anger" in the individual, the eyes remained hard and fixed, but the mouth opened further and normally the gums and teeth were shown and the lips curled back. These latter two expressions described by Schaller (1963, 1965) have been termed, respectively, the staring open-mouthed face and the staring bared-teeth scream face by van Hooff (1967); both are also found in the chimpanzee. Van Hooff states, however, that normally the latter has a large flight as well as aggressive component in it, though this may be untrue for the great apes, as certainly the staring bared-teeth scream face in chimpanzees can be given by dominant individuals as often as subordinates.

Schaller (1963, 1965) lists four (split into five, 1965) main aggressive behaviors:

1. Stare (contained in the previously mentioned two facial expressions). This is a fixed, unwavering stare at another individual, sometimes with a furrowed brow. When found in an intragroup context (the most frequent one), the stares are of a shorter duration than in an intergroup confrontation. These are generally accompanied by short grunts.
2. Jerk of head, or snap. This is a jerk or thrust of head toward the antagonist, occasionally ending with an audible snap from the closing jaws.

3. Forward lunge of body. This is interpreted by Schaller as an incipient charge, and usually consists of an abrupt movement of two or three steps toward the opponent.
4. Bluff charges. The literature on this behavior directed toward man is considerable, from Du Chaillu (1861) onward. This is the charging display which has enabled many people to label the gorilla as very dangerous and aggressive, the opposite of the truth (see Fossey, 1971). Schaller (1963, 1965) reported that this behavior was rarely observed; it could take place at any distance from 10 to about 80 ft, and consisted of the animal's charging the opponent (usually man) on all fours, either silently or with roaring.
5. Physical contact. This consists of grabbing and grappling, with biting and screaming, and occurred rarely in Schaller's observations.

Two other displays have also been seen regularly, particularly the chest beating, mentioned by Yerkes (1929), Bingham (1932), Yerkes and Donisthorpe (1958), Kawai and Mizuhara (1959), Schaller (1963, 1965), and Fossey (1970). Schaller (1963, 1965) has split the chest-beating sequence into nine parts: hooting; "symbolic feeding" (plucking a leaf and eating); rising (bipedal stance); throwing vegetation into the air; chest beating with rapid, alternate, open-handed slapping; leg kicking (one leg into air); running sideways; slapping and tearing at vegetation; and ground thumping. This may take as long as 30 sec to complete, and it can be seen that many components of the chimpanzee explosive display are present.

The second display is called the strutting walk by Schaller (Fig. 1). In this the animal chooses an open area and walks quadripedally very stiffly. The arms are bent outward at the elbow, the body held very stiff and erect, and only brief glances made to the opponent. The steps are unusually short. This has also been observed in captive mountain gorillas (Schaller, 1963). The author has observed it as well in the captive lowland gorillas in the Basel Zoo, sometimes combined with a "lip-in" expression in which the lips are firmly compressed and drawn into the mouth (a similar expression occurs in humans—Grant, 1969). The strutting walk (without lips in) was often followed by a running sequence, in which the animal may hit the wall or another gorilla with its shoulder. Schaller describes this as "often seemingly playful" (1963, p. 291), and found it to occur more commonly in zoo animals.

Many of these behaviors have been described also for the lowland gorilla, but usually in zoo conditions (e.g., see Carpenter, 1937 and Riess *et al.*, 1949). However, Imanishi (1958) describes the bluff charge in a wild adult male (silver-backed) lowland gorilla, in which "again

Fig. 1 On right, silver-back male, lowland gorilla Stephi at Basel Zoo, in stance preliminary to strutting walk.

and again he made a mock charge toward us, with a furious cry, then he disappeared [1958, p. 76]." The vocalizations of the gorilla in aggressive situations consist of:

1. Harsh grunt. Schaller describes this as a series of two to five short grunts in succession, and Fossey (1971) probably refers to this call as a pig grunt.
2. Bark. A loud, often high-pitched sound, possibly the "Hwat, wat" of Kawai and Mizuhara (1959).
3. Roar. A loud and very intense roar made only by silver-backed and large black-backed males (Schaller, 1963). It is two- or three-toned.
4. Harsh scream. The small black-backed males and females give this loud call. It is short and harsh, and usually given in a series. This is perhaps the "kien" of Kawai and Mizuhara (1959).

In addition to these calls, Schaller lists four calls given by displaying animals, of which hooting is the most common, given mainly by silver-backed males during chest beating. This occurs in a series which begins slowly and then speeds up so that the last vocalizations form a nearly continuous growl. Females, however, give a series of "panting ho-ho" calls when chest beating. Fossey (1972) has published a list of vocalizations which agrees largely with that of Schaller, except that she states he has subdivided some vocalizations, the "roar" and "wraagh," for example, into four distinct calls and the pig grunt into two calls. Fossey also notes that the silverback male was found to call more frequently and had a larger repertoire than any other age/sex class.

The Orangutan

The orang, found only in the islands of Borneo and Sumatra in Indonesia, is the pongid in greatest danger of extermination. The large adult males (Fig. 2) look very spectacular with their orange to brown fur, large dewlaps, and highly developed side pieces on the head. They are greatly endangered by logging activities that reduce the area of forest open to them (de Silva, 1971; Davenport, 1967).

This least known of the great apes has been studied in the wild by Carpenter (1938), Harrisson (1960, 1962, 1963), Schaller (1961), Davenport (1967), de Silva (1971), and Mackinnon (1971). In addition, Rodman has worked with these animals, but as yet no material has been published, and Jantschke (1972) has worked extensively with captive orangutans in Frankfurt and Nürnberg zoos. Unfortunately, few of these

Fig. 2 Adult male orangutan Niko at Basel Zoo.

publications (except Schaller, Davenport, Mackinnon, and Jantschke) contain detailed behavioral observations, as the animals appear to be the least social of all pongids, and occur at an estimated density of between one per 1.5 square miles (Yoshiba, 1964) to two to three per square mile (Mackinnon, 1971).

The aggressive behaviors shown seem in many ways to be similar to those of other pongids. Aggression is rare, even in the crowded conditions of the rehabilitation area at Sepilok (Mackinnon, 1971). However, Jantschke (1972) does state that it is impossible for two adult males to be kept together in zoo conditions. His description of fighting behavior is by far the most detailed, and so this will be taken as the basic description, even though it was not observed in naturally occurring groups.

1. Staring. This direct and steady look, as in all pongids, is the first noticeable threat behavior.
2. Hitting away. This is a quick raising of the hand or beat in the air in the direction of the opponent. Mackinnon describes this behavior in a young animal at Sepilok, "brushing away" an approaching cat. He states that it is very similar to the "arm raise" and "soft bark" threat of chimpanzees.
3. Attacking. Directed run with look at and hair erect. The movements are exaggerated and jerky.
4. Grappling and biting. This has rarely been reported in the wild, except in the semicaptive conditions at Sepilok. Brandes (1939) reported extensive wounds, present on older captured males, indicating that such behavior does occur.

Threat displays have also been reported. Mackinnon (1971) lists eight vocalizations commonly heard during threat and warning displays.

1. Kiss squeak (smacking, Davenport, 1967). A sharp intake of air through pushed-out lips.
2. Raspberry (from Harrisson, 1963). In this the air is blown out, with the lips in the same position as above.
3. Chomping. Rapid closing of mouth to produce loud chewing sound.
4. "Lork" noises. These are described as "loud repetitive noises . . . made by animals frustrated by my persistent observation [Mackinnon, 1971, p. 173]."
5. "Ahoor" calls. Long, loud exhalations.
6. Barks. Single, sharp exhalations.
7. "Grumphs" (two-toned sounds, Davenport, 1967). Deep throaty grunts, the head often being turned from side to side. These sounds may be intermingled with or followed by, kiss squeaks.

8. Complex calls. Strings of deep guttural noises, repetitive and sometimes combinations of the above sounds.

One very common threat behavior listed by nearly all authors is that of stick dropping or throwing. Often occurring as a response to human presence, adult animals will break off small twigs and drop them onto observers below. De Silva (1971) noted that the throwing gesture was invariably directed backward with an underarm motion; he once saw (or was subjected to) defecation and urination from a female orang who first maneuvered herself into a suitable position. Kiss squeaks, barks, and grumphs often accompany the threat (Davenport, 1967; Mackinnon, 1971). Schaller and Davenport, who both describe the throwing, suggest that the intensity of the display depends on the proximity of the observer in the horizontal rather than vertical plane, though Mackinnon mentions animals climbing up trees away from the observer, rather than moving from tree to tree. Harrisson (1962) mentions branch shaking in an adolescent she observed. It gave a deep, short bellow on noticing the observer, grunted, and standing on all fours, branch shook, threw a twig, and then moved off.

Adult male orangs will sometimes also approach human observers on the forest floor with a threat display (de Silva, 1971).

The display observed in zoo animals by Jantschke (1972) differs considerably from that observed in the wild. Throwing is rare, for there are few projectiles available, although throwing of feces has been seen. In the wild the twigs are not picked from the ground, but from branches and are dropped rather than thrown, whereas in the zoo, the orangs frequently used objects or part of the cage to beat loudly and rhythmically. This was often done while hanging sideways or with the head down, and hair was sometimes raised.

AGGRESSION IN THE CONTEXT OF GROUP STRUCTURE

Within the Group

In the foregoing paragraphs, aggressive and threat behavior has been considered as a series of patterns which individual animals can display. We must now turn our attention to the relevance of these patterns in the social life of the animal.

RANK ORDER AND GROUP STRUCTURE

Most, if not all, primates in some part of their life cycle form social groups, and within these groups much of the behavior to be observed is agonistic. It has been stated above that, at times, aggressive behavior can be maladaptive in that severe damage to individuals, particularly females and young, limits the breeding success of a population. The importance of limiting aggressive behavior within some form of controlling framework has been stressed by most workers in the field (Rioch, 1967), and one mechanism observed in many primates is that of rank orders, or status hierarchies (for an extensive review of this phenomenon in primates, see Bernstein, 1970). Many primate and especially pongid groups, however, seem to be extremely flexible in organization—what has been called open and adaptive (Crook, 1970).

Reynolds (1966), particularly, has stressed the importance of open groups in relation to human evolution, backed up by the results of his field work on chimpanzees (Reynolds and Reynolds, 1965). He observed groups of chimpanzees in the Budongo forest region of western Uganda. The groups observed were not closed social groups, but were "constantly changing membership, splitting apart, meeting other groups and joining them, congregating or dispersing [Reynolds and Reynolds, 1965, p. 396]." They referred to these parties of individuals as "bands," and Reynolds (1968) refers to the population of one area as the "local regional population." Four different types of bands were observed:

1. Adult bands, with adults of both sexes but no mothers with dependent young.
2. Male bands with only adult males.
3. Mother bands, with mothers and infants and sometimes other females.
4. Mixed bands.

The adult and mixed bands were most frequently composed of large (more than seven individuals) bands with about equal frequency of all in small parties (six or less). Thus it would appear that the adult and mixed bands form loose aggregations from different bands fairly often. Van Lawick-Goodall (1965a, 1968a) also refers to the temporary nature of the association of chimpanzees in a "group" in the deciduous woodland area of the Gombe stream reserve in Tanzania. Itani (1966) pointed out that chimpanzee groups "do not always move as a unified whole [Izawa, 1970, p. 1]," and Itani and Suzuki (1967) state that changeability in the social grouping occurs frequently but that there is a basic large group found in any area, which is more or less closed (cf. Reynolds local regional population). Sugiyama (1967, 1968, 1969) worked with the Budungo forest population and found virtually exactly the same situation as had been described by the Reynolds. Nishida (1967), however, working with the population of savannah chimpanzees in western Tanzania reported that the population is much more closed and organized. The population density of the region seems to be much lower than that of the other areas investigated, and it is the membership of the regional population that is stable.

The implication of this work is that a strict dominance hierarchy cannot exist because of the fluctuations within the population. Eibl-Eibesfeldt (1970) maintains that the rank order is established by a series of fights, and maintained by ritualized aggression. Van Lawick-Goodall (1968a) states that the "temporary nature of the chimpanzee groups results in an *apparently* loose social structure [p. 212]." The regular observations of individuals in an area made it possible to determine dominance positions with a fair degree of accuracy. Nevertheless, it remains possible that the artificial provisioning of her chimpanzees resulted in a tightening of the social structure to form more closely defined social groups. It is interesting to note in this context that in the case of the two authors who support the idea of a more rigid social structure, van Lawick-Goodall (1965a, 1968a) and Nishida (1967, 1968), both artificially provisioned the chimpanzees in their study areas.

Certainly, in the case of the gorilla, the group is much more defined and closed. Schaller (1963) reports that the group size varies from 2 to 30 animals, with all groups having at least one silver-backed male, which is a mature male over about 12 years of age (Kawai and Mizuhara, 1959; Schaller, 1963) having a gray or silver-colored coat on the back. The groups studied were fairly stable throughout the observation period, most members being added to or lost by the processes of birth or death. Most groups had more females than males, a higher proportion of silver-backed to black-backed males, fewer juveniles than infants. Blower (1956) states that the gorilla group is a family, and Kawai and Mizuhara (1959) that it is one polygamous family. Fossey (1971) has observed a "bachelor band" of one silver-backed and four black-backed males; there was, however, an old female with this group at one time who later died. That these individuals then continued their bachelor existence as a group shows to some extent their closed nature.

Dominance interactions have been noted within the group, mainly taking the form of slight touches from one animal to another, leading to physical displacement, such as on a narrow trail, or at a sitting place (Schaller, 1963). Silver-backed males were dominant over all group members, and if more than one silver-backed male was present, the dominant animal was easily recognizable as the focal point for all females and young. However, dominance interactions were recorded by Schaller (1963) only 0.23 time per hour of observation. In chimpanzees,

Reynolds and Reynolds (1965) saw only 25 dominance interactions in over 300 hr of observation, an even lower frequency.

Dominance relationships in wild orangutans are virtually unknown, as these animals are the most solitary of all the great apes. Mackinnon (1971) saw 143 subgroups, the average size of which was 1.83; 84% of these contained only two animals, usually a mother with dependent young. Jantschke (1972), however, has observed dominance relations in groups of subadult orangs in captivity. The organization here was unclear, but he states that direct interaction was much rarer than the avoidance of subordinates on the approach of a dominant to a play object. These relationships are obviously promoted to a large extent by the confined zoo conditions.

Thus, dominance interactions would seem to be neither frequent nor obvious in the Pongidae. Albrecht and Dunnett (1971) could not find a rank order among the chimpanzees they studied in Guinea. However, van Lawick-Goodall (1968a) does state that they exist in a complex fashion in the Gombe stream chimpanzees, and furthermore that these are influenced by kinship and dependent rank (Kawai, 1965), such that a high-ranking mother will raise the apparent rank of a juvenile. Personal relationships also play a part, as when a high-ranking male will aid a lower-ranking "friend." However, these dominance relationships are not necessarily based directly on aggressive behavior as Reynolds and Luscombe (1969) have shown in the captive colony of Holloman Air Force Base chimpanzees. The animals which initiate most aggressive interactions are not the ones at the top of the rank order, as measured by access to preferred foods.

What, then, is the nature of aggressive behavior? Very few violent aggressive incidents were observed by any author. Van Lawick-Goodall (1968b) relates some of those which did occur to the high density around the artificial feeding site, and Sugiyama (1969) reports a seasonality in severe attacks, which occur more frequently in the dry season between December and February when the chimpanzees are more active, moving from place to place. Van Lawick-Goodall states that mature males are responsible for the initiation of many more attacks than either adolescent males or females, and mature females were the object of attack more often than any other age–sex class. This could be due to the prevalence of redirected aggression among group members. Females, being generally lower-ranking than mature males, were frequently the object of such attacks, particularly by lower-ranking males during charging displays, an exclusively male behavior. Mature males also attacked subordinates approaching with submissive gestures, or who were screaming. Van Lawick-Goodall also describes aggression based on kinship or personal relationships in which one animal came to the aid of another when it was being attacked, or when dominant males attacked individuals already fighting. True dominance fights were also observed in males and females and between individuals who were close in rank. These fights, however, frequently took the form of extended threat displays such as bipedal swaggering, branch shaking, and ground slapping. Nishida (1970) reported a similar pattern of interactions in his group of chimpanzees.

In the gorilla, interactions of any sort are less frequent. The two most common behaviors were rank-order agonistic behavior and grooming. In 110 dominance interactions recorded by Schaller (1963), 51 were by silver-backed males, half of which were directed to females, and another 25% to other silver-backed males.

Interactions of any kind were rare between silver-backed and black-backed males. Schaller reports a tendency for males low in the rank order, either black- or silver-backed, to assume peripheral positions, and suggests that the spatial organization to some extent resembles the centric or centripetal one of many macaques (Imanishi, 1957; Chance and Jolly, 1970). He also states that one reason why the level of aggression is so low is that the subordinate individuals are usually aware of the behavior of the dominant, and take care to "circumvent the issues before they materialize [Schaller, 1963, p. 345]."

In orangs, too, avoidance is often the means by which possible aggressive encounters are prevented. On hearing the approach of another male, for example, the animal may climb higher into his tree, or simply remain hidden. At other times passersby are completely ignored (Davenport, 1967). The adult males also have a "long-call," which signals the fact that they are in a disagreeable mood, and so best left alone (Mackinnon, 1971). No general pattern of aggressive behavior has yet emerged from the orang studies.

The relationship between aggression and dominance behavior is unclear. Dominance relationships certainly do exist in the pongids, but a strict dominance hierarchy, or rank order, perhaps does not. As Nishida (1970) says, it is impossible in these animals to use the simple "peanut" or "sweet potato" test of Itani (1954) to decide dominance between two chimpanzees; equally, purely agonistic relationships do not fit the pattern either; for example, a high-ranking male may run off or "grimace" (a submissive gesture) when a lower ranking male shows an explosive display (Nishida, 1970). In gorillas the direction of aggression seems much more often to follow that of the putative dominance order, but the level of such interactions is very low indeed. The attentiveness of subordinates to dominants might suggest that Chance's (1967) concept of attention structure would be of greater relevance here.

SEXUAL BEHAVIOR

Aggression within the courtship display, in a ritualized form, is a feature of many animal species, and the great apes are no exception to this. Van Lawick-Goodall (1968a) describes four acts which occur also in aggressive context as being typical of the male courtship display in chimpanzees. These are:

1. The sitting hunch
2. The bipedal swagger
3. Branching
4. Glaring (this rarely)

These are described in the preceding section on the repertoire of aggressive behavior—the chimpanzee. Females responded mostly by "presenting" for copulation, either remaining where they are or running to the male. On some occasions the female retreated from the male, screaming, and then the male either chased until the female presented, or branch shook (a threat display) and then gave up. On some occasions a female was "forced" into a form of consort relationship with a male when she was in estrus, even though chimpanzees are normally completely promiscuous. The male behavior, which was the same in all cases, consisted of walking a few steps, turning, and looking at the female and branch shaking if she was not following. If she still did not follow, he chased her, and this occasionally led to an attack (van Lawick-Goodall, 1968a). This behavior seems similar to that observed in the hamadryas baboon (Kummer and Kurt, 1963), but seems to serve little function in the chimpanzees, as they do not form the hamadryas type of one-male group.

Copulations usually were allowed to proceed undisturbed, especially if between a mature male and female. On occasion, in copulation adolescent males (and on one occasion a high-ranking male) were charged or threatened; in the former case the adolescents dismounted. It is thus the males which are the object of attack (van Lawick-Goodall, 1968a). Sugiyama (1969), however, states that it is the female, and not the male (who adopts a "defensive position") who is attacked.

Observations of copulation in free-living gorillas are very few. Schaller (1963) saw it only on two separate occasions during his study. In both cases the male involved was a subordinate silver-back, and the courtship sequences were not seen. In the first case, the dominant male did not interrupt the copulation. In the second, the dominant twice approached the couple, upon which the male desisted and retreated. Mating activities were shortly resumed after the dominant had moved away the first time, and the second approach was after ejaculation by the copulating male. In captive groups, however, the male often first approaches with the strutting walk display, with head averted. Wrestling, which Schaller describes as ritualistic, but would seem to have a high agonistic content from personal observation at the Basel Zoo, also plays a prominent part, and finally, running or chasing around the cage has also been described by Schaller. Interruption, in the form of a running charge with slapping at or bumping into, can be carried out by subordinate males even when the silver-backed male is copulating (personal observation).

Aggression in a sexual context in orangutans has also been observed in the wild. Mackinnon (1971) reported that out of eight observed copulations, seven were rapes. The females tried to escape but were pursued, caught, and sometimes struck and bitten. The females often screamed as did their dependent young, who also struck and bit the male. The male grasped the female by her thighs or waist with his feet, and continued to thrust pelvically whether the female was dorsal or ventral side up as she pulled herself along by her arms. Zoo observations suggest that these rapes are the exception rather than the rule. Mackinnon states that in captivity the pairs can get to know each other first, dispelling the fear in the female, thus suggesting that successfully mating pairs normally form a pair-bond first.

The curious form of aggressive mating behavior of the orangutan is somewhat different from the aggression shown in chimpanzees and gorillas, where it mainly takes the form of ritualized threat displays during the courtship sequence. Such behavior presumably relates to the quite different social structure of the orangutan, which is usually solitary and forms short-term pair-bonds—in the area studied by Mackinnon at least. So far, in these animals, the exact nature of the social structure remains obscure.

MOTHER–INFANT RELATIONSHIPS

Mother–infant relationships in all three great apes seem to be remarkably free from aggression, at least on the mother's side. Reynolds and Reynolds (1965) state that mother chimpanzees tolerate a good deal of bother from their young without signs of aggression, and Schaller (1963) talks of "mild rebuffs" from mother to young. This consists mainly of holding an infant's hand back, or brushing it away gently with the forearm. During the juvenile stage, the social break between the mother and young gorilla takes place without any great trauma; as the juvenile begins to seek the company of other troop members more and more, the rebuffs of the mother become unnecessary, and the two processes become complementary. Schaller states that he once saw a female "swatting at" a juvenile who was associated with her, and once saw one female snap at her juvenile. By the time they are 4½ to 5 years old, all ties seem to be broken.

Van Lawick-Goodall (1968a) talks of the tolerance of a mother for her infant, and says that mothers play a denying rather than aggressive role with the infants. The infants and young juveniles show occasional "temper tantrums," which grow less as they grow older. These consist of throwing themselves on the ground, screaming loudly, and writhing about, often hitting surrounding objects. This may be associated with weaning, and seems to have a large "frustration" component in it. This behavior can be controlled. Yerkes (1943) mentions the infant in a temper tantrum "glancing furtively" at the mother or caretaker.

Orangutan mothers also seem to be remarkably tolerant and show virtually no aggression toward infants (Mackinnon, 1971). One mother was observed grunting (two-toned sound) at her infant, who immediately returned to her. He later went off again and did not return even after repeated grunts by the mother, who then turned to move away, upon which the infant rushed back to her (Davenport, 1967). Concurrent threatening between mother and infant, in which the mother was always the more vigorous, has also been reported by Davenport.

One further aspect of mother–infant aggression must be mentioned here. Mother chimpanzees will threaten other individuals with whom their young are interacting (van Lawick-Goodall, 1968a). This means that youngsters with high-ranking mothers will obtain a "lift" in the hierarchy owing to their kinship, as has been reported for Japanese macaques (Kawai, 1965) and rhesus macaques (Sade, 1967).

Between Groups

The relationship between groups of chimpanzees seems obscure. The large-sized groups (Itani, 1966; Izawa, 1970) or (local) regional populations (Reynolds, 1966; Sugiyama, 1968) are split into smaller subunits, termed variously bands, small-sized groups, and parties by the various authors. Nishida (1967, 1968) as already discussed, considers these large groups to be closed and having dominant–subordinate relationships between them. Reynolds and Reynolds (1965), Sugiyama (1968), and Izawa (1970) report that the populations can intermingle with very little agonistic behavior. There is, however, a great deal of display behavior observable at their coming together. The Reynolds give a detailed description of what they term a "carnival," where two different (they assume) parties of chimpanzees gathered at one spot, usually a food source. Stamping and fast running, with very loud calling and buttress drumming, went on for several hours on the six occasions that this behavior was heard. Observation proved impossible because of the rapid movement of the chimpanzees.

"Normal" drumming bouts were heard more often, and seemed to be associated with both the coalescence and splitting up of groups. The bouts occurred when food was relatively abundant and the groups were not widely scattered, and the Reynolds suggest that it serves a communicatory function to other bands in the area, which may form part of the regional population. Branch shaking, ground slapping, and sapling breaking occurred in this context with the calls and drumming; in some contexts these behaviors seem to have an aggressive or threatening connotation.

Van Lawick-Goodall (1968a) also mentions the display elements on the coming together of two parties. Males stamped their feet, hooted, drummed on tree trunks, and broke and shook branches. Females and youngsters often ran out of the way, screaming loudly. She points out that the amount of daily activity in general also varied according to group size, the activity (and distance traveled) increasing with the number of individuals in a group. The display is almost identically described by Sugiyama (1968) who also mentions that the groups tended to split up on traveling and congregate on feeding.

The dominance–subordinance relationships of groups mentioned by Nishida (1967, 1969) seem to be based on threat displays and displacement of one group by another, rather than actual aggression. Izawa (1970) states that groups he observed in the Kasakati Basin in Tanzania, an area of savannah woodland, seemed deliberately to avoid one another. Even when the ranges overlapped, one group never moved into the main nomadic area of the other. The main difference between the areas studied is in population density and type of environment (forest or savannah woodland) and these obviously change the forms of interaction between groups, the high-population density forested areas having a looser social unit. Azuma and Toyoshima (1961–1962) also consider the variation in grouping to be a result of ecological factors.

The mountain gorillas studied by Schaller (1963, 1965) and Fossey (1970, 1971) definitely live in much more closed groups. Though they tend to stay within one large area, the nomadism is such that interactions between groups are not infrequent. No serious aggressive encounters were recorded, and on several occasions night nests were made fairly close together. One bluff charge by the dominant male from one group at the dominant of the other, preceded by chest beating, was recorded. They then stood staring at each other, nearly touching, until they parted and began to feed. Fifteen minutes later the charge was repeated twice, then the groups parted. One other encounter led to considerable chest beating, and a juvenile from the one group bluff-charged a juvenile from the other. A black-backed male later threatened a female of the other group, and was chased off. Thus the amount of intergroup aggression seen was low, and the response of a particular group to the approach of another varied from the incidents described above to no visible response at all.

The dominant males of the groups were always the most excited group members and determined the form and duration of contact (all from Schaller, 1963). Groups can and do influence each others' ranges of travel. One group of bachelors, after the death of their solitary female, began following more closely the neighboring groups (Fossey, 1971). One of these, with a new leader male (because the previous dominant had died) moved south away from the bachelors. The latter then concentrated their attention on the northern group, and their home ranges merged. This caused an increase in aggressive displays such as branch breaking, bluff charging, and chest beating, not only between males of opposing groups but also within each group. The two groups moved gradually northward together, while the southern group regained its old area.

As the basic group in orangutans seems to be the lone adult male or the female with dependent young, intergroup as opposed to interindividual aggression does not really exist. Suffice it to say that most interactions consist of avoidance, flight or ignoring, with the addition of the sexual and rape contacts between males and females.

AGGRESSION TOWARD OTHER SPECIES

Nonhuman Species

Aggression toward other animal species takes two main forms: that of predator defense, and that of aggression against competing species. A third type of behavior, predatory behavior, will also be discussed.

Most responses of pongids to nonpredatory species seem to be "accidental," i.e., a response due to the sudden sound or sight of other creatures. Orangutans have been known to branch shake at gibbons in the same tree, or vocalize at red-leaf monkeys or hornbills who were being noisy. Males were seen to give the long-call in response to calls of pig and deer (Mackinnon, 1971). Similarly, gorilla males have been recorded roaring at ravens and a hawk. The one exception to this accidental nature is the relationship of chimpanzees to baboons.

Chimpanzees and baboons live together in many parts of the chimpanzees' range. It has been suggested (Marler, 1965) that such sympatric species would benefit by responding to each others' alarm signals, and this they seem to do (van Lawick-Goodall, 1968a). Chimpanzees are always alerted by baboon alarm barks, and numerous reciprocal examples have been seen. However, being truly sympatric, they are also in competition for food. Thus some aggressive interaction has been seen, though they usually coexist peacefully. The frequency of aggressive incidents has increased since the setting up of the artificial feeding regime at Gombe stream. In 1963 and 1964, out of 198 aggressive encounters between the species, only 11 resulted in physical combat, with male baboons attacking and biting feeding chimpanzees, or when a female chimpanzee twice hit a baboon threatening her young. The other situations were solved either by the chimpanzees moving away, or by threat gestures on either or both sides. Stick and stone throwing has also increased; Köhler (1917, 1931) noted that in captivity, throwing behavior developed slowly after capture, and thus it may be that the increase in this form of aggression is due to a learning procedure rather than simply an increase of agonism between the two species.

Another interaction between chimpanzees and baboons is that of the predator–prey relationship. Chimpanzees have been observed actively hunting animal prey (Izawa and Itani, 1966; Kawabe, 1966; van Lawick-Goodall, 1965a, 1968a; Teleki, 1973) as well as eating insects (Reynolds and Reynolds, 1965; van Lawick-Goodall, 1965a, 1968a; Suzuki, 1966). Other primate species, such as the red-tailed monkey (*Cercopithecus ascanius*) (Kawabe, 1966) and red colobus (*Colobus badius*) (van Lawick-Goodall, 1965a) have been seen to be hunted as well as the baboon. However, it does seem that this behavior is *not* related to aggression, as would be true also for any other predator–prey relationship, for Teleki (1973) states that predatory episodes are unknown when the chimpanzees are interacting among themselves, or acting with other species, aggressively. This must be pointed out also for the recent evidence on cannibalism in this species (Bygott, 1972). In one example, a group of males suddenly attacked, very violently, a strange female who was probably carrying a baby. They chased her away and some minutes later one of the males was observed to be carrying a 6-month-old infant chimpanzee which he then proceeded to kill by hitting against a stone. This corpse was later partly eaten by members of the group. It is not unknown for sudden attacks like this to occur; also, van Lawick-Goodall (1968a) notes that young infants may be picked up and used as branches or other objects in part of the explosive display. This killing is not aggression directed toward the young, but merely making use of a convenient display object nearby. This meat-eating behavior has not been reported for the gorilla or the orang.

The other interspecific form of aggressive behavior is that directed at predators. For the orangutan, the only predator of any importance is man, and this will be discussed in the next section. Both the chimpanzee and the gorilla were reported to be predated by leopards, however. For the gorilla, this has been noted by Burbridge (1928), Geddes (1955), Zahl (1960), Dart (1961), and Baumgartel (1961). Tobias (1961) [quoted from Schaller (1963)], reported on the killing of two adult gorillas by a leopard. No evidence of aggression by gorillas toward leopards exists, however, and in fact Merfield and Miller (1956) relate an incident they saw in which a leopard jumped to the ground in full view of a group of gorillas, which did not respond in any observable way.

The chimpanzee is somewhat different, there being a fully documented series of reports by Kortlandt and his colleagues on its aggressive relationship to leopards (Kortlandt and Kooij, 1963; Kortlandt and van Zon, 1969; van Zon and van Ortshoven, 1967; Albrecht and Dunnett, 1971). Kortlandt *et al.* have experimented in the wild by presenting stuffed leopards to groups of chimpanzees in both savannah and forest environments. The most common reaction was loud vocalizations and a scattering of individuals on exposure of the dummy. The individuals grouped, performed "reassurance" behavior (touching, pseudocopulation, embracing), and screamed. The "reassurance frenzy" lasted for some time, then individuals began to approach the leopard, and throw objects. Quadripedal charges, branch shaking, and aimed throw all took place, and then bipedal charge with an object. Some considerable variation is seen between the savannah and forest populations, with the savannah animals showing much more stick use and club making, more cooperative attacks and vocal support, and a longer duration and greater frequency of bipedalism. However, no actual attacks on real leopards have yet been seen, although alarm calls caused by a leopard have been heard by van Lawick-Goodall (1968a) and Izawa and Itani (1966). In neither of these cases was any notice taken by other members of the group. Nishida (1968) has seen an adult male chimpanzee in a tree suddenly start branch throwing and shaking, then climb down, whereupon a leopard some distance away turned into the undergrowth. Thus, the actual reaction of the chimpanzee to a leopard differs considerably from the extremely aggressive reaction to the model, behavior which is not often observed in any other circumstance.

Humans

All members of the pongidae react readily to the presence of human beings. "Although orangutans seemed to pay little attention to other animals in the jungle they seemed very interested in me [Mackinnon, 1971, p. 183]." The specific reaction varied from individual to individual, but immediate flight was rare. Commonly, a threat display of twig dropping, combined with "grumphs" and kiss squeaks would result. Apparently directed urination and defecation have been observed by de Silva (1971), but as this pattern seems to occur with haphazard movements in frightened animals, it may just have been a coincidental. Mackinnon has also seen males dropping to a hanging position below the branch with one arm and leg which Jantschke (1972) states is a threat posture in captive orangs. Large males have even been known to pursue a man on the ground.

The reaction of chimpanzee is essentially similar to that of the orang. Stick throwing is common (Mackinnon states that at one time or another all members of the Gombe stream research station have been hit by objects thrown by the chimpanzees), and Sugiyama in his list of displays (1969) states that they are mainly used against humans. These include slapping the ground, branch throwing, exaggerated eating, buttress beating and branch shaking, and the whooping display. All of these, except exaggerated eating, can be seen as part of the normal threat display repertoire of the chimpanzee, and such behaviors have also been noted by Reynolds and Reynolds (1965), and van Lawick-Goodall (1965, 1968a,b).

In the gorilla, Schaller (1963) divides the response to man into the categories of being hunted and quietly observed. The early literature (Du Chaillu, 1861; Owen, 1859; Gatti, 1936) creates the impression of a large dangerous beast, attacking without provocation. The displays and bluff charges undoubtedly do occur, but in hunting, after repeated charges, the male usually is killed or flees, and the females then make no attempt at retaliation (Merfield and Miller, 1956). If an individual stands his ground when bluff-charged, the animal usually stops, only pursuing and biting a fleeing man. (Orangutans have also been reported to bite, but this is usually in captivity.) The attacks usually consist of a lunge forward, brief contact, and retreat, with no prolonged mauling. The response to an observer sitting quietly is very often chest beating and perhaps roaring on the part of the dominant male. With habituation to the observer's presence this may become no more than an initial annoyed grunt, and they will tolerate the observers at a closer distance. Females will also let infants wander from them. Fossey (1970, 1971) has managed to get even closer to some groups by adopting submissive postures, such as arm folding.

CONCLUSIONS

Aggressive behavior in pongids seems to be basically similar in most aspects. The component patterns, with great reliance on threat behavior rather than on attack, are similar; for example, the occurrence of object throwing in all species even though in the gorilla it seems to be little directed. The displays in response to human intruders are also similar in many ways. However, there are large differences in the social organization of the three types, and as Hall (1964) says, "The apes are difficult to place in comparison with the monkeys [p. 62]." The organization of the chimpanzee is essentially open and adaptive in that to a certain extent movement between groups, and certainly the splitting and recombination of subgroups, is unique within the primate kingdom. This form of behavior would place enormous strain on macaque or savannah baboon social organization, which is much more rigid and structured to a large extent around a central hierarchy (but see Kummer, 1971 for a description of the "fission–fusion" society in hamadryas baboons). The control mechanism involved then cannot be simply a relaxation of the rank order, with less emphasis on actual aggression and more on threat behavior. There is also a great increase in the number and modes of contact behavior with reassurance gestures. Reynolds and Luscombe (1969) have shown how dominance in a captive chimpanzee colony relates to positive social behavior and not to amount of aggression [van Lawick-Goodall (1968a) also states that the amount of aggression shown by one dominant male (Mike) reduced slowly after he became dominant], and that full display was only shown by fully adult males. The most frequent event following Mike's display was approach by other animals followed by associative acts. Chance and Jolly (1969) call this sort of social structure a hedonic one; this is really only a description of what is happening, but it does at least point out the nature of the difference between these behavioral mechanisms and those of the lower primates, which are essentially agonistic. Gartlan (1968) makes an interesting attempt to clarify the situation by talking of roles in the social structure, but in my mind this essentially fails in this case because he is then establishing a functional criterion, whereas the difference seems essentially to be a structural one.

The gorilla has an apparently more "normal" closed group society, with central core and peripheral members. Yet even here there are some large if subtle distinctions from the centripetal primate society in that the level of aggression (indeed all interactive behaviors) is very low; Schaller (1963) states that this could be due to the anticipation of the dominants' behavior by the subordinates. The society is still flexible, however, with the acceptance of lone black- or silver-backed males into the group (Fossey, 1970). This difference in behavior can perhaps best be related to the ecological situation of the animal; he is an eater of well-distributed, bulky, less nourishing foods such as plant stems and leaves, as opposed to the fructivorous (and carnivorous) chimpanzee and leaf- and fruit-eating orangutan.

The orang seems to be a relic population which at one time spread over the Pleistocene rain forest as far as South China. The arboreal mode of life is obviously highly specialized, and the solitary nature of the orang's existence does not necessitate a large behavioral repertoire. The aggressive component of behavior may well not be markedly reduced in comparison to other more solitary primates, such as prosimians, for the rape chases by males involve biting and hitting.

Thus the behavior of the chimpanzee alone seems to occupy a unique position in the primate field. Kortlandt and his co-workers (e.g., 1962, 1963, 1969) have proposed the interesting but as yet unsupported theory that the social behavior of the chimpanzee is related to a conflict with man at a protohomini stage, the idea being that the early savannah chimps were driven back into the less hospitable forest environment by early savannah man, based on the evidence of fossil chimpanzee distribution and the difference in behavior between savannah and forest chimpanzees. In whatever way it may have derived, aggression in chimpanzee groups has become less important in relation to the social structure. The same may be true for the gorilla and orangutan, but for clarification, further field work is necessary.

REFERENCES

Albrecht, H., and Dunnett, S. C. (1971). "Chimpanzees in Western Africa." Piper, München.

Azuma, S., and Toyoshima, A. (1961–1962). Progress report of the survey of chimpanzees in their natural habit, Kabogo Point area, Tanganyika. *Primates* 3, 61–70.

Baumgartel, M. W. (1958). The Muhavura gorillas. *Primates* **1**, 79–83.

Baumgartel, M. W. (1961). The gorilla killer. *Uganda Wild Life Sport* **2**, 14–17.

Bernstein, I. S. (1970). Primate status hierarchies. *In* "Primate Behaviour: Developments in Field and Laboratory Research" (L. A. Rosenblum, ed.), Vol. 1, pp. 71–109. Academic Press, New York.

Bingham, H. C. (1932). Gorillas in a native habitat. *Carnegie Inst. Washington Publ.* **426**, 1–66.

Blower, J. (1956). The mountain gorilla. *Uganda Wild Life Sport* **1**, 41–52.

Brandes, G. (1939). "Buschi. Vom Orang-Säugling zum Backenwülster." Quelle u. Meyer, Leipzig.
Burbridge, B. (1928). Gorilla N.Y.
Bygott, J. D. (1972). Cannibalism among wild chimpanzees. *Nature (London)* **238**, 410–411.
Carpenter, C. R. (1937). An observational study of two captive mountain gorillas (*Gorilla beringei*). *Hum. Biol.* **9**, 175–196.
Carpenter, C. R. (1938). A survey of wildlife conditions in Atzeh, North Sumatra. *Netherlands Comm. Int. Prot.* **12**, 1–33.
Carpenter, C. R. (1940). A field study in Siam of the behaviour and social relations of the gibbon (*Hylobates lar*). *Comp. Psychol. Monogr.* **10**, 1–168.
Carpenter, C. R. (1954). Tentative generalizations in the grouping behaviour of non-human primates. *In* "The Non-human Primates and Human Evolution" (J. A. Gaven, ed.), Wayne Univ. Press, Detroit, Michigan.
Carpenter, C. R. (1965). The howlers of Barro Colorado Island. *In* "Primate Behaviour." (I. DeVore, ed.), pp. 250–291. Holt, New York.
Carthy, J. D., and Ebling, F. J., eds.). (1964). "The Natural History of Aggression." Academic Press, London and New York.
Chance, M. R. A. (1967). Attention structure as the basis of primate rank orders. *Man N.S.* **2**, 503–518.
Chance, M. R. A., and Jolly, C. J. (1970). "Social Groups of Monkeys, Apes and Men." Cape, London.
Crook, J. H. (1970). Social organization and the environment: Aspects of contemporary social ethology. *Anim. Behav.* **18**, 197–209.
Dart, R. A. (1961). The Kisoro pattern of mountain gorilla preservation. *Curr. Anthrop.* **2**, 510–511.
Davenport, R. K. (1967). The Orang-utan in Sabah. *Folia Primat.* **5**, 247–263.
Donisthorpe, J. (1958). A pilot study of the mountain gorilla (*Gorilla g. beringei*) in S.W. Uganda. *S. Afr. J. Sci.* **54**, 195–217.
Du Chaillu, P. B. (1861). "Explorations and Adventures in Equatorial Africa." Murray, London.
Eibl-Eibesfeldt, I. (1970). "Ethology: the Biology of Behaviour." Holt, New York.
Fossey, D. (1970). Making friends with mountain gorillas. *Nat. Geogr. Mag.* **137**, 48–67.
Fossey, D. (1971). More years with mountain gorillas. *Nat. Geogr. Mag.* **140**, 574–585.
Fossey, D. (1972). Vocalizations of the Mountain Gorilla (*Gorilla gorilla beringei*). *Anim. Behav.* **20**, 36–53.
Garner, R. L. (1896). "Gorillas and Chimpanzees." Osgood, London.
Gartlan, J. S. (1968). Structure and function in a primate society. *Folia Primat.* **8**, 89–120.
Gatti, A. (1936). "Great Mother Forest." London.
Geddes, H. (1955). "Gorilla." Frankfurt.
Grant, E. C. (1969). Human facial expression. *Man N.S.* **4**, 525–536.
Hall, K. R. L. (1964). Aggression in monkey and ape societies. *In* "Natural History of Aggression" (J. D. Carthy and F. J. Ebling, eds.), Academic Press, New York.
Hamburg, D. A. (1971). Aggressive behaviour of chimpanzees and baboons in natural habitats. *J. Psychiat. Res.* **8**, 385–398.

Harrisson, B. (1960). A study of orang-utan behaviour in semi-wild state. *Sarawak Mus. J.* **15–16**, 422–447.

Harrisson, B. (1962). "The Orang-utan." Doubleday, Garden City, New York.

Harrisson, B. (1963). The education of young orangs to live in the wild, *Oryx* **7**, 108–127.

Hooff, J. van (1967). The facial displays of the Catarrhine monkeys and apes. *In* "Primate Ethology." (D. Morris, ed.). Weidenfeld and Nicolson, London.

Hooff, J. van (1971). "Aspecten von het Sociale Gedrag en de Communicatie bij Humane en Hogere Niet-Humane Primaten." Bronder-Offset, Rotterdam.

Imanishi, K. (1957). Social behaviour in Japanese monkeys, *Macaca fuscata. Psychologia* **1**, 47–54.

Imanishi, K. (1958). Gorillas: a preliminary survey in 1958. *Primates* **1**, 73–78.

Itani, J. (1954). "The Monkeys of Takasakiyama" (Transl. by S. L. Washburn). Ford Foundation.

Itani, J. (1961). "Gorilla to Pigmie no Mori." Iwanami, Tokyo.

Itani, J. (1966). Social organisation of chimpanzees. *Shizen* **21**, 17–30.

Izawa, K. (1970). Unit groups of chimpanzees and their nomadism in the savannah woodland. *Primates* **11**, 1–46.

Izawa, K., and Itani, J. (1966). Chimpanzees in Kasakati Basin, Tanganyika. (1) Ecological studies in the rainy season. *Kyoto Univ. Afr. Stud.* **1**, 77–156.

Jantschke, F. (1972). "Orang-Utans in Zoologischen Gärten." Piper, München.

Kawabe, M. (1966). One observed case of hunting behaviour of wild chimpanzees living in the savannah woodland of western Tanzania. *Primates* **7**, 393–396.

Kawai, M. (1965). On the system of social ranks in a natural troop of Japanese monkeys. I. Basic Rank and Dependent Rank. *In* "Japanese Monkeys" (S. A. Altmann, ed.). Publ. by editor.

Kawai, M., and Mizuhara, H. (1959). An ecological study on the wild mountain gorilla. *Primates* **2**, 1–42.

Köhler, W. (1917). "Intelligenzprüfungen an Anthropoidean. I (Einzelausgabe)." Verlag Kön Akad. Wiss. Berlin.

Köhler, W. (1921). Aus der Anthropoidenstation auf Teneriffa. V. zur Psychologie des Schimpansen. *Akad. Wiss.* **39**, 686–692.

Köhler, W. (1931). "The Mentality of Apes." Harcourt, New York.

Kortlandt, A. (1962). Chimpanzees in the wild. *Sci. Amer.* **206**, 128–138.

Kortlandt, A. (1967). Experimentation with chimpanzees in the wild. *In* "Neue Ergebnisse der Primatologie (Progress in Primatology)" (D. Stark, R. Schneider, and H. J. Kuhn, eds.) pp. 208–224. Fischer, Stuttgart.

Kortlandt, A., and Kooij, M. (1963). Protohominid behaviour in primates (Preliminary Communication). *Symp. Zool. Soc. London* **10**, 61–88.

Kortlandt, A., and van Zon, J. C. J. (1969). The present state of research on the dehumanization hypothesis of African ape evolution. *Proc. Int. Congr. Primat., 2nd, Atlanta.* Karger, Basel.

Kummer, H. (1971). "Primate Societies." Aldine, Chicago, Illinois.

Kummer, H., and Kurt, F. (1963). Social units of a free-living population of hamadryas baboons. *Folia primat.* **1**, 4–19.

Lawick-Goodall, J. van (1965a). Chimpanzees of the Gombe Stream Reserve. *In* "Primate Behaviour" (I. DeVore, ed.). Holt, New York.

Lawick-Goodall, J. van (1965b). New discoveries among Africa's chimpanzees. *Nat. Geogr. Mag.* **138**, 802–831.

Lawick-Goodall, J. van (1967a). "My Friends, the Wild Chimpanzees." Nat. Geogr. Soc., New York.

Lawick-Goodall, J. van (1967b). Mother-infant relations in chimpanzees. *In* "Primate Ethology" (D. Morris, ed.). Weidenfeld and Nicolson, London.

Lawick-Goodall, J. van (1968a). The behaviour of free-living chimpanzees in the Gombe Stream Reserve. *Anim. Behav. Mono.* **1**, 165–311.

Lawick-Goodall, J. van (1968b). A preliminary report on expressive movements and communication in the Gombe stream chimpanzees. *In* "Primates, Studies in Adaption and Variability" (P. C. Jay, ed.). Holt, New York.

Mackinnon, J. (1971). The orang-utan in Sabah today. *Oryx* **11**, 141–191.

McQuitty, L. L. (1966). Similarity analysis by reciprocal pairs for discrete and continuous data. *Educ. Psychol. Measur.* **26**, 825–831.

Marler, P. (1965). Communication in monkeys and apes. *In* "Primate Behaviour" (I. DeVore, ed.). Holt, New York.

Marler, P. (1969). Vocalizations of wild chimpanzees, an introduction. *Proc. 2nd Int. Cong. Primatol. Atlanta* 1968. Karger, Basel.

Merfield, F. G., and Miller, H. (1956). "Gorilla Hunter." New York.

Napier, J. (1963). The locomotor functions of hominids. *In* "Classification and Human Evolution" (S. L. Washburn, ed.), Vol. 37, pp. 178–189. Viking Fund Publ. Anthropol.

Nishida, T. (1967). Savannah chimpanzees. *Shizen* **22**, 31–41.

Nishida, T. (1968). The social group of the wild chimpanzees in the Mahali Mountains. *Primates* **9**, 167–224.

Nishida, T. (1970). Social behaviour and relationship among wild chimpanzees of the Mahali Mountains. *Primates* **11**, 47–87.

Nissen, H. W. (1931). A field study of the chimpanzee. *Comp. Psychol. Monogr.* **8**, 1–105.

Owen, R. (1859). "On the Classification and Geographical Distribution of the Mammalia." London.

Pitcairn, T. K. and Grant, E. C. (1972). A Dyadic Analysis of Human Behaviour. Paper to Assoc. Study Anim. Behav., Lond. Nov. 1972.

Reynolds, V. (1966). Open groups in hominid evolution. *Man* **1**, 441–452.

Reynolds, V. (1968). Kinship and the family in monkeys, apes and men. *Man* **3**, 209–223.

Reynolds, V., and Reynolds, F. (1965). Chimpanzees of the Budongo Forest. *In* "Primate Behaviour" (I. DeVore, ed.). Holt, New York.

Reynolds, V., and Luscombe, G. (1969). Chimpanzee rank order and the function of displays. *Proc. Int. Congr. Primat., 2nd, Zurich,* 1968. Karger, Basel.

Riess, B. F., Ross, S., Lyerly, S. B., and Birch, H. G. (1949). The behaviour of two captive specimens of the lowland gorilla, *Gorilla gorilla gorilla* (Savage and Wyman). *Zoologica* **34**, 111–118.

Rioch, D. (1967). Discussion of agonistic behaviour. *In* "Social Communication Among Primates" (S. A. Altmann, ed.). Univ. Chicago Press, Chicago, Illinois.

Sade, D. S. (1967). Determinants of dominance in a group of free-ranging rhesus monkeys. *In* "Social Communication Among Primates" (S. A. Altmann, ed.). Univ. Chicago Press, Chicago, Illinois.

Schaller, G. B. (1961). The orang-utan in Sarawak. *Zoologica* **46**, 73–82.

Schaller, G. B. (1963). The mountain gorilla. "Ecology and Behaviour." Chicago Univ. Press, Chicago, Illinois.

Schaller, G. B. (1965). The behaviour of the mountain gorilla. *In* "Primate Behaviour" (I. DeVore, ed.). Holt, New York.

Schaller, G. B., and Emlen, J. T. (1963). Observations on the behaviour of the mountain gorilla. *In* "African Ecology and Human Evolution" (F. C. Howell, ed.). Viking Fund Publ. Anthrop.

Scott, J. P. (1958). "Aggression." Univ. Chicago Press, Chicago, Illinois.

Silva, G. S. de (1971). Notes on the orang-utan rehabilitation project in Sabah. *Malay Nat. J.* **24,** 50–77.

Sugiyama, Y. (1967). Forest living chimpanzees. *Shizen* **22,** 18–29.

Sugiyama, Y. (1968). Social organisation of chimpanzees of the Budongo Forest, Uganda. *Primates* **9,** 225–258.

Sugiyama, Y. (1969). Social behaviour of the chimpanzees in the Budongo Forest, Uganda. *Primates* **10,** 197–225.

Suzuki, A. (1966). On the insect-eating habits among wild chimpanzees living in the savannah woodland of western Tanzania. *Primates* **7,** 482–487.

Tobias, P. V. (1961). The work of the gorilla research in Uganda. *S. Afr. J. Sci.* **57,** 297–298.

Teleki, G. (1973). The omnivorous chimpanzee. *Sci. Amer.* **228,** 32–42.

Walther, F. (1968). "Verhalten der Gazellen." Ziemsen, Wittenburg.

Wilson, W. L., and Wilson, C. C. (1968). Aggressive interactions of captive chimpanzees living in a semi-free-ranging environment. *Techn. Rep. ARL. TR-68-9, 6571st. Aeromed. Res. Lab. New Mexico.*

Yerkes, R. M. (1943). "Chimpanzees, a Laboratory Colony." Yale Univ. Press, New Haven, Connecticut.

Yerkes, R. M., and Yerkes, A. W. (1929). "The Great Apes." Yale Univ. Press, New Haven, Connecticut.

Zahl, P. (1960). Face to face with gorillas in Central Africa. *Nat. Geogr. Mag.* **117,** 114–137.

Zon, J. C. J. van, and Ortshoven, J. van (1967). Enkele Resultaten van de Zesde Neederlandse Chimpanzee-Expeditie. *Vakbl. Biol.* **47,** 161–166.

COMPARATIVE PHYSIOLOGICAL DATA

ANDROGENS AND AGGRESSION: A REVIEW AND RECENT FINDINGS IN PRIMATES

ROBERT M. ROSE

Boston University School of Medicine

IRWIN S. BERNSTEIN

University of Georgia and Yerkes Regional Primate Research Center of Emory University

THOMAS P. GORDON

Yerkes Regional Primate Research Center, Field Station and University of Georgia

SHARON F. CATLIN

Boston University School of Medicine

INTRODUCTION

There has been a renewal of interest in the relationship between androgens, the class of male sex hormones, and sexual and aggressive behavior. However, much of the nature of this interaction remains unclear. We have learned much from the study of rodents, but only recently have parallel investigations in nonhuman primates begun. Assessment of hormonal determinants of behavior in various nonhuman primate groups is further hampered by the fact that there are significant nonendocrine influences on sexual and aggressive behavior, e.g., social influences, past experience of the individual, etc. This is not to imply that endocrine influences are not relevant, for indeed, they may well function to establish clear behavioral propensities, as well as to facilitate the appearance of specific aggressive and sexual behavior. Rather, their influence appears to be more subtle than in nonprimate mammals, and thus it is increasingly difficult to separate endocrine from experiential factors.

Before we examine the relevant studies, it might be useful to digress a bit and present a very brief overview of relevant endocrinology. The term "androgen" refers to a group of individual hormones, all of which exert similar biological effects. They are responsible for the development in the adult of male secondary sex characteristics, such as increased muscle mass, larger physical size, development of canine teeth, and a host of other sexually dimorphic characteristics. In the developing fetus, androgens are responsible for embryological differentiation of the male urogenital system, e.g., penis, vas deferens, etc. As we shall discuss, it also appears that androgens may function to differentiate and establish some male behavioral propensities.

Testosterone is the most potent of the androgens in most of the effects exerted by this class of hormones. In the intact male, it is secreted almost entirely by the testes. In both sexes, the adrenal glands secrete significant amounts of less potent androgens, mostly Δ^4-androstenedione and dehydroepiandrosterone. In the female, small amounts of circulating testosterone are derived by conversion from Δ^4-androstenedione. However, males produce approximately 20 times as much testosterone as females, thereby significantly differentiating the sexes endocrinologically.

In human and probably in nonhuman primates, adrenal androgens serve as the major source of androgenic hormones in females (Rose, 1972). Consequently, adrenalectomy removes most androgens from females, while ovariectomy is more effective in removing estrogens and progesterone.

Traditionally, most of our knowledge about the effects of hormones has been via administration of specific hormones, either in the intact animal at various stages of development, or administration following removal of the endocrine organ responsible for secretion of the hormone under study. Most of the studies reviewed in this chapter utilize some variation of this basic experimental design.

In recent years, we have become increasingly aware of the importance of psychological influences in the control of endocrine secretion. A very large literature has developed documenting that exposure of the organism to potentially noxious or stressful events results in significant increases in cortisol, epinephrine, and growth hormone secretion (Mason, 1968). Most recently, we have also found that following exposure to potentially stressful environments, there is a fall in plasma testosterone (Rose, 1969; Rose *et al.*, 1971, 1972).

These observations raise an interesting possibility. Androgens appear to affect certain aspects of sexual and aggressive behavior. Psychological stimuli may function to regulate the secretion of androgens. It is therefore possible that alterations in behavior following exposure to defeat, for example, may be mediated by changes in testosterone secretion. The defeat inhibits testosterone secretion, and the subsequent fall in plasma testosterone makes it less likely for the animal to engage in future aggressive behavior. This potential interaction between environmental events and endocrine secretion will be discussed in greater detail in the last section of the chapter.

The first section reviews the relevant studies in rodents, as they more or less set the stage for the work undertaken with nonhuman primates. We will then go on to a review of the studies involving the influence of androgens in the fetal and neonatal animal, especially in terms of how administration of testosterone may affect future behavior. The next stage of development where androgens may significantly affect behavior is during puberty. There are very little data to date that systematically study this particular interaction. However, the small amount of information available, along with anecdotal observations, does suggest that this may be a crucial period of hormonal–behavioral interactions in primates. Seasonal influences are reviewed next, and it appears that these variables may function to influence significantly the secretion of testosterone, with the concomitant changes observed in sexual and aggressive behavior. Observations on male castrates are then reviewed, followed by a presentation of recent work done by our group.

STUDIES IN RODENTS

Most studies of the relationship between androgens and aggressive behavior have been done in rodents. This is directly related to various important technical considerations, such as the availability of large numbers of animals, ease of handling in the laboratory, and a long history of investigation of biological variables in the laboratory rat. However, rats, especially those such as are usually housed in the laboratory, exhibit a very narrow range of social behavior. Observations of aggressive behavior in this situation are thus severely limited by (*1*) the rats being housed individually or in small groups, (*2*) their severely limited access to space, compared to the home range of the wild rat, (*3*) the usual practice of studying only the social interactions of pairs of animals, and (*4*) the less complex pattern of behavior and social interaction of the rat, as compared to that of the primate. In addition, as Beach (1947, 1958, 1970) has pointed out, as one moves from rodents to other mammals such as dogs or cats to the study of nonhuman primates, to man, sexual behavior becomes increasingly independent of hormonal influences, and the influence of learning and past experience becomes much more important in determining individual responses. This probably also applies to aggressive behavior. Nevertheless, many important studies of the interaction between aggression and hormonal influences have been done in rodents. They have served to shape our thinking about the pertinent questions to be studied in various primate species and therefore should be reviewed briefly at this time.

The studies fall into several categories, as illustrated by the list which follows. These categories are applicable not only to rodents, but also outline investigations with various nonhuman primate groups.

A. Differences in the aggressive behavior of normal males and females
   1. Shock-elicited aggression, usually done with pairs of animals which simultaneously receive foot shock and subsequently attack one another
   2. Pup-killing behavior of adult animals
B. Manipulation of the endocrine status of the adult animals, e.g., castration and replacement by androgens or estrogens and observations of subsequent changes in aggressive behavior
C. Manipulation of the endocrine status of the fetal or neonatal animals during critical periods of brain development and observation of any subsequent changes in behavior during maturation or adulthood

To date, however, in contrast to work with primates, there is no work with rodents measuring differences in the levels of testosterone in individual animals as a function of differences in social rank or the effects of various psychological stimuli. This is in part related to the difficulty of obtaining repeated samples on individual rodents, as well as to the restricted social repertoire of the laboratory animal as he is usually studied.

Conner and Levine (1969) found a significant difference in the shock-induced fighting response of males and females, with males exhibiting higher mean number of fighting responses, greater tendency to maintain the fighting position, and a tendency to fight sooner when exposed to a series of shocks. When they castrated adult males, however, they found no change in this fighting response within a month after castration. They also found no change when they injected both normal, as well as ovariectomized, females with testosterone propionate. However, they did find significantly less fighting in male animals who were castrated at weanling age (22 days), and further decreases in fighting when males were castrated neonatally. Hutchinson *et al.* (1965), however, have reported a gradual decline in fighting frequency in castrated males after more extended testing, irrespective of whether they were castrated before or after puberty, although the effect was not immediate when the operation was performed after maturation. In addition, Hutchinson *et al.* (1965), found that the probability of the shock-induced fighting response increased as a direct function of age. Powell and Creer (1969) also found that shock-induced aggression increased with age, but suggested, based on their investigation, that this relationship is partially due to the confounding of age and size of the enclosure, as smaller enclosures are known to increase the frequency of shock-induced aggression. These findings would be compatible with the idea that testosterone levels, higher in males than females and higher in adult males than prepubertal males, influence aggressive behavior.

Similar findings have been made in studies of the golden hamster (*Mesocricetus auratus*). Vandenbergh (1971) and Payne and Swanson (1972) found that castrated males showed significantly less aggressive behavior than the intact males with which they interacted and that they exhibited a significant increase in aggression with administration of testosterone propionate (TP).

Powell *et al.* (1971) did a study in which they injected neonatal rats with testosterone propionate in vegetable oil and found that the mean overall fighting by these subjects was greater than for those injected with oil alone. In addition, they found that of the rats castrated at 35 days of age, those with prior fighting experience fought as often as sham-operated subjects, whereas those without experience fought at similar rates for a time, but then less often than the sham castrates. This suggests that prior shock-induced fighting and castration interact in a complex manner to determine level of fighting behavior later on.

A recent study done with trained male mouse fighters matched fighters with inexperienced castrates, half of which had been treated with testosterone propionate and half with oil. Lee and Brake (1972) found that the fighters attacked testosterone propionate subjects more often and for a longer period of time than oil-treated subjects. They suggest that a possible reason for this observation is that the TP-treated castrates may produce an aggression-promoting pheromone. They did not observe a greater degree of aggression on the part of the TP-treated castrates than on the part of the oil-treated castrates.

Rosenberg *et al.* (1971) performed a series of experiments in which they compared the pup-killing behavior of normal adult rats. In general, male rats were observed to kill rat pups placed with them, whereas females did not. In one study, they demonstrated that castrating males in adulthood, and therefore significantly lowering plasma testosterone, had no effect on pup-killing behavior. In a subsequent study, they castrated neonatal males and found pup-killing behavior significantly reduced (almost to the level of incidence of female pup killing) when these animals became adult. Testosterone injections to these rats in adulthood did not increase their killing rate. They also castrated young rats at 15 days of age and found that they had as low an incidence of killing as the day 2 castrates. Females which were administered testosterone propionate at age 2 days, still during the period of critical brain development in rats, showed no increase in pup-killing behavior as adults, compared to controls. Davis and Gandelman (1972), however, found that the administration of high doses of testosterone propionate to both ovariectomized and intact adult virgin female mice elicited pup-killing behavior. This behavior disappears within approximately 3 weeks after cessation of treatment. Gandelman (1972) also found that administration of testosterone propionate to ovariectomized females in infancy led to a significant increase in the number of females that exhibited pup killing.

Another behavior which has been used to study the potential influence of androgens is the frequency of the mounting response, which is widely recognized as being sexually dimorphic in rats. Males display significantly more mounting behavior and females demonstrate significantly more lordosis. The relative frequencies of these behaviors, however, have been altered experimentally by manipulating the hormone levels by several different methods. Gerall and Ward (1966) found that when they injected pregnant Sprague-Dawley females with testosterone propionate, they obtained female offspring which demonstrated varying degrees of masculinization of their external genitalia. In addition, when these females were injected with testosterone propionate as adults, they demonstrated approximately the same frequency of mounting behavior as normal males, and significantly more mounting behavior than females who had received injections of testosterone in adulthood, but had not been androgenized *in utero*. These same prenatally treated females were also observed to have significantly fewer estrous responses than normal females. Similar observations have been made in guinea pigs by Gerall (1966) and support the hypothesis that prenatal androgens may exert a permanent organizational effect on the central nervous system.

Ward and Renz (1972) found that both pre- and postnatal testosterone propionate enhanced the incidence of masculine sexual behavior in female rats, but that prenatal TP was more effective. In addition, it has been shown recently that administration of testosterone propionate to female rats during either prenatal development or up to age 10 days after birth results in their being acyclic in adulthood (the androgen-sterilized rats) (Harris, 1964; Bradbury, 1941; Barraclough, 1961; Harris and Levine, 1962).

In a similar light, Ward (1972a) did a study in which she exposed pregnant female rats to restraint and light stress, and observed changes in the behavior of the male pups. Those males whose mothers were stressed prenatally showed lower frequencies of copulatory behavior and higher frequencies of lordosis than controls. Similarly, Ward (1972b) also found that female rats which were injected with a potent antiandrogen (cyproterone acetate) had male pups which showed significant feminization of their adult sexual responses. Both of these studies are based on experiments which appear to have lowered the level of functional testosterone in the developing fetus. In the first study, Ward suggests that stress lowers the level of testosterone secreted from the gonads, while increasing the secretion of the weaker adrenal androgen, Δ^4-androstenedione. In the second study, the administration of cyproterone acetate blocks the action of endogenously secreted testosterone. This work serves to support the interpretation that androgens significantly affect the adult behavior of animals when administered to rodents early in development and during this crucial period of brain maturation.

In summary, two general categories of behavior, aggressive and sexual, have been examined in order to determine the effects on the individual of pre- and postnatal hormones. Shock-induced fighting is significantly reduced by castration performed early in life. In addition, this fighting may be reduced if castration is done in adulthood, but appears to be heavily influenced by fighting experience and other factors yet to be investigated. More studies are needed in order to confirm and extend the findings of Powell *et al.* (1971). Studies of pup-killing behavior are somewhat inconclusive, as findings have been contradictory in some areas; but they suggest that higher androgen levels, at least in the neonate, and perhaps also in the adult, lead to significantly increased pup killing.

Observations of sexual behavior show that both pre- and postnatal administration of testosterone propionate to females increases the incidence of male sexual behavior (mounting), but that prenatal administration is more effective. In males, lowering the level of functional testosterone in the developing fetus results in lower frequencies of copulatory behavior and higher frequencies of lordosis than those exhibited by controls.

These studies suggest that, in the absence of androgens during fetal development, the maturing brain retains its more basic female predisposition, resulting in lower levels of aggressive behavior in adulthood, along with decreased mounting and increased frequency of lordosis. The presence of androgen during this crucial early period of brain development appears to exert an organizational effect leading to a more masculine behavioral predisposition. The administration or removal of androgen in adulthood may also be effective, but the effect, in general, is not as profound.

Although the evidence for these endocrine influences on behavior in rodents is very persuasive, it should be noted that there are significant exceptions, along with some contradictory evidence. Nevertheless, these data are conclusive enough to support parallel investigation of endocrine determinants of behavior in nonhuman primates.

PRIMATE STUDIES

Pre- and Neonatal

The study of the relation between aggression and testosterone in primates, as compared with rats, is complicated by such factors as primates' greater behavioral repertoire and the increased importance of learning. Harlow (1965) attempted to control for this influence in a series of peer group rearing experiments in which rhesus monkeys were taken from their mothers at birth and lived in individual cages with cloth surrogates, except for a portion of each day, which they spent in an elaborate playpen in groups of four. He observed several differences in the behavior of males and females. Infantile sexuality, expressed by brief, incomplete sexual reactions, such as rubbing and thrusting responses directed toward a second infant, seemed to appear earlier and with greater frequency in the male, but the data were not conclusive. From 73 to 159 days, however, males definitely exhibited significantly more infantile sexual responses, as described above.

In addition, males exhibit from 2½ months on a significantly higher frequency of threat response, which, as defined by Harlow, involves a stiffening of posture, staring at the other monkey, flattening the hair on top of the head, retracting the lips and baring the teeth. Females, after the first few weeks, exhibit a significantly higher frequency of passivity, in which little or no response is made to an approaching animal. Females also typically exhibit patterns of withdrawal or rigidity.

Play patterns, also, appear to be sexually dimorphic. From the first appearance of rough-and-tumble play, males initiate this pattern more frequently than females, and from 100 days on, engage more frequently in contact play. Following these peer-group rearing experiments, in which only surrogate mothers were used, another series of studies was done, employing a different experimental situation. In general, similar results were obtained in this second test situation. In this set of studies, the infants had contact with both surrogates and real mothers. The animals were housed in a playpen, divided by removable partitions into four units, each 30 in by 30 in. The playpen was surrounded by four living cages, each housing a mother monkey or surrogate and her babies. A small opening allowed each baby to come and go at will into a play area. The partition between pairs of playpens was raised for a portion of each day, allowing the small monkeys to interact with each other and one or more of the mothers. Data from these groups supplemented the data from the playpen group. In this situation, these workers found similar differences in the play behavior of male and female infants as observed when the infants had contact only with peers.

In another study, Jensen *et al.* (1966) found that male infant pigtail monkeys (*Macaca nemestrina*) raised with their mothers in a highly simplified and controlled laboratory condition were more "independent" than females; that is, they spent more time in the most separated position and became relatively less manipulative of their mothers. Also, under conditions of privation, that is, barren cages with smooth walls located alone in soundproof rooms, male infants were reported to be more punitive toward their mothers than female infants.

These studies provide important information on differences in behavioral propensity or predisposition between infant male and female monkeys. By excluding contact with adults of both sexes, except for mothers, one may study the play behavior of infants which more accurately reflects the potential influence of different hormonal environments during fetal development. However, one cannot extrapolate from these studies to differences in the behavior of infant and juvenile animals in normal social groups.

There are relatively little data available on the differences in play behavior of infants and juveniles in normally constituted social groups. Kaufmann (1966) reported on a free-ranging band of rhesus on Cayo Santiago. He studied two females and their infants intensively for 4 months and, in addition, 70 hr were spent observing other females and their infants. All observations were made during daylight hours, mostly in the early morning. He was unable to obtain a great deal of information regarding the differences between males and females and their rate of development. Generally, the infants ride on their mothers' backs after they relinquish the normal ventral-to-ventral position seen during early development, and he reported that female offspring tended to ride on their mothers' backs earlier than males. This could be interpreted as evidence of a slightly increased rate of development. He concluded, however, that in most activities he could not discriminate clear sexual differences in play behavior or in rate of development.

In another study which aimed at a compromise between a natural setting and a laboratory environment, Hinde and Spencer-Booth (1967) observed two groups of young rhesus living with their mothers in pens, each containing an adult male, two to four females, and their young. Each pen had an outside run 18 ft × 18 ft × 8 ft and an inside room 6 ft × 4½ ft × 7½ ft. The main group consisted of eight young observed over their first 2½ years of life, and the early group consisted of eight young observed for the first 24 to 30 weeks. Social play was recorded as rough-and-tumble if physical contact occurred and as approach–withdrawal if it did not. These researchers report that they found no clear differences between the sexes in the time of appearance of play, in the changes with age, the amounts of locomotor play (leaps and gallops), or in the amounts of passive rough-and-tumble and approach–withdrawal social play. However, males showed somewhat more active social play of both types than females, but the difference was not significant.

Observations of sexual behavior in this study revealed that male mounting attempts preceded female mounting attempts considerably. In both sexes, these early attempts were incomplete and disoriented. In most males, first mounting occurred between weeks 12 and 46, and in most females was much later. Only one female in the early group was seen to mount (weeks 13 to 18) and one of the three in the main group did not mount during the course of the 2½ years. The other two first mounted in weeks 34 and 97, respectively. In addition, male young in the main group showed significantly more mounting than females, but there was no clear difference between the sexes in the amount they were mounted. Females were mounted more often than they mounted and the reverse was true for four out of the five males.

Branch shaking in both sexes peaked in the middle of the second year and was more common in males than females. It was also more common in the dominant male than in any other male.

These two studies, although far from conclusive, strongly suggest the differences in play behavior reported by Harlow (1965) may not be directly applicable to animals raised in social groups. It is not clear what may modify the possible differences in behavioral propensities shaped by increased testosterone levels in the fetal male. It is possible that the presence of other adults of both sexes, besides the mother, plays a crucial role in shaping the infant's behavior. Nevertheless, before these questions may be answered, detailed observations of the play behavior of infants and juveniles in normal social groups are necessary.

With these qualifications, it is instructive to examine some recent work on the effects on future behavior of experimentally manipulating the early endocrine environment. In order to evaluate the potential significance of such manipulations, it would be relevant to examine differences in endocrine levels in normal fetal development. Resko (1970), in a study of hormone levels in the rhesus fetus and neonate, measured testosterone in plasma from the umbilical artery and in the fetal gonads. The amount of testosterone in umbilical artery plasma was significantly greater in male fetuses than female fetuses. The small amounts which are found in the female could be explained by peripheral conversion of androstenedione to testosterone. In addition, high levels of testosterone were found in the testes of fetal males, diminishing from 3560 ng/gm on day 96 to 380 ng/gm on day 150, approximately 2 weeks before birth. Testosterone was not detectable in the fetal ovaries. These measurements support the view that the fetal testes secrete testosterone during the second half of gestation.

The sex differences in the amount of plasma and gonadal testosterone in the rhesus indicate that the developing anlagen of the genetic male are exposed to larger quantities of it than those of the genetic female. As the quantities of androstenedione did not differ in fetal plasma and gonads with sex, one would suspect that this hormone is not responsible for masculine differentiation. During the neonatal period, plasma and testes testosterone levels drop off sharply in the male, and plasma levels remain low until about 3 to 4 years of age (Resko, 1967).

In order to study the effects of fetal androgen on adult rhesus behavior, Goy and Phoenix (1972) studied the behavior of young individuals under conditions similar to those of Harlow, that is, taking the young from their mothers at birth and providing them with surrogates or leaving them with their own mothers and removing them at the weanling stage (90–100 days). Their observations of peer group interactions support Harlow's conclusions: They found that male rhesus monkeys performed threat, play-initiation gestures, rough-and-tumble play, and pursuit play 5–10 times as frequently as females. They then treated genetic females with testosterone propionate from the 39th through the 70th or 105th day of gestational age, thereby obtaining pseudohermaphrodites, and studied them under comparable conditions. Their average frequency of performance is between that of males and normal females on every measure. They also found mounting behavior markedly sexually dimorphic—relatively frequently displayed by about 80% of young males and rarely or never displayed by females. The pseudohermaphroditic females treated prenatally with testosterone propionate displayed mounting frequencies close to those of normal males.

Both males and females gonadectomized at birth developed sexually dimorphic behavior typical of intact animals of the same sex, suggesting that the presence of the gonad after birth is not essential to the sexually differentiating effects of social experience in the testing situation. This is consistent with Resko's observation of the drop in testosterone level shortly after birth.

In addition, Goy and Phoenix performed another experiment in which female rhesus monkeys were injected with testosterone propionate three times a week for 8 months, beginning at 6 months of *postnatal* age. This had marked effects on rate of body growth and dominance status, but did not vary the frequencies of threat, play initiation, rough-and-tumble play, pursuit play, and mounting behavior from those of normal females or of the same females prior to treatment. Apparently the effects of pre- and postnatal androgen on the genetic female are distinctive and have significantly different effects.

A major distinction between the response of the rhesus monkeys and that of the rats to administration of testosterone propionate is in ovarian cyclicity (Karsch *et al.*, 1973). In rats, this cycling is frequently suppressed and sterility induced by a single injection of androgen administered during the first 10 postnatal days. Treloar *et al.* (1972) attempted to duplicate this in the rhesus by injecting newborn females with testosterone (35 ng/kg). Menstrual cycle histories, gross ovarian morphology, and plasma progestin levels reflecting ovulation and corpus luteum development failed to reveal any differences between injected and control animals. The authors suggest, however, that the dose could have been too low or that the type of androgen preparation could have been responsible for the absence of differences, as it was different from that most frequently used in studies of other species.

Also, Goy and Resko (1972) reported the establishment of stable and regular menstrual cycles typical of normal females of comparable age in their pseudohermaphrodites which had been treated with testosterone propionate from 30 to 82 days, beginning at 38–40 days of gestational age. Biochemical and morphological evidence showed that ovulation and normal corpus luteum formation was occurring, and direct evidence was obtained in five of the seven pseudohermaphrodites upon ovariectomy on day 18 of the cycle.

All this evidence suggests that female cyclicity in primates is not suppressed by exogenous testosterone propionate, either pre- or postnatally, but it is possible that a normal male hormonal environment may not yet have been faithfully replicated. Such cyclicity, however, has been successfully suppressed in rats under comparable circumstances, whether or not the normal male hormonal environment was replicated.

Developmental

Little has been done on the relationship between maturation and aggression in primates. Schaller (1963), in his book on the mountain gorilla, suggests that silver-backed males (fully mature) are the most aggressive and are dominant. However, he offers no quantified analysis of agonistic behavior in that species, and there have been no measures of testosterone level made.

Van Lawick-Goodall (1973), writing about the chimpanzees of the Gombe National Park in Tanzania, says that for the male, adolescence often seems to be a stressful period. The male goes through a period of rapid growth, after which he tends to become more aggressive, especially toward females. In late adolescence, he begins to threaten and occasionally attack low-ranking males of the adult hierarchy. Shirley McCormack (1971) has investigated plasma testosterone concentrations in the male chimpanzee and has found a large rise in testosterone level in the adolescent (ages 6–9 years) and another large rise in the adult. The level found in infants was 4.7 ± 6.2 ng/100 ml, 263 ± 144 ng/100 ml in adolescents, and 495 ± 143 ng/100 ml in adult males. This suggests that testosterone levels rise during maturation, possibly synchronously with the increase in normal aggressive behavior.

Resko (1967) reports that in the rhesus, testosterone is not detectable prepubertally in the peripheral plasma, but that at approximately 3–4 years of age its concentration reaches early adult levels. Similar findings have been made in boars: Carlson *et al.* (1971) report that the maximal rate of increase of testosterone production occurs during puberty between 78 and 130 days of age in Poland China boars. However, quantified measures of changes in aggressiveness at puberty are not available for either species.

Clearly, there is the need for a good deal of work in determining the relationship of increasing levels of testosterone and frequency of aggression at the time of and following puberty. Conclusions cannot be drawn at this time.

Seasonal

Examination of the literature relating to primate breeding suggests another link between testosterone and aggressive behavior. Breeding in many primate species appears to be seasonal (Lancaster and Lee, 1965), and a higher level of aggression and wounding seems to be associated with this season. In addition, there is strong indirect evidence that testosterone levels in males are higher at this time than at other times of the year, although until very recently there have been no direct measurements of plasma testosterone variations with breeding season.

Southwick *et al.* (1961) found indications of seasonal breeding in rhesus monkeys in India, and Imanishi (1960) came to similar conclusions about native Japanese macaques. Conaway and Koford (1965) observed sexually mature monkeys in the largest of six social groups on Cayo Santiago, a small island (37 acres) off Puerto Rico. The island was stocked in late 1938 with monkeys from India. A distinct annual reproductive cycle has been observed at least since 1956, when the National Institutes of Health assumed responsibility for the colony, and systematic record-keeping began. Mating appears to be confined to the period from July to January, peaking in September and October.

In addition, they found the breeding season of the female clearly indicated by the occurrence of distinct estrous periods during which consort relation, copulation, and associated behavior occur, but they did not note an increase in aggression among males. Vandenbergh and Vessey (1968), however, found evidence that aggression among individuals is higher during the breeding season than at other times of the year, but there was no indication (Vessey, 1968) of an increase in group interactions accompanied by aggression.

Kaufmann (1967) observed a free-ranging colony of rhesus on Cayo Santiago and found seasonal differences in the proportion of agonistic versus nonagonistic and aggressive versus submissive behavior of males. In the mating period, although there was relatively less agonistic behavior than in the rest of the year, this agonistic behavior consisted of a higher proportion of aggressive as opposed to submissive acts. He suggests that this could be influenced by the typical male–aggressive, female–submissive nature of the consort relation, as males have a higher proportion of their interactions with females during this time.

Lindburg, in his study (1969) of five rhesus groups in northern India, found that adult males occasionally transferred from one social group to another. The criterion for such a shift was simply the sighting of a male in the vicinity of a group other than that of his origin. In every case, however, a male which was counted as having shifted clearly altered his patterns of traveling, foraging, and resting to conform with his new group. The frequency of such shifts corresponded to the breeding season, and none were observed between the end of January and mid-October when the mating season begins. These transfers were accompanied by an increase in aggressive behavior between the intruding male and the new group, consisting mostly of branch shaking and exchange of threats. Lindburg suggests, however, that actual fighting, although observed only once, was probably more frequent, as several severe wounds were noted in adults of both sexes.

Further observations were made of increased aggressive behavior by Wilson and Boelkins (1970) observing on Cayo Santiago. As noted earlier, the mating season begins around July to August and is over by the end of December. Wilson logged over 1400 hr of direct observation time between July 1965 and August 1967 and notes that death and wound counts for July–August are probably low as a result of interruptions and difficulties in observing. They arbitrarily defined a wound as any visible cut of 4 cm or longer. The males incurred significantly more wounds during the mating season than at any other time. In females, there was slightly more wounding in the birth season (January through May); but the difference was not statistically significant, and the authors suggest that it may have been due to the irascibility of some females in late pregnancy. Male deaths, too, were significantly higher in mating season with 86.6% of all male deaths occurring at that time. Female deaths were higher, but not significantly so, during the birth season. This may be related to complications developing during labor and the postpartum period, although this is not known. These data are strong evidence for a seasonal increase in aggressive behavior in the male rhesus at breeding time.

The squirrel monkey (*Saimiri sciureus*), too, shows seasonal morphological and behavioral changes. DuMond and Hutchinson (1967) observed a Miami colony living in a seminatural state and found that during the mating season the males change in appearance, becoming heavy and fluffy about the shoulders and arms—the so-called "fatted male phenomenon." They become aggressive, volatile, and vocal as compared with the retiring, more passive behavior of the nonfatted male during the birth season and when females are actually lactating. The investigators felt that these changes in behavior and appearance and in actual mating indicated cyclical production of androgen and therefore tested for the existence of a spermatogenic cycle. They took testis samples from animals newly imported from their natural environment. During the mating season, the fatted males are in a spermatogenic phase, and when the males are not fatted, the testes are in an inactive phase with intermediate stages between the two extremes.

It seems clear, then, that breeding in several primate species is a seasonal affair, and there are clear indications that it is accompanied by an increase in aggressive behavior. The question now becomes whether and how this is related to testosterone production and levels in the plasma. Until recently, there has been no direct evidence of a seasonal rise in hormone level, but studies have been done on testis size and sex-skin coloration which strongly indicate such variation.

Sade (1964), for example, observing the rhesus on Cayo Santiago, measured testis size over time and found that the cycle of large testes and small testes corresponds to the cycle of true copulations and births. He also found that the color of the sex skin is more intense at the time of the year of larger testes.

Vandenbergh (1965), studying rhesus monkeys removed from free-ranging colonies on Cayo Santiago and La Cueva, attempted to determine the hormonal basis of such sex skin changes, to quantify these changes, and to reproduce them by hormone administration. He found that castration of the adult rhesus results in a loss of redness and edema (but not complete disappearance) of the sex skin. Administration of estrogen to the castrates led to a sex skin resembling that of a pubertal female. Low doses of testosterone produced coloration of the sex skin resembling that of a free-ranging male during the nonbreeding season, and higher doses produced sex skin like that of a breeding male. This strongly suggests, although it does not prove, that testosterone levels are higher during breeding season, when aggressive behavior and wounding are peaking, and that these higher levels cause morphological changes.

These results generated the further question of whether the seasonal variation in male hormone levels was simply coincidental with the female breeding cycle or if, indeed, it could be induced by estrous females. Vandenbergh, in a later study (1969), found that by bringing females artificially into estrus in the nonbreeding season, morphological changes could be induced in the males exposed to the treated females. Following exposure to the estrogen-treated females, testicular size, seminiferous tubule diameter, and sex skin redness all increased significantly, thereby suggesting that (*1*) females communicate their endocrine status to males, and (*2*) there is a system for the synchronization of mating activities between the sexes.

Recent studies indicate that not only is sexual behavior seasonal in both the male and female rhesus, but that the seasonal sexual arousal in the male is dependent upon their contact with estrous females. Gordon and Bernstein (1973) divided 35 sexually mature male rhesus at Yerkes Regional Primate Research Center in Lawrenceville, Georgia, into three groups. One ($n = 7$) was housed with 46 females and 23 juveniles and infants as part of a breeding group; the second ($n = 8$) had visual and olfactory contact with the breeding group, but was housed separately; and the third group of males ($n = 20$) had no contact with either the breeding group or the other all-male group. Each group was maintained in a separate 38.1-m square outdoor compound with an attached indoor area. Observations were made during the summer and fall of 1971.

No true sexual behavior was observed during the nonbreeding season. With the advent of mating season, the animals in the breeding group resumed sexual activity and showed evidence of increased testosterone secretion in terms of heightened coloring in facial and perianal regions. The all-male group which had access to visual contact with the breeding group also evidenced hormonal activation and engaged in frequent mounting activity. Mount rates in this male group increased 50-fold, with some mounts terminating in ejaculation. In the group which had no contact with the mixed group, there was no evidence of sexual arousal. There was no variation in their mounting frequency and no evidence of hormonal arousal.

Very recently, we have been able to analyze plasma samples from males both during the breeding season as well as during the birth season. Findings from a small number of males indicate that there is indeed a significant two- to threefold increase in plasma testosterone during the breeding season (Gordon *et al.*, 1973).

These data provide strong evidence that female breeding activity is seasonal and that the females in some way communicate their state of sexual receptivity to the males. This communication then induces a state of sexual arousal in the male.

Somewhat similar observations have been made in the red deer (*Cervus elaphus*), the roe deer (*Capredus capreolus*), and the Asiatic elephant (*Elephas maximus*).

Work done by Lincoln *et al.* (1970) indicates that the period of intense sexual activity of the red deer stag, referred to as rutting, is largely determined by rising levels of testosterone. Aggressive behavior, however, appears to be influenced by the effect of rising testosterone levels on the size and condition of the antlers. Stags lose, or cast, their antlers every second year at about the end of March and begin growing new ones. The new antlers are covered with soft growth, which is referred to as velvet. In general, the older the stag, the larger his antlers grow. When testosterone levels begin their seasonal rise in July and August (as measured by determinations done on ether extracts of stag testes), the velvet is rubbed off the antlers and the stags are referred to as being in hard horn. At this time, following the cleaning of the velvet, there is an abrupt change in aggressive behavior within 2–5 days. When the horns are in velvet, the forefeet are the primary means of offense and defense, whereas once the velvet is removed, antlers become the primary means of fighting. Often the animal with the largest antlers becomes dominant, and the dominance hierarchy becomes linear. By mid-September, the older stags move to their individual territories where they herd and mate with the females. Testosterone levels fall considerably by November and rut comes to a gradual end. By March, testosterone is usually undetectable and antlers are again cast. These findings suggest that the type of aggressive behavior is determined by antler state which, in turn, appears to be controlled by changes in testosterone levels. In the roe deer, Bramley (1970) also found that increases in territorial behavior coincided with peak androgen secretion.

The male Asiatic elephant periodically comes into "musth," a time which is characterized by a very aggressive temperament. Jainudeen *et al.* (1972) found that during this time plasma testosterone levels rise to a high of from 2960 to 6540 ng/100 ml, as compared with levels of from <20 to 140 ng/100 ml during nonmusth periods. Musth, although it is commonly believed to be a form of sexual activity, is not a well-understood phenomenon.

Additional findings concerning the relationship of transitional rises in plasma testosterone levels and copulatory activity have been made recently by several groups of researchers. Significant increases in plasma testosterone in male rabbits after copulation have been reported by Haltmeyer and Eik-Nes (1969) and Saginor and Horton (1968). This may be analogous to reflex ovulation in the female and is probably induced by release of gonadotropin.

Castration

There have been several studies done with primate castrates in an attempt to determine the effect of administration of various hormones on sexual and aggressive behavior and dominance. The results of these studies, taken as a whole, are inconclusive, because the examination of aggression is mostly in the context of sexual interactions, and further work clearly needs to be done. Nevertheless, these results are relevant and should be reported here.

The castrate research may be divided conceptually into two groups—the papers dealing primarily with sexual behavior and with aggression in the context of sexual activity, and those dealing with dominance. Michael and Zumpe (1970), examining the relationship between changes in agonistic behavior between males and females and changes in the endocrine status of the female, observed opposite-sexed pairs over time with both sexes intact and with females ovariectomized. They found that ovariectomy resulted in a significant decrease in female receptivity, a slight nonsignificant increase in female aggression, and a marked increase in number of aggressive episodes by males. Male mounting attempts remained constant; however, they state that the great majority of male aggressive episodes with ovariectomized females was not associated with such mounting attempts. The increased female aggression was, in 21 out of 40 episodes, a response to male aggression, and in 12 was associated with male mounting attempts. The possibility, however, that the increased aggression observed on the part of both sexes was only a reflection of the changes in female receptivity must be considered.

Trimble and Herbert (1968), in their investigation of the effects of testosterone propionate and estradiol monobenzoate on the sexual behavior of the ovariectomized female rhesus, found that giving either estrogen or low doses (1 mg/day) of testosterone resulted in a marked increase of sexual presentations made by the female. When estrogen alone was given, there was an increase in the males' mounting activity. This was not found when testosterone was given alone. This suggests that the testosterone-treated females were not sexually attractive to the males. Higher doses of testosterone decreased the females' presentation and increased their aggressiveness toward males.

A more recent study was done by Everitt *et al.* (1972), in which they studied the effects of adrenal androgens on the sexual receptivity of the female rhesus monkey. They employed females which had previously been ovariectomized and were maintained on 25 μg/day of estradiol replacement therapy. They measured the number of presentations by the females, number of mount refusals, female ejaculatory responses, and female acceptance ratio. All these measures fell following adrenalectomy of the female. This fall in various measures of sexual receptivity was reversed following the administration of large doses of one of the adrenal androgens (200–400 μg/day of Δ^4-androstenedione; dehydroepiandrosterone was ineffective). This level of androgen administration did not affect agonistic interaction between the males and females studied.

These data support the interpretation that in female primates, as opposed to other mammals, androgens are more essential for female receptivity than either estrogens or progesterone (for general review, see Rose, 1972).

In the adult male rhesus, recent research suggests that castration and the consequent large decrease in plasma testosterone level results in a significant reduction of sexual activity. Wilson *et al.* (1972) found in observations of eight males that the number of mounts, intromissions and ejaculations declined significantly in each case within 6 weeks of operation. However, in four of the animals, ejaculation continued for 20, 27, 41, and 109 weeks, respectively. Injections of testosterone propionate resulted in significant increases in sexual activity within 2–8 weeks.

Resko and Phoenix (1972) also found sexual behavior in males greatly reduced or eliminated by castration, but 3 of their 10 castrates displayed the full complement of sexual behavior, including ejaculation, 1 year after castration. Level of sexual behavior after castration did not correlate with the low levels of plasma testosterone still present, which were assumed to be of adrenal origin.

The other major area of investigation involving castrates is the work relating to dominance. Early research was done by Birch and Clark with both male and female castrate chimpanzees. In the male-castrate study (Clark and Birch, 1945), two animals were used, one a 9½-year-old prepubertal castrate, and the other a 10½-year-old intact male of approximately the same weight. Their behavior was evaluated in a competitive feeding situation. Drugs used were methyl-testosterone and alpha-estradiol administered orally to the castrated animal. Before hormone therapy, the castrate was significantly but not completely dominant over the other animal. Upon administration of 50 mg testosterone daily, the castrate became clearly dominant. This relationship continued for about 2 months after cessation of treatment. With estrogen administration to the castrate, the other animal became dominant.

In the female-castrate study (Birch and Clark, 1946) three postpubertally ovariectomized and hysterectomized females were used. These animals, upon initial testing, showed a stable linear dominance–subordination hierarchy. The competitive feeding situation was again used to determine dominance. Birch and Clark came to the following conclusions: (*1*) female dominance status was enhanced by raising the estrogen level, (*2*) female dominance status was enhanced by raising the androgen level, (*3*) rise in dominance induced by estrogen paralleled the course of sexual swelling and disappeared with detumescence, and (*4*) improvements in dominance induced by androgen are not accompanied by swelling and appear to be more persistent.

While these are very interesting results, it is important to keep in mind that they are predicated on the observations of few animals and their findings have not been duplicated. It is also important to remember that cross-species generalizations should not be made without adequate evidence.

Mirsky (1955) investigated the effects of hormone administration in gonadectomized but immature rhesus monkeys. He used in the first part of his experiment a group of five males and a group of five females, determined the initial linear hierarchy, and then administered hormones (first androgen, later estrogen) to selected animals (numbers 2 and 5 in the hierarchy) by means of pellet implantation. In no case did he find a change in the dominance–subordination relationship existing between two animals prior to such treatment.

In the second part of the experiment, he used four like-sexed pairs of gonadectomized monkeys and administered hormone (androgen or estrogen) to the subordinate of each pair. The score changes that occurred showed little or no consistency and the hierarchies remained unchanged. Mirsky suggests that a possible explanation of the difference between these findings and those of Birch and Clark could be that any hormonal effect on dominance might be masked in the rhesus by the greater degree of aggressiveness characterizing its social behavior, as compared with the chimpanzee. It should be noted that the animals studied by Mirsky were still adolescent and had not reached sexual maturity. Consequently, the males had not experienced the potential influence of rising testosterone levels on the development of normal adult aggressive behavior. It may be argued from analogy with sexual studies that the effects of replacing testosterone following castration differ in mature and immature castrates. Testosterone is more effective in restoring sexual behavior in animals which have a greater history or repertoire of sexual behavior. Consequently, the potential effects of testosterone on aggressive behavior and dominance might differ if it were administered to mature adult males which had already passed puberty. Nevertheless, this work raises serious questions about a simple unitary relationship of testosterone with dominance.

One final set of observations should be noted here, and these involve the results of castration on the free-ranging male rhesus. Wilson and Vessey (1968) report on four infant castrates and one 4-year-old castrate on Cayo Santiago and on five males on La Parguera castrated at ages 3 to 7 years. All animals were free ranging at the time of the study. Wilson and Vessey found several long-term effects of castration: (*1*) 6 of the 10 gradually fell in dominance rank several years postoperatively, (*2*) the castrates generally showed less sexual behavior than other males, (*3*) 8 of the 10 preferred to associate with other castrates, (*4*) 2 of the 10 dominated males twice their weight. As there were no animals which were studied as controls, it is difficult to draw firm conclusions from this study. As only 6 out of 10 castrates fell in dominance, this might have occurred solely as a result of aging. Castration appears to have more clearly affected future sexual behavior, and its long-term effects upon aggressive behavior remain unclear.

The results of the studies reviewed in this section which investigate the effects of castration and hormone replacement on sexual behavior suggest that such behavior may be greatly reduced or eliminated by castration. In some individuals, however, the full complement of sexual behavior may be retained after gonadectomy, possibly indicating that hormonal influence may be overridden by learning or sexual experience.

Studies of castrates relating to dominance are somewhat less conclusive as they are few in number, involve different species, and in the case of Birch and Clark, use very small numbers of animals. Testosterone levels probably relate to position in a dominance hierarchy in some complex fashion, which will be elucidated only with further work.

Yerkes Studies

In the summer of 1968, a joint program was started between the Walter Reed Army Institute of Research and the Yerkes Regional Primate Research Center. This program was undertaken to further our understanding of the effects of psychological stress on rhesus macaques living in social groups. The development of facilities and personnel necessary to carry out the related behavioral and endocrine measurements was supported by the U.S. Army Medical Research and Development Command.[1]

An initial study was undertaken to clarify some of the issues surrounding the relationship between testosterone, dominance rank, and aggressive behavior in the male rhesus. Thirty-four males maintained in a 125 ft × 125 ft outdoor compound were studied (Rose *et al.*, 1972). Food and water were freely provided, and the monkeys had access to enclosed temperature-regulated shelter. During the first month of the experiment, the dominance hierarchy was established and stabilized. In months 2 through 4, behavioral data were collected, and in months 5 and 6, the animals were adapted to the collection procedure. During the seventh month, blood samples were taken for plasma testosterone and cortisol determination. Of all observed behavior, only a very small portion (1.85%) was aggressive in nature, and of that, only one-fifth was contact aggression. This suggests that dominance rank, once established, is maintained largely by threat gestures and submissive responses.

The authors found that both aggressive behavior and being submitted to correlated positively and significantly with plasma testosterone level. A high number of submissive responses, however, did not have a strong relationship with low testosterone concentration, suggesting that an animal could be highly aggressive, have a relatively high level of plasma testosterone, and still exhibit many submissive responses. Dominance rank, also, was positively correlated with testosterone concentration. The animals in the highest quartile of the dominance hierarchy had significantly higher plasma testosterone than those in the lower three quartiles, but there were no significant differences between quartiles two, three, and four.

[1] Contract No. DADA 17-69-C-9014.

Although animals high in either dominance rank or in frequency of aggressive behavior tended to have high levels of plasma testosterone, a high degree of aggressive behavior does not necessarily signify a high rank in the hierarchy or vice versa. The alpha male, for example, was twelfth in frequency of aggressive behavior, and the most aggressive animal was tenth in dominance rank.

Having found a significant relationship between testosterone level and frequency of aggressive behavior, the next question to be addressed was whether this high hormone level in aggressive and dominant animals preceded their attainment of dominant and/or aggressive status or whether it was a reflection of the effect of their social environments. Rose *et al.* (1972) deliberately altered the social environment of four male rhesus to determine what effect, if any, these changes would have on testosterone levels. When these animals were housed with adult females, some of which were in estrus, they engaged in frequent sexual activity, each became the dominant (alpha) animal, and their testosterone level significantly increased. They remained with the females 2 weeks. Within 1 week after removal and return to an individual cage, mean testosterone levels fell back toward base line.

They then introduced each male individually to a well-established group of 30 adult males, with the expectation that the group would be intolerant of a newcomer. This proved to be the case, and each of the four suffered a sudden and profound defeat. Every male was removed within 2 hr of his introduction. All exhibited a marked decrease in plasma testosterone by the end of the week, averaging an 80% drop from base-line levels. Two of these males still exhibited significantly depressed levels 6 and 9 weeks, respectively, after defeat. These same two animals were, then, at 9 and 15 weeks, respectively, reintroduced to the female group and their testosterone levels showed significant increases, reaching or surpassing within 4 days those observed originally with the females.

The observation that plasma testosterone falls after defeat in all four animals is somewhat complicated by the fact that two males suffered wounding extensive enough to require sutures. Although two did not, it is possible that the fall in levels during the first week after defeat could be due at least in part to physical injury.

Another study was undertaken to examine systematically the effect of loss of dominance status on plasma testosterone, without confounding the results by physical injury. Two groups of rhesus were introduced to one another.[2] One group was large and well established, consisting of many juveniles and females. However, the only two adult males in the group were older and had had their canines removed surgically years ago.

[2] See chapter by Bernstein *et al.*, this volume.

The second group consisted of 4 adult males and 13 females, but had been formed only several months previously. Approximately 1 week following the merger of the two groups, the older, more well-established group clearly became dominant over the smaller, more recently established group containing the four males. There was no serious wounding or injuries. In the week following the defeat of the smaller group, all four males dropped in plasma testosterone, and levels continued to fall in the following 6 weeks. After this time, taking samples for plasma testosterone was discontinued. By the sixth week, the plasma testosterone for the four males decreased to approximately 70% of their values prior to group merger (Rose *et al.*, 1973). This study strongly suggests that plasma testosterone is suppressed following loss of dominance status. The fall in testosterone paralleled the observation that the four males developed a peripheral status in the group, interacting infrequently with resident animals, and failing to compete for preferred food, even with the juveniles of the well-established group.

These findings support the interpretation that social and environmental stimuli can function as important regulators of testosterone secretion. Testosterone levels are increased following access to estrous, receptive females, concomitant with assumption of dominance status by the male in the new group. Further work is in progress to separate the potential influence of these two stimuli in increasing plasma testosterone.

Defeat and loss of dominance status appear to function as powerful stimuli inhibiting testosterone secretion. This effect appears to persist for many weeks following the actual event. Whether the defeated males are housed individually in cages or remain in the new group as peripheral animals, testosterone levels remain depressed for at least 6 weeks after defeat and loss of status. Preliminary evidence suggests that the adult females of the group may have a crucial role in integrating the new males into the group structure. Concomitant with this integration, it may be that plasma testosterone returns to normal or even elevated levels, such as those seen during the breeding season. Further work is in progress to define the relationship between the presence and behavior of the adult females, the integration of recently introduced males, and their plasma testosterone levels.

CONCLUSION

In this chapter, we have attempted to review the evidence on the relationship between androgens and aggressive behavior. The data generally support the conclusion that, in rodents, testosterone has an important role in regulating many aspects of sexual and aggressive behavior.

However, one cannot extrapolate to other species, especially primates, save in a limited way. Even the studies of the effects of androgens on aggressive behavior in rodents are limited primarily to relatively artificial situations. It is still unclear what relationship shock-elicited fighting bears to the agonistic behavior of animals living in social groups and in their usual home range. Furthermore, as Beach and others have pointed out, as one moves from studies of rodents to other nonprimate mammals such as cats and dogs, to primates, hormonal influences on behavior are increasingly modified by experiential variables. Nevertheless, there is some convergence of work done with various primate groups which strongly suggests that endocrine factors significantly shape some aspects of sexual and aggressive behavior.

The evidence for the role of androgens in regulating sexual behavior is less ambiguous. Castration in males or removal of adrenal androgen in females significantly depresses sexual behavior. Replacement with certain biologically active androgens, such as testosterone, and to a lesser extent, Δ^4-androstenedione, reverses this loss of sexual functioning. However, it should be noted that removal of androgens is not immediately followed by a complete cessation of sexual behavior. Indeed, it may take as long as 2 years for the effects of castration to become evident, possibly reflecting the important influence of experiential factors.

In our discussion of endocrine influences on aggressive behavior in primate groups, we have restricted our review to those studies dealing with intraspecific aggression. Furthermore, we are interested primarily in the aggressive behavior that occurs in the context of maintaining the dominance hierarchy, preserving dominance status, defending the group against the presence or introduction of strange animals, competition for preferred food, or increased fighting associated with the breeding season. We have thus excluded other categories of aggressive behavior, such as that associated with hunting for prey, which, although rare, occurs in some primate groups, or the aggressive behavior of a mother defending her offspring. Whatever influence androgens have on aggressive behavior, it is more likely related to the agonistic responses associated with dominance status, attacking strange animals, etc., rather than hunting or defense of young. As these modes of aggressive behavior often appear more frequently in females, it is unlikely that they relate directly to levels of circulating androgens.

Our data on the relationship between androgens and aggressive behavior in nonhuman primate groups are really quite limited. Most of the studies provide information showing that plasma testosterone may be increased or decreased under certain circumstances. It is highly likely that it is increased during the breeding season, or after males are permitted access to receptive females after some period of isolation, or maybe even after assumption of dominance. We have also observed one instance in which testosterone increased in a dominant rhesus male when he successfully defended his group after it was confronted by the introduction of another strange group. As we have also noted, defeat and loss of dominance status are followed by a fall in testosterone. However, the fact that testosterone levels change subsequent to important environmental events does not document the degree to which testosterone or other androgens influence aggressive behavior. It might be argued teleologically that the rise in testosterone during the breeding season supports the increase in fighting behavior and sexual activity observed during this period. In a parallel fashion, the fall in testosterone following defeat might be considered adaptive. As aggressive threats or challenges are severely punished by the dominant members of a group, higher levels of testosterone that might stimulate such behavior would be inappropriate. Nevertheless, these observations fall far short of demonstrating that appearance of aggressive behavior is controlled by the level of circulating androgens.

The types of studies that would provide more direct information on this question are few in number and far from conclusive. The short or long-term effects of castration and hormonal replacement on aggressive behavior are still unclear. The presence of testosterone during fetal development does suggest at least a permissive effect on the future behavior of the animal, i.e., it may be necessary, but not sufficient for the appearance of certain patterns of behavior in the male monkey. These observations must be extended to animals born and living in normally constituted social groups before more definitive conclusions may be drawn.

Perhaps one of the most important studies that can be done to clarify these issues is to focus on the relationship between rising levels of testosterone and the changes in behavior that occur during normal development and puberty. We lack the most rudimentary data on any parallel between the increased frequency of threat, challenge, or fighting behavior seen in the adolescent or adult male and the rise in plasma testosterone that occurs during this period. Studies on the development of the social behavior of animals who were castrated early in life are under way at the Caribbean Primate Research Center facility near La Parguera, Puerto Rico (Loy, 1973) and will provide important information on this issue.

As an overview, it is highly likely that androgens do play some role in the expression of aggressive behavior in various primate groups. Perhaps this effect is only permissive; perhaps hormones function to establish some behavioral propensity, which may be highly modified or overshadowed by the influence of life experiences. As a guide to our thoughts and future experiments, it is well to keep in mind that there are two periods in development where the sexes differentiate the most endocrinologically, *in utero* and during puberty. The systematic study of these periods may provide the clearest means of discerning the endocrine influences on aggressive behavior.

REFERENCES

Barraclough, C. A. (1961). Production of anovulatory, sterile rats by single injections of testosterone propionate. *Endocrinology* **68**, 62–67.

Beach, F. A. (1947). Evolutionary changes in the physiological control of mating behavior in mammals. *Psychol. Rev.* **54**, 297–315.

Beach, F. A. (1958). Evolutionary aspects of psychoendrocrinology. *In* "Behavior and Evolution" (A. Roe and G. G. Simpson, eds.), pp. 81–102. Yale Univ. Press, New Haven, Connecticut.

Beach, F. A. (1970). Some effects of gonadal hormones on sexual behavior. *In* "The Hypothalamus" (L. Martini, M. Motta, and F. Fraschini, eds.), pp. 617–639. Academic Press, New York.

Birch, H. G., and Clark, G. (1946). Hormonal modification of social behavior: II. The effects of sex-hormone administration on the social dominance status of the female-castrate chimpanzee. *Psychosom. Med.* **8**, 320–331.

Bradbury, J. T. (1941). Permanent after-effects following masculinization of the infantile female rat. *Endocrinology* **28**, 101–106.

Bramley, P. S. (1970). Territoriality and reproductive behaviour of the Roe Deer. *J. Reprod. Fertil. Suppl. II*, 43–70.

Carlson, I. H., Stratman, F., and Hauser, E. (1971). Spermatic vein testosterone in boars during puberty. *J. Reprod. Fertil.* **27**, 177–180.

Clark, G., and Birch, H. G. (1945). Hormonal modifications of social behavior: I. The effects of sex-hormone administration on the social status of a male-castrate chimpanzee. *Psychosom. Med.* **7**, 321–329.

Conaway, C. H., and Koford, C. (1965). Estrous cycles and mating behavior in a free-ranging band of rhesus monkeys. *J. Mammal.* **45**, 577–588.

Conner, R. L., and Levine, S. (1969). Hormonal influences on aggressive behavior. *In* "Aggressive Behaviour" (S. Garattini and E. B. Sigg, eds.), pp. 150–163. Excerpta Medica Foundation, Amsterdam.

Davis, P. G., and Gandelman, R. (1972). Pup-killing produced by the administration of testosterone propionate to adult female mice. *Horm. Behav.* **3**, 169–173.

DuMond, F. V., and Hutchinson, T. C. (1967). Squirrel monkey reproduction: The "fatted" male phenomenon and seasonal spermatogenesis. *Science* **158**, 1467–1470.

Everitt, B. J., Herbert, J., and Hamer, J. D. (1972). Sexual receptivity of bilaterally adrenalectomised female rhesus monkeys. *Physiol. Behav.* **8**, 409–415.

Gandelman, R. (1972). Induction of pup-killing in female mice by androgenization. *Physiol. Behav.* **9**, 101–102.

Gerall, A. A. (1966). Hormonal factors influencing masculine behavior of female guinea pigs. *J. Comp. Physiol. Psychol.* **62**, 365–369.

Gerall, A. A., and Ward, I. (1966). Effects of prenatal exogenous androgen on the sexual behavior of the female albino rat. *J. Comp. Physiol. Psychol.* **62**, 370–375.

Gordon, T. P., and Bernstein, I. S. (1973). Seasonal variation in sexual behavior in all-male rhesus troops. *Am. J. Phys. Anthrop.* **38,** 221–226.

Gordon, T. P., Bernstein, I. S., and Rose, R. M. (1973). Seasonal changes in sexual behavior and plasma testeosterone levels of group living monkeys. *Anatom. Record* (abstract) (in press).

Goy, R. W., and Phoenix, C. H. (1972). The effects of testosterone propionate administered before birth on the development of behavior in genetic female rhesus monkeys. *In* "Steroid Hormone and Brain Function" (C. Sawyer and R. Gorski, eds.), pp. 193–201. Univ. of California Press, Berkeley, California.

Goy, R. W., and Resko, J. A. (1972). Gonadal hormones and behavior of normal and pseudohermaphroditic nonhuman female primates. *Rec. Progr. Horm. Res.* **28,** 707–733.

Haltmeyer, G. C., and Eik-Nes, K. B. (1969). Plasma levels of testosterone in male rabbits following copulation. *J. Reprod. Fertil.* **19,** 273–277.

Harlow, H. F. (1965). Sexual behavior in the rhesus monkey. *In* "Sex and Behavior" (F. A. Beach, ed.), pp. 234–265. Wiley, New York.

Harris, G. W., and Levine, S. (1962). Sexual differentiation of the brain and its experimental control. *J. Physiol.* **163,** 42P–43P.

Hinde, R. A., and Spencer-Booth, Y. (1967). The behaviour of socially living rhesus monkeys in their first two and a half years. *Anim. Behav.* **15,** 169–196.

Hutchinson, R. R., Ulrich, R. E., and Azrin, N. H. (1965). Effects of age and related factors on the pain-aggression reaction. *J. Comp. Physiol. Psychol.* **59,** 365–369.

Imanishi, K. (1960). Social organization of subhuman primates in their natural habitat. *Curr. Anthropol.* **1,** 393–407.

Jainudeen, M. R., Katongole, C. B., and Short, R. V. (1972). Plasma testosterone levels in relation to musth and sexual activity in the male asiatic elephant, *Elephas Maximus. J. Reprod. Fertil.* **29,** 99–103.

Jensen, G. D., Bobbitt, R. A., and Gordon, B. N. (1966). Sex differences in social interaction between infant monkeys and their mothers. *In* "Recent Advances in Biological Psychiatry" (J. Wortis, ed.), pp. 283–293. Plenum Press, New York.

Karsch, F. J., Dierschke, D. J., and Knobil, E. (1973). Sexual differentiation of pituitary function: Apparent difference between primates and rodents. *Science* **179,** 484–486.

Kaufmann, J. H. (1967). Social relations of adult males in a free-ranging band of rhesus monkeys. *In* "Social Communications Among Primates" (S. A. Altman, ed.), pp. 73–98. Univ. of Chicago Press, Chicago, Illinois.

Lawick-Goodall, J. van (1973). The behavior of chimpanzees in their natural habitat. *Am. J. Psychiat.* **130,** 1–12.

Lee, C. T., and Brake, S. C. (1972). Reaction of male mouse fighters to male castrates treated with testosterone propionate or oil. *Psychon.* Sci. **27,** 287–288.

Lincoln, G. A., Youngson, R. W., and Short, R. V. (1970). The social and sexual behavior of the Red Deer stag. *J. Reprod. Fertil. Suppl. II,* 71–103.

Lindburg, D. G. (1969). Rhesus monkeys: Mating season mobility of adult males. *Science* **166,** 1176–1178.

Loy, J. (1973). Personal communication.

Mason, J. W. (1968). Organization of psychoendrocrine mechanisms. *Psychosom. Med.* **30,** 565–808.

McCormack, S. A. (1971). Plasma testosterone concentration and binding in the chimpanzee; Effect of age. *Endocrinology* **89,** 1171–1177.

Michael, R. P., and Zumpe, D. (1970). Aggression and gonadal hormones in captive rhesus monkeys (*Macaca mulatta*). *Anim Behav.* **18**, 1–10.

Mirsky, A. F. (1955). The influence of sex hormones on social behavior in monkeys. *J. Comp. Physiol. Psychol.* **48**, 327–335.

Payne, A. P., and Swanson, H. H. (1972). The effect of sex hormones on the agonistic behavior of the male golden hamster (*Mesocricetus auratus* Waterhouse). *Physiol. Behav.* **8**, 687–691.

Powell, D. A., and Creer, T. L. (1969). Interaction of developmental and environmental variables in shock-elicited aggression. *J. Comp. Physiol. Psychol.* **69**, 219–225.

Powell, D. A., Francis, J., and Schneiderman, N. (1971). The effects of castration, neonatal injections of testosterone, and previous experience with fighting on shock-elicited aggression. *Commun. Behav. Biol.* **5**, 371–377.

Resko, J. A. (1967). Plasma androgen levels of the rhesus monkey: Effects of age and season. *Endocrinology* **81**, 1203–1212.

Resko, J. A. (1970). Androgen secretion by the fetal and neonatal rhesus monkey. *Endocrinology* **87**, 680–687.

Resko, J. A., and Phoenix, C. H. (1972). Sexual behavior and testosterone concentrations in plasma of the rhesus monkey before and after castration. *Endocrinology* **91**, 499–503.

Rose, R. M. (1969). Androgen responses to stress: I. Psychoendocrine relationships and assessment of androgen activity. *Psychosom. Med.* **31**, 405–417.

Rose, R. M. (1972). The psychological effects of androgens and estrogens—A review. *In* "Psychiatric Complications of Medical Drugs" (R. Shader, ed.), pp. 251–295. Raven Press, New York.

Rose, R. M., Holaday, J. W., and Bernstein, I. S. (1971). Plasma testosterone, dominance rank and aggressive behaviour in male rhesus monkeys. *Nature* (London) **231**, 366–368.

Rose, R. M., Gordon, T. P., and Bernstein, I. S. (1972). Plasma testosterone levels in the male rhesus: Influences of sexual and social stimuli. *Science* **178**, 643–645.

Rose, R. M., Bernstein, I. S., and Gordon, T. P. (1973). (in preparation).

Rosenberg, K. M., Denenberg, V. H., Zarrow, M. X., and Frank, B. L. (1971). Effects of neonatal castration and testosterone on the rat's pup-killing behavior and activity. *Physiol. Behav.* **7**, 363–368.

Sade, D. S. (1964). Seasonal cycle in size of testes of free-ranging *Macaca mulatta*. *Folia primat.* **2**, 171–180.

Saginor, M., and Horton, R. (1968). Reflex release of gonadotropin and increased plasma testosterone concentration in male rabbits during copulation. *Endocrinology* **82**, 627–630.

Schaller, G. B. (1963). "The Mountain Gorilla." Univ. of Chicago Press, Chicago, Illinois.

Southwick, C. H., Beg, M. A., and Siddiqi, M. R. (1961). A population survey of rhesus monkeys in northern India: II. Transportation routes and forest areas. *Ecology* **42**, 698–710.

Treloar, O. L., Wolf, R. C., and Meyer, R. K. (1972). Failure of a single neonatal dose of testosterone to alter ovarian function in the rhesus monkey. *Endocrinology* **90**, 281–284.

Trimble, M. R., and Herbert, J. (1968). The effect of testosterone or oestradiol upon the sexual and associated behavior of the adult female rhesus monkey. *J. Endocrinol.* **42**, 171–185.

Vandenbergh, J. G. (1965). Hormonal basis of sex skin in male rhesus monkeys. *Gen. and Comp. Endocrinol.* **5**, 31–34.

Vandenbergh, J. G. (1971). The effects of gonadal hormones on the aggressive behaviour of adult golden hamsters (*Mesocricetus auratus*). *Anim. Behav.* **19,** 589–594.

Vandenbergh, J. G. (1969). Endocrine coordination in monkeys: Male sexual responses to the female. *Physiol. Behav.* **4,** 261–264.

Vandenbergh, J. G., and Vessey, S. (1968). Seasonal breeding of free-ranging rhesus monkeys and related ecological factors. *J. Reprod. Fertil.* **15,** 71–79.

Vessey, S. H. (1968). Interactions between free-ranging groups of rhesus monkeys. *Folia primat.* **8,** 228–239.

Ward, I. L. (1972a). Prenatal stress feminizes and demasculinizes the behavior of males. *Science* **175,** 82–84.

Ward, I. L. (1972b). Female sexual behavior in male rats treated prenatally with an anti-androgen. *Physiol. Behav.* **8,** 53–56.

Ward, I. L., and Renz, F. J. (1972). Consequences of perinatal hormone manipulation on the adult sexual behavior of female rats. *J. Comp. Physiol. Psychol.* **78,** 349–355.

Wilson, A. P., and Boelkins, R. C. (1970). Evidence for seasonal variation in aggressive behavior by *Macaca mulatta*. *Anim. Behav.* **18,** 719–724.

Wilson, A. P., and Vessey, S. H. (1968). Behavior of free-ranging castrated rhesus monkeys. *Folia primat.* **9,** 1–14.

Wilson, M., Plant, T. M., and Michael, R. P. (1972). Androgens and the sexual behavior of male rhesus monkeys. *J. Endrocinol.* **52,** 11.

COMPARATIVE PRIMATE NEUROANATOMY OF STRUCTURES RELATING TO AGGRESSIVE BEHAVIOR

ORLANDO J. ANDY
University of Mississippi Medical Center

HEINZ STEPHAN
Max-Planck-Institut für Hirnforschung

INTRODUCTION

Aggressive behavior may be defined as a sensorimotor response integrated as an emotional drive to attack (Andy *et al.*, 1973). Based on an arbitrarily restricted form of this definition, it was decided to comparatively describe those brain structures in which direct stimulation and/or ablation would elicit or attenuate aggressive behavior. According to this criterion, the structures most closely related to aggressive behavior were the amygdala (Egger and Flynn, 1962, 1963, 1967; Fonberg, 1963; Fonberg *et al.*, 1962; Gastaut *et al.*, 1951; Hilton and Zbrozyna, 1963; Ingram, 1952; Kaada and Bruland, 1960; Kido *et al.*, 1967; Schreiner and Kling, 1953; Ursin, 1965; Wood, 1958; Woods, 1956; Yoshida, 1963; Zbrozyna, 1963; Fernandez de Molina and Hunsperger, 1959), hippocampus (Andy, 1961; MacLean and Delgado, 1953), septum (Andy *et al.*, 1957; Brady and Nauta, 1953, 1955; Brugge, 1965; King and Meyer, 1958), hypothalamus (Akert, 1961; Bard, 1928; Egger and Flynn, 1962, 1963; Hunsperger, 1963; Hunsperger and Bucher, 1967; Hunt, 1967; Karli and Vergnes, 1964; Levinson and Flynn, 1965; Nakao, 1958, 1967; Spiegel *et al.*, 1940; Wasman and Flynn, 1962; Wheatley, 1944; Yasukochi, 1960(, mesencephalon (Abrahams *et al.*, 1962; Flynn, 1962, 1967; Skultety, 1963), and thalamus (Ingram, 1958; Schreiner *et al.*, 1953). Most of the investigations relating these structures to aggression were performed on species other than the primate. However, there appeared sufficient consistency among species to assume that homologous structures participated in primate aggression. It is significant that in the human, the highest primate, lesions in the hypothalamus (Sano, 1958, 1966; Sano *et al.*, 1970), thalamus (Andy, 1966, 1970; Andy and Jurko, 1972; Dieckmann and Hassler, 1971; Hassler and Dieckmann, 1967, 1971; Poblete *et al.*, 1970; Schvarcz *et al.*, 1972), and amygdala (Balasubramaniam, 1972; Balasubramaniam and Ramamurthi, 1969, 1970; Blasubramaniam *et al.*, 1969, 1970; Heimburger *et al.*, 1966; Narabayashi, 1972; Narabayashi *et al.*, 1963) have been effective in alleviating aggressive behavior. Whereas the limbic, diencephalic, and brainstem structures may serve as the core through which an integrated aggressive attack is mobilized and expressed (Barnett, 1967; Fernandez de Molina and Hunsperger, 1962), the neocortex and other allocortical structures may serve as the monitor of aggressive behavior (Bard and Mountcastle, 1948; Spiegel *et al.*, 1940). It is thus considered pertinent to include also those cortical structures with their related basal ganglia in presenting the comparative primate neuroanatomy of structures relating to aggressive behavior.

TECHNIQUE

Quantitative volumetric studies have been utilized to determine differential degrees of development of various brain parts. In evaluating development it is imperative that the reference point be an indifferent structure, such as body size, to which the brain bears a constant relationship. Snell (1892) and Dubois (1897) used the body weight as an indifferent reference and found a mathematical expression for the brain–body size relationship within narrowly related forms. This method of comparison (known as the "allometric method") made possible the phylogenetic interpretation of comparative studies of the primate brain by using insectivores from which primates were thought to have evolved (Stephan, 1960, 1961). Through such studies it was subsequently noted that a structure such as the septum which was thought to become progressively smaller in phylogeny actually became larger in primate evolution and attained its greatest development in the human brain (Andy and Stephan, 1966, 1967, 1968; Stephan and Andy, 1962, 1964, 1969). By utilization of this technique therefore, differential degrees of development of various brain parts which relate to aggressive behavior are discussed. Volumetric determinations were made by outlining photographic enlargements of various brain parts, cutting them out, and weighing them (Fig. 1).

RESULTS

Neocortex

The neocortex in the simian revealed a development 45 times greater than in the basal insectivore. Within these groups, however, the various species have a wide range (Figs. 2 and 3). In contrast, the neocortex in the prosimian only averaged 15 times that in the basal insectivores. It should be stressed that the marked difference of increased neocortical volume in the simian over that of the prosimian was not shared by any of the other brain structures (Fig. 3).

A surprisingly very marked neocortical development was found in the human brain (Fig. 2). The human neocortex was 156 times that of the basal insectivores. In the chimpanzee the neocortex was approximately 60 times that of the basal insectivores. It should be noted that the developmental step from the basal insectivores to the chimpanzee is only one-third of that to the human. The extremely marked neocortical development in man undoubtedly represented the morphologic substrate for the very high and complex functional capacity of the human brain.

In comparison to the neocortex, the progression of all other brain structures in phylogeny, with the exception of the olfactory bulb, was relatively small but definitely evident.

Striatum

The second highest progression, following that of the neocortex, was striatum. However, the degree of progression was much less and its development in the human was closely similar to that in the lower primates (Figs. 3 and 4). With reference to the basal insectivores, it was five times enlarged in prosimians and nine times enlarged in simians. The progression of the striatum maintained an almost constant proportion to that of the neocortex, the enlargement of the neocortex being five times that of the striatum. This close relation in size would seem to indicate a functional relation between the two structures. However, there were specific species among lower primates such as the aye-aye (*Daubentonia madagascariensis*) which had a relatively high striatal value in contrast to the gorilla, chimpanzee, and human.

Hippocampus (Archicortex)

The hippocampus showed a definite enlargement in the prosimian of 2.0 to 2.4 times that of the basal insectivore. In progressing from the prosimian to the simian, there was no change in absolute size of the hippocampus (Fig. 3). The average progression index for the simians was 2.3. It was of interest to note that specific animal species in the prosimians and simians varied with respect to their relative hippocampal development. For example, comparatively low values were found among the prosimians in *Loris* and among the simians in the *Gorilla.* In contrast, comparatively high values were found among the prosimian in *Lemur* and *Daubentonia* and among the simians in *Homo* (Fig. 5). The wide scattering of the hippocampal indices for similar neocorticalization indicated further that there was no close size relationship between hippocampus and neocortex. The human had a progression index of 4.2, which meant that the hippocampus of the human was four times as large as that of a hypothetical basal insectivore of the same body weight. These comparative volumetric developments between hippocampus and neocortex revealed that both structures underwent enlargement in primate development but the hippocampus enlarged at a much slower rate than the neocortex which was 156 times the basal insectivore.

The hippocampus appeared to undergo a slight rotation and change in position relative to other structures during development from the lowest primate to the highest primate, including man (Fig. 6).

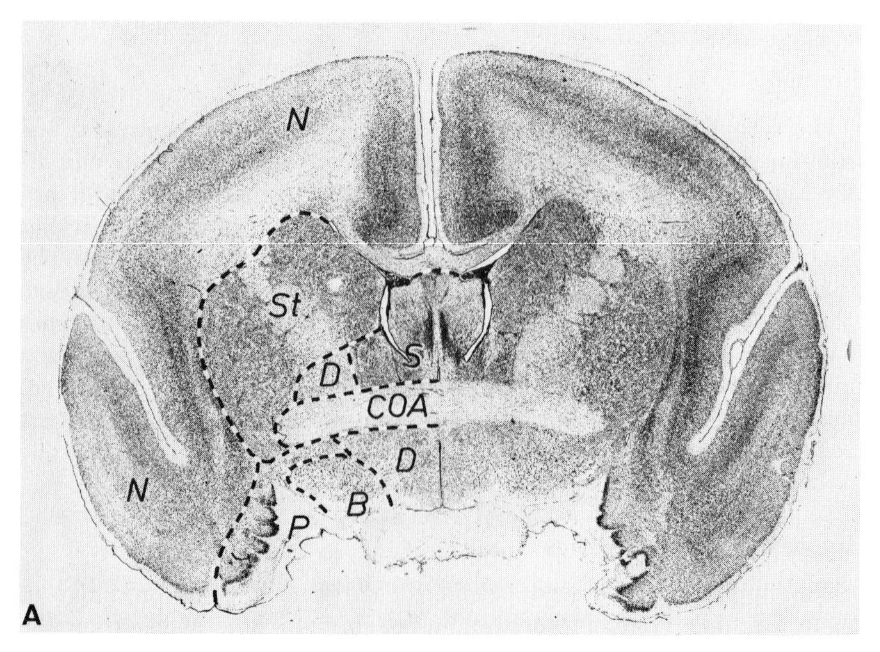

Fig. 1, part A See legend and part B on facing page.

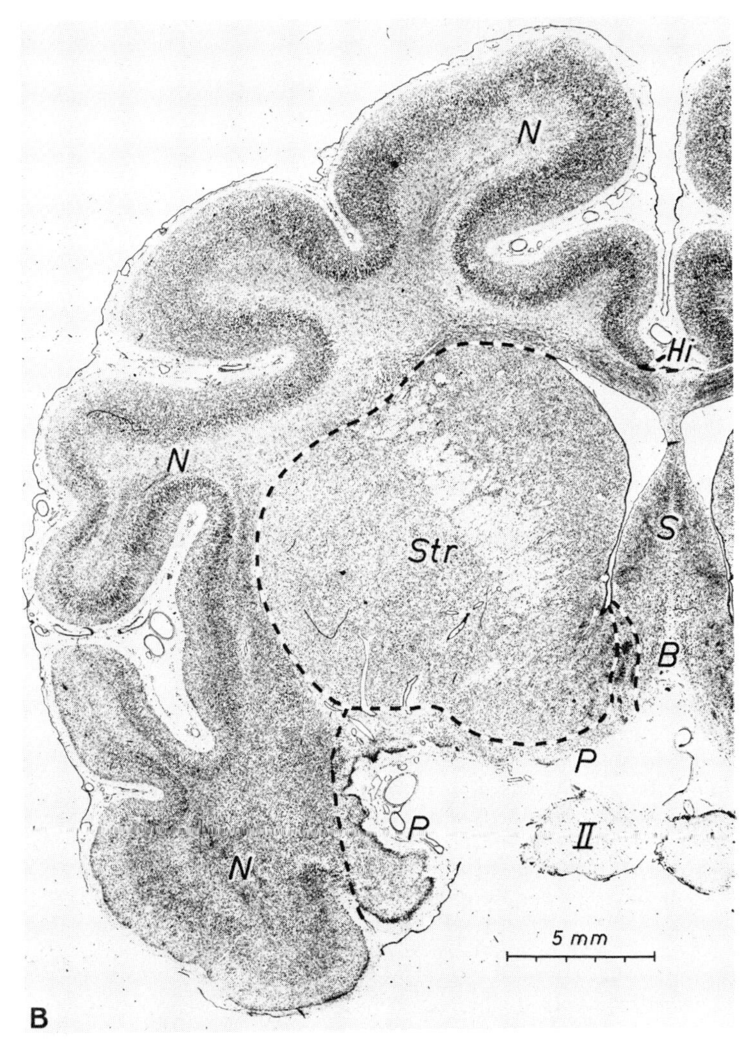

Fig. 1 Various brain parts are outlined to facilitate cutting them out and were weighed to derive the volumetric measurements. A: *Ateles paniscus*. B: *Leontocepus oedipus*. Abbreviations: B, diagonal band of Brocha; COA, anterior commissure; D, diencephalon; Hi, hippocampus; N, neocortex; P, paleocortex; S, septum; St and Str, strianum; II, optic tract.

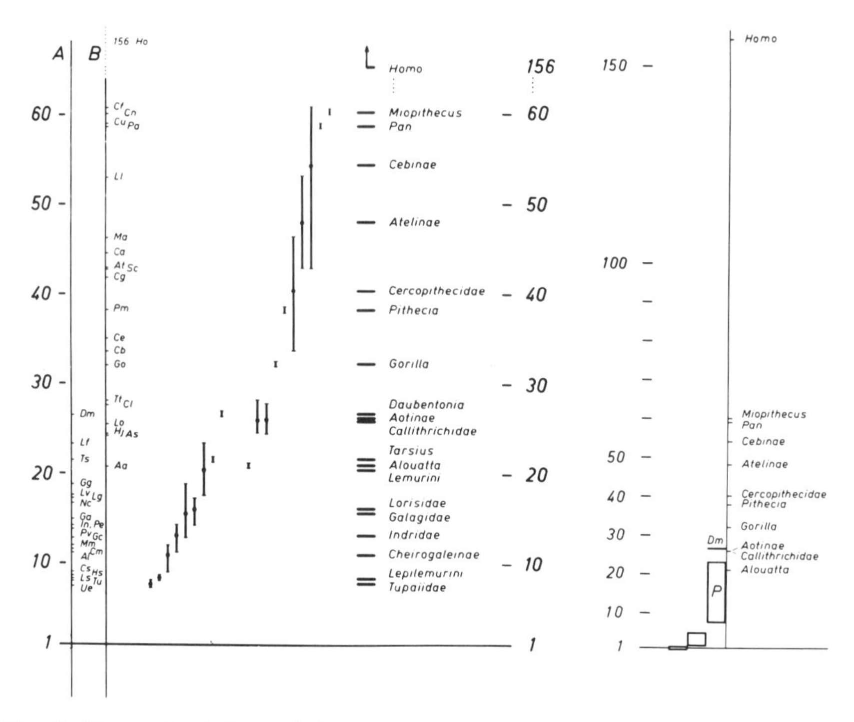

Fig. 2 Progression indices of the neocortex. They express the enlargement multiple of the neocortex in relation to the average basal insectivore of equal body weight. A: The scale for prosimians. B: The scale representing simians. Horizontal bars represent averages of the indices within the indicated systematic group. The vertical bars represent the respective variabilities within each of the systematic groups of primates. The vertical bars are arranged from left to right according to increasing neocortical progression which is considered to represent the ascending primate scale. The horizontal base line is arbitrarily given a value of 1 and represents the average value obtained from basal insectivores. On the right side of the figure there is a reduced scale to demonstrate the actual position of the human neocortical development among the primates. The following is a list of abbreviations used in Figs. 2, 3, 4, 5, 8, and 11 to identify the various species. A—Prosimians: *Al, Avahi laniger; Cs, Cheirogaleus medius; Cm, Cheirogaleus major; Dm, Daubentonia madagascar; Ga, Galago senegalensis; Gc, Galago crassicaudatus; Gg, Galago demidovii; Hs, Hapalemur simus; In, Indri indri; Lf, Lemur fulvus; Lg, Loris gracilis; Ls, Lepilemur ruficaudatus; Lv, Lemur variegatus; Mm, Microcebus murinus; Nc, Nycticebus cougang; Pc, Perodicticus potto; Pv, Propithecus verreauxi; Tx, Tarsius syrichta; Tu, Tupaia glis; Ue, Urogale everetti.* B—Simians: *Aa, Alouatta seniculus; As, Aotes trivirgatus; At, Ateles paniscus; Ca, Cercopithecus ascanius; Cb, Colobus badius; Ce, Cercopithecus mitis; Cf, Cebus albifrons; Cg, Cercocebus albigena; Cl, Callicebus moloch; Cn, Cercopithecus (Miopithecus) talap.; Cu, Cebus* sp.; *Go, Gorilla gorilla; Hj, Callithrix jacchus; Ho, Homo sapiens; Li, Lagothrix lagotricha; Lo, Leontocebus oedipus; Ma, Macaca mulatta; Pa, Pan troglodytes; Pm, Pithecia monacha; Sc, Saimiri sciureus; Tt, Saguinus tamarin.*

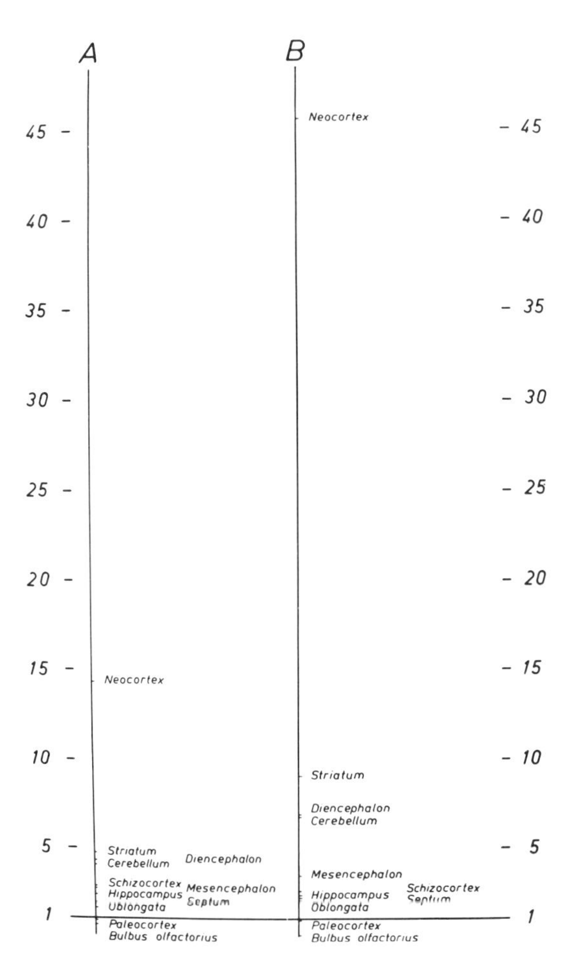

Fig. 3 Average development of various brain structures among 20 prosimians (A) and 21 simians (B). The base line of reference is given an arbitrary value of 1 and represents the average progression index of those various structures in the basal insectivores. It should be noted that the amygdala is included in the paleocortex.

Schizocortex (Periarchicortex)

The schizocortex was predominantly periarchicortex and consisted of regio entorhinalis, regio perirhinalis, and regio prae- and parasubicularis. It also included the retrosplenial, supracallosal, and subgenual areas of the periarchicortex. The schizocortex paralleled the hippocampus and septum in its phylogenetic development. This was very evident from the regression line which demonstrated a very close size correlation between those structures (Stephan and Andy, 1964, 1969). There was definite evidence of a progression in the prosimians, but there was a definite lack of progression in the simians. In fact, there may have been a slight regression. *Alouatta* and *Gorilla* showed a particularly low value whereas

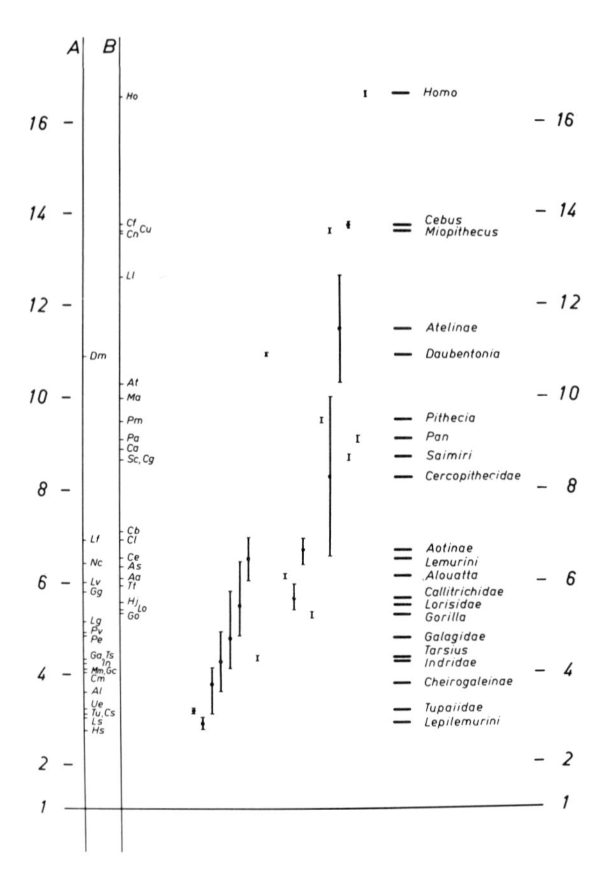

Fig. 4 Striatum development in primates. The vertical lines represent prosimians (A) and simians (B). The horizontal bars represent the average for each of the identified groups of primates. The vertical bars represent variabilities within each of the identified species and are arranged from left to right in the order of the progressively increasing neocorticalization. The base line of reference is given a value of 1 and represents the average size of striatum in basal insectivores.

Daubentonia, a prosimian, revealed a value of 6.2 times enlargement with reference to the basal insectivores. In the human, there existed a 5.3 times enlargement. These volumetric developmental changes for the schizocortex were very closely similar to those found for the hippocampus. In contrast to the hippocampus, the schizocortex possessed a more highly differentiated cortical structure and thus appeared to be more progressive. The regio entorhinalis was by far the most extensively developed structure making up the schizocortex (Fig. 6). It was of interest to note that the regio entorhinalis was undifferentiated in the insectivores whereas in

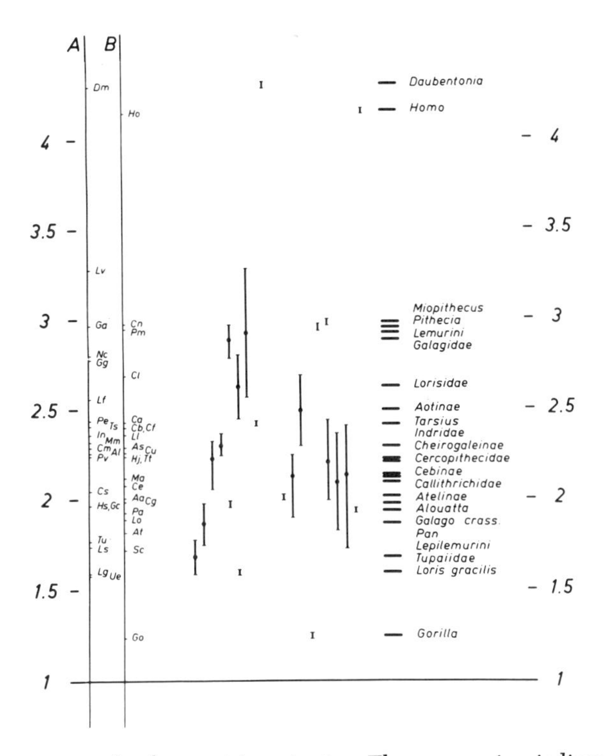

Fig. 5 Hippocampus development in primates. The progression indices for volumes of the hippocampus in relation to body weight is represented for prosimians and simians in relation to the average size of the hippocampus in the basal insectivores which are utilized as a reference. The size of the hippocampus in the basal insectivores is arbitrarily assigned a value of 1. The vertical column A represents the prosimians and the vertical column B the simians. The horizontal bars represent the average values for each of the indicated groups. The vertical bars indicate the variability within each group; they are arranged from left to right in the order of progressively increasing level of neocorticalization in the primate scale. This is based on the relative degrees of neocortical development as shown in Fig. 2.

the *Lepilemur* (a primitive prosimian) there was a clearly differentiated external zone which appeared as two-cell layered. Although in *Homo* the development of this area was not as marked as in the lower primates, the differentiation could have been more advanced, especially in the rostromedial region, close to the amygdala (Fig. 7). In *Homo,* the entorhinal area actually could be subdivided into a series of well-defined subareas according to Rose (1927a,b). Similar developmental and differentiational changes have taken place in the regio praesubicularis (Fig. 6). This region was very well developed and differentiated in primates but could hardly be recognized in subprimate forms.

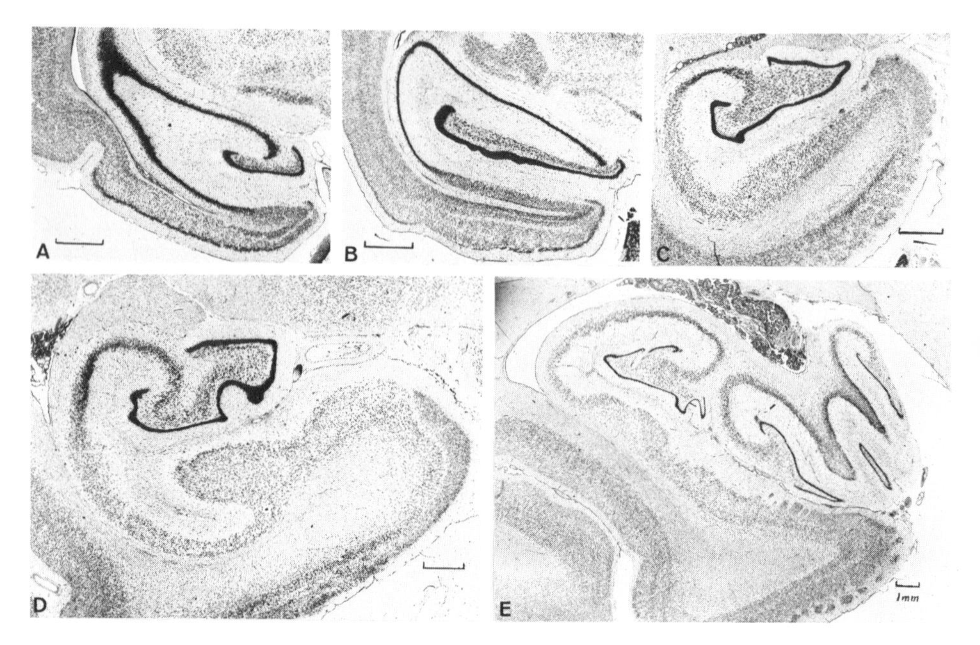

Fig. 6 Hippocampus, ventral component in A: *Tupaia glis* 969, section 810, 6.1×. B: *Galago demidovii* A 102, section 596, 6.2×. C.: *Cercopithecus ascanius* A 219, section 1590, 5.4×. D: *Pan troglodytes* A 280, section 2901, 5.2×. E: *Homo sapiens* 1044, section 3530, 2.9×.

Septum

Size alterations of the septum were very similar to those of the hippocampus (Figs. 2 and 3). These two structures appeared to correlate very closely in their development and thus appeared not to vary independently of one another. The obvious close size correlation in their development was evident from their regression coefficients (Stephan and Andy, 1969). Percent size increase of the septum as a whole, progressing from prosimians to simians, was very small, approximately 30%. For example, in the prosimian *Galago* there was a 1.7 times enlargement and in the simians *Colobus, Cercopithecus, Callithrix, Aotes,* and *Leontocebus* there were 1.4 to 2.5 times enlargements. However, there was a 100% increased size in progression from higher simians to human (Fig. 8). The shape and structural relationships of the septal nuclei underwent a change in primate phylogeny (Figs. 9 and 10). Nucleus septalis triangularis was the only septal nucleus which underwent definite regression in phylogeny.

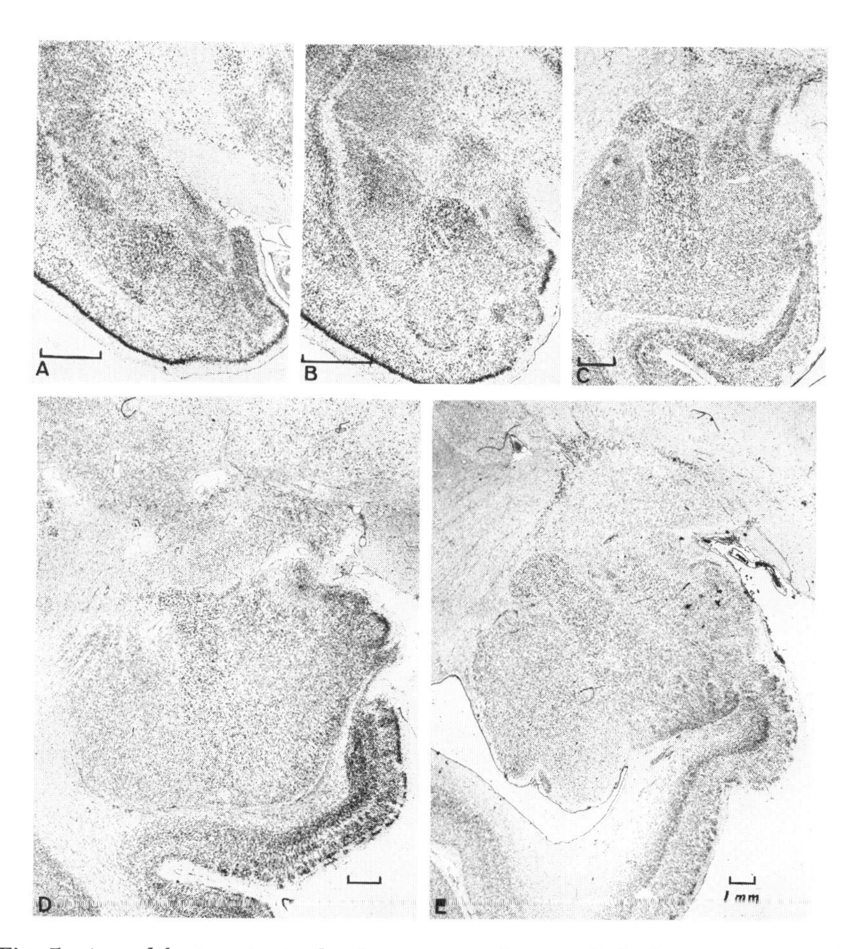

Fig. 7 Amygdala in primate development reveals a marked change in structural organization and relationships in progressing from prosimian to simians and to *Homo*. A: *Tupaia,* section 685, 8.1×. B: *Galago,* section 500, 9.1×. C: *Cercopithecus,* section 1180, 4.9×. D: *Pan,* section 2200, 4.3×. E: *Homo,* section 3060, 3.2×.

Histologically, the nuclei of the septum were divided into four major groups which were very well defined and differentiated from one another. This nuclear organization was best demonstrated in the lower primates, especially the prosimians. The major groups were (*1*) nucleus septalis dorsalis, containing four subdivisions; (*2*) nucleus septalis lateralis, containing two subdivisions; (*3*) nucleus septalis medialis and nucleus of the diagonal band of Broca, each containing two subdivisions; and (*4*) a caudal group consisting of a nucleus septalis fimbrialis, septalis triangu-

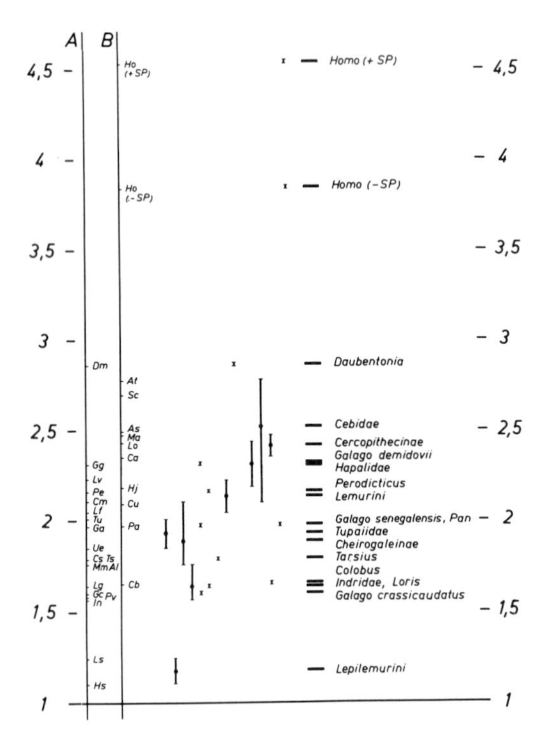

Fig. 8 Septum development in primates. The volume of the septum in the prosimians (vertical line A) and simians (vertical line B) is presented in relation to the average of the septum and basal insectivores which is arbitrarily given a value of 1 and utilized as the base line of reference. The horizontal bars represent average values for each group. The vertical bars represent variability within each group and are arranged from left to right in the order of increased neocorticalization. *Homo* (+SP) septum telencephali, *Homo* (—SP) septum verum. (e.g., without septum pellucidum).

laris, bed nucleus of the anterior commissure and bed nucleus of the stria terminalis which contained three subdivisions. In progressing up the primate scale, these various nuclei tended to become less well differentiated from one another, although they were specifically identified (except for nucleus septalis triangularis) in all forms studied, including the human brain. It should be noted that in the human brain and higher primates the total septum consisted of two major parts which were easily differentiated histologically. One part contained the well-developed nuclei similar to those described in lower forms and was identified as the septum verum (the true septum). The remaining part, which was in the dorsal position, extended from the septum verum to the corpus callosum where it was attached. It was primarily made up of fiber tracts and glia, appearing like a thin vellum and devoid of nuclei. This structure was identified as the "septum pellucidum." It should be emphasized that in spite of the appearance of the septum pellucidum in the human and higher primates, the major increase in the developmental size of the septum in those forms was due to the marked nuclear development of the septum verum (Fig. 8).

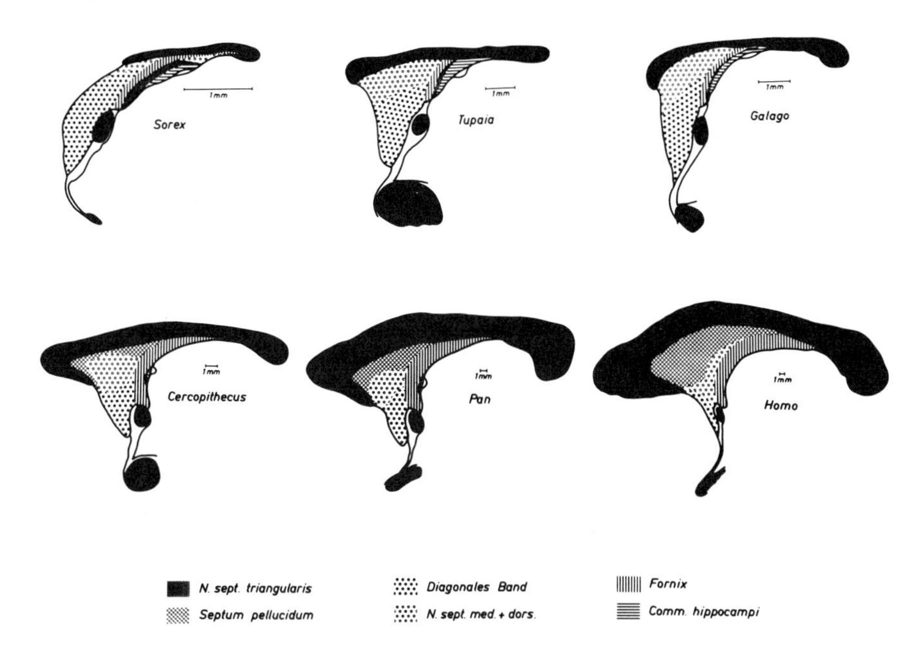

Fig. 9 Sagittal view of the septum in primate evolution. Note that the septum pellucidum begins to make its appearance in the simians and is not present in the prosimians. It progressively becomes more extensive and is largest in *Homo.*

Amygdala

In primates it was not possible to histologically establish a definite borderline between the nucleus corticalis amygdala and the nucleus basalis and lateralis amygdala (Fig. 7). However, it was relatively easier to establish a border between the nucleus corticalis and medialis amygdala. Consequently, the nucleus corticalis amygdala was incorporated with nucleus basalis and lateralis amygdala and together identified as the corticobasolateral group. The remaining amygdaloid nuclei were collectively identified as the centromedial group which included the nucleus centralis, nucleus medialis, nucleus of the lateral olfactory tract, and the anterior amygdaloid area. This classification represented a deviation from the more commonly accepted classification in which the nucleus corticalis was considered as a component of the medial group.

The amygdaloid complex as a whole underwent enlargement in phylogeny from insectivores to higher primates (Stephan and Andy, 1973). However, the degree of enlargement from prosimians to simians was relatively small. Some amygdaloid components such as the centromedial

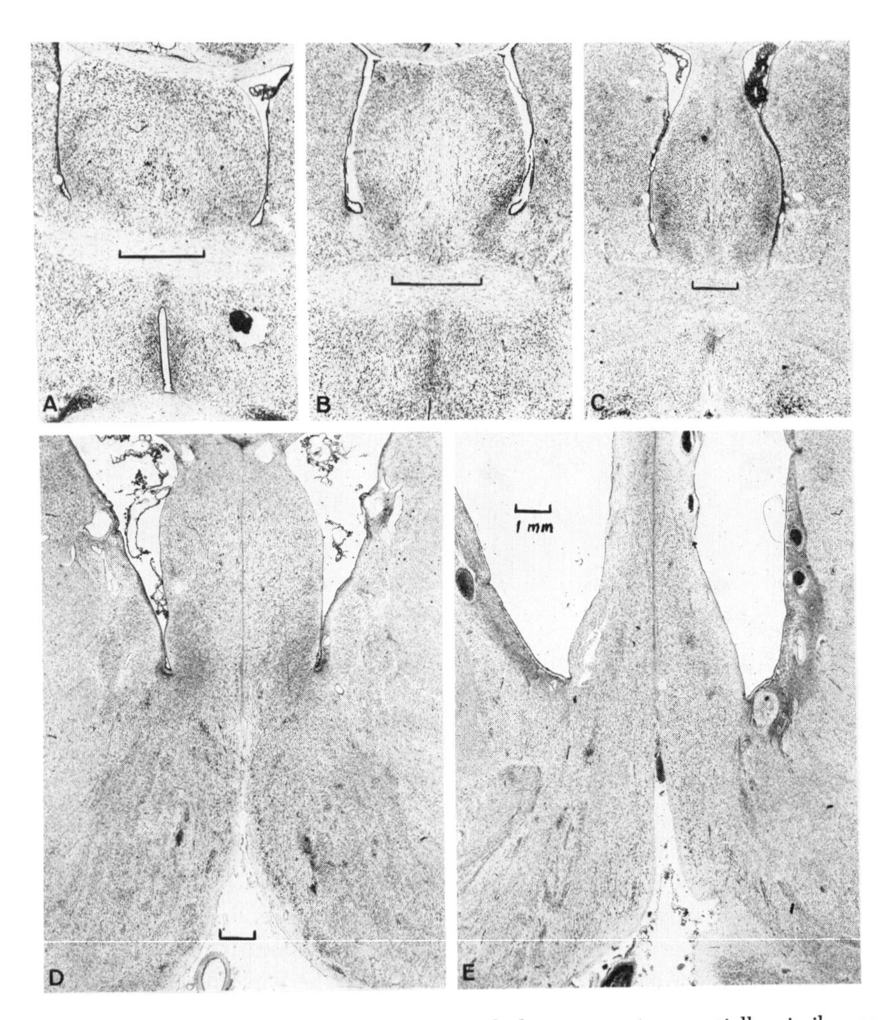

Fig. 10 Septum. The nuclear composition of the septum is essentially similar as one progresses from prosimians to simians and to humans. In primate evolution the structural conformation of the septum as a whole undergoes a change. The dorsal component becomes very thin in contrast to the ventral component. This is thought to be secondary to the marked enlargement and caudal extension of the corpus callosum to which it is attached. The dorsal component which does not contain nuclei is identified as septum pellucidum. The largest mass of the septum is predominantly made up of the septum verum. A: *Tupaia,* section 625, 11.6×. B: *Galago,* section 444, 11.9×. C: *Cercopithecus,* section 1030, 5.9×. D: *Pan,* section 1800, 4.6×. E: *Homo,* section 2690, 4.6×.

group did not increase or decrease in size in progressing from prosimians to higher primates. The corticobasolateral group underwent a definite increased size from prosimians to simians and humans. In relation to the basal insectivores the enlargement indices of the total amygdala were: prosimians 2.4 times, simians 3.0 times, and human 3.5 times. It is of interest to note that the enlargement in the corticobasolateral group was predominantly due to the cortical, lateral, and the small-celled basal components. The large-celled basal component of this group underwent no developmental change. In the higher primates, the area of greatest development involved those cells making up the ill-defined transitional area between the nucleus corticalis and basalis amygdala. These relatively recent cellular collections were identified as accessory nuclei and considered in their classification under the basal group. Within the centromedial group, the nucleus of the lateral olfactory tract underwent an absolute regression in primate phylogeny. This was undoubtedly associated with the marked regression of the olfactory apparatus in primate evolution (Fig. 11). It should be noted that the amygdala underwent a marked rotational change in phylogeny from the lowest to the highest primate, including man. The greatest transition appeared to be between the prosimians and simians in the primate scale (Fig. 7). The predominant factors influencing this change were the marked development of the various temporal neocortical areas.

Paleocortex

The paleocortex appeared to undergo a regressive change in primate phylogeny (Fig. 1). In higher primates and the human it underwent a reduction which most likely resulted from the marked regression of the olfactory apparatus. The combined paleocortex–amygdala volumetric measurements showed no change in phylogenetic development because the amygdala underwent a progression in primate phylogeny and thus compensated for the paleocortical regressive changes (Fig. 3).

Bulbus Olfactorius

The phylogenetic changes of the olfactory system in primate evolution revealed an absolute regression. In relation to the basal insectivores which serve as a point of reference, the olfactory bulb in primate evolution underwent a progressive reduction (Fig. 11), as evident from the following developmental indices. In prosimians the developmental index was 1.1 to 1.3. In the simians (Old World and New World monkeys) the indices were 0.06 to 0.13. In the leaf-eating guerezas (*Colobus*) and

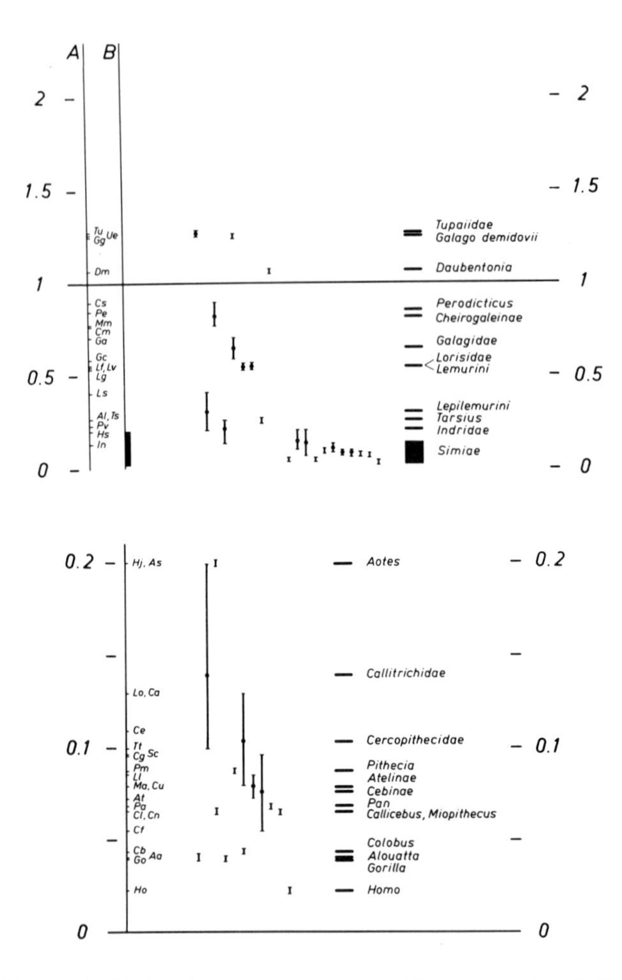

Fig. 11 Olfactory bulb development in primates. The progression indices of development for the olfactory bulb are given in relation to the average size in basal insectivores which is assigned an arbitrary value of 1 and used as a base line (horizontal bar). Prosimians (A) and simians (B) are represented along the vertical lines. Note that in simians the very marked reduction in development of the olfactory bulb with progression indices values of less than 0.3 and the location of the simians along the vertical lines is represented by a thickening of the line at the lower end. This end of the figure is enlarged in order to illustrate the relative locations of various species among the simians and to demonstrate the lowest value that is being represented by *Homo* (lower figure). The horizontal bars represent the averages for each of the groups. The vertical bars represent variabilities within each group and are arranged from left to right in the order of their progressively increased neocorticalization.

howler (*Alouatta*) and in the *Gorilla* the average index was 0.04. The human, among all primates, revealed by far the greatest reduction of the olfactory bulb: The index was 0.023 with reference to the basal insectivores.

Diencephalon

The thalamus and hypothalamus were the two major components of the diencephalon which in prosimians was 3.5 to 4.5 times larger than in basal insectivores. In simians it was 5.5 to 6.0 times enlarged (Fig. 3). The enlargement of the diencephalon was predominantly due to the increased growth rate of the thalamus. Comparatively, the hypothalamus did not progress as rapidly in its phylogenetic development. However, with respect to the relatively slow rate of hypothalamic development, it was of interest to note that the ventromedial nucleus of the hypothalamus was just as well developed in the prosimian as it was in the simians, higher primates, and human (Fig. 12). With regard to the faster growth rate of the thalamus, the center median nucleus represented a thalamic nucleus which apparently underwent a marked increase in size from the primitive primate (*Tupaia*) to the highest primate, *Homo* (Fig. 13). Note the small center median nucleus in the *Tupaia* in relation to the medial dorsal nucleus above it. In the galago, a prosimian of higher development, there was a relatively much greater enlargement of the center median nucleus. In progressing through the simians and higher primates to man, the size of the center median nucleus appeared to be approximately the same, although it may have been slightly larger in the human brain than in other higher primates.

Mesencephalon

Although the mesencephalon was definitely larger in prosimians in contrast to the basal insectivores (2.0 to 3.5 times enlarged), there was no marked size progression from prosimians to simians, which underwent a 3.7 times enlargement in the ascending primate scale (Fig. 3). The central gray, from which rage can be elicited by direct stimulation, was well developed throughout the primates from *Tupaia* to *Homo*. The nucleus of Gudden, viewed in frontal sections and demonstrated in various species of primate phylogeny, served as a representative nucleus of the central gray which has remained well developed and has not undergone major phylogenetic changes (Fig. 14).

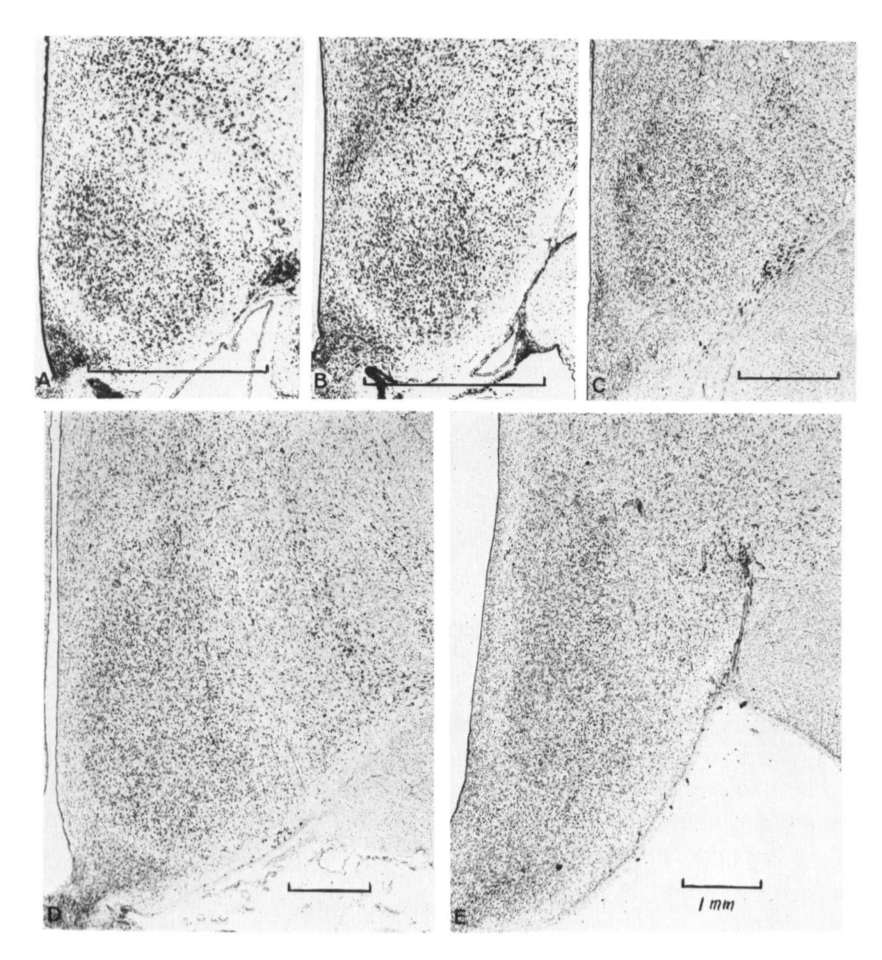

Fig. 12 Ventromedial nucleus of the hypothalamus in A: *Tupaia,* section 720, 24×. B: *Galago,* section 536, 23.4×. C. *Cercopithecus,* section 1210, 13.3×. D: *Pan,* section 2161, 10.9×. E: *Homo,* section 3060, 10.6×.

DISCUSSION

In studying developmental trends of the brain and its parts, erroneous impressions may be obtained from comparing specific brain structures with reference to one another or to the total brain. Note that all brain structures, except for the neocortex, appeared to become smaller in primate phylogeny (Fig. 15). This error stems from their possessing relatively different rates of development. Structures such as the neocortex which develop at a faster rate than others will appear to have enlarged,

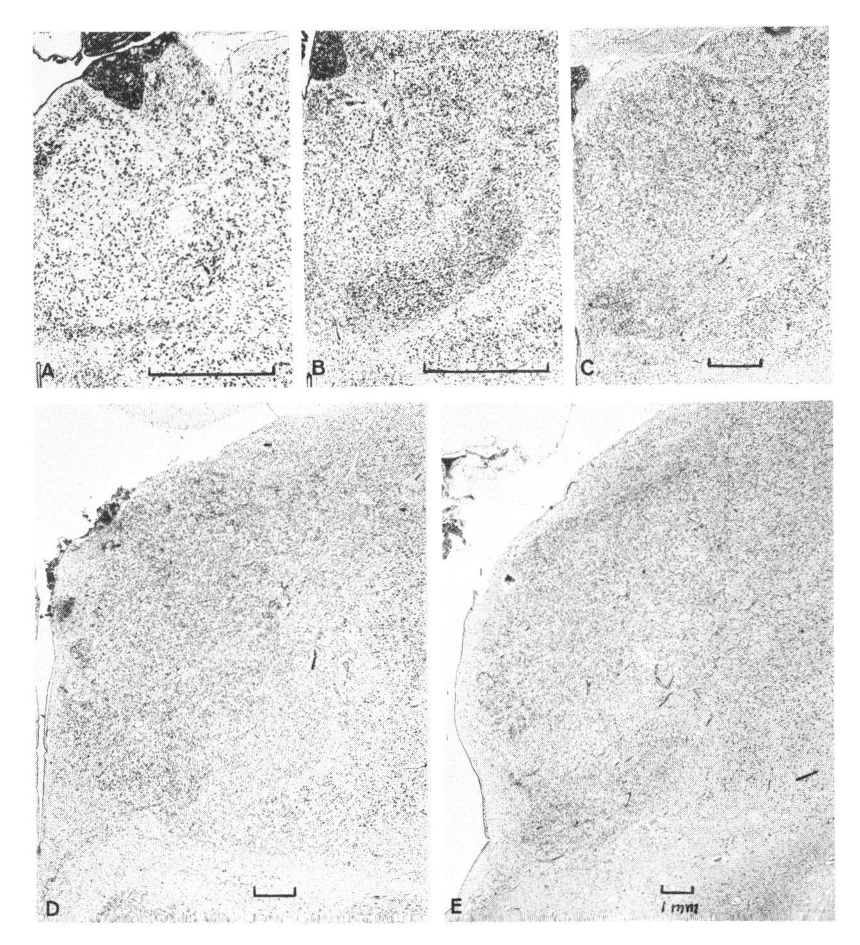

Fig. 13 Center median nucleus and medial dorsal nucleus in A: *Tupaia,* section 825, 16.8×. B: *Galago,* section 588, 16.6×. C.:*Cercopithecus,* section 1430, 7.0×. D: *Pan,* section 2560, 5.5×. E: *Homo,* section 3710, 4.5×.

and vice versa, structures such as the septum and hippocampus with a slower rate will appear to have become smaller. In reality, both may be enlarging. Consequently, in order to have accurate developmental measurements in phylogeny, an indifferent point of reference must be used as a base line. It has been well accepted that body size served as an excellent point of reference in order to evaluate developmental changes of the brain and its various parts. This technique of measurement, known as the allometric method, was developed by Snell (1892) and Dubois (1897).

The following discussion is an attempt to relate the anatomical findings

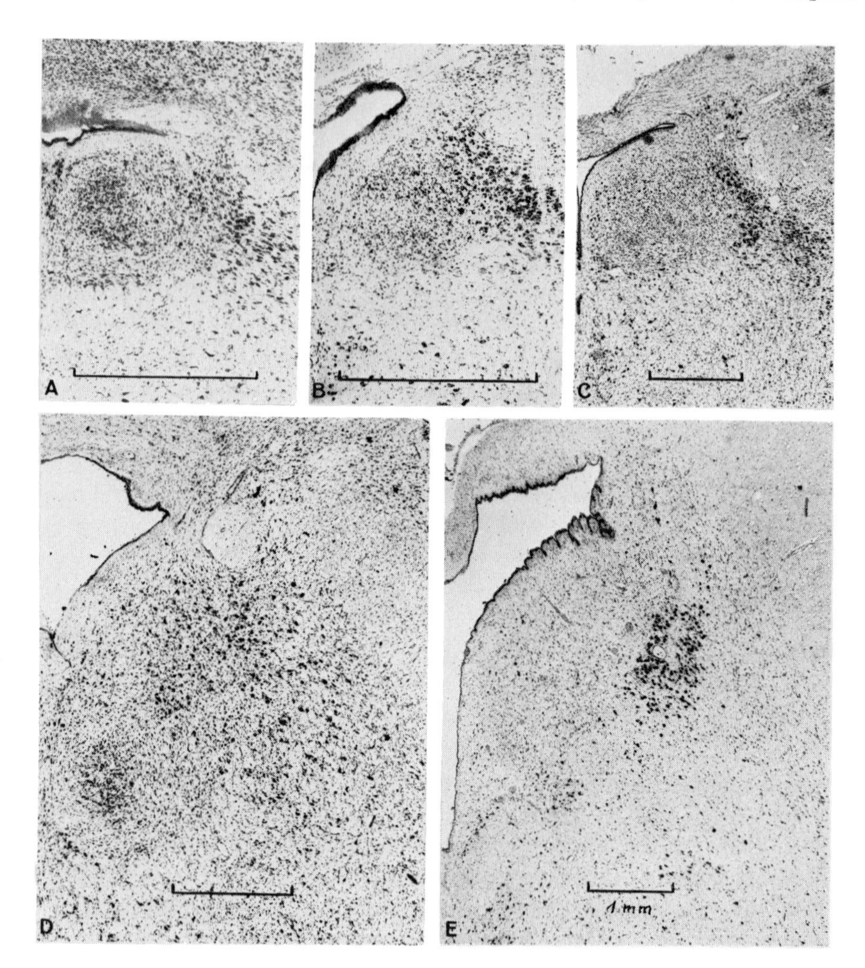

Fig. 14 Nucleus of Gudden in primate development of the central gray reveals continued histologic differentiation and development which appears essentially unchanged in the primate scale. A: *Tupaia,* section 1055, 24.6×. B: *Galago,* section 772, 26.6×. C: *Cercopithecus,* section 1820, 12.2×. D: *Pan,* section 3200, 16.0×. E: *Homo,* section 4510, 11.3×. Note the dark-stained cells of the nucleus coeruleus, which also is essentially unchanged in primates, just lateral to the nucleus of Gudden.

to the functional significance, especially with respect to aggressive behavior.

Based on the allometric method of evaluation, it was found that among brain structures related to aggression, all but one have undergone a progressive enlargement in primate phylogeny. The one exception is the olfactory bulb. It should be noted that there was a very marked and absolute regression of the olfactory system in progressing from lower to

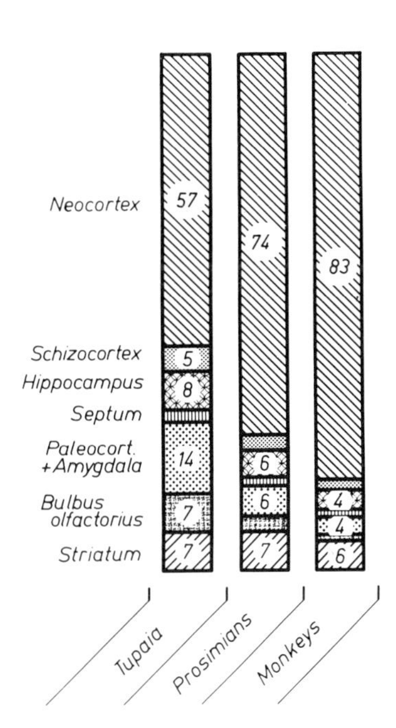

Fig. 15 Volumes of subdivisions of the telencephalon expressed as percent of total telencephalon. Except for the neocortex, note that all other structures and especially those representing the limbic system undergo apparent progressively decreasing development in phylogeny from prosimians to simians. This type of relative comparison leads to erroneous conclusions because the neocortex undergoes a much faster rate of development in phylogeny than other brain structures. Thus, when compared to the telencephalon in which the neocortex is a part, other brain structures appear to progressively become smaller. *Tupaia* is included in the primate evolutionary scale and may be thought of as representing a transitional form of primate linking insectivores and primates.

higher primates, including man. This was evident from the marked volumetric reduction of the olfactory bulb and nucleus of the olfactory tract. However, the very marked regression of the olfactory system in primate phylogeny had no apparent effect upon the overall functional homeostatic balance of the remaining allocortical and other brain structures subserving aggressive behavior. This finding was of importance since it was observed that disruption of the olfactory nerves in the rat resulted in increased aggression whereas disruption of olfaction by eliminating the peripheral olfactory epithelium did not (Alberts, 1972). Consequently, the presence or absence of aggression was not dependent upon the presence of olfactory function, as also concluded by Spiegel *et al.* (1940). This is in accord with the observation that the septum, hippocampus, and amygdala which were believed to subserve aggression and which were once thought to be dependent upon olfaction for their existence, also were found to be independent of that sensory function in their development. It also may be observed that although interruption of the olfactory nerve in the rat resulted in aggressive behavior, there has been no similar reaction in primates, including man.

Brain structures relating to aggressive behavior possessed different rates of enlargement. They were divided into three groups, based on their differential rates of development. Group 1 contained structures which revealed the least development in primate phylogeny, the mesencephalon and diencephalon. In the second group, consisting of the hippocampus, septum, and amygdala, the rate of enlargement tended to be greater than in the first. The third group, consisting of the neocortex and striatum, underwent the greatest rate of enlargement in primate evolution. It should be stressed that the neocortex far exceeded the striatum in its rate of progression and went from 15 times in the prosimians to an index of 156 times in the human.

These observations of differential development rates were of special interest because those three anatomic and phylogenetic differentiated groups also appeared to relate differentially to aggressive behavior. That is, the most completely integrated aggressive attack was most easily elicited from electrical and chemical stimulation of the least progressive structures, the diencephalon and mesencephalon. In contrast, electrical stimulation of the amygdala, hippocampus, and septum tended to elicit fragments or attenuated forms of aggression. The neocortex, on the other hand, representing the most progressive group, did not elicit aggression when stimulated, but tended to facilitate aggression when ablated. With regard to cortical inhibition of aggression, it was of interest that electrical stimulation of the cortex, such as the cingulate, did not inhibit spontaneous aggression in the cat, whereas stimulation of the hippocampus (archicortex) did inhibit the aggressive behavior (Andy, 1961).

SUMMARY

It is significant that the most completely integrated forms of aggression are elicited from the phylogenetically least progressive brain structures. It is suggested that the more progressive brain structures exerted varying degrees of control and influence upon those less-developed parts from which the aggressive state emanated. The neocortex, the most highly developed brain structure, exerted the greatest controlling influence. That the neocortex is involved becomes evident from the observation that a conditioned auditory stimulus evoked either a directed or nondirected aggression of hypothalamic origin (Andy *et al.,* 1973; Nakao, 1958, 1960). These combined anatomic and physiologic observations imply that the complex physiologic and behavioral changes which together make up aggression were predominantly integrated through rudimentary brainstem and diencephalic structures which revealed very little change in phylogenetic development from lower to higher primates, including man. The progressively increased control of the aggressive state, as the primate evolved, was predominantly due to the increased neocorticalization in phylogeny. However, in spite of the markedly advanced neocortical development among the higher primates, especially in the human, a developmental abnormality or lesion implicating those rudimentary structures may activate physiologic mechanisms, resulting in aggressive states which cannot be controlled by inhibitory mechanisms of neocortical origin. If, in addition, the neocortex were malfunctioning, one would expect the acme of the aggressive state to develop.

REFERENCES

Abrahams, V. C., Hilton, S. M., and Malcolm, J. L. (1962). Sensory connections to the hypothalamus and midbrain, and their role in the reflex activation of the defence reaction. *J. Physiol.* **164**, 1–16.

Akert, K. (1961). Diencephalon. *In* "Electrical Stimulation of the Brain" (D. E. Sheer, ed.), Univ. Texas Press, Austin, Texas.

Alberts, J. R. (1972). Olfactory bulb removal but not anosmia increases emotionality and mouse killing. *Nature (London)* **238**, 454–455.

Andy, O. J. (1961). Personal Observations.

Andy, O. J. (1966). Neurosurgical treatment of abnormal behavior. *Amer. J. Med. Sci.* **252**, No. 2, 132–138.

Andy, O. J. (1970). Thalamotomy in hyperactive and aggressive behavior. *Confina Neurol.* **32**, 322–325.

Andy, O. J., and Jurko, M. F. (1972). Thalamotomy for hyperresponsive syndrome. *In* "Psychosurgery" (E. Hitchcock, L. Laitinen, and K. Vaernet, eds.), pp. 127–135. Thomas, Springfield, Illinois.

Andy, O. J., and Stephan, H. (1966). Septal nuclei in primate phylogeny. *J. Comp. Neurol.* **126**, 157–170.

Andy, O. J., and Stephan, H. (1967). "Phylogeny of the Primate Septum Telencephali." Thieme Verlag, Stuttgart.

Andy, O. J., and Stephan, H. (1968). The septum in the human brain. *J. Comp. Neurol.* **133**, 383–410.

Andy, O. J., Chinn, R. McC., and Bonn, P. (1957). Seizures from the septal region. Behavioral and electrical study in the cat. *Trans. Amer. Neurol. Ass.* 128–129.

Andy, O. J., Giurintano, L., and Laing, J. W. (1973). Conditioned aggressive behavior. IRCS reprint number (73-4) 39-2-6.

Balasubramaniam, V. (1972). Surgery for Behavioural Disorders. *Inst. Neurol. Madras Proc.* **2**, No. 1, 1–6.

Balasubramaniam, V., and Ramamurthi, B. (1969). Stereotaxic amygdalotomy in behaviour disorders. *Symp. Int. Soc. Res. Stereoenceph., 4th New York. Confin. Neurol.* **32**, 367–373.

Balasubramaniam, V., and Ramamurthi, B. (1970). Stereotaxic amygdalotomy in behavior disorders, *Advan. Stereoenceph.* **5**, 367–373.

Balasubramaniam, V., Kanaka, T. S., Ramanujam, P. B., and Ramamurthi, B. (1969). Sedative neurosurgery—A contribution to the behavioral sciences. *J. Indian Med. Ass.* **53**, 377–381.

Balasubramaniam, V., Kanaka, T. S., Ramanujam, P. B., and Ramamurthi, B. (1970). Sedative neurosurgery. *Neurology (India)* **18**, 46–52.

Bard, P. (1928). A diencephalic mechanism for the expression of rage with special reference to the sympathetic nervous system. *Amer. J. Physiol.* **84**, 490–513.

Bard, P., and Mountcastle, V. B. (1948). Some forebrain mechanisms involved in expression of rage with special reference to suppression of angry behavior. *Res. Publ. Ass. Nerv. Ment. Dis.* **27**, 362–404.

Barnett, O. (1967). "Aggression and Defense." *In* "Brain Function" (C. D. Clemente and D. B. Lindsley, eds.). Univ. Calif. Press, Berkeley and Los Angeles, California.

Brady, J. V., and Nauta, W. J. H. (1953). Septal lesions rat emotion. *J. Comp. Physiol. Psychol.* **46**, 339–346.

Brady, J. V., and Nauta, W. J. H. (1955). Subcortical mechanisms in emotional behavior: The duration of affective changes following septal and habenular lesions in the albino rat. *J. Comp. Physiol. Psychol.* **48**, 412–420.

Brugge, J. F. (1965). An electrographic study of the hippocampus and neocortex in unrestrained rats following septal lesions. *Electroenceph. Clin. Neurophysiol.* **18**, 36–44.

Dieckmann, G., and Hassler, R. (1971). Relief from compulsion and obsession by combined intralaminarmedial thalamotomy. *Eur. Congr. Neurosurg., 4th, Prague* 28 June–2 July (Abstr.).

Dubois, E. (1897). Uber die Abhangigkeit des Hirngewichts von der Korpergrosse bei den Saugetieren. *Arch. Anthrop.* **25**, 1–28.

Egger, M. D., and Flynn, J. P. (1962). Amygdaloid suppression of hypothalamically elicited attack behavior. *Science* **136**, 43–44.

Egger, M. D., and Flynn, J. P. (1963). Effects of electrical stimulation of the amygdala on hypothalamically elicited attack behavior in cats. *J. Neurophysiol.* **26**, 705–720.

Egger, M. D., and Flynn, J. P. (1967). Further studies on the effects of amygdaloid stimulation and ablation on hypothalamically elicited attack behavior in cats. *Progr. Brain Res.* **27**, 165–182.

Fernandez de Molina, A., and Hunsperger, R. W., (1959). Central representation of affective reactions in forebrain and brain stem: electrical stimulation of amygdala stria terminalis, and adjacent structures. *J. Physiol.* **145**, 251–265.

Fernandez de Molina, A., and Hunsperger, R. W. (1962). Organization of the subcortical system governing defense and flight reactions in the cat. *J. Physiol.* (*London*) **160**, 200–213.

Flynn, J. P. (1967). "Neurophysiology and Emotion" (D. Glass, ed.), pp. 40–60. Rockerfeller Univ. Press, New York.

Flynn, J. P. (1972). Biting attack elicited by stimulation of the ventral midbrain tegmentum of cats. *Science* **177**, 364–366.

Fonberg, E. (1963). The inhibitory role of amygdala stimulation. *Acta Biol. Exp.* **23**, 171–180.

Fonberg, E., Brutkowski, S., and Mempel, E. (1962). Defensive conditioned reflexes and neurotic motor reactions following amygdalectomy in dogs. *Acta Biol. Exp.* **22**, 51–57.

Gastaut, R., Vigouroux, R., Carriol, J., and Badier, M. (1951). Effets de la stimulation electrique (par electrodes a demeure) du complexe amygdalien chez le chant non narcose. *J. Physiol.* (*Paris*) **43**, 740–746.

Hassler, R., and Dieckmann, G. (1967). Stereotaxic treatment of compulsive and obsessive symptoms. *Confin. Neurol.* **29**, 153–158.

Hassler, R., and Dieckmann, G. (1971). Violence against oneself and against others as a target for stereotaxic. *Eur. Congr. Neurosurg., 4th, Prague, 1971* (Abstr.).

Heimburger, R. F., Whitlock, C. C., and Kalsbeck, J. E. (1966). Stereotaxic amygdalotomy for epilepsy with aggressive behaviour. *J. Amer. Med. Ass.* **198**, 741–745.

Hilton, S. M., and Zbrozyna, A. W. (1963). Amygdaloid region for defense reactions and its efferent pathway to the brainstem. *J. Physiol (London)* **165**, 160–173.

Hunsperger, R. W. (1963). Comportements affectifs provoques par la stimulation electrique du gronc cerebral et due cerebral anterieur. *J. Physiol. (Paris)* **55**, 45–97.

Hunsperger, R. W., and Bucher, V. M. (1967). Affective behavior produced by electrical stimulation in the forebrain and brain stem of the cat. *Progr. Brain Res.* **27**, 103–127.

Hunt, H. F. (1967). Aggression. *Ann. N. Y. Acad. Sci.* **10**, 712.

Ingram, W. R. (1952). Brain stem mechanisms in behavior. *Electroenceph. Clin. Neurophysiol.* **4**, 397–406.

Ingram, W. R. (1958). Modification of learning by lesions and stimulation in the diencephalon and related structures. *In* "Reticular Formation of the Brain" pp. 535–544. (H. Jasper, L. Proctor, R. Knighton, W. Noshay, and R. Costello, eds.), Little, Brown & Co., Boston, Massachusetts.

Kaada, B. R., and Bruland, H. (1960). Effect of chlorpromazine on the "attention" (orienting), fight and anger response elicited by cerebral stimulation. *Acta Physiol. Scand. Suppl.* **50**, 175.

Karli, P., and Vergnes, M. (1964). Nouvelles donnees sur les bases neurophysiologiques du comportement d'aggression interspecifique rat-souris. *J. Physiol. (Paris)* **56**, 384.

Kido, R., Hirose, K., Yamamoto, K. I., and Matsushita, A. (1967). Effects of some drugs on aggressive behavior and the electrical activity of the limbic system. *Progr. Brain Res.* **27**, 365–387.

King, F. A., and Meyer, P. M. (1958). Effects of the amygdaloid lesions upon septal hyperemotionality in the rat. *Science* **128**, 655–656.

Levinson, P. K., and Flynn, J. P. (1965). The objects attacked by cats during stimulation of the hypothalamus. *Anim. Behav.* **13**, 217–220.

MacLean, P. D., and Delgado, J. M. R. (1953). Electrical and chemical stimulation of frontotemporal portion of limbic system in the waking animal. *EEG Clin. Neurophysiol.* **5**, 91–100.

Nakao, H. (1958). Emotional behavior produced by hypothalamic stimulation. *Amer. J. Physiol.* **194**, 411–418.

Nakao, H. (1960). Hypothalamic emotional reactivity after amygdaloid lesions in cats. *Folia Psychiat. Neurol. (Japan)* **14**, 357–366.

Nakao, H. (1967). Facilitation and inhibition in centrally induced switch-off behavior in cats. *Progr. Brain Res.* **27**, 103–127.

Narabayashi, H. (1972). Stereotaxic amygdalotomy. *In* "The Neurobiology of the Amygdala" (B. E. Eleftheriou, ed.), pp. 459–483. Plenum Press, New York.

Narabayashi, H., Nagao, T., Saito, Y., Yoshida, M., and Nagahata, M. (1963). Stereotaxic amygdalotomy for behaviour disorders. *Arch. Neurol.* **9**, 1–16.

Poblete, M., Palestini, M., Figueroa, E., Gallardo, R., Rojas, J., Covarrubias, M. I., and Doyharcabal, Y. (1970). Stereotaxic Thalamotomy (Lamella Medialis) in Aggressive Psychiatric Patients. *Advan. Stereoenceph.* **5**, 326–331.

Rose, M. (1927a). Der Allocortex bei Tier und Mensch. I Teil *J. Psychol. Neurol.* **34**, 1–111 (1926).

Rose, M. (1927b). Die sog. Riechrinde beim Menschen und beim Affen. *J. Psychol. Neurol.* **34**, 261–401.

Sano, K. (1958). *Folia Psychiat. Neurol. (Japan)* **12**, 152–176.

Sano, K. (1966). Sedative stereoencephalotomy: fornicotomy, upper mesencephalic reticulotomy and postero-medial hypothalamotomy. *Progr. Brain Res.* **21-B**, 350–372.

Sano, K., Mayanagi, Y., Sekino, H., Ogashiwa, M., and Ishijima, B. (1970). Results of stimulation and destruction of the posterior hypothalamus in man. *J. Neurosurg.* **33**, 689–707.

Schreiner, L., Rioch, D. McK., Pechtel, C., and Masserman, J. H. (1953). Behavioral changes following thalamic injury in cat. *J. Neurophysiol.* **16**, 234–246.

Schreiner, L., and Kling, A. (1953). Behavioral changes following rhinencephalic injury in cat. *J. Neurophysiol.* **16**, 643–659.

Schvarcz, J. R., Driollet, R., Rios, E., and Betti, O. (1972). Stereotactic hypothalamotomy for behavior disorders. *J. Neurol. Neurosurg. Psychiat.* **35**, 356–359.

Skultety, F. M. (1963). Stimulation of periaqueductal gray and hypothalamus. *Arch. Neurol.* **8**, 608–620.

Snell, O. (1892). Die Abhangigheit des Hirngewichts von dem Dorpergewicht und den geistigen Fähigkeiten. *Arch. Psychiat.* (*Berlin*) **23**, 436–446.

Spiegel, E. A., Miller, R. R., and Oppenheimer, M. J. (1940). Forebrain and rage reactions. *J. Neurophysiol.* **3**, 538–548.

Stephan, H. (1960). Methodische Studien Uber den quantitativen Vergleich architektonischer Struktureinheiten des Gehirns. *Z. Wiss. Zool.* **164**, 143–172.

Stephan, H. (1961). Vergleichend-anatomische Untersuchungen an Insektivorengehirnen. V. Die quantitative Zusammenstzung der Oberflachen des Allocortex. *Acta Anat.* (*Basel*) **44**, 12–59.

Stephan, H., and Andy, O. J. (1962). The Septum. *J. Hirnforsch.* Bd. 5, Heft 3.

Stephan, H., and Andy, O. J. (1964). Quantitative comparisons of brain structures from insectivores to primates. *Amer. Zool.* **4**, 59–74.

Stephan, H., and Andy, O. J. (1969). Quantitative comparative neuroanatomy of primates: An attempt at a phylogenetic interpretation. *Ann. N. Y. Acad. Sci.* **167**, 370–387.

Stephan, H., and Andy, O. J. (1974). Quantitative comparison of the amygdala in insectivores and primates. Submitted for publication.

Ursin, H. (1965). The effect of amygdaloid lesions on flight and defense behavior in cats. *Exp. Neurol.* **11**, 61–79.

Wasman, M., and Flynn, J. P. (1962). Directed attack elicited from hypothalamus. *Arch. Neurol.* **6**, 220–227.

Wheatley, M. D. (1944). The hypothalamus and affective behavior in cats. *Arch. Neurol. Psychiat.* **52**, 296–316.

Wood, C. D. (1958). Behavioral changes following discrete lesions of temporal lobe structures. *Neurology* **8**, 215–220.

Woods, J. W. (1956). 'Taming' of the wild Norway rat by rhinencephalic lesions, *Nature* (*London*) **178**, 869.

Yasukochi, O. (1960). Emotional responses elicited by electrical stimulation of the hypothalamus in the cat. *Folia Psychiat. Neurol.* (*Japan*) **14**, 260–267.

Yoshida, M. (1963). Effects of amygdaloid stimulation on emotional responses produced by hypothalamic stimulation in cats. *Psychiat. Neurol.* (*Japan*) **65**, 863–879 (English summary, pp. 71–72).

Zbrozyna, A. W. (1963). The anatomical basis of the patterns of autonomic and behavioural response effected via the amygdala. *Progr. Brain Res.* **3**, 50–68.

PROBLEMS IN THE ANALYSIS OF AGONISTIC BEHAVIOR IN THE SQUIRREL MONKEY: AN ATTEMPT AT A SOLUTION BY MEANS OF TELESTIMULATION TECHNIQUE, TELEMETRY, AND STATISTICAL METHODS [1]

MANFRED MAURUS

Max-Planck-Institut für Psychiatrie

INTRODUCTION

Identification of Problems

The squirrel monkey has found its place in the scientific laboratory to an increasing extent, particularly within the past two decades. This animal proved itself suitable at the outset, especially for neurophysiological investigations and for space research (e.g., MacLean and Delgado, 1953; Carmichael and MacLean, 1961; Beischer and Furry, 1964). The widespread use of this species as a laboratory animal intensified the interest in its behavior toward working out the basis for successful animal care and breeding on the one hand and out of ethological interest on the other. Apart from the first behavioral experiments carried out with the squirrel monkey by Klüver (1933), more extensive behavioral investigations began first during the 1960s. Behavioral studies in the laboratory (e.g., Ploog, 1961, 1963, 1967; Ploog and MacLean, 1963; Kirchshofer, 1963; Ploog *et al.*, 1963, 1967; MacLean, 1964; Winter *et al.*, 1966; Bowden *et al.*, 1967; Hopf, 1967, 1970, 1971; Castell and Ploog, 1967; Castell and Maurus, 1967; Latta *et al.*, 1967; Rosenblum, 1968;

[1] Supported by the Deutsche Forschungsgemeinschaft, Bad Godesberg.

Winter, 1968, 1969; Castell, 1969; Castell and Heinrich, 1970; Peters, 1970; Mayer, 1971; Schott, 1972) and field studies (e.g., Hill, 1938; Thorington, 1967; DuMond and Hutchinson, 1967; Baldwin, 1968a,b, 1969, 1970, 1971; DuMond, 1968; Thorington, 1968; Winter, 1972) furnished comprehensive knowledge about the behavior of squirrel monkeys. The field studies were occupied mainly with questions of ecology, group dynamics, and the observation of particular behavior patterns as shown by the animals in natural or seminatural environments. The laboratory inquiries resulted in a comprehensive qualitative description of the behavior units within agonistic and nonagonistic behavior and in quantitative investigations concerning several of these behavioral modes. Among the behavior units described up to now, one finds 30–40 visually recognizable behavior units which occur in agonistic interactions and about 10–20 different types of vocalizations frequently associated with agonistic behavior. The visually recognizable units are set up according to morphological aspects (i.e., a description of postures and movements). They comprise movement sequences which appear repeatedly in quite similar form and which distinguish themselves with sufficient clarity from other units (see Table 1). The delimitations between units is partially made possible through the nonoccurrence of fluid transitions between them (e.g., 12, 15, 18, 20). The limits must be arbitrarily laid out where fluid transitions occur among the units (e.g., touching of the partner with the hand: 05, 06, 07, etc.). Seemingly plausible criteria were used in determining the limits (e.g., body-touching delimitation corresponds to the anatomical points touched, as for example, the head 05, back 07, and so on). The vocal units are arranged by means of time–frequency diagrams and through subjective acoustic impressions of the observers. There are clear-cut vocalization types, but the delimitation among several types is difficult, because continuous transitions occur quite frequently (Schott, 1972). In using this method to mark off the units, the question whether the cataloging of categories of behavior shown by the animals "labelled in terms of physical description of the movements involved [Hinde, 1966, p. 291]" agrees with the meaning which they have as communication agencies for the conspecific partner remains unanswered. There is much which indicates the lack of this conformity with respect to the squirrel monkey. This becomes plain through the fact that only a few pairs, each made up of two successive events, could be discovered whose transition frequencies exceed chance level, enabling us to comment on an association of a signal with a behavioral response (e.g., Hopf, 1972; Winter *et al.*, 1966; Schott, 1972). We can recognize the significance of the signals for the partner only in a few cases (e.g., warning cries, contact sounds). In all

TABLE 1 **List of Behavior Units**

No. of behavior unit	Description of behavior units
04	Vocalization (without differentiating types of calls)
05	Touching partner's head
06	Touching partner's hip
07	Touching partner's back
08	Lolling, sprawling
09	Touching partner's tail
10	Pulling partner's tail
11	Touching partner's extremities
12	Advancing mouth toward partner's neck
13	Bending forward upper part of auricle
14	Straightening body in front of partner
15	Thrusting chin toward partner
16	Uncovering teeth
17	Circling around partner
18	Genital display[a]
19	Inspection of partner's genitals
20	Back rolling with presentation of ventral view[b]
21	Mounting
22	Copulatory movements
23	Jumping onto partner
24	Beating partner
25	Biting partner
26	Fighting
27	Running away from
28	Pushing partner away
29	Running to
30	Touching partner's ventral side
31	Advancing nose toward partner's sitting place

[a] Ploog and MacLean (1963).
[b] Castell *et al.* (1969).

other cases we do not know why the behavior units occur nor why in the succession observed.

In principle, the lack of determined transitions from one action to the following ones can have different reasons: It is possible that the units employed, obtained through a physical description of the movements involved, are incorrectly chosen inasmuch as the distribution into the designated units does not represent the distribution into the different informations which the partners mutually convey. Expressed otherwise, each of the various behavior units conveys not just one definite item of information, nor is each information conveyed through only one definite behavior unit.

Moreover, further problems which quite generally tangle up an analysis of communication sequences must be considered. The observer cannot always recognize whether a signal is actually directed to one partner and to just this one partner. In the same way he cannot know whether a signal was received from the corresponding partner (MacKay, 1972). That tells us that the dependence of an animal's action on the preceding action of another animal can never be stringently proven (Cullen, 1972). Additional prior signals can exert an unrecognizable effect (Cullen, 1972). Each activity, i.e., sitting, running, eye contact, etc., can or cannot be a relevant signal depending on the context, where the human observer would not be in a position to recognize this. Theoretically, a random amount of unrecognizable signals can occur between any two recognizable signals. We have demonstrated such unrecognizable socially relevant visual signals in the squirrel monkey (Maurus and Ploog, 1969, 1971). These considerations sufficiently explain why a behavioral analysis based exclusively on the observation of behavior units following one another as they occur in spontaneous behavior cannot lead to a sufficient understanding of the communication processes of the squirrel monkey.

The use of dummies is one of the established methods employed in ethological research in analyzing social behavior of other animal species. The introduction of dummies whose appearance more or less resembles that of the companion species leads to quite constant reactions on the part of the experimental animal with many species. It responds to the displayed social signal (the dummy) with a reaction in which one or more behavior events are contained, with frequency exceeding chance level.

According to experiences in our laboratory, it is difficult to find a suitable dummy for squirrel monkeys. Dummies of conspecifics have no clear-cut effect on squirrel monkeys; no reproducible reactions can be incited even when the dummies are taxidermic models of conspecific partners. The same holds true for the utilization of tape recordings of most vocalizations peculiar to the species. Obviously only conspecifics which move about naturally and emit natural sounds can be repeatedly used to initiate communication processes (e.g., mirror display, Ploog *et al.*, 1963; Ploog and MacLean, 1963; MacLean, 1964). For the time being, the telestimulation technique offers the most refined method of releasing naturally appearing behavior in freely moving animals. An animal can be "forced" to perform certain social behavior units through electrical stimulation of certain brain structures which induce response in the nonstimulated group members (Delgado, 1966; Robinson *et al.*, 1969; Maurus and Ploog, 1969). The behavior "forced" on the stimulated animal can be repeated almost as frequently as desired. A large number of communication processes can be initiated in this manner. All of these processes are set off by the same event, i.e., the behavior of the stimulated animal. Just about every animal in the group can be made into such a dummy. In this way we get an ideal "dummy" through which we can repeat a socially relevant signal quite frequently. We possibly gain the chance to recognize the function of the signals by this means.

Based on the experiences with other animal species, one would also expect from the squirrel monkeys that the repeated display of the dummy would be followed by one or a few definite behavior units with above-average frequency. But even in the most ideal case, where the frequent repetition of a constant stimulus response of one animal (= unchanging dummy) is always answered only by the same partner, varying behavioral events continually occur in diverse sequences in the answer. This holds true in the case that the experimental group consists of four or five animals and in the case that the group consists of only two animals. The resulting variability is so great that it seems to the observer that the behavior units occur randomly. This impression can be only a deception, however, because a communication system must be constructed along certain rules in order to function. The partner must be able to understand the information contained in the signal (Flechtner, 1966) in order to react to it in the proper way. Where a signal is misunderstood, it will cause problems for the animals communicating until these animals have learned to behave according to the underlying rules of the communication system (Kummer *et al.*, 1970). Consequently, just one explanation remains for the randomly distributed occurrence of the squirrel monkeys' behavior units. The behavior units obtained so far do not coincide with the units transmitted as information. As these information units are unknown to us, and as we cannot observe any other units, an analysis of the social behavior must begin on the basis of the described behavior units. The behavior units must be arranged in classes so that all units combined in a class have the same or at least a similar function in the communication processes, i.e., so that they convey the same or very similar information. The units combined in another class have another function. As the behavioral repertoire of the animals was broken down into behavior units to and not beyond an extent thought reasonable in terms of physical description of postures and movements involved, a class can consist of one or of several behavioral units.

Because of the high variability in the occurrence of the behavioral events, promising approaches to such a classification can be attained only on the basis of statistical procedures. Large quantities of data are beneficial for statistical calculations. These large amounts of data can be accumulated with the aid of telestimulation. This method offers the following advantages:

1. Electrical brain stimulation can be repeated at just about any chosen interval and very frequently if required. Thus a sufficient quantity of data can be gained for statistical computation in a relatively short time. Undesired fundamental changes in the group structure of the experimental animals are eliminated as a possible source of error.
2. We can limit all animals' behavior to the desired behavioral category through the choice of suitable electrode positions (e.g., agonistic behavior of an intensity excluding serious injuries of the animals).
3. The investigator determines the release time of social interaction through brain stimulation. This permits recording on motion picture film at an economically acceptable expense. The advantage of filmed records is self-evident.

TELESTIMULATION

Technique

The following conditions must be met in applying technical aids in investigations of social behavior:

1. The device the animal has to carry must be sufficiently small and light and fastened in such a way as to hinder the carrier animal as little as possible and alter its appearance so little that it will be accepted by its partners as a normal species companion.
2. Possible necessary work on these devices (e.g., battery recharging) should be required only over long time spans so that the animal does not become disquieted too much through frequent capture and manipulation.
3. The brain stimulation incited by the device must be brought about in such a way that it causes as little irreversible injury as possible.

The telestimulator devices developed for rhesus monkeys by other authors (Delgado, 1963; Robinson *et al.*, 1964) do not meet the conditions of numbers (1) and (3) in use on the squirrel monkeys in that they are too large and produce a monophosic stimulation. Therefore, a new telestimulator had to be developed (Maurus, 1967a,b). The receiver of the telestimulation equipment (i.e., the part that must be carried by the animal) adequately meets all conditions laid down:

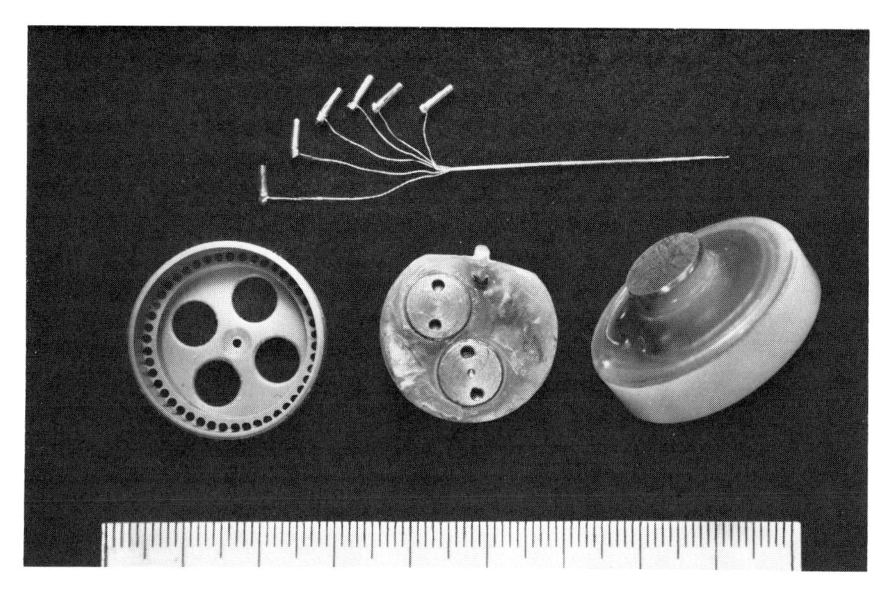

Fig. 1 Equipment carried by the animal for telestimulation and telemetry. Above: multilead electrode. Below: from left to right: plastic box for containing the electrode plugs and the telestimulation receiver, telestimulation receiver, lid as transmitter for telemetry.

1. Its weight comes to about 5 gm, including the batteries (modified form of 1967 device). Its outer dimensions are: 23 mm diameter, height 8 mm (see Fig. 1). This receiver is housed in a cylindrical plastic container (diameter 25 mm, length 12 mm) which is fastened to the animal's skull. This construction does not seem to disturb the animals; the conspecific partners ignore it. Only very young animals occasionally reach for this construction during the first hours following implantation on the other animals. All members of several groups of monkeys have lived for more than 4 years with this device and have borne and successfully raised infants during this time.
2. A manipulation at the receiver is necessary only at intervals of several weeks or months. The receiver works with a residual current of about 2 μA. A battery charge (35 mAh) is adequate for ca. 50–100 hr of uninterrupted brain stimulation, depending on the stimulation parameters.
3. Monophase current impulses are unsuitable for frequently repeated brain stimulations because they cause considerable electrolytic lesions in the brain with just relatively small total current amounts (see Fig. 2). Biphasic impulses reduce these injuries considerably with

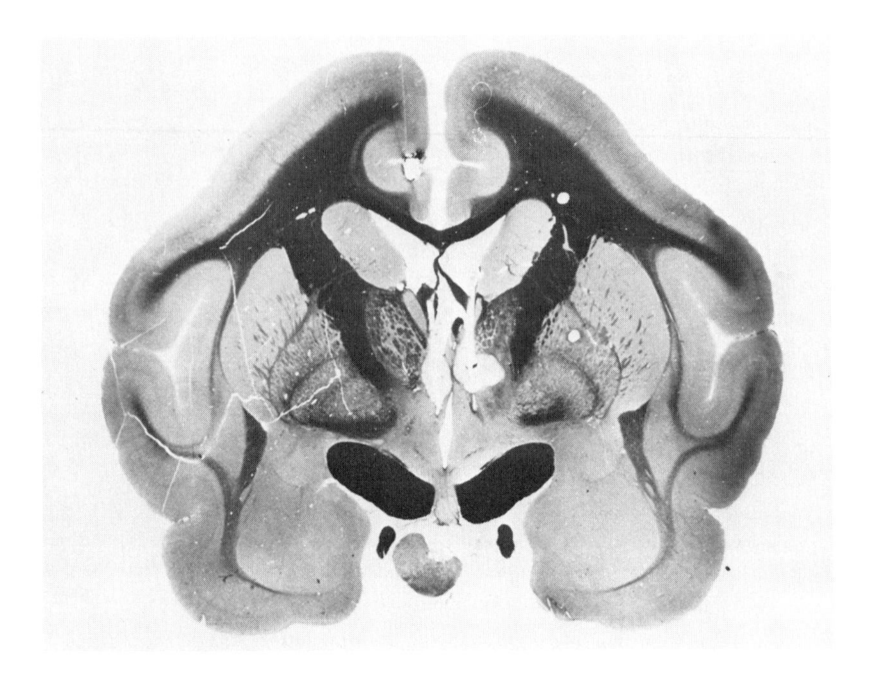

Fig 2 Frontal section through squirrel monkey brain. Two electrode tracts with lesions at the electrode tips. The lesions are due to stimulation with monophasic rectangular impulses. Total current: left electrode ca. 4.2 mC, right electrode ca. 34 mC.

still the same stimulation effect (Lilly, 1961; Mickle, 1961; Terasawa and Sawyer, 1969). The receiver developed for the squirrel monkeys furnishes a biphasic rectangular pulse with the following parameters: repetition rate, 10–100 Hz; pulse duration, 0.1–1.5 msec; amplitude, 5V over a series resistance of 5 kΩ per electrode. Each negative impluse is followed by a positive impulse with the same pulse width and amplitude at an interval of 3–6 msec. The residual current over the electrodes between impulses is less than 0.5×10^{-13} A. We have not found any damage following application of these technical elements.

Preparation of the Animals

We began our studies with groups of four or five squirrel monkeys. These groups consisted of two males and two females or of two males and three females each. The groups were combined so that the male rank order was determined by a considerable difference in weight. All females were lower in rank than the males. For several months these groups were allowed to get accustomed to each other and to their cages (270 cm × 170 cm × 100 cm). Following cessation of spontaneously occurring visible aggressive interactions (the so-called social stable groups), electrodes were implanted in the animals. Multilead electrodes with five to seven terminals and an external diameter of about 0.4 mm were used (Fig. 1). The five to seven individual leads of the electrodes were arranged in a staggered fashion so that a vertical tip separation of 1 or 1.5 or 2 mm was obtained. The insulation of the tips was removed for 1, 1.5, and 2 mm, respectively. Individual leads measured 0.12 mm in diameter. These electrodes were implanted mainly in medial structures of the limbic system. Each exterior end of the individual leads was provided with a subminiature female plug (Fig. 1). These plugs were inserted concentrically into a plastic case fixed to the skull. The indifferent electrode consisted of four stainless steel screws within the skull, short circuited and connected to a male plug protruding from the center of the box. The telereceiver (Fig. 1) intended for insertion into the box was provided with a female plug at bottom center and a male plug at its margin. Any chosen lead could be connected in this way to complement the indifferent electrode. Leads were switched by slightly lifting, turning, and reinserting the receiver into the plug of another lead. When the animals were not stimulated, they wore dummy receivers of equal weight so as to stay accustomed to the receiver weight.

DATA RECORDING

Technique

The completion of the telestimulation technique and the preparation of the animals for telestimulation are not the only technical preconditions. The behavior of the squirrel monkeys unfolds with a speed requiring two persons for the direct observation of the spontaneous behavior of just one animal in the group and for recording the events sent and received by him (Hopf, 1972). If all the events in the behavior of all animals in a group are to be noted, it is possible only through use of technical aids allowing a repeated and variable speed rendering for later evaluation. Visually recognizable signals can be filmed by a movie camera.

A similarly simple method for recording the acoustic signals is inapplicable. A tape recording, accompanied by motion picture, cannot attribute the sound to the vocalizing animal individually, since many sounds are made with the mouth closed. Only a telemetrical system with separate transmission channels for each individual enables definite attributing of all emitted calls. Each animal wears the telestimulator box. The lid of the box (see Fig. 1) was rebuilt into a telemetrical transmitter of acoustical signals (see Fig. 3). The sound waves conducted via bone and

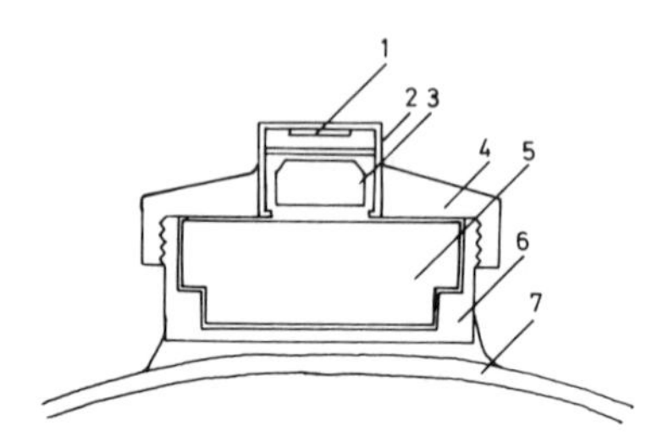

Fig. 3 Systematic cross-section of the construction attached to the skull of the animal (actual size). 1: Piezoelectric crystal; 2: stainless steel cylinder; 3: battery; 4: lid with the molded-in electronic transmission equipment; 5: telestimulation receiver; 6: plastic box; 7: skull.

plastic box reach a piezoelectric transducer that modulates the frequency of the transmitter. Each animal wears its own transmitter. Each transmitter has its own carrier frequency and its own receiver, demodulator, amplifier, and assigned track on a multitrack tape recorder.

Still another very significant factor in social behavior is the distance between the animals (Maurus and Ploog, 1969, 1971). Particularly the agonistic behavior shows strong dependence on distance between the animals involved. It is not only the distance between the animals which plays the role of momentarily directing agonistic signals reciprocally, but also the distance from these two animals to other members of the group, particularly to animals of higher rank. The distance between the animals could be gotten from the film by manual interpretation. A manual interpretation would have a quite considerable delaying effect on this film analysis, as all distances between the members of the group must be included in the evaluation. For example, it would come to 10 distance values for five animals in addition to each event. In this case the quantity of data recorded by hand would be multiplied 11-fold, and an equally manifold delay of the analysis would ensue. As the film evaluation already constitutes the main time problem in the plan of the investigation, even where distance values are not being considered, a more efficient procedure for extracting the distance values has to be applied. We have decided on an automatic position location procedure.

Position location (PL) is done by means of ultrasonic sounds of 50 kc/sec (Riebling, 1970). There is an ultrasonic source located on the ceiling of the test cage in three corners. These sources emit short ultrasonic impulses of 1 msec duration with an interval of about 30 msec in sequence. This series, consisting of three ultrasonic impulses, is repeated with every time interval (that is, 100 msec). The transmitters worn by the animals as well as the receivers of the data recording system for acoustical signals also serve ultrasonic PL. This is possible because the piezoelectric transducer for the transmitter modulation is constructed so that it has its resonance frequency at 50 kc/sec. Thus the piezoelectric transducer is sufficiently sensitive to sound waves conducted via air at this single frequency. The delay caused by the ultrasonic impulse passing from the ultrasonic source to the animals' transmitters is in linear proportion to the distance between the respective animal and the corresponding source. We get a distance value per animal per 1/10 sec for each of the three ultrasonic sources. The conversion of time delay into distance values occurs within an accuracy of 1.7 cm. The frequency of the repetitions is sufficient (10 per sec) to guarantee registration of all locomotions of animals. The incoming distance values are stored in blocks on digital tape. (The automatic PL was not developed just for the simplification of the analysis, it also serves as precondition for an automatic situation-dependent stimulus application; i.e., a stimulus application takes place automatically only under predetermined conditions, appearing in the behavior of the monkeys. See conclusions.)

For the evaluation it is necessary to link the recorded events retained by the three recording systems to one another timewise. Accordingly, all three recording systems are steered from a common timer. It produces a time interval of 100 msec. These time intervals are counted each day, beginning in the morning. The 24-hr day has 864,000 time intervals of this kind. These enumerated time intervals are so inserted into each of the recording systems that the beginning and end of each event can be read with an exactness of 1/10 sec from the respective recordings. Sufficient precautions are built into the system with respect to error avoidance. (Detailed accounts of data recording are found in Maurus *et al.*, 1972.)

Execution of Experiments and Data Conversion

The "socially stable" group of animals prepared for telestimulation is put into a cage identical in size and arrangement to the acclimatization cage (see Fig. 4). In an "explorative study" the electrodes are activated one after the other in all animals with different stimulus parameters controlled from the telestimulation transmitter. Only one animal is stimulated at a time. Stimulus responses are tested for reproducibility in all electrode positions by several repetitions of the stimulus. In these explorative studies (Maurus and Ploog, 1971) the following behavior units could be incited up to now as stimulus response to stimulated animals: 04 (various

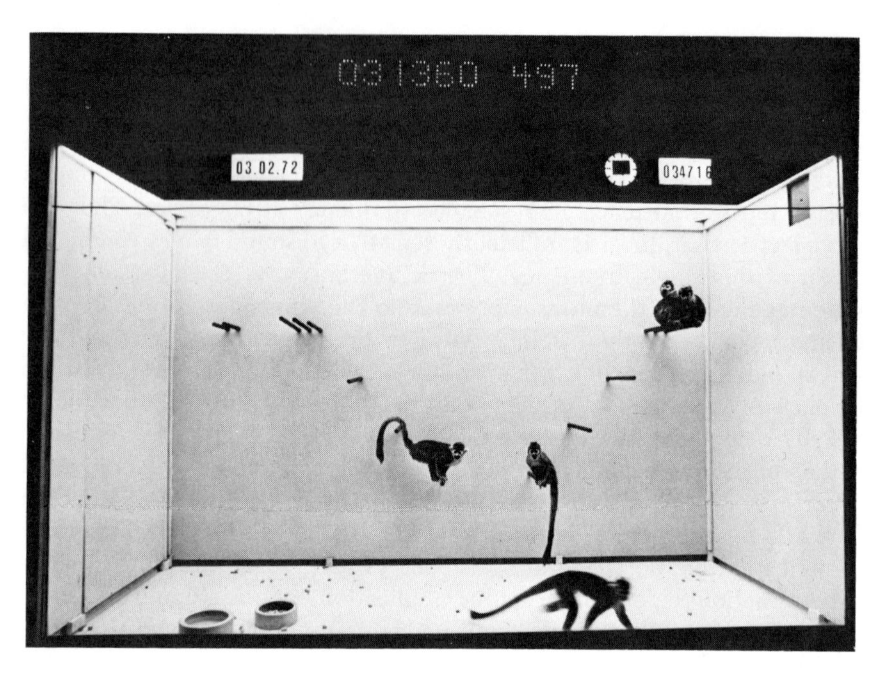

Fig. 4 Experimental cage with monkeys. The cage size is 270 × 170 × 100 cm. At the upper margin: Time markings in 1/10 sec. Below data: Clock times and number of the stimulated electrode. All monkeys are provided with the telestimulation devices.

forms of vocalization), 05, 06, 07, 09, 10, 11, 12, 13, 14, 15, 21, 23, 24, 25, 26, 29, and 30 (see Table 1). The behavior units mentioned can occur alone or form the stimulus response together with others. Other group members react reproducibly to these stimulus responses. (Of all observable, visually recognizable behavior units, numbers 08, 16, 17, 18, 19, 20, 27, 28, and 31 have not yet been induced as stimulus response to the stimulated animal.)

The investigator has an enormously variable dummy at his disposal capable of inciting behavior units useful as aids in the analysis of the intraspecific communication. This dummy is an arbitrarily manipulated group member. Its behavior appears as natural to the observer. It cannot be differentiated from spontaneously occurring behavior units. As nonstimulated group members cannot perceive the electrical brain stimulation, and yet react to the behavior of the stimulated animal, it can be assumed that the electrically induced response appears as normal behavior to the partners as well. It certainly does not appear, however, at the proper time point in the continuing communication process. It therefore disturbs the existing state of balance in the socially stable group. This state is finally reestablished again (i.e., no evident agonistic interaction during the pauses between the stimulus-induced interactions). One can conclude that reproduction of the state of balance is attained simply through processes following the rules of communication processes as well. In other words, everything shown following the stimulus response of the stimulated animal by nonstimulated partners is "natural" behavior, even if it occurs at the "wrong" point in time.

The brain stimulations were done automatically in order to avoid disturbing the animals through the presence of a superfluous observer (see earlier discussion) and to obtain the necessary large mass of data. A device was developed for these automated experiments which repeated the stimulus automatically. This device switched on the film camera and the transmitter within predetermined time intervals so as to record the initial situation before onset of the stimulation and the sequence of all consecutive events caused by the stimulus. An experiment consisted of a certain number of stimulus repetitions. The number of repetitions could exceed some thousands. Within one experiment the stimulated animal remained the same; the electrode, all stimulus parameters, and the time interval from stimulus to stimulus were kept constant. The time between two repetitions was not less than 5 min. In most cases it was 20 min. The animals were exposed to a light–dark cycle: 12 hr light and 12 hr darkness. Stimuli were administered only during the 12-hr day.

Each stimulus repetition leads to a sequence of short events following one another. This sequence follows a pause in which almost no visually discernible interaction occurs among the animals (social stable group). This pause is ended with the succeeding stimulus repetition which initiates a new sequence. Each sequence consists of the stimulus response and the consecutively occurring reactions of the group members caused by the stimulus response. The events were coded into so-called actions. An action consists of a sender animal (a), a recipient animal (b), and a behavioral unit (u) which is directed from the sender animal to the recipient one. The animals were numbered according to their rank within the group: number 1, the dominant male; number 2, the subordinate male; numbers 3–5, the females. If a behavior unit is not definitely directed toward a certain partner, we write $b = 0$. The list of behavior units observable in the experiments is shown in Table 1. Each action (a, b, u) is coded into a four-digit number. For example, if the dominant male ($a = 1$) showed genital display ($u = 18$) toward the subordinate male ($b = 2$), this action was coded into 1218. The set of all different actions forms the behavioral catalog.

In the experiments cited in the following discussion, the recording is limited to the visually recognizable actions, which require the least technical expenditure, so that the experiments could be taken up before the completion of the technique involved in recording the remaining data. The time intervals separating the actions of a sequence were not taken into consideration, as the visually discernible actions all follow one another at short intervals, and coordination with vocal signals and distance values was not possible. That means that a large portion of the communication remains disregarded. It can be assumed that an analysis based exclusively on the transition probability from one action to the next cannot be promising due to previously mentioned reasons. For this reason, procedures in which the lack of single events in the sequence can be neglected have to be used for the analysis.

The film is replayed slowly, and individual scenes are reshown as frequently as required for the clear recognition of all actions of all animals in the evaluation of the film records. Owing to this care, there are practically no omissions in contrast to direct observation (Hopf, 1972), rather an error rate of about 4%, predominately a result of difficulty in indentifying certain actions. (Either the animals are partially concealed by partners or are in an inconvenient position for the camera.) The poorly developed facial expressions of the squirrel monkeys is particularly convenient here. The observed actions of each of the enumerated sequences are noted in the order of occurrence in the earlier mentioned code. The following sequence serves as an example. The stimulated subordinate male puts his hand on the head of female number 3 (stimulus response); the dominant male thrusts chin toward subordinate male, as a result the subordinate male runs away from the dominant male, whereupon the dominant male shows back rolling with presentation of his ventral view. This sequence would appear in the record in the following form: 2305, 1215, 2127, 1020. The records can be produced by hand and then transferred for electronic data processing onto paper tapes or punch cards. They can also be transferred directly onto paper tape via the evaluation device developed for this purpose (Höhne and Maurus, in preparation). The beginning and end periods of the actions are given automatically on the same paper tape with this method.

Along with the construction of the technical apparatus for the performance of the telestimulation experiment and the procedures for collecting the data, we have established a necessary precondition for the analysis of the communication system. The next problem lies in the question of how these data can be processed so that they lead to biologically relevent statements. As we see it, there is no way other than establishing the mathematical methods which can be applied to the data from a mathematical point of view and checked for relevant biological results.

An example of these procedural means will be presented in the visually recognizable signals of the squirrel monkeys.

STATISTICAL ANALYSIS

Variability of Quantities

An abundance of behavior unit sequences arises through the frequent repetition of brain stimulation. It is not the goal of the investigation to use technical aids, such as automatic stimulus repetition and a data recording system, to gather this abundance of sequences and describe them in the form of episodes. The data extracted should serve as a basis for recognizing rules on which intraspecific communication is built. It is necessary for recognition of rules to find the suitable quantities first of all. These quantities must be invariant for the range where the rules are valid [i.e., at least be invariant within the investigated population type; (that is, gothic arch type, MacLean, 1964)]. To be invariant means they must be independent of individual characteristics of the animals or of the animal groups, and from experimental conditions.

Each quantity must be checked to see whether it meets the stated requirements. This check shows clear results for the underlying units, the visually recognizable units. These behavior units are invariant. They fulfill all conditions mentioned, inasmuch as the statement relates exclusively to morphological aspects (the physical description of the postures and movements involved). Possibly occurring individual deviations in performing certain signals cannot presently be recognized by the observer. In the case of their existence, the interindividual variability lies obviously in a value order within the intraindividual variability. (Such problems are discussed for acoustic signals in Schott, 1972.) From the fact that no significant, individual variability exists in the execution of the behavior units, it follows that the actions are invariant as well.

It is much more difficult to answer the question of invariability with the quantitative values than with the qualitative values of "behavior unit" and "action." For the quantitative values it is easier to prove that they are not invariant since it is sufficient to show that a quantity is different in at least two experiments. In order to demonstrate conclusively that a value is invariant, it would have to remain constant in many experiments, with many different animals, with many different monkey groups, and under many different experimental conditions. It will take many years to carry out these numerous experiments. For the time being, the verification of quantities is confined to some experiments when brain stimulation is carried out on different electrode positions on the same animal, on different animals in the same group, and on different groups under different experimental conditions as in the following:

Monkey group number 0 (= GO), where a female was stimulated at one brain site (GO ES1).

Monkey group number 1 (= G1), where one female (number 3) was stimulated in one brain site and another female (number 4) was stimulated in two different brain sites (that is, two different stimulus responses) (G1 ES1).

Monkey group number 2 (= G2), where the subordinate male was stimulated in a brain site in several experiments with constant interval between the stimulus repetitions: (G2 ES1) and in another series of experiments on the same brain site, where the interval between the stimulus repetitions was different for each experiment. 20 min: (E8), 10 min: (E9), 5 min: (E10), 20 min: (E11), no females were present in this experiment), 20 min: (E12), 20 min: (E13). A series of experiments in which a female was stimulated at three different brain sites (in this case three identical stimulus responses) (G2 ES2). (For more details see Maurus *et al.*, 1973a,b).

Quantities remaining constant under all these conditions are considered invariant, with the limitation that they are to be provisionally understood as invariant. We want to limit ourselves to the investigation of the simplest statistical value: frequency. The following frequencies vary from experiment to experiment:

1. The absolute and relative (relative in regard to the number of all the actions observed per experiment) frequency where behavior units are actually sent or received by each of the animals pro experiment.
2. The absolute and relative (relative in regard to the number of all the actions observed per experiment) frequency with which each of the different behavior units occurs pro experiment.
3. The absolute and relative frequency (relative in regard to the number of all the actions observed per experiment) with which each of the different actions occurs pro experiment [e.g., with which each of the behavior units (u) of each animal (a) is directed toward each partner (b) pro experiment].

Also variable, as mentioned above, are the sequences of actions following a constant stimulus response. (Compare p. 377 in Maurus and Pruscha, 1972.) Statements derived from these variable quantities cannot be regarded as generally valid for the population or the species.

Invariant values can be found only if the animals are numbered for each experiment according to the following criteria:

Under I are all those males ranking higher than the stimulated animal.

Under II we find the animal electrically stimulated (i.e., the animal disturbing the state of balance) in the experiment (male or female).

Under III are combined all females unstimulated in the experiment.

With this codification, for example, action 1318 would be written as I II 18 if animal 3 were stimulated electrically, and I III 18 if animal 2, 4, or 5 were stimulated.

One can count how often a certain behavior unit is carried out by each animal and how often it is directed by him toward each of his partners. If this procedure is performed for all different behavior units, then it becomes apparent that there are behavior units which are not directed from certain animals toward certain partners. This means certain animals (a) "omit" performing certain behavior units (u) toward certain animals (b), or stated otherwise, certain (a,b) combinations are omitted for certain behavior units (u). These omitted (a,b) combinations are not the same for all behavior units. But for each behavior unit there are (a,b) combinations uniformly omitted in all the experiments. We can presently view this finding as an invariant quantity. For the behavior units occurring more frequently, Table 2 shows all (a,b) combinations that did not occur in any of the experiments. For example, behavior unit 18 (that

TABLE 2 **List of (*a,b*) Combinations That Do Not Occur in Any Experiments**[a]

	(I, 0)	(I, II)	(I, III)	(II, I)	(II, III)	(III, I)	(III, II)	(III, III)
05	0					0	0	
06	0					0	0	0
07	0					0		
08		0	0	0	0	0	0	0
12	0					0	0	0
14	0					0	0	0
15	0		0		0	0	0	0
18						0		0
19	0			0		0	0	
20		0	0	0	0	0	0	0

[a] 0 = (a,b) combinations that do not occur in any of the experiments. (Frequencies ≤ 2 were given the value 0.)

is, genital display) is "omitted" only between the nonstimulated females. Behavior unit 08 and behavior unit 20 are only shown by animal I and never directed toward a definite partner (i.e., the observer cannot recognize the partner toward which the two behavior units are directed). The behavior units 08 and 20 distinguish themselves clearly from all other behavior units in this respect. A differentiation of these other units from one another is impossible by means of the simple criteria sufficient for distinguishing 08 and 20, although it is certainly a possibility as soon as a further question is answered.

The question is as follows: With which relative frequencies is a behavior unit divided up into its different (a,b) combinations? Or stated otherwise: If a behavior unit occurs, then to which division will it be conducted from I toward 0 (for explanation of 0 see p. 343), from I toward II, from I toward III, and so on? Where one enters these numbers for all (a,b) combinations of a certain behavior unit, one gets a distribution of the relative frequencies of the (a,b) combinations of this behavior unit. This distribution can be calculated for each behavior unit. We want to restrict ourselves to those behavior units which have occurred in each experiment with sufficient frequency. These behavior units are 06, 07, 12, and 18. We structure the distribution of the relative frequencies of the (a,b) combinations for each of these behavior units in each experiment (experimental series) and compare these distributions with one another. We take f^2 (f^2 = chi square divided by the frequency), a measurement of the similarity or homogeneity between the distributions being compared (Cramer, 1954, p. 445). If $f^2 = 0$, then the compared frequency distributions are completely proportional; $f^2 = 1$ stands for maximum difference.

Following computation of the values of f^2, we can ascertain that in all experiments (or experimental series) f^2 is least in the comparison of the distributions of the behavior units 06, 07, and 12 among one another; that f^2 increases if behavior units 06, 07, and 12 (combined) are compared with behavior unit 18; that f^2 increases again between behavior unit 18 and behavior unit 20, and is largest between the behavior units 06, 07, and 12 on the one hand, and behavior unit 20 on the other (see Fig. 5). This succession in the similarity of the distribution of the relative frequencies of (a,b) combinations of behavior units is apparently a further quantity we can consider as invariant.

We want to waive the presentation of further invariant quantities. (For more invariant quantities see Maurus *et al.*, 1973a,b.) In this paper, instead, using these two quantities we want to test the significance that such quantities obtained through computation have for biological statements.

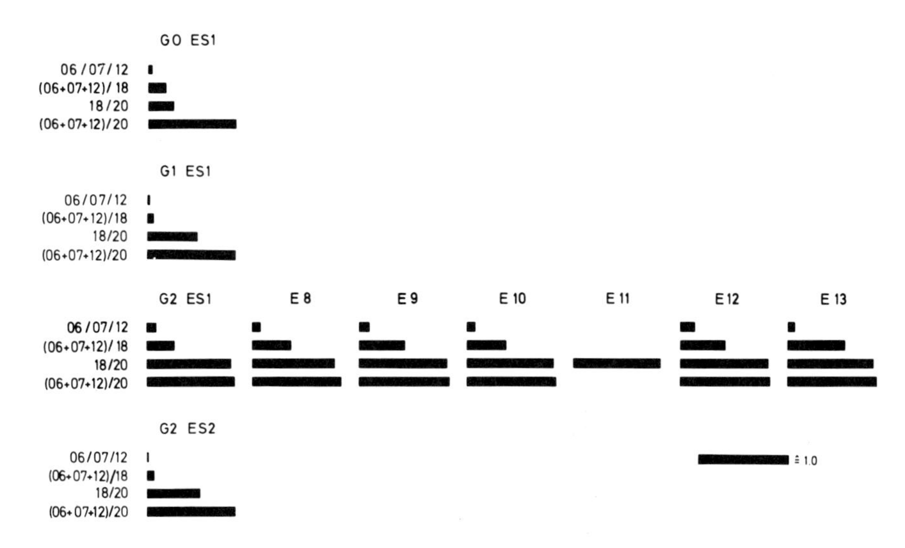

Fig. 5 Values (f^2) giving the similarities between the distributions of (a,b) combinations of behavior units within the experiments E8–E13. The length of the bar is proportional to the calculated similarity value (f^2) between the behavior units listed on the left side. Behavior units separated by / are compared with one another; behavior units in ( ) are pooled.

Relevance of the Quantities

If it is established that a quantity is invariant, a necessary condition for the recognition of rules is met. Meeting this condition alone, however, is insufficient, i.e., it must also be demonstrated that the values obtained are biologically relevant as an important constituent of the intraspecific communication and useful toward the solution of the classification problems addressed here (see p. 335).

Both invariant quantities are based on the question of how frequently certain behavior units from certain animals are directed toward each of their partners. It seems reasonable that the meaning of a signal (i.e., the information which carries the signal to the partner) is connected in some way with whether or not it is performed by a specific animal toward a specific partner and how often toward each of these partners. For example, the behavior units 06, 07, and 12 are directed mainly from dominant males toward nonstimulated females. Behavior unit 15 is exchanged only between the dominant male (I) and the stimulated animal (II). None of the other animals ever used this signal nor is it ever directed toward any other animal. Behavior unit 18 is directed from all animals toward almost every partner. It is difficult to imagine that all these behavior units could have the same meaning in the communication process. Differentiating the application of these signals would be meaningless unless accompanied by a differentiation of their function in the communication process. These arguments can suffice for the time being in order to attempt a provisional classification of the behavior units appropriate to the distribution of the frequencies of their (a,b) combinations. We can obtain the four classes, namely a class consisting of behavior units 06, 07, and 12; a class consisting of behavior unit 15; a class consisting of behavior unit 18; and a class consisting of behavior units 08 and 20.

In order to demonstrate that this classification is correct, it must be shown that it results in an increase of predictability in regard to the succession of the behavioral events (i.e., classes). The answer to the question whether a classification causes this increase or not can be found in the following two ways:

1. A procedure can be applied which allows calculation and valuation of a quantitative value of predictability for a given classification. A procedure which permits us to say whether this value lies above or below the average of all possible classifications was developed (Pruscha and Maurus, 1973). However, this value does not tell us if there is a further classification other than the tested one, which results in a better value, nor tell us how this classification must be set up. In order to obtain this knowledge, the value for each single classification from all possible ones must be calculated. In the number of about 120 different actions found so far in agonistic interactions there arises a quantity of possible classifications so large that the amount of calculation would hardly be manageable.

2. Another procedure does not derive from a proposition testing a previously given classification, but rather from a proposition employing the measurement for predictability (that is, correlation measurement) as a criterion for classification. This classification procedure is carried out successively in many similar steps. The first step consists in selecting two actions from the set of all different actions (that is, catalog) in such a way that through their combination the newly formulated catalog causes a maximal predictability in regard to the succession of the behavioral events. In the next step the now available catalog, newly formed, is changed through combining once again two events with the same goal, as in the preceding procedure. This procedure is repeated until all of the actions are combined into one single class. The result of this "agglomerative cluster method" (Socal and Sneath, 1963; Orloci, 1967, 1968; Cole, 1969) is a series toward a hierarchy of classes steadily becoming more and more comprehensive. It can be graphed in the form of a dendrogram (see these dendrograms in Maurus and Pruscha, 1973). The statistical correlation measure T (in information theory also called mutual infor-

mation, information rate, rate of transmission of information) forms for both methods the criterion for the degree of predictability. In case of a nominal scale, T has the same meaning as has the correlation coefficient (more exactly the empirical covariance) in case of an interval scale. This property of quantity T is clearly seen at its extreme value: T is 0 where the sequence of actions occurs stochastically independent. T is at its maximum when each action is fully determined by the one preceding it.

In three experimental series carried out in two different monkey groups, this agglomerative clustering method was applied (Maurus and Pruscha, 1973). In all three experimental series the results were by and large identical: The first steps of this cluster analysis lead to a classification ending in a "lumping" of the actions according to the animals involved. This subsequently confirms the necessity of including the animals involved (sender animal = a and recipient animal = b) in the codification of the actions. Furthermore, it is an indication that this cluster method gives us a classification with biologically relevant facts. This primary result means that the social function of behavior units (u) depends essentially on the combinations of the animals involved (a,b combination). Further steps lead to a classification of the actions according to the behavior units involved. Summing up, four classes of behavior units can be delimited from one another (see Maurus and Pruscha, 1973): Behavior units 05, 06, 07, 09, 11, 12, and 21 usually fall together into one class (number 91). Behavior units 14 and 15 usually fall together into another class (number 92). Behavior units 18, 19, and 31 fall together into a third class (number 93). And behavior units 08 and 20 fall together into a fourth class (number 94).

When one compares the standing of the units in the classes built on the basis of the invariant values, then it is demonstrable that both procedures lead to classifications not reciprocally contradicting one another. For example, the classification based on the invariant quantities of the (a,b) combinations leads to a clear increase of predictability in regard to the succession of the behavioral event (see further detail in Pruscha and Maurus, 1973).

CONCLUSIONS

Based on the invariant quantities, the following statements concerning agonistic behavior in the squirrel monkeys can be derived.

1. The behavior units set up according to morphological aspects (that is, a physical description of postures and movements involved) are not directed from each animal toward each partner in random frequency.

Rather, each behavior unit is characterized in its distribution among the partners in a relative strictly fixed frequency relation.

2. Behavior units resembling each other in regard to this frequency relation have similar functions in the communication processes (that is, communicate similar information).

3. Behavior units clearly distinguishable in regard to this frequency relation have different functions in the communication processes.

4. From 2 and 3 follows that not only what is done is of importance, but also who does it and toward whom it is done.

5. Who "who" is, is determined through the hierarchical rank of the animal.

6. The behavior units can be combined according to 2 and 3 in classes with the result that based on these classes the predictability in regard to the succession of the behavioral events (that is, classes) is significantly increased.

7. The classes found with the help of the proscribed methods correspond with morphological characteristics of the postures and movements involved. The behavior units usually falling into class 91 are behavior units in which the partners act in short distance (05 equals touching partner's head, 06 equals touching partner's hip, 07 equals touching partner's back, 09 equals touching partner's tail, 11 equals touching partner's extremities, 12 equals advancing mouth toward partner's neck, and 21 equals mounting).

The second class (92) consists of behavior units (14 equals straightening body and 15 equals thrusting chin toward partner) which are performed toward the partner from a greater distance.

The third class (93) is made up of behavior units which are related to the genital organs. In behavior unit 18 (that is, genital display) and 19 (that is, inspection of partner's genitalia) the genital organs directly participate. Behavior unit 31 (advancing nose toward partner's sitting place), also belonging to this class, can be assumed to be connected with olfactory stimuli originating from the genital organs.

Class 94 includes behavior units 08 (that is lolling, sprawling) and 20 (back rolling with presentation of ventral view).

The following morphological criteria are relevant in this connection:

For the first class (91), short distance to the partner.

For the second class (92), greater distance between the partners.

For the third class (93), reference to genitalia of partners.

For the fourth class (94), contact of larger body regions with the environment (cage).

It cannot be deduced from this result that in the end the morphological criteria are relevent and that therefore the mathematical analyses have been superfluous because:

a. It is not evident a priori that morphological properties of social signals are associated with functional properties, so that social signals with similar morphological properties have similar functions.

b. Even if the existence of such an association is hypothetically assumed, it is still unknown which of the many possible morphological properties are relevant.

c. The mathematical analysis carried out not only solved the problems in (a) and (b) but also gave us further information. It has been demonstrated that this connection can differ depending on the sender–recipient combination. I II 91 is, for example, easy to distinguish from I II 92, while II I 91 cannot be distinguished from II I 92. Behavior units from class 91 shown by the dominant male require other reactions in the group than behavior units from class 92 shown by him. On the contrary, no differences in the group reactions are recognized if behavior units from both classes 91 and 92 are shown by the subordinate male (see table 5 in Maurus and Pruscha, 1973). Here we have an example that shows that behavior units originating from different classes can be cataloged under certain (a,b) combinations in one class. Beyond this, we cannot exclude the possibility that the one or the other behavior unit within a certain (a,b) combination appears in one class and in a different (a,b) combination in another class. It follows from this that we cannot a priori conclude that the final classification must exclusively consist of a combining of behavioral units, even though this be the case with the classes recognized up to now.

8. The correctness of our classification proves itself not only through the increase in predictability in regard to the succession of the behavioral events, but also through other results. We mention here the run of the individual behavior units' frequencies as an example in experiments E8, E9, E10, E11, E12, and E13 in the monkey group G2. In this experimental series the interval between the stimulus repetitions was altered from experiment to experiment (all other parameters remained constant). Figure 6 shows that the run of the frequencies is similar for all behavior units belonging to the same class and is different for behavior units from other classes.

The results obtained thus far relate exclusively to visually recognizable behavior units occurring most often. The other visually recognizable signals, all acoustic signals, and also further important parameters such as duration of the actions, time interval between the actions, and the dis-

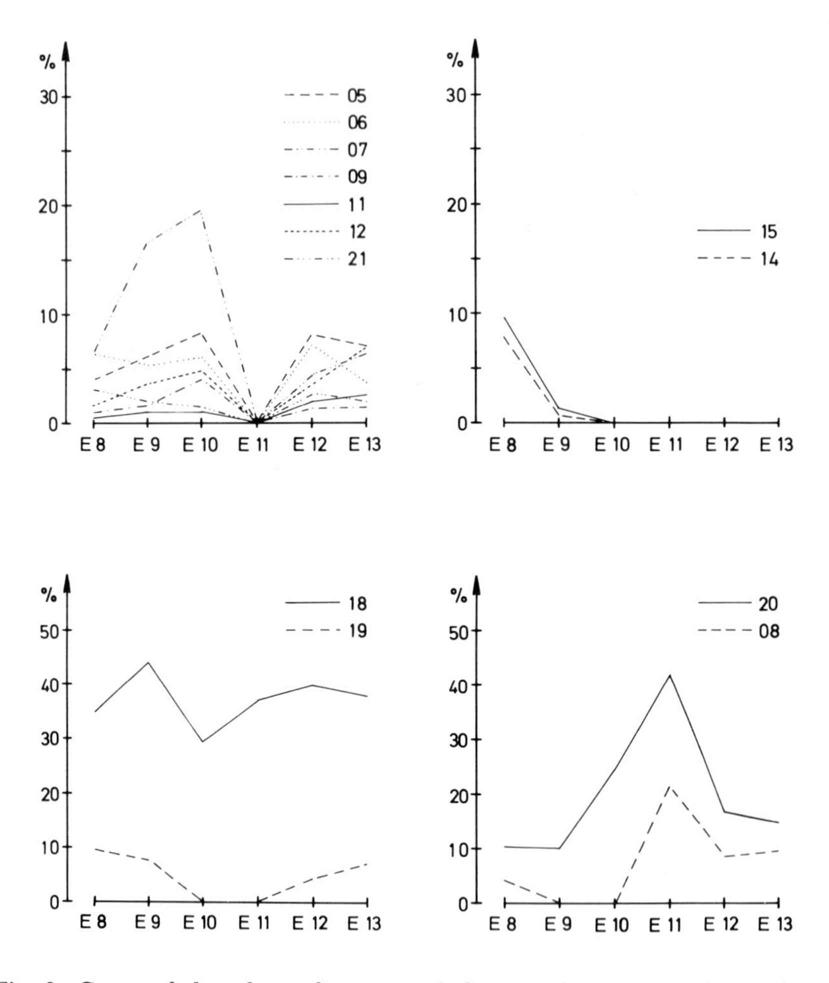

Fig. 6 Course of the relative frequencies (relative with respect to the number of all actions per experiment) of the behavior units in experiments E8–E13. The combination of the behavior units in the four diagrams is the consequence of the classification derived.

tance separating the animals, were neglected due to technical reasons. That we obtained usable results even when neglecting all these important factors is an argument for the usefulness of the methods of analysis applied. Herewith we have achieved the intention of our contribution, namely to demonstrate a few examples of a new method that allows us to gain more insight into the communication processes of squirrel monkeys than was possible up to now using other methods.

This success justifies the use of necessary technical aids in order to include the remaining communication signals in the analysis. These aids, such as the data recording system, enable not only the recording of all registrably relevant signals, but form the basis for the application of the electrical brain stimulus that takes place automatically only under predetermined conditions appearing in the behavior of the monkeys. This automatic situation-dependent stimulation is based on the automatic recording of the locomotion (Riebling, 1970) and following completion of the automatic on-line recognition of the vocalization types of the monkeys (Peetz, 1969; Maurus *et al.*, 1970; Schott, 1972), which is still in the developmental stage. With this, the variability of the takeoff situation in which the stimulus is applied and consequently the variability in the communication processes initiated is reduced. The reduction of variability will bring notable advantages to mathematical analysis. Moreover, the situation-dependent stimulation sets up the technical preconditions for tracking targeted questioning. Thereby it will be possible to insert precisely defined behavioral events in exactly defined situations.

The scope of the analyses remains restricted not only to the classification problems treated extensively in this contribution. Classification is a necessary first step. Other questions arise. For example, what does the classification of acoustic signals look like? Which acoustic expressions are signals as such; and which and how do acoustic expressions modify visual signals? What role does the distance among the animals play? What influence do previously occurring events have? What influence do learning processes have? Which signals depend on the context? How does this dependence function, and so on. There are still many questions to be answered before an analysis of the communication system will have succeeded to the extent which will enable us to predict behavioral events. If an analysis of the complex communication system of the squirrel monkey should succeed in this way, then we certainly have an important success in ethological research. Moreover, we have proof that the methods applied are correct. In this case they could also be used successfully on other communication systems of similar complexity.

The knowledge of the communication system along with the data recording aids at our disposal provide a promising basis for research of the central nervous control of social behavior in suitable combination with brain stimulation and lesion experiments. Here the manipulability of the animals' behavior will serve as a tool to initiate social interactions. The knowledge of the communication system, along with the data recording aids, will help to detect more subtle behavioral effects caused by manipulations in the central nervous system. The more sensitive detectors of changes in an animal's social behavior are the conspecific partners rather than the human observer.

ACKNOWLEDGMENTS

The author wishes to thank D. Ploog for continuous support, H. Pruscha for his advice in mathematical matters and for placing his computer programs at our disposal, E. Hartmann and B. Kühlmorgen for their competent assistance with electrode implantation, evaluation of the motion pictures, and data processing.

REFERENCES

Baldwin, J. D. (1968a). The social behavior of adult male squirrel monkeys (*Saimiri sciureus*) in a seminatural environment. *Folia Primat.* **9,** 281–314.

Baldwin, J. D. (1968b). A study of the social behavior of a semi free-ranging colony of squirrel monkeys (*Saimiri sciureus*). *Diss. Abstr.* **28,** 4423B.

Baldwin, J. D. (1969). The ontogeny of social behavior of squirrel monkeys (*Saimiri sciureus*) in a semi-natural environment. *Folia Primat.* **11,** 35–79.

Baldwin, J. D. (1970). Reproductive synchronization in squirrel monkeys. *Primates* **11,** 317–326.

Baldwin, J. D. (1971). The social organization of a semi free-ranging troop of squirrel monkeys (*Saimiri sciureus*). *Folia Primat.* **14,** 23–50.

Beischer, D. E., and Furry, D. E. (1964). *Saimiri sciureus* as an experimental animal. *Anat. Rec.* **148,** 615–624.

Bowden, D., Winter, P., and Ploog, D. (1967). Pregnancy and delivery behavior in the squirrel monkey (*Saimiri sciureus*) and other primates. *Folia Primat.* **5,** 1–42.

Carmichael, M., and MacLean, P. D. (1961). Use of squirrel monkey for brain research, with description of restraining chair. *Electroenceph. Clin. Neurophysiol.* **13,** 128–129.

Castell, R. (1969). Communication during initial contact: A comparison of squirrel and rhesus monkeys. *Folia Primat.* **11,** 206–214.

Castell, R., and Heinrich, B. (1971). Rank order in a captive female squirrel monkey colony. *Folia Primat.* **14,** 182–189.

Castell, R., and Maurus, M. (1967). Das sogenannte Urinmarkieren bei Totenkopfaffen (*Saimiri sciureus*) in Abhängigkeit von umweltbedingten und emotionalen Faktoren. *Folia Primat.* **6,** 170–176.

Castell, R., and Ploog, D. (1967). Zum Sozialverhalten der Totenkopfaffen (*Saimiri sciureus*). Auseinandersetzung zwischen zwei Kolonien. *Z. Tierpsychol.* **24,** 625–641.

Castell, R., Krohn, H., and Ploog, D. (1969). Rückenwälzen bei Totenkopfaffen (*Saimiri sciureus*): Körperflege und soziale Funktion. *Z. Tierpsychol.* **26,** 488–497.

Cole, A. J. (1969). "Numerical Taxonomy." Academic Press, New York.

Cramer, H. (1954). "Mathematical Methods of Statistics." Princeton Univ. Press, Princeton, New Jersey.

Cullen, J. M. (1972). Some principles of animal communication. *In* "Non Verbal Communication" (R. A. Hinde, ed.). Cambridge Univ. Press, London and New York.

Delgado, J. M. R. (1963). Telemetry and telestimulation of the brain. *In* "Bio-Telemetry" (L. Slater, ed.), pp. 231–249. Pergamon, Oxford.

Delgado, J. M. R. (1966). Aggressive behavior evoked by radio stimulation in monkey colonies. *Amer. Zool.* **6,** 669–681.

DuMond, F. V. (1968). The squirrel monkey in a semi natural environment. *In* "The

Squirrel Monkey" (L. A. Rosenblum and R. W. Cooper, eds.). Academic Press, New York.

DuMond, F. V., and Hutchinson, T. C. (1967). Squirrel Monkey Reproduction: The "fatted" male phenomenon and seasonal spermatogenesis. *Science* **158**, 1067–1070.

Flechtner, H. J. (1966). "Grundbegriffe der Kybernetik." Wissenschaftliche Verlagsgesellschaft mbH, Stuttgart.

Hill, W. C. O. (1938). A curious habit common to Lorisoids and Platyrrhine monkeys. *Ceylon J. Sci., Sect. B.* 21–65.

Hinde, R. A. (1966). "Animal Behaviour." McGraw-Hill, New York.

Höhne, A., and Maurus, M. (in preparation). A device for semiautomatical conversion of behavioral data recorded on film to paper tape suitable for direct computer access.

Hopf, S. (1967). Neue Ergebnisse der Primatologie. *In* "Progress in Primatology," *Congr. Int. Primatol. Soc., Frankfurt, 1966* (D. Starck, D. Schneider, and H.-J. Kuhn, eds.). Fischer Verlag, Stuttgart.

Hopf, S. (1967). Notes on pregnancy, delivery, and infant survival in captive squirrel monkeys. *Primates* **8**, 323–332.

Hopf, S. (1970). Report on a hand-reared squirrel monkey. *Z. Tierpsychol.* **27**, 610–621.

Hopf, S. (1971). New findings on the ontogeny of social behavior in the squirrel monkey. *Psychiat. Neurol. Neurochir.* **74**, 21–34.

Hopf, S. (1972). Sozialpsychologische Untersuchungen zur Verhaltensentwicklung des Totenkopfaffen. Diss. im Fachbereich Psychologie, Univ. Marburg.

Hopf, S. (1972). Study of spontaneous behavior in squirrel monkey groups: Observation techniques, recording devices, numerical evaluation and reliability tests. *Folia Primat.* **17**, 363–388.

Kirchshofer, R. (1963). Einige bemerkenswerte Verhaltensweisen bei *Saimiris* im Vergleich zu verwandten Arten. *Z. Morph. Anthrop.* **53**, 77–91.

Klüver, H. (1933). "Behavior mechanisms in monkeys," p. 387. Univ. Chicago Press, Chicago, Illinois.

Kummer, H., Goetz, W., and Angst, W. (1970). Cross-species modifications of social behavior in baboons. *In* "Old World Monkeys" (J. R. Napier, and P. H. Napier, eds.) pp. 351–363. Academic Press, New York.

Latta, J., Hopf, S., and Ploog, D. (1967). Observation on mating behavior and sexual play in the squirrel monkey. *Primates* **8**, 229–246.

Lilly, J. C. (1961). Injury and excitation by electric currents. *In* "Electrical Stimulation of the Brain" (D. E. Scherer, ed.). Univ. Texas Press, Austin, Texas.

MacKay, D. M. (1972). Formal analysis of communication processes. *In* "Non-verbal Communication" (R. A. Hinde, ed.). Cambridge Univ. Press, London and New York.

MacLean, P. D. (1964). Mirror display in the squirrel monkey (*Saimiri sciureus*). *Science* **146**, 950–952.

MacLean, P. D., and Delgado, J. M. R. (1953). Electrical and chemical stimulation of frontotemporal portion of limbic system in the walking animal. *Electroenceph. Clin. Neurophysiol.* **5**, 91–100.

Maurus, M. (1967a). A new telestimulation technique for the study of social behavior of the squirrel monkey. *In* "Progress in Primatology," *Congr. Int. Primatol. Soc., Frankfurt, 1966*. Fischer Verlag, Stuttgart.

Maurus, M. (1967b). Neue Fernreizapparatur für kleine Primaten. *Naturwissenschaften,* **22**, 593.

Maurus, M., and Ploog, D. (1969). Motor and vocal interactions in groups of squirrel monkeys, elicited by remote-controlled electrical brain stimulation. *Rec. Advan. Primatol.* 3, 59–63.

Maurus, M., and Ploog, D. (1971). Social signals in squirrel monkeys: Analysis by cerebral radio stimulation. *Exp. Brain Res.* **12**, 171–183.

Maurus, M., and Pruscha, H. (1972). Quantitative analyses of behavioral sequences elicited by automated telestimulation in squirrel monkeys. *Exp. Brain Res.* **14**, 372–394.

Maurus, M., and Pruscha, H. (1973). Classification of social signals in squirrel monkeys by means of clusteranalysis. *Behavior* **47**, 106–128.

Maurus, M., Peetz, H.-G., and Jürgens, U. (1970). Elektronische Lauterkennungsschaltung zur automatischen Steuerung von Fernreizversuchen an Totenkopfaffen *Saimiri sciureus*). *Naturwissenschaften* **57**, 141.

Maurus, M., Höhne, A., Peetz, H., and Wanke, J. (1972). Technical requirements for the recording of significant social signals in squirrel monkey groups. *Physiol. Behav.*, **8**, 969–971.

Maurus, M., Hartmann, E., and Kühlmorgen, B. (1973a). Invariant quantities in the communication processes of squirrel monkeys. *Primates* **15**, in press.

Maurus, M., Kühlmorgen, B., and Hartmann, E. (1973b). Concerning the influence of experimental conditions on social interactions initiated by telestimulation in squirrel monkey groups. *Brain Research* **64**, 271–280.

Mayer, W. (1971). Gruppenverhalten von Totenkopfaffen unter besonderer Berücksichtigung der Kommunikationstheorie. *Kybernetik* **8**, 59–68.

Mickle, A. W. (1961). The problems of stimulation parameters. *In* "Electrical Stimulation of the Brain" (D. E. Scheer, ed.), Univ. Texas Press, Austin, Texas.

Orloci, L. (1967). An agglomerative method for classification of plant communities. *Ecology* **55**, 193–205.

Orloci, L. (1968). Information analysis in phytosociology: Partition, classification and prediction. *J. Theoret. Biol.* **20**, 271–284.

Peetz, H.-G. (1969). Entwurf und Bau einer Erkennungsschaltung für den Affenlaut Kakeln. Dipl.-Arbeit, TU München.

Peters, M. (1970). Mouth to mouth contact in squirrel monkeys (*Saimiri sciureus*). *Z. Tierpsychol.* **27**, 1009–1010.

Ploog, D. (1961). Untersuchungen am Totenkopfaffen: Die cerebrale Lokalisation der männlichen Genitalfunktion und die Bedeutung dieser Funktion für das soziale Verhalten. *Klin. Woschr.* **39**, 657.

Ploog, D. (1963). Vergleichend quantitative Verhaltensstudien an zwei Totenkopfaffen-Kolonien. *Z. Morph. Anthrop.* **53**, 92–108.

Ploog, D. W. (1967). The behavior of squirrel monkeys (*Saimiri sciureus*) as revealed by sociometry, bioacoustics, and brain stimulation. *In* "Social Communication among Primates" (A. S. Altmann, ed.). Univ. of Chicago, Chicago, Illinois.

Ploog, D., Hopf, S., and Winter, P. (1967). Ontogenese des Verhaltens von Totenkopfaffen (*Saimiri sciureus*). *Psych. Forsch*, **31**, 1–41.

Ploog, D. W., and MacLean, P. D. (1963). Display of penile erection in squirrel monkey (*Saimiri sciureus*). *Anim. Behav.* **11**, 32–39.

Ploog, D. W., Blitz, J. and Ploog, F. (1963). Studies on social and sexual behavior of the squirrel monkey (*Saimiri sciureus*). *Folia primat.* **1**, 29–66.

Pruscha, H., and Maurus, M. (1973). A statistical method for the classification of behavior units occurring in primate communication. *Behavioral Biology* **9**, 511–516.

Riebling, W. (1970). Ortung beweglicher Objekte in einem abgeschlossenen Raum. Dipl.-Arbeit, TU München.

Robinson, B. W., Warner, H., and Rosvold, H. E. (1964). A headmounted remote-controlled brain stimulator for use on rhesus monkeys. *Electroenceph. Clin. Neurophysiol.* **17**, 200–203.

Robinson, B. W., Alexander, M., and Bowne, G. (1969). Dominance reversal resulting from aggressive responses evoked by brain telestimulation. *Physiol. Behav.* **4**, 749–752.

Rosenblum, L. A. (1968). Mother–infant relations and early behavioural development in the squirrel monkey. *In* "The Squirrel Monkey" (L. A. Rosenblum, and R. W. Cooper, eds.). Academic Press, New York.

Schott, D. (1972). Quantitative Analyse der Vokalisationen von Totenkopfaffen (*Saimiri sciureus*). Dissertation, Univ. München.

Socal, R. R., and Sneath, P. H. A. (1963). "Principles of Numerical Taxonomy." Freeman, San Francisco, California.

Terasawa, E., and Sawyer, C. H. (1969). Electrical and electrochemical stimulation of the hypothalamo-adenohypophysial system with stainless steel electrodes. *Endocrinology* **84**, 918–925.

Thorington, R. W., Jr. (1967). Feeding and activity of *Cebus* and *Saimiri* in a Colombian forest. *In* "Progress in Primatology" (D. Starck, R. Schneider, and H. J. Kuhn, eds.), pp. 180–184. Fischer, Stuttgart.

Thorington, R. W., Jr. (1968). Observation of squirrel monkeys in a Colombian forest. *In* "The Squirrel Monkey" (L. A. Rosenblum and R. W. Cooper, eds.). Academic Press, New York.

Winter, P. (1968). Social communication in the squirrel monkey. *In* "The Squirrel Monkey" (L. A. Rosenblum and R. W. Cooper, eds.), pp. 235–253. Academic Press, New York.

Winter, P. (1969). The variability of peep and twit calls in captive squirrel monkeys (*Saimiri sciureus*). *Folia primat.* **10**, 204–215.

Winter, P. (1972). Observations on the vocal behaviour of free ranging squirrel monkeys (*Saimiri sciureus*). *Z Tierpsychol.* **31**, 1–7.

Winter, P., Ploog, D., and Latta, J. (1966). Vocal repertoire of the squirrel monkey (*Saimiri sciureus*), its analysis and significance. *Exp. Brain Res.* **1**, 359–384.

ALTERATIONS OF SOCIAL BEHAVIOR WITH NEURAL LESIONS IN NONHUMAN PRIMATES [1]

ARTHUR KLING

College of Medicine and Dentistry of New Jersey—Rutgers Medical School

ROSLYN MASS

College of Medicine and Dentistry of New Jersey—Rutgers Medical School

INTRODUCTION

The study of localized brain lesions (as well as other manipulations, e.g., electrical stimulation) and social behavior in nonhuman primates provides the behavioral scientist with opportunities, as of now largely unexploited, to investigate the interrelationships between neural systems and the dynamics of the essential behavior repertoires common to the various primate species. Primate social systems can afford models for studying those complex behaviors which appropriately occur only within the social context.

The variability that exists between different species, and indeed between individuals, can result in difficulties in generalizing about lesion effects; nevertheless, species-specific behaviors and highly specialized adaptations can be of great value in understanding the biological basis for these adaptations. These differences in utilization of sensory inputs, communication patterns, intensities of social bondings, dominance patterns, and reproductive behavior may be advantageous in elucidating underlying neural mechanisms.

[1] Supported by the Behavioral Science Foundation, Boston, Massachusetts.

While much is now known about the cognitive, sensory and perceptual deficits occurring after selective ablations of brain structures, we have yet to understand how these deficits affect the adaptation of individuals within the social group. Further, how do genetic, environmental, and group structural factors interact with the brain damage to alter the final expression of these adaptations?

Following is a review of those studies relevant to the relationship between brain function and social behavior in nonhuman primates. The studies are categorized according to brain area involved and are arranged along the artificial–natural dimension as regards the setting and group composition. Since this research area is of relatively recent origin and still in the exploratory stage, there is considerable spread of methodologies, species used, experimental settings, and group compositions. Consequently, this review is intended to generate interest and to explore the methodological problems rather than to draw firm conclusions.

AMYGDALOID NUCLEI

In the individually caged monkey, bilateral ablation of the amygdaloid nuclei and surrounding temporal lobe structures, or ablation of the amygdaloid nuclei alone, has been observed to result in a syndrome of behavioral changes which can be summarized as follows: (*1*) a decrease in belligerence and a reduction of fear toward normally fear-inducing objects, including man; (*2*) a tendency to investigate orally and generally contact orally inedible objects including coprophagia and uriposia; (*3*) increased and inappropriate sexual behavior; (*4*) "hypermetamorphosis" (Kluver and Bucy, 1939; Schreiner and Kling, 1956).

Variability in the intensity and duration of all or some of the symptoms has been documented (especially with respect to age, sex, and species differences, Kling, 1966). However, studies of these preparations in social groups in the laboratory and free-ranging environments have revealed additional behavioral deficits and major modifications in the syndrome which have previously not been observed and bear directly on the role of this nuclear group in the regulation of affective behavior.

Dyadic Interactions

The interactions of amygdalectomized juvenile, male *Macaca mulatta* were compared with interactions of comparable normal pairs. Observations were conducted in a partitioned cage, during which time-based frequencies of social interactions were scored. During the intertest periods, the subjects were kept in individual cages. As contrasted with the normal pairs, the lesioned pairs showed fewer aggressive interactions than did the normals, but significantly more rough-and-tumble play, attempted and appropriate mounts, and grooming bouts. No clearly dominant animal appeared in the lesioned pairs in contrast to the normal pairs in which one of the dyad was clearly dominant. The operates also displayed frequent mutual solicitation for mounting and grooming with persistent erection and inappropriate and excessive oral behavior (Kling, 1968).

Miller (1968) paired operated, juvenile female *Macaca speciosa* with a normal male peer as well as with each other. Postoperatively, she found a drop in the frequency and duration of grooming in both sets of dyads (see Fig. 1). A postoperative increase in aggression by the male toward the

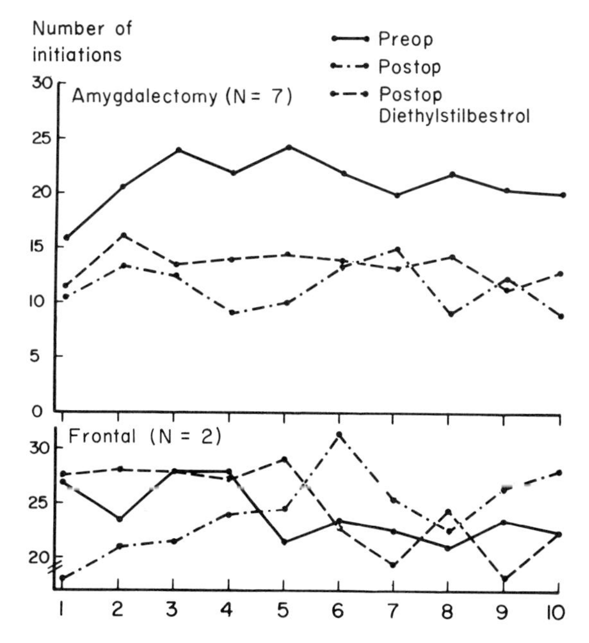

Fig. 1 Averages of daily grooming interactions of amygdalectomized and frontal lesioned juvenile *Macaca speciosa*. (From Miller, 1968.)

females was reported to be the result of the operates inappropriately challenging the male over food and cage position. A slight postoperative increase in sexual behavior was seen in these females who became more receptive to the male, and also displayed more female–female mountings.

Thompson *et al.* (1969) studied social fear responses in infant female *Macaca mulatta* and compared dyads of lesioned animals, dyads of control animals, and dyads of lesioned and control animals together. The operates showed more social fear and performed less social exploration (grooming, sitting together, clasping) than normals. However, the operates showed less fear toward novel situations. These authors concluded that their operates showed heightened fear responses when in contact with normal peers, while other behaviors were grossly normal.

In adult male dyads of *Macaca fuscatta* (Iwato and Ando, 1970), temporally lobectomized subjects became subordinate with respect to a normal adult male. However, in those dyads composed of lesioned subjects, rank became indeterminant. Their operates also displayed hypersexual behavior between them.

Small Laboratory Groups

Rosvold *et al.* (1954) observed a group of subadult *Macaca mulatta* of both sexes in a large cage and reported that amygdalectomy resulted in alterations of dominance relations and reduction in group aggression. One operate, a female, rose in rank. They also noted that some subjects were more aggressive to man when housed individually than in the group.

In a group of six *Cercopithecus aethiops* (Kling *et al.*, 1968), three males and three females ranging from juvenile to adult were observed primarily for dominance relations, spatial positions within the cage, and group behavior. After amygdalectomy, all operates regardless of age or sex fell in rank and exhibited decreased social interactions. The operates would huddle together in a corner of the cage apart from the normal animals. The operates also disregarded the previously established feeding order and were occasionally attacked by more dominant subjects. No pre- or postoperative copulations were observed, but hyperoral and copraphagic behavior was noted.

Plotnik (1968) using adult male *Saimiri sciureus* also found the operates to fall in rank with respect to normal peers after anterior temporal lobectomy.

In still another species, *Macaca speciosa*, Kling and Cornell (1971) observed a group of six juvenile-to-adult male and females. (See Fig. 2.) The juvenile male operate was found to fall in rank, but the dominant male retained his rank, as did an adult female. As a group, there was a quantitative decrease in both aggressive behavior and positive social interactions, the most marked changes being due to the operates in the group. The increase in threat behavior was due entirely to the adult operated female. No hypersexual or persistent hyperoral behavior was noted. The effects of amygdalectomy on similarly composed groups of *Macaca mulatta* and *Macaca ira* were found to result in somewhat different effects than in *Macaca speciosa*. In lesioned *M. mulatta*, the dominant male was

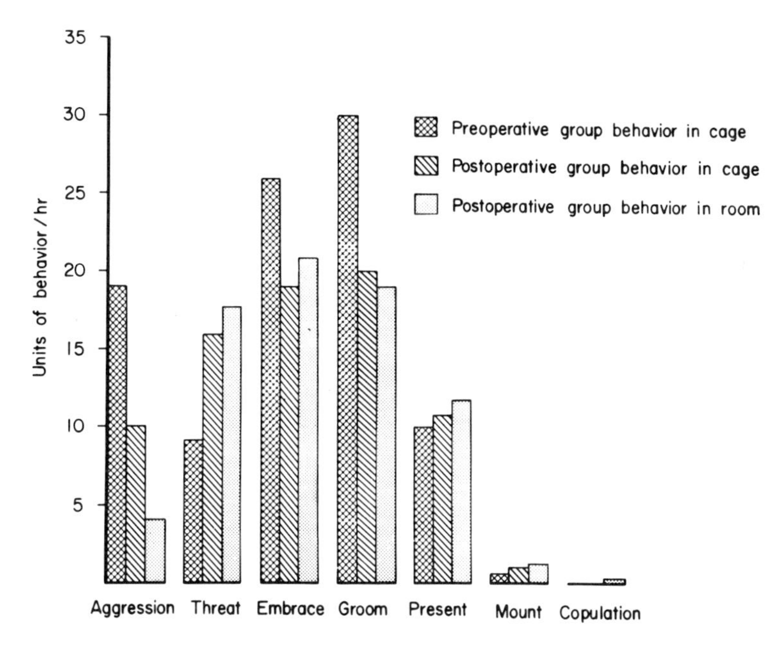

Fig. 2 Mean behavioral interactions per hour for group of six *Macaca speciosa.* Increase in threat was due solely to behavior of adult female. (From Kling, A. and Cornell, R. P.: Amygdalectomy and social behavior in the caged stump-tailed macaque (*Macaca specioso*). *Folia Primat.* **14,** 190–208 (Karger Basel, 1971). By permission.)

repeatedly attacked and eventually killed by the adult female. Because of her hyperaggressiveness, the group could not be maintained for further study.

Semi-Free–Ranging Setting

Dicks *et al.* (1969) observed the effects of complete and partial amygdalectomy in the semi-free–ranging colony of *Macaca mulatta* of the Cayo Santiago, Puerto Rico. In this setting (1968–1969), some 850 rhesus, naturally divided into a number of bands of various sizes, freely roamed the 40-acre island. The monkeys were artificially provisioned by food and water stations around which considerable inter- and intragroup agonism occurred.

Seven males from one group (E) of 85 were trapped, operated, and released after recovery. (See Fig. 3.) Postadolescent subjects or those with complete ablations displayed social indifference, became solitary, were frequently attacked by normal group members, and all eventually died of wounds or inanition. Two juvenile subjects did not reveal gross disturb-

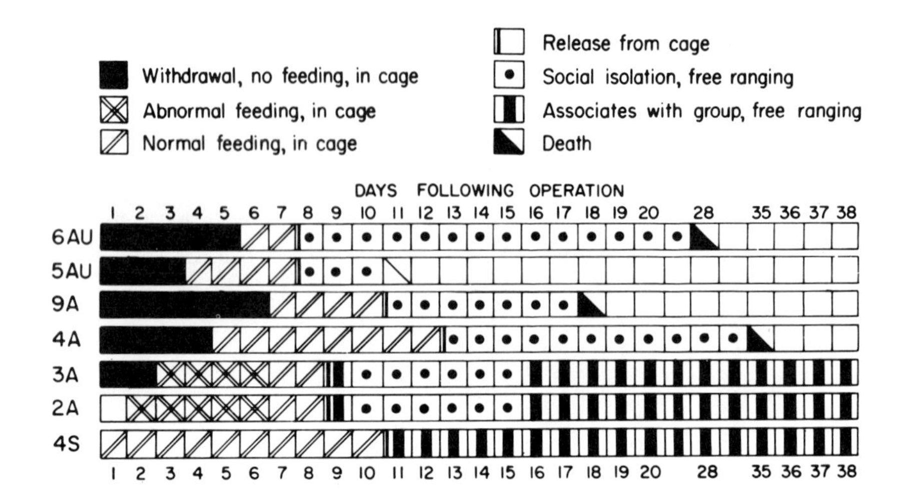

Fig. 3 Chronological diagram of postoperative behavior for six operates (6AU, 5AU, 9A, 4A, 3A, 2A) and one control (4S). AU signifies removal of amygdala plus uncinate cortex, bilaterally; A signifies removal of amygdala, bilaterally. Numerals signify age of animal, in years.

ances in their social interactions. Unfortunately, little information was obtained from those subjects who went solitary and eventually died.

Myers and Swett (1970), in the same setting, found that lesions of the inferior temporal lobe, sparing the amygdala, in five adults also resulted in solitary behavior and eventual death of the operates. Upon release near their group, all subjects ran through the band and disappeared in the underbrush. Survival times were estimated to range from 1 to 32 weeks during which time they lived as isolates. They also exhibited diminished emotional reactions and aggressiveness to other monkeys and human observers.

Free–Ranging Setting

In a completely free field setting along the Zambesi River in Zambia, Kling *et al.* (1970) observed the effects of bilateral amygdalectomy in 7 subjects within a group of 42 *Cercopithecus aethiops.* All operates, regardless of age or sex, failed to resocialize after operation. All normal controls who were trapped, kept in captivity, and later released eventually resocialized. While in captivity, the behavior of the operates was similar to the typical syndrome as initially described, but changed dramatically when freed. (See Figs. 4 and 5.) They rejected and withdrew from positive social gestures by normal group members who attempted to investi-

Fig. 4 Amygdalectomized *Cercopithecus aethiops* in cage prior to being released within the group. Livingston, Zambia, 1969.

gate, groom, or play with them. The operates either climbed to the highest branches of nearby trees or hid in dense brush. They tended to remain motionless and moved only when group members approached. They were not seen to eat, drink, investigate their surroundings, or attend to the vocalizations of their group. Unlike their behavior in cages, in the field they withdrew from human contact. With one exception (an adult female), none displayed agonistic gestures when closely followed or investigated by group members. Unlike the rhesus of the Cayo Santiago, the vervet operates were not attacked by intact group members; instead the normals repeatedly attempted to socialize and play with the operates. None of the operates could be followed for more than a day in the field setting, and it is assumed that they were eventually taken by predators. One striking exception was the case of an adult male who disappeared after being released within his own group. However, he was seen again several months later sitting in a school yard in Livingston, about 6 miles from where he was released. While he most likely remained an isolate, he was obviously able to feed, drink, and avoid predators during the intervening period.

Fig. 5 Operated *Cercopithecus aethiops* (rear) maintaining distance from normal group member shortly after being released into the group.

Comment

The results of the studies reported to date suggest that (*1*) the expression of the syndrome as described by Kluver and Bucy (1939) is highly dependent on the social–environmental setting in which the subject is observed; (*2*) the characteristic increase in oral and sexual behavior tends to be suppressed when the operates are housed with normal conspecifics, and has not yet been observed in natural settings; (*3*) increasing social environmental complexity tends to result in increased social fear, withdrawal from social interactions, and social isolation in free-ranging settings; (*4*) a decrease in aggressive behavior and a fall in social rank seems common to all settings, but paradoxical heightened aggression and elevation in rank has been seen in adult females (this issue has been dealt with in greater detail elsewhere, Kling, 1972b); (*5*) rather than the originally described decreased response to threat, the operates tend to show withdrawal from social interactions with normal conspecifics and man; (*6*) species-specific differences in social bondings and aggressive behavior may significantly affect the potential for resocialization of the operates.

It has been previously suggested (Kling, 1972a) that a major factor in lesion–environment interaction may be related to the role of the amygdala in visual perception through its connections with inferior temporal cortex, or possibly by cells intrinsic to the nuclear group. Gross *et al.* (1972) have described single units in the inferior temporal cortex of rhesus monkey that respond only to specific complex visual stimuli, while Jacobs and McGinty (1972) have recently reported single units in cat basolateral amygdala that respond to specific complex auditory stimuli. It is not unlikely then that the behavioral disturbances seen after amygdalectomy are related in part to the absence of temporal lobe structures necessary for the interpretation of complex stimuli. Some support of this hypothesis is that inferior temporal lobe ablation, sparing amygdala, was also reported to result in excessive social fear and solitary behavior in *Macaca mulatta* (Myers and Swett, 1970).

Accordingly, as the complexity of the environment is increased, the affected subject may become increasingly flooded with information which it is unable to sort or selectively inhibit, resulting in fear, withdrawal, and at times, immobility. In the visually simple cage situation, this deficit may be expressed by the increased investigatory and oral behaviors in an attempt to comprehend its environment.

While amygdalectomy usually results in an affective change toward greater placidity or diminished aggressiveness, female operates have been reported to show increased aggressiveness after operation. This sex difference is probably related to alterations of hypothalamic–pituitary function via amygdala hypothalamic projection systems to preoptic and midtuberal regions (Zolovick, 1972).

PREFRONTAL CORTEX

The characteristic syndrome resulting from ablation of prefrontal cortex in monkeys includes (*1*) hyperreactivity, circling and pacing, (*2*) cognitive deficits marked by failure to master delayed response and delayed alternations, (*3*) affective changes toward increased belligerence.

While the major interest in frontal lobe function in monkeys has focused on understanding the cognitive deficits, there has not been comparable progress in the analysis of its role in the regulation of social–affective behavior.

Dyadic Interactions

Perhaps the first published descriptions of the behavior of monkeys sustaining frontal lobe ablation was by David Ferrier in 1875. He added a normal monkey to cages housing two frontal-lobectomized subjects. One of the operates took little or no notice of the companion, whereas preoperatively, he took great interest in any companion placed nearby. The other operate showed no interest in its cage mate at first, although the companion examined the operate. Gradually, however, the operate got closer to the companion and eventually sat embracing him. The companion would frequently tug or bite the operate and seemed "annoyed at its occasional restless movements" and "to lose patience with its waywardness." Ferrier was impressed with the loss of active interest in surroundings of the operates, even though they were still capable of exhibiting various emotions, and with their indifference which varied with "restless and purposeless wanderings to and fro."

Warden and Galt (1945) using *Cebus capucina, Macaca mulatta,* and *Cercocebus fuliginosus* found that unilateral extirpation of prefrontal lobes markedly decreased time spent grooming; bilateral ablations were followed by not a single instance of grooming being observed. Dominance relations were unaffected by unilateral ablation and only minor changes in dominance followed the bilateral ablations.

Deets *et al.* (1970) paired young *Macaca mulatta* who had sustained bifrontal lobectomy with a stimulus animal who was a stranger to them. Control animals were paired with stimulus animals also. The authors concluded that the operates were more socially withdrawn and more distressed than their controls and that these differences made the stimulus animals more "ill at ease" in their presence than in the presence of the control animals. One of their categories of distress was pace-stereotypy, a behavior characteristic of monkeys after lobectomy but not lobotomy (Wade, 1952). While these investigators concluded that the operates were less aggressive than their controls, they felt that the operates also initiated inappropriate aggressions. Of the 11 operates, 9 were females, and whereas the controls threatened females more than males, the operates threatened males more than females.

Laboratory Groups

Brody *et al.* (1952) were interested in the influence of prefrontal lobotomy on the social dynamics of a group of six *Macaca mulatta.*

Occasionally food was manipulated or animals who were strangers were introduced in order to intensify social interactions. All operates became more active. Two showed an increase in food-getting attempts "sometimes snatching food from under the noses of higher status monkeys [p. 411]." One subject, second lowest in rank, had a very unstable position in the hierarchy postoperatively; he often aggressed higher ranking animals who did not always retaliate. The female operate who ranked lowest preoperatively was once seen to attack a higher ranking animal. These authors saw the effect of lobotomy as a reduction of rigidity and stability of the social organization because of the "disappearance or marked diminution of learned avoidance responses in the low-status animals [Brody *et al.*, 1952, p. 415]."

At the Rutgers laboratories, we have been studying a group of eight juvenile male *M. mulatta* with ablations limited to the dorsolateral frontal cortex. In contrast to lobectomized preparations, after all eight had been operated, the stability of the group was not disrupted. Each monkey, except the most dominant, spent the most time with the same monkey he had spent time with preoperatively. Time spent grooming increased when four or eight animals were operated. When the number of operates increased from four to eight, the four who had already undergone surgery showed an increase in aggression, while the four newly operated monkeys showed a decrease in aggression. There was a marked diminution of threat behavior despite the increase in physical aggression. Hyperactivity, pacing, and circling were characteristic of all operates. While it tended to diminish by several weeks after surgery, social stress within the group reactivated the hyperactivity. It appeared that the physical interactions, both agonistic and nonagonistic (together, play, grooming) increased, while noncontact communications (threat and submissive gestures) decreased as a result of the lesion. (See Table 1, Fig. 6.)

Mass (1972) utilized a more naturally composed group of 11 *Macaca speciosa* of both sexes. After 100 hr of preoperative observations, 5 of the 11 sustained dorsolateral frontal cortex ablations. During the 2 months following surgery there was a statistically significant increase in aggression. All operates paced and circled inside the perimeter of the cage, especially during commotion or when strangers entered the room. After 2 months, the three female operates (formerly ranked 3, 5, and 7) fell in rank; they fell in inverse relationship to their preoperative ranks. Two of these operates, 3 and 5, fell after violence erupted which was directed at them primarily by other females. The fall in dominance was measured by displacement behavior and was found to permeate several behavioral categories. No inappropriate upward aggressions or inappropriate feed-

TABLE 1 **Comparison of Group Social Interactions of Eight *Macaca mulatta* before and after Ablation of Dorsolateral Frontal Cortex**

Behavior	x̄ Preop	x̄ Postop	P
Aggression	0.6	1.81	0.05
Self-grooming	3.56	1.54	n.s.[a]
Grooming	1.14	4.55	0.01
Mounting	2.01	1.31	n.s.
Play	2.88	3.55	n.s.
Threat	1.38	1.09	n.s.
Together	2.61	15.52	0.01

[a] n.s. indicates not significant.

ing orders were noticed for the operates, contrary to reports of other investigators.

The effect of lesions of the other major subdivision of the prefrontal cortex, the orbital surface, was studied by Snyder (1970). He formed a permanent matrix colony of four male rhesus monkeys. Within a few weeks, a linear dominance hierarchy had formed and nearly all aggression and submission was unidirectional and stable. Experimental (operated) animals were introduced into this stable group.

Each of the monkeys replaced the dominant, or alpha male when they were introduced into the matrix group preoperatively. Following orbital cortex ablation, 3 of the 4 subjects retained their dominant position for 1, 4, and 6 months, respectively, at which time they rapidly fell to the bottom of the dominance hierarchy. To explain this delayed fall in status, Snyder looked at the relationship between the level of stress or threat and the time of the orbital monkey's fall in dominance. The orbital monkeys did not receive equally intense challenges from the matrix group dominant male after postoperative introduction into the colony. The orbital monkey who fell first from social dominance to total submission was attacked more frequently and for a longer time by the displaced dominant male. Furthermore, the fall of the other two orbital monkeys was coincidental with a rise in aggression directed at them from the displaced alpha male.

Other evidence (Snyder, 1970) suggests that (*1*) decreased aggression, (*2*) inappropriate or lack of properly signaled submission, and (*3*) an inability to perceive threat signals lead to repeated attacks on the orbital monkeys by the matrix members which, in turn, reinforce the orbital monkeys' new position at the bottom of the hierarchy.

In another study on orbital cortex, a group of 14 *Macaca speciosa*

Fig. 6 Dorsolateral frontal lesioned lobectomized male juvenile *Macaca mulatta* sitting together and huddling. Flattening of facial expression was typical after operation.

comprised of adults, subadults, and juveniles of both sexes were established in our laboratory enclosure (Notkin, 1972). Following 100 hr of quantitative measures of their social interactions, social ranking, and sexual behavior, three subjects matched for age and sex were removed from the group. One sustained an orbital frontal gyrus lesion, one operative control, a superior temporal gyrus lesion, and the third remained as an unoperated but separated control. After full recovery, the group of three was returned to the group. After 2 weeks of observation, a second group of three was removed and treated in a similar fashion. Eventually, there were four subjects with orbital cortex lesions, four with superior temporal lesions, and four who were unoperated controls. Two remaining juveniles were both unoperated and not removed from the group. Among the subjects with orbital lesions was the dominant male.

Postoperatively, 3 of the 4 orbitals were lethargic and apathetic for several weeks; they were hyporeactive to human observers and oblivious to bouts of aggression or other commotion among others in the group. Two adults spent most of the observation time alone, engaged in self-

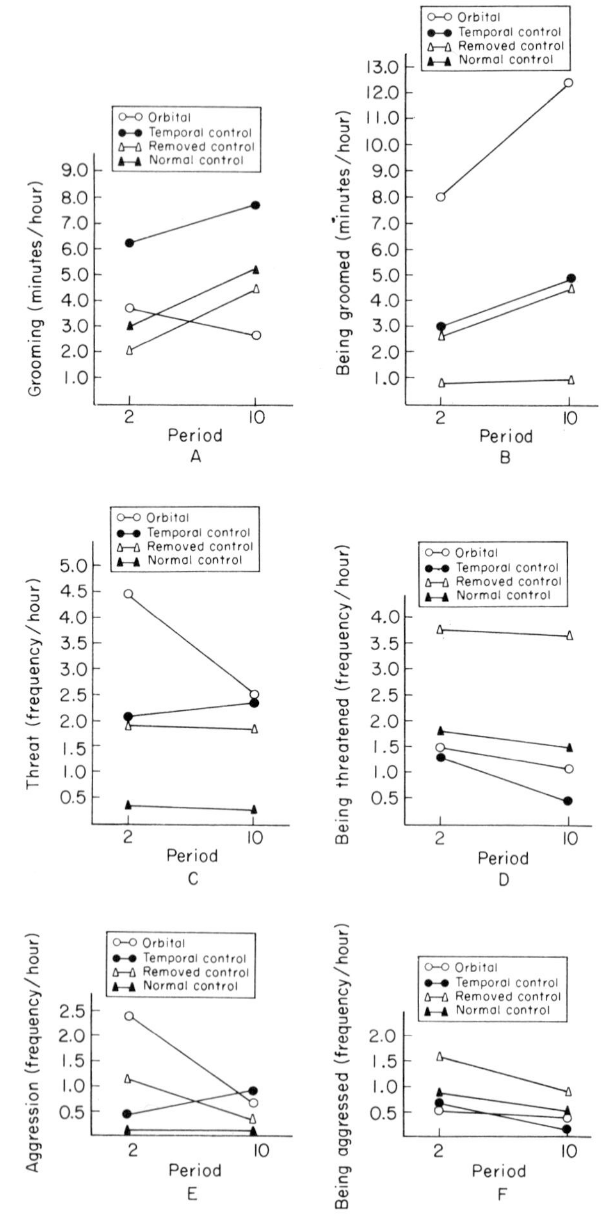

Fig. 7 Legend on facing page.

grooming in a stereotyped fashion, picking at one small area of the skin. Interactions with group members were as frequent as before, but brief. After operation, the dominant male particularly tended to wander aimlessly, pacing and circling about the enclosure, but no hyperreactivity was seen in any of these operates.

Orbitals showed a marked decrease in active grooming, but were groomed and joined by others more than preoperatively. Only one subject, the alpha female who ranked second to the dominant male in the hierarchy, fell in status (to third place). While the dominant male retained his position, he was repeatedly challenged by the next ranking adult male, sometimes resulting in violent physical aggression between them. Nevertheless, there was an overall decrease in threat and aggression by the operates. Both heterosexual and homosexual copulations decreased as a result of the orbital lesion, particularly among two adult females who preoperatively were frequently observed engaged in homosexual copulatory-like behavior often culminating in orgasm. The subjects sustaining the control lesion (superior temporal gyrus) showed none of the changes seen in the orbitals. (See Figs. 7 and 8.)

After completing the laboratory observations, the entire group was transferred to a half-acre enclosure in Puerto Rico; 1 month later the dominant male fell to number 2. The social ranks of the other operates remained unchanged. The group remained spatially cohesive and none of the operates became solitary. Quantitative scores of social interactions in the enclosure showed a decrease in threat and aggressive interactions, but joining, groomings, and huddlings remained at the same frequencies as in the smaller laboratory enclosure.

Semi-Free–Ranging

Miller *et al.* (1971) trapped five monkeys from the semi-free–ranging group of *Macaca mulatta* of the Cayo Santiago and subjected them to prefrontal ablations. Of the five frontal operates, all but the youngest showed an increase in activity as measured in an enclosure before release, and the same four did not rejoin their social group nor any other social group upon release. The operate rejected him as did another operate mother of a 10-month-old normal male. Both infants rejoined

---

Fig. 7 Changes in grooming, threat, and aggression after orbital frontal lesions in *Macaca speciosa*. For each of the four experimental groups as a function of periods is shown: A: Mean grooming time. B: Mean being groomed time. C: Mean threat frequency. D: Mean being threatened frequency. E: Mean aggression frequency. F: Mean being aggressed frequency. (From Notkin, 1972.)

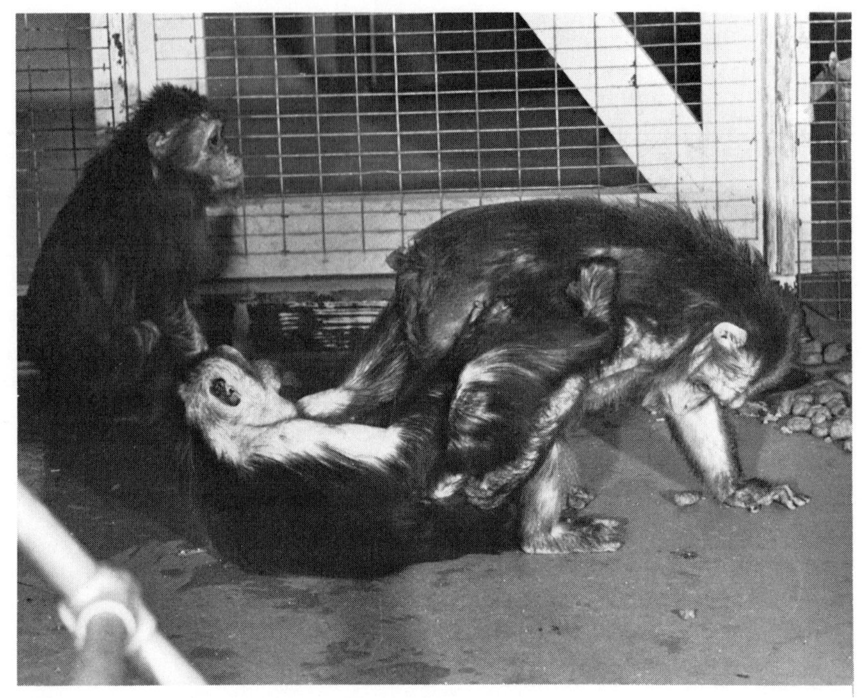

Fig. 8 Female homosexual behavior in *Macaca speciosa*. Initiator on her back pulling the subordinate female into position for vaginal contact.

their group upon release and were adopted by normal group members. Some operates were attacked before they became solitary and others ran through their group upon release and became solitary without being attacked. Control animals (scalp incisions, superior temporal lobe lesions, pinealectomies) returned to their respective groups. (See Table 2, Figs. 9 and 10.)

Comment

The effects of selective ablations of prefrontal cortex, either the dorsolateral surface or orbital cortex, on social behavior differ markedly from the larger lobectomy or lobotomy. In those studies, using the rhesus, the larger lesions appear to cause more intense hyperreactivity, more immediate and severe disruption in the hierarchy, more inappropriate aggression, a decrease in grooming, and degrees of social withdrawal depending upon the setting. Free-ranging subjects are reported to become solitary.

The effects of lesions of orbital cortex by Snyder (1970) and Notkin

TABLE 2 **Effects of Prefrontal and Cingulate Lesions on the Social Affinities of Rhesus Monkeys**

Lesion	Age	Sex	Days removed	Remained solitary	Rejoined group
Prefrontal	1[a]	m	14		x
	2	m	16	x	
	4[a]	f	14	x	
	5	f	2	x	
	5[a]	f	8	x	
Cingulate	2	f	20		x
	4	f	36		x
	4	m	17		x
	5	m	8	x	
Control	1	f	8		x
	1	m	8		x
	3[b]	f	12		x
	3[b]	m	7		x
	4[c]	m	10		x
	5[d]	f	8		x
	5[d]	m	2		x
	5	m	1		x
	5	m	3		x
	6+	m	46		x

[a] Activity pattern observed in enclosure setting.
[b] Bilateral bone flap.
[c] Scalp incision only.
[d] Pinealectomy.

(1972) show certain common features. Namely, the initial apathy, lack of attention to surroundings, a general decrease in aggression and over time, a fall in rank for some operated subjects. While only four subjects showed the latter effect in Notkin's study, group composition, species, and experimental design were quite different, especially since Snyder introduced a stranger to an already established group composed of similar ages and sex. Repeated challenges by subordinate to more dominant subjects eventually caused the fall in rank of subjects in both studies. It would be safe to assume that if there had been no normal adult male in Notkin's group, there would not have been any effective opponent to the dominant male and he undoubtedly would have retained his rank.

In the laboratory studies on dorsolateral ablations, social rank does not seem to be as affected as with orbital lesions. Although in Mass' study, three females fell in rank several months after operation, after

Fig. 9 Frontal lobectomized *Macaca mulatta* being chased by normal. Cayo Santiago, 1969.

being attacked by subordinate females, the male operates retained their rank. Similarly, in the study by Miller *et al.* (1971) using a group of all-male juvenile *Macaca mulatta,* even after all had been operated, no significant shifts in the preoperative hierarchy occurred. This is despite the general increase in aggression, gross disturbances of hyperactivity, pacing, and decrease of facial gestures associated with threat. In the two groups of *Macaca speciosa* who were transferred to a half-acre enclosure, none of the dorsolateral or orbital-lesioned subjects became solitary. It is not clear, however, whether this was due to a species effect or because of the more limited lesions.

As with amygdalectomy, major differences in social affective behavior after prefrontal lesions are seen when the setting, species, age and sex, lesion locus, and group composition are varied. While the significance of these studies to the understanding of frontal lobe function must await further analysis, it appears that frontal lesions in general seriously interfere with motor expression of affective gestures while intensifying both

Fig. 10 Frontal lobectomized *Macaca mulatta* who escaped to an offshore rock, remained as an isolate, and eventually died.

positive and agonistic physical contact. Their obvious bizarre hyperactivity and apparent lack of attention to surroundings does not appear to cause serious concern among normal group members in enclosure settings, except when challenges to their rank occur by appropriate group members. Stereotyped behaviors, pacing and circling, resulting from the lesion seem to diminish or even disappear in more natural settings.

CINGULATE GYRUS

The role of the cingulate cortex in behavior regulation is yet poorly understood. Anatomically, it is closely related to the hippocampus and other limbic structures as well as to the prefrontal cortex. Somatomotor, autonomic, and electroencephalographic effects have all been described from electrical stimulation studies (Crosby *et al.*, 1962).

Bilateral ablation of the anterior cingulate region (area 24) of monkeys has generally been shown to result in reduction of fear or shyness toward man and temporary tameness (Smith, 1944; Ward, 1948; Glees *et al.*, 1950; Mirsky, 1951; Showers, 1959). In addition, Glees *et al.* (1950) found hyperactivity and restlessness. In the study by Smith (1944), some operates displayed increased cutaneous sensitivity and piloerection. The behavioral effects noted by the above studies have been considered transient.

Dyadic Interactions

Glees *et al.* (1950) found their *Cebus capucina* and *Macaca mulatta* with bilateral and unilateral lesions to treat a cage mate as an inanimate part of the cage, "leaping over it, knocking into it, sitting on it, and taking food from its hands [p. 188]." The operates would retire briefly when the cage mate showed signs of resentment but would then repeat the offending act. "All this makes a striking contrast to its own preoperative behavior or to that of a normal animal [p. 188]."

Pribram and Fulton (1954) paired a mature male *M. mulatta* with another adult male. The operate remained dominant over the cage mate but there was somewhat less prolonged fighting over food. When the same operate was paired with an adult female, no operative effects were seen. The operate was appropriately protective when a baby was born.

Showers (1959) used *M. mulatta* and found no essential difference in the operates when placed with a cage mate; operates were equally adept in competition for food. The behavior of the operates toward the cage mate was described as "indifferent."

Laboratory Group

Ward (1948) used *Macaca mulatta* and found his operates to show a loss of mimetic activity and to show no grooming or "acts of affection" toward others. He reports that the operates behaved as though the others were inanimate, walking over them or even sitting on them. The operates would take food from the others and "appeared surprised" when they retaliated, yet this never led to a fight, for it was neither pugnacious nor even aggressive, seeming merely to have lost its "social conscience."

Anterior cingulectomy was not observed to alter the social behavior of a colony of preadolescent *M. mulatta* (Pribram and Fulton, 1954).

Kennard (1955) reported her cingulate lesioned *M. mulatta* to be uncompetitive and retiring and to vocalize excessively and inappropriately.

Mirsky *et al.* (1957) used young *M. mulatta,* some immature males, and some ovariectomized females. In none of their several measures of group-cage behavior was there any significant group effect of surgery.

Semi-Free-Ranging

Miller *et al.* (1971) operated four *M. Mulatta* from a semi-free–ranging group on the Cayo Santiago. He found one operate, the 5-year-old male, to rush through the group upon release without interacting. He was judged to be healthy and vigorous at time of release; his body was found 3 days later in a decomposing state. The other three operates (a 2-year-old female, a 4-year-old female, a 4-year-old male) all rejoined their group.

THALAMIC LESIONS

Of the brain regions discussed so far, all represent studies of lesions of cortical or limbic structures. One experiment has now been done which observed the effects of medial thalamic lesions on social behavior of a group of juvenile male *Macaca mulatta* (Case, 1972). Three of the group sustained stereotaxically placed lesions of centromedian parafasicular complex. The most striking change in their behavior was the increase in positive social interactions. The lesioned animals mounted more frequently and were more frequently invited to groom; there was no significant change in who was mounted or who did the inviting to be groomed. The operates huddled more and threatened less (these differences were not quite significant). The author concludes that there was an increase in sociability on the part of the operates.

If these effects can be replicated in further studies, it would represent the only lesion so far that results in the facilitation of positive social interactions along with a reduction of agonistic behavior.

Comment

Compared to amygdaloid or frontal lesions, cingulectomy does not seem to result in marked disturbances in social behavior. Ward (1948) suggests that cingulectomy may result in social indifference and a lack of appropriate responsiveness to social communications. Agonistic behaviors as well as positive interactions seem to be diminished, but social rankings remain unchanged. In some respects, the behavior of the cingulate monkeys resembles frontally lesioned animals, especially those with lesions limited to the orbital cortex (Notkin, 1972). The study by Miller on the rhesus of the Cayo Santiago suggests that for subadults at least, reintegration within the band, maintenance of social bonds and retention of social rank are compatible with lesions of the cingulate gyrus, even though the animals may display diminished interactions.

DISCUSSION

The field studies done so far suggest that monkeys (*Macaca mulatta* and *Cercopithecus aethiops*) sustaining bilateral ablation of amygdala, prefrontal lobectomy, and possibly inferior temporal lobectomy become social isolates whether they were apparently aggressed upon, as in the rhesus of the Cayo Santiago, or given positive communications, as in *C. aethiops* (amygdalectomy only). Nevertheless, the neural deficits resulting from amygdala and frontal cortex surgery are severe enough to shatter essential social bondings even between familially related subjects. Whether these deficits are primarily perceptual, as has been suggested for amygdala lesions, or a motor expressive disorder (frontal lesions) remains highly speculative. Of equal interest is that operates were not observed to exhibit many of the behaviors occurring under caged conditions.

While the importance of environmental–social factors on the expression of behavioral deficits after localized ablations is evident, the essential task is to understand the underlying mechanisms. This will require the combined efforts of formal testing procedures and field observation. Unfortunately, it has not yet been possible to carry out continued observations of subjects who leave the group and become isolates. It would be extremely important to know if and how these isolates survive, find food, avoid predators, or perhaps make attempts to resocialize. Suitable radio tracking devices might be employed to solve this problem. Studying small groups in large enclosures of one-half acre or more may afford operates the opportunity to withdraw and yet remain under observation; this setting would also allow for long-term observation under controlled conditions. Appropriate predators could be introduced as well as alternative food sources.

In the two studies using groups of *M. speciosa* which were initially composed and studied in a laboratory enclosure and later transferred to such an enclosure, the five subjects with dorsolateral ablations and five with orbital lesions remained with the group, although they showed lower levels of interaction than normals. None showed the excessive fear and wandering seen in the rhesus of the Cayo Santiago. However, previous studies suggest *Macaca speciosa* may have more intense social bondings than *Macaca mulatta* or *Cercopithecus aethiops*. Comparable experiments in the latter two species have not yet been done. (See Fig. 11, A and B.)

The effects of brain manipulations on social bondings in different species may prove to be useful to those concerned with the organization of primate social systems as well as to brain researchers. As more field and laboratory studies increasingly demonstrate the variability in social

Fig. 11 A and B: Social group of *Macaca speciosa* in half-acre enclosure. Caribbean Primate Research Center, Sebana Seca, Puerto Rico, 1972.

organization and intragroup behavior of the various species, it becomes increasingly necessary to be cautious in making generalizations based on one or two species. An important factor which may confound the results of studies using artificially composed groups in the laboratory is that the subjects are not related by familial or group bonds and we do not know to what extent this affects their social interactions and subsequent lesion effects. Nor do we know how long it takes for relationships which nearly resemble the natural condition to develop, if ever. In such artificial groups, age and sex distribution are crucial to determining the effect of a lesion on dominance relations, changes in rank, and bondings between subjects.

Age at the time of insult is an important factor in the expression of symptoms of brain injury. This has been documented for motor, sensory, cognitive, and affective functions (Kling and Tucker, 1968, 1971; Kennard, 1936; Benjamin and Thompson, 1959; Tucker *et al.*, 1968). While none of the studies reviewed in this report utilized subjects younger than 1 year, Dicks *et al.* (1969) found that several amygdala lesioned juveniles rejoined their mothers and remained within the group, while older subjects became isolates. In the study of free-ranging vervets (Kling *et al.*, 1970), the only operate to show any interest in and follow the group even for a short time was the youngest, an 18-month-old female. As for long-term studies of operated subjects within social groups, there is a need to study the development of neonatally damaged infants to observe whether certain deficits noted in adult operates will eventually appear, and if so, at what ages. Laboratory studies suggest that nursing females will retain and provide good maternal care to operated infants (Kling and Tucker, 1968). Conversely, both amygdala and frontally lesioned females who had infants or juveniles with them failed to show maternal behavior and abandoned their offspring (Kling, 1972a).

While field studies may eventually provide answers to questions related to survival functions, social bondings, and appropriate responses to the environment, questions related to the details of social interactions can probably best be studied in enclosure-type settings. Innovative techniques as described by Menzel (1967) may also provide the means to study cognitive functions under natural or seminatural conditions.

REFERENCES

Benjamin, R. M., and Thompson, R. F. (1959). Differential effects of cortical lesions in infants and adult cats on roughness discrimination. *Exp. Neurol.* **1**, 305–321.

Brody, E. B., and Rosvold, H. E. (1952). Influence of prefrontal lobotomy on social interaction in a monkey group. *Psychosom. Med.* **14**, 406–415.

Case, R. B. (1972). Personal communication.

Crosby, E. C., Humphrey, T., and Lauer, E. W. (1962). Cingulate Region. *In* "Correlative Anatomy of the Nervous System," pp. 473–476. Macmillan, New York.

Deets, A. C., Harlow, H. F., Singh, S. D., and Bloomquist, A. J. (1970). Effects of bilateral lesions of the frontal granular cortex on the social behavior of rhesus monkeys. *J. Comp. Physiol. Psychol.* **72**, 452–461.

Dicks, D., Myers, R. E., and Kling, A. (1969). Uncus and amygdala lesions: effects on social behavior in the free-ranging rhesus monkey. *Science* **165**, 69–71.

Ferrier, D. (1875). The Croonian Lecture: experiments on the brain of monkeys (second series). *Phil. Trans. Roy. Soc. London* **165**, 433–488.

Glees, P., Cole, J., Whitty, C. W. M., and Cairns, H. (1950). The effects of lesions in the cingular gyrus and adjacent areas in monkeys. *J. Neurol. Neurosurg. Psychiat.* **13**, 178–190.

Gross, C. G., Rocha-Miranda, C. E., and Bender, D. B. (1972). Visual properties of neurons in inferotemporal cortex of the macaque. *J. Neurophysiol.* **35**, 96–111.

Iwato, K., and Ando, Y. (1970). Socio-agonistic behavior in temporal lobectomized monkeys. Pre-print.

Jacobs, B. L., and McGinty, D. J. (1972). Participation of the amygdala in complex stimulus recognition and behavioral inhibition: evidence from unit studies. *Brain Res.* **36**, 431–436.

Kennard, M. A. (1936). Age and other factors in motor recovery from precentral lesions in monkeys. *Amer. J. Physiol.* **115**, 138–146.

Kennard, M. A. (1955). The cingulate gyrus in relation to consciousness. *J. Nerv. Ment. Dis.* **121**, 34–39.

Kling, A. (1966). Ontogenetic and phylogenetic studies on the amygdaloid nuclei. *Psychosom. Med.* **28**, 155–161.

Kling, A. (1968). Effects of amygdalectomy and testosterone on sexual behavior of male juvenile macaques. *J. Physiolog. Psychol.* **65**, 466.

Kling, A. (1972a). Effects of amygdalectomy on social-affective behavior in non-human primates. *Neurobiol. Amygdala* **2**, 511–536.

Kling, A. (1972b). Differential effects of amygdalectomy in male and female non-human primates. Presented at the *Int. Congr. Primatol., 4th, Portland, Oregon.*

Kling, A., and Cornell, R. (1971). Amygdalectomy and social behavior in the caged stump-tailed macaque (*Macaca speciosa*). *Folia Primat.* **14**, 190–208.

Kling, A., and Tucker, T. (1968). Sparing of function following localized brain lesions in neonatal monkeys. *In* "The Neuropsychology of Development" (R. Isaacson, ed.). Wiley, New York.

Kling, A., and Tucker, T. (1971). Motor and behavioral development after neo-decortication in the neonatal monkey. *Med. Primatol., 1970, Proc. Conf. Exp. Med. Surg. Primat., 2nd, New York, 1969,* pp. 376–393. Karger, Basel.

Kling, A., Dicks, D., and Gurowitz, E. M. (1968). Amygdalectomy and social behavior in a caged group of vervets (*C. aethiops*). *Proc. Int. Congr. Primat., 2nd, Atlanta, Georgia* **1**, 232–241. Karger, Basel.

Kling, A., Lancaster, J., and Benitone, J. (1970). Amygdalectomy in the free-ranging vervet (*Cercopithecus aethiops*). *J. Psychiat. Res.* **1**, 191–199.

Kluver, H., and Bucy, P. (1939). Preliminary analysis of functions of the temporal lobes in monkeys. *Arch. Neurol. Psychiat.* **42**, 979.

Mass, R. (1972). The effects of dorsolateral frontal ablations on the social behavior of a caged group of eleven stumptail macaques. Doctoral dissertation, Rutgers Univ.

Menzel, E. W., Jr. (1967). Naturalistic and experimental research in primates. *Hum. Develop.* **10**, 170–186.

Miller, M. H., Myers, R., and Swett, C. (1971). Personal communication.
Miller, R. (1968). Effects of amygdalectomy on sexual behavior in juvenile female monkeys (*M. speciosa*). Masters thesis, Illinois Inst. of Technol.
Mirsky, A., Rosvold, H. E., and Pribram, K. H. (1957). Effects of cingulectomy on social behavior in monkeys. *J. Neurophysiol.* **20**, 588–601.
Myers, R. E., and Swett, C. (1970). Social behavior deficits of free-ranging monkeys after anterior temporal cortex removal. A preliminary report. *Brain Res.* **18**, 551–556.
Notkin, E. (1972). Effects of orbital lesions on the social behavior of the caged stump-tailed macaque (*Macaca speciosa*). Masters thesis, Rutgers Univ.
Plotnik, R. (1968). Changes in social behavior of squirrel monkeys after anterior temporal lobectomy. *J. Comp. Physiol. Psychiat.* **66**, 369.
Pribram, K. H., and Fulton, J. F. (1954). An experimental critique of the effects of anterior cingulate ablations in monkeys. *Brain* **77**, 34–44.
Rosvold, H. E., Mirsky, A. F., and Pribram, K. H. (1954). Influence of amygdalectomy on social behavior in monkeys. *J. Comp. Physiol. Psychol.* **47,** 173.
Showers, J. J. (1959). The cingulate gyrus: additional motor area and cortical autonomic regulator. *J. Comp. Neurol.* **112,** 231–302.
Schreiner, L., and Kling, A. (1953). Behavioral changes following rhinencephalic injury in cat. *J. Neurophysiol.* **16,** 643–659.
Smith, W. K. (1944). The results of ablation of the cingular region of the cerebral cortex. *Fed. Proc.* **43,** 42–43.
Snyder, D. R. (1970). Fall from social dominance following orbital frontal ablation in monkeys. *Proc. Annu. Convention, APA,* 78th 235–236.
Thompson, C., Schwartzbaum, J. S., and Harlow, H. F. (1969). Development of social fear after amygdalectomy in infant rhesus monkeys. *Physiol. Behav.* **4,** 249.
Tucker, T., Kling, A., and Scharlock, D. (1968). Sparing of photic frequency and brightness discriminations after striatectomy in neonatal cats. *J. Neurophysiol.* **31,** 818–832.
Wade, M. (1952). Behavioral effects of prefrontal lobectomy, lobotomy and circumsection in the monkey (*Macaca mulatta*). *J. Comp. Neurol.* **96,** 179–207.
Ward, A. A., Jr. (1948). The cingular gyrus: area 24. *J. Neurophysiol.* **11,** 13–23.
Warden, C. J., and Galt, W. (1943). A study of cooperation, dominance, grooming and other social factors in monkeys. *J. Genet. Psychol.* **63,** 213–233.
Zolovick, A. J. (1972). Effects of lesions and electrical stimulation of the amygdala on hypothalamic-hypophyseal-regulation. *Neurobiol. Amygdala* **2,** 643–683.

COMPARATIVE APPROACHES

BRAIN STIMULATION AND AGGRESSION: MONKEYS, APES, AND HUMANS

ROD PLOTNIK

San Diego State University

INTRODUCTION

Each time the experimenter flicked a switch, the tiny device on the monkey's head caused current to flow through a fine wire implanted in the monkey's brain. The current flow in this wire caused the brain cells or neurons, located at the tip of this wire, to become active. Each time these particular neurons were activated, the monkey would attack another monkey in the cage. From these results, the experimenter concluded that he had activated neural centers, or triggers, or circuits for aggression in the brain. This finding and this interpretation have been reported a number of times (Delgado, 1955, 1959, 1962, 1963, 1964, 1965a,b, 1966, 1967, 1968), leading to the common notion that there are neural triggers for aggression in the brain. And there may well be. But, there is another interpretation for these results. Instead of eliciting aggression directly, brain stimulation may elicit painful sensations and these sensations, in turn, elicit aggression. In this chapter, reports of brain stimulation eliciting aggression will be examined to determine which of these two interpretations is supported.

BACKGROUND: INNATE CIRCUITS OR NEURAL TRIGGERS

If a monkey's brain is electrically stimulated and this results in aggression, one of the underlying issues is whether an innate neural circuit or trigger has been located. The terms "trigger" and "circuit" assume an organization of neurons whose activation results in a certain behavior. Given the brain's complexity in number of neurons (10 billion) and interconnections, there are endless ways a circuit or trigger could be organized. Speculating on the various neural combinations that might compose such a circuit would be premature at this point. Thus, when these terms are used here, they simply refer to groupings of neurons that, when activated, produce a specified, well-organized, and directed behavioral response.

During the past few years, we have been bombarded with a new litany of innate aggressive instincts, each assuming an innate neural circuit. Those who believe in aggressive instincts and, thus, innate neural circuits, can marshal evidence from either behavioral or neurological data, or both. Using primarily behavioral data, a number of authors—Ardrey (1961, 1966), Eibl-Eibesfeldt (1970), Lorenz (1966), Morris (1968), and Storr (1968)—have argued that man is equipped with an innate aggressive drive or instinct. These conclusions about man are usually based on observations of animals in environments that preclude learning. These authors contend that since the existence of unlearned patterns of behavior for various aggressive and sexual responses and displays have been identified for many species of animals, it is reasonable to generalize aggressive instincts to man. Not everyone would agree on the reasonableness of this generalization.

There are many authors convinced that man should not be included in the catalog of animals having aggressive instincts (Montagu, 1968). The identification of instinctual patterns or drives, also called fixed action patterns, demands some kind of environmental manipulation that precludes learning of the responses. This could be isolation or rearing with another species. Since experiments requiring isolation or rearing with other species are impossible with humans, and since isolation greatly changes the response patterns of monkeys (Harlow and Harlow, 1962), it may be very difficult to determine whether innate aggressive neural circuits exist in primates. Thus, this one question of whether circuits are innate will probably never be answered to our satisfaction for primates. Since lower species can be isolated with less drastic effect on behavior, it is important to determine if there is evidence for innate aggressive circuits in these animals. If neurological circuits for aggression can be demonstrated in lower species, then we might be able to draw limited conclusions for their possible existence in higher species.

NEURAL CIRCUITS FOR AGGRESSION IN NONPRIMATES

Prototype Experiment

The following experiment would demonstrate the existence of innate neural circuits. Isolate an animal from birth. When the animal is mature, implant in its brain fine wires (electrodes) which can be used for electrical stimulation, or tiny tubes which can be used for injection of chemicals. With the application of minute electrical currents to these electrodes or the introduction of small amounts of chemicals, neurons in the brain can be excited. After placing the animal in an appropriate environment for testing, its brain can be stimulated electrically or chemically. If there is an innate neural circuit and your electrode or tube is in this area, and the environment is appropriate, the stimulation should cause the animal to perform some response. For example, electrical or chemical stimulation could activate a neural circuit for aggression and, "presto," you would see an organized, directed, aggressive response. If the "presto" did not work and no aggressive response were produced, there are several good explanations for failure. Always with us is the possibility that the electrode was in the wrong brain area; there are millions of places to stimulate. Second, the animal may have been so changed by isolation that the environment was inappropriate for aggression. Third, there were no innate circuits to begin with and so there were none to activate.

It would also be possible to study innate circuits without using electrical or chemical stimulation, but that, too, has its problems. Suppose an animal were raised in some environment which prevented learning of a response, and then at a later time, the animal were tested for display of this response. If the response were present, then we would conclude that the animal had the necessary innate neural circuitry. If the response were not present, we would have to consider the following examples before arriving at any conclusion. There are some domesticated rats which do not show the predatory behavior that most wild rats display. There are some cats (household variety) that do show predatory behavior while others do not. Instead of concluding that some rats and cats have the neural circuitry and others do not, we might conclude that all rats and cats have the circuitry for predation, but changes in their environment have suppressed these responses in some. Or we might conclude that these animals have been selectively bred so that the predatory neural circuits are still present, but other neural systems keep them in check. Several experiments have indicated that the latter is the case. Rats and cats that normally show no predatory behavior will kill after electrical or chemical stimulation (Karli *et al.,* 1968; Roberts and Berquist, 1968; Smith *et al.,* 1970). Thus, chemical or electrical stimulation of the brain can elicit responses that might otherwise be absent in the animal's repertoire. In this case, brain stimulation has indicated the existence of innate neural circuits that behavioral studies might have been unable to demonstrate. These circuits have been for predatory behavior, which is not the same as aggression (see the following section). To date, there are no experiments in which electrical or chemical stimulation has elicited aggressive responses in animals that have been prevented from learning these responses. Therefore, *neurological* evidence for innate (unlearned) aggressive circuitry is lacking in lower as well as higher species.

Predation versus Aggression

Some of the scientists who study predatory behavior will take exception to the distinction between predation and aggression. They call predatory responses—killing an animal of a different species as part of food-getting behavior (cats killing rats; rats killing mice; lions killing gazelles)—aggressive responses. Others (Carthy and Ebling, 1964; Davis, 1964; Eibl-Eibesfeldt, 1967; Scott, 1971) disagree. Before resolving the question of whether predation is the same as aggression, let us review briefly the large body of research studying the neurological basis of cats killing rats and rats killing mice (Flynn *et al.,* 1971; Karli *et al.,* 1968; MacDonnell and Flynn, 1964; Panksepp, 1971; Woodworth, 1971). These are representative papers, not an exhaustive listing. In an excellent series of experiments by John Flynn and his associates (Flynn, 1967), the location of the cat-killing-rat neural circuit was localized in the brain. With this information, other investigators (Roberts and Berquist, 1968) implanted electrodes in the brain of an adult cat that had been raised in isolation. Having never seen a rat before, the cat upon stimulation exhibited a fairly normal rat-killing pattern. Similarly, predatory behavior can be elicited by electrical and chemical stimulation in rats and cats that do not show killing without stimulation. These experiments have shown that the latter animals do have an innate neural circuit for predation, which, for reasons of breeding or environment is normally (without brain stimulation) not displayed.

If predatory behavior were defined as aggression, then the above experiments would represent evidence for the presence of innate aggressive circuits. If predation is not aggression, a large body of neurological evidence for innate aggression vanishes. There are good reasons for saying predatory behavior is not aggression. Predation is the normal pattern of behavior the animal engages in to seek its food—it is part of consummatory behavior. Animals use different patterns of behavior when preying than when aggressing. By analogy, a cat killing a rat or a rat killing a mouse is no more aggressing than a person butchering a chicken or a hunter shooting a duck. Nevertheless, most articles on predatory behavior are titled with the words "attack" or "aggression," leading to the impression that neural circuits for aggression have been located in the brain.

Instead of saying predation simply is or is not aggression, one author has defined it as one kind of aggression; other kinds are intermale, fear-induced, irritable, territorial defense, maternal, instrumental, and sex-related (Moyer, 1968). (And, in my experience, trying to define aggression makes people aggressive.) Defining aggression on the basis of eliciting stimuli (Moyer, 1968) would seem to be a reasonable approach, provided one does not revert to the general term. After identifying various kinds of aggression, it can be confusing and misleading to use the general term "aggression." For example, using this multidefinitional system, one can say that there is evidence for the existence of innate neural circuits of *predation*, which is considered one kind of aggression. Since predation is only one kind of aggression, it is then misleading to conclude that there is evidence for the existence of innate neural circuits of *aggression*.

Moyer concluded that there is evidence for innate circuits of aggression, but the evidence given for this comes primarily from studies on predation, which many would not equate with aggression. Thus, it is confusing to have many kinds of aggression and then use the general term "aggression" to refer to all or any one of them.

Although there is good evidence for innate neural circuits for predation, data from brain stimulation in nonprimates to date offer no evidence for innate neural circuits for aggression. Thus, for one of the underlying questions of whether there is any neurological evidence from brain stimulation for the existence of innate aggressive circuits, the answer is "no." If an author reports that such evidence exists, then he or she has included predation under aggression.

Definition of Aggression

Having excluded predation from aggression, it is time to say what aggression does include. A frequently used definition of aggression is the inflicting of, or attempting to inflict, or threatening to inflict damage on others of the *same* species (Carthy and Ebling, 1964; Scott, 1971). The behaviorist may particularly like the following definition: Aggression is the presentation of, or attempt to deliver, noxious stimuli to another organism (Buss, 1971; Ulrich and Symannek, 1968). An ethological definition of aggression is any act that leads to spacing or subordination, even a display (Eibl-Eibesfeldt, 1971). And, finally, there must be the usual and valid reminder that there are many kinds of aggressive behavior, and we will have to wait for the behavioral observations to remove all the excess meaning (Johnson, 1972).

For the sake of communication, this article defines aggression as the presenting or attempting to present noxious stimuli to another of the same species. Whether the stimuli are "noxious" is determined by the animal's reaction to these stimuli. In other words, the animal's response, not the observer's preconceptions, defines which stimuli are noxious. This definition, specifying same species, excludes predation or food seeking from aggression. In some experiments, brain stimulation has elicited responses against inanimate objects or animals of other species, and these responses have been called aggressive. In this review, such responses will be considered aggressive if the brain stimulation has also been shown to elicit aggressive responses against the animal's own species. Before reviewing the results of brain stimulation and aggression in primates, it is necessary to discuss some methodological problems which can make results difficult to interpret.

BRAIN STIMULATION AND AGGRESSION IN PRIMATES

Methodological Considerations

It is relatively easy to restrain a monkey in a chair, electrically stimulate its brain, and study the responses. This is the least desirable and reliable procedure for studying aggression. Because the monkey is restrained, it may be difficult to interpret the elicited response. For example, the rhesus monkey has a very stereotyped facial gesture of submission, called the grimace. This response is made to a threatening or attacking monkey. The grimace consists of lip retraction, which exposes the teeth. In contrast, the facial gesture for dominance consists of raising the eyebrows, opening the mouth without exposing the teeth, and possibly bobbing the head at the target monkey. When a monkey is stimulated while restrained in the chair, it is not always possible to tell whether an elicited response that looks like a grimace or threat is, in fact, a social–emotional response or whether it is simply a motor response (Delgado,

1969). To make this distinction, the animal must be stimulated while unrestrained and in the presence of another monkey. If it is a motor movement, it will not be well directed against the second animal, and the target monkey may not respond to it. If it is not a motor, but rather an aggressive response, the stimulated animal will direct the gesture against the target monkey, which will, in turn, show some behavior indicating reception of this signal.

Another problem in studying restrained monkeys is that the elicited responses may be peculiar to that condition and not generalizable to other conditions and targets. For example, some brain stimulation will elicit aggressive responses against humans when the monkey is restrained; but the same stimulation fails to elicit this aggression against humans when the monkey is unrestrained and in the presence of other monkeys (Delgado, 1965a; Plotnik *et al.*, 1971). Thus, studying only the restrained animal would have led to the erroneous conclusion that this stimulation elicited aggression against humans when the effect was peculiar to the restrained condition. Another study showed that responses elicited from the restrained monkey could not be used to predict what responses would be elicited when the monkey was free and in the presence of other monkeys (Robinson *et al.*, 1969). Thus, if a monkey is restrained in a chair, brain stimulation may elicit responses that look like aggressive responses but are, in fact, motor movements; or stimulation may elicit aggressive responses that are peculiar to the restrained condition; or stimulation may fail to elicit aggressive responses that would be obtained in the freely moving animal with an appropriate target. For these reasons, aggressive responses should be studied in a group of unrestrained monkeys. Accordingly, studies in which all data have been taken from restrained monkeys will not be included in this paper.

Appropriate Stimuli

A major finding from brain stimulation research is that the elicited response is as much a function of the environment as it is of the stimulation itself. The environment may determine the occurrence and kind of aggressive response elicited. For example, brain stimulation elicited few dominant responses from a monkey of low social ranking in a colony. When dominant monkeys were removed and submissive monkeys added, thus raising the animal's ranking, a higher number of aggressive responses were elicited (Delgado, 1967; Maurus and Ploog, 1968). An even more dramatic example of environmental importance came from a study which paired the same animal with either a dominant or a submissive animal (Plotnik *et al.*, 1971). The same electrode and electrical current, in the same monkey, would elicit dominant responses if the stimulated animal was paired with a submissive monkey; would elicit submissive responses if the stimulated animal was paired with a dominant partner; would elicit an aggressive response if a stuffed toy was present; and would elicit no aggressive response if no target was present. In this experiment, brain stimulation was not begun until the social interactions between the paired monkeys were very stable (1 hr of interactions per day for 15 days). In this stable interaction, brain stimulation never elicited attack or threat by a submissive monkey against a dominant one.

In another study (Robinson *et al.*, 1969), monkeys had been observed together for a very short period of time (two 45-min sessions) before brain stimulation. Following this shorter period of interaction, brain stimulation caused the submissive monkey to attack its dominant partner. These studies indicate that there are two important variables in brain stimulation research. First, the occurrence and kind of aggressive responses elicited depends upon the kind of environment the animal is tested in. This indicates that, if neural circuits are activated, they do not act independently of the environment. Second, the more established the social relationships (dominant–submissive) the more likely it is that these relationships or patterns will be maintained during stimulation-elicited behavior. In other words, if there are neural circuits for aggression, established social hierarchies will limit the activation of these circuits.

Finally, another methodological difficulty derives from the strong individual differences seen in primates. Different monkeys may prefer to attack different targets (Plotnik *et al.*, 1971). In this study, monkeys were stimulated in the presence of different stimuli. For one monkey, brain stimulation elicited the highest percentage of attacks against a toy tiger, the next highest percentage against an unfamiliar human (stranger), the next highest against a submissive monkey, and the next against a familiar human (regular monkey tester). This monkey never threatened a dominant monkey or a mirror in which it could see a reflection. For another monkey, the highest percentage of attacks elicited by brain stimulation was against a mirror, followed by a toy tiger and a submissive monkey. This monkey never threatened a human, familiar or unfamiliar, or a dominant monkey. These relationships held whether the monkey was stimulated while free in a cage or restrained in a chair. Thus, frequency of aggressive responses elicited by brain stimulation depends upon a monkey's preference for targets, and this preference may differ for different monkeys. The most consistent preference across monkeys for a target to attack was a submissive monkey. The least consistent preference was for a human target. This is another compelling reason why aggression should be studied within the species rather than across species.

Brain Stimulation in Humans—Problems

Since it is very difficult to determine what a monkey is feeling during brain stimulation, it would be advantageous to have a human subject who could report these sensations. But often, what advantage is gained from self-reports during brain stimulation in humans is lost by the limited control the investigator has in selecting, treating, and observing these patients postoperatively. With regard to selection, brain stimulation is only done in patients who have some severe neurological problem—epileptic seizures, intractable pain, tumor—or some severe behavioral problem with suspected neurological causes—assaultive behavior, uncontrollable outbursts. With regard to treatment, the surgeon does not have the freedom to explore many different brain areas because of possible damage to the brain, or repeat the stimulation since responses may be elicited which would be dangerous to the patient or staff; nor can the surgeon always arrange for an appropriate target stimulus. With regard to postoperative followup, further study may be terminated by discharge from the hospital. There is usually no staff available to observe the patient in his/her normal environment after discharge. Thus, data we have from brain stimulation in humans has come from a very restricted population with severe neurological and/or psychological problems, a very small number of brain areas explored, very few reported cases in which the stimulation could be repeated, few if any test situations with the appropriate target stimuli present, and very little postoperative followup of the patient in his/her normal environment.

The previous discussion (see Appropriate Stimuli) of how the environment and stimulation interact to determine the elicited response in monkeys can give us some idea of how inappropriate an operating room would be to study aggressive responses. Lacking an appropriate object to attack, brain stimulation causes chickens to peck the ground and monkeys to walk about their cage. Similarly, we might expect brain stimulation of humans in the hospital setting, where the examining physician is not an appropriate target, to elicit any behavior other than rage or anger.

Methodological Summary

In the course of this chapter, studies which report aggression elicited by brain stimulation will be reexamined in the light of the above methodological considerations. In deciding whether or not brain stimulation elicited aggression, we must consider whether the subject was restrained or free; whether appropriate targets were present; the subject's preference for targets; and the length of social interaction between subject and target.

Instrumentation

Descriptions of the instrumentation needed for studying free-moving monkeys or humans have been published (Delgado, 1964, 1968; Maurus, 1967; Robinson, 1969). Studying brain stimulation in unrestrained monkeys takes exceeding patience, competent electronic staff, and cooperative monkeys. Typically, 30–50 electrodes are implanted in the monkey's brain. Then, the usual procedure is to explore all the electrodes in the restrained monkey. This is to ensure that all the electrodes are functioning and to select, by some criterion, those electrodes that will be studied in the unrestrained monkey. A radio stimulator has the capacity, depending on the model, to electrically activate from 1 to 12 different electrodes in the freely moving monkey. After the first 1–12 electrodes have been explored, the monkey must be restrained again and 1–12 different electrodes connected to the radio stimulator; this process is repeated until all 30–50 electrodes have been tested.

The stimulator is in two separate parts. One part, the receiver, fits on the monkey's head or on a collar. When the receiver is triggered, it sends out a tiny current which travels down the fine wire (electrode) implanted in the brain. The tip of the wire is uninsulated for 3–4 mm and this is the area through which the current flows to excite the surrounding neurons. The other part of the stimulator, called the transmitter, can be as much as 100 ft from the receiver on the animal. The transmitter sends a radio signal to start or stop the receiver mounted on the monkey. Since the receiver is what actually provides the current to excite the neurons, it is commonly called a stimulator. This is analogous to a radio station (transmitter) broadcasting a signal and a portable radio receiving it. In the same way that a portable radio must be tuned to get the best reception, the stimulator on the monkey must be tuned and interference minimized.

Great advances have been made in radio stimulator technology. The size is now reduced to the size of a book of matches, and the stimulator is totally implantable under the skin. There is also a slightly larger version that can both stimulate the brain and record brain waves. Almost all of these advances are traceable to the research of Jose Delgado. The ability to stimulate and record remotely has been used successfully in human patients who were implanted with electrodes for therapeutic reasons (Delgado *et al.*, 1968).

DATA ON BRAIN STIMULATION AND AGGRESSION

Overview of Results

The neurobiologist has implanted electrodes into the brains of many species: rats, cats, monkeys and, less frequently, chimpanzees and humans. From other readings studies reporting the existence of neural centers, triggers, or circuits for aggression may be recalled. With this popular belief in mind, the following facts might come as a shock. There has never been a study on rats or cats which demonstrated conclusively that brain stimulation produced aggression. Of all the studies with monkeys, there has been exactly one monkey in which brain stimulation can be said to have elicited aggression (Robinson *et al.*, 1969); in all other cases there were alternative interpretations. There are no studies on chimpanzees in which brain stimulation can be said to have elicited aggression. There is, with qualification, one study on a human in whom aggression was elicited by brain stimulation (Heath *et al.*, 1955; King, 1961). There are studies on chickens (von Holst and von Holst, 1962), ducks (Phillips, 1964), opossums (Roberts *et al.*, 1967), pigeons (Akerman, 1966), and cats (Delgado, 1955; Adams, 1968), which report aggressive behavior with brain stimulation, but for all of these data there are at least two interpretations.

Two Interpretations of Data

There are many studies on cats and monkeys which report that brain stimulation elicited aggressive responses. This very straightforward-sounding statement has been the basis for much confusion about the existence of neural triggers, centers or circuits for aggression.

When an electrode in the brain carries current, the neighboring neurons are excited. These neurons may, indeed, be involved in the start or execution of an aggressive response. On the other hand, they may *not* be involved in the start or execution of an aggressive response. These neurons may, instead, be involved in the mediation of pain, or noxious or aversive sensations. In this case, each time the electrode is stimulated, the animal will have a noxious sensation, and this noxious sensation may be what elicits the aggressive response. Thus the statement, "Brain stimulation elicited aggressive responses," has two interpretations. First, brain stimulation may have elicited aggression by exciting neurons that are involved directly in the start or execution of an aggressive response. Second, brain stimulation may have elicited aggression by exciting neurons that are involved in pain or noxious sensations, and these noxious sensations, in turn, elicited aggression.

The first interpretation, often quoted in the popular press, is very dramatic, since it indicates the existence of neurons directly involved in aggression. The second interpretation is not very dramatic since it indicates what may be an axiom—pain can elicit aggression. Each time we see the statement, "Brain stimulation elicited aggression," we must ask whether the brain stimulation was painful or noxious to the animal or human. If the stimulation was noxious, then the interpretation of pain causing aggression cannot be ruled out. If the stimulation was *not* painful, then the interpretation of exciting neurons involved directly in the start or execution of aggression has greater credibility. Thus, when nonpainful brain stimulation produces aggression, it has been called primary aggression. In contrast, when painful or noxious stimulation elicits aggression, it has been called secondary or pain-mediated aggression (Plotnik *et al.*, 1971).

Brain Stimulation and Noxiousness

We have seen that one criterion for primary aggression is that the brain stimulation *not* be aversive or noxious or painful. With humans, this information can be obtained with self-report. With animals, certain conditioning procedures can be used to test the properties of brain stimulation. One procedure is to observe whether an animal will press a lever to administer brain stimulation to itself. If the animal does press the lever, that is, works to obtain the stimulation, then the stimulation is said to be rewarding. If the animal presses a lever to avoid or terminate the stimulation, then this stimulation is said to be aversive, noxious, or possibly painful. In animals it is difficult to determine whether a stimulus is actually painful, since the animal's performance may be the same in avoiding stimuli that are noxious or aversive but not painful. For this reason, it is only possible to say that the animal works to avoid or terminate the stimulation, and this indicates that the stimulation is aversive, or noxious and possibly painful.

A recent technique used to assess properties of brain stimulation has been to compare an animal's lever responses for known rewarding (food), punishing (shock), or neutral stimuli (nothing) with its responses for brain stimulation (Plotnik *et al.*, 1972). If the animal's lever responses resemble those for food reinforcement, then the brain stimulation is termed rewarding; if the animal's lever responses to brain stimulation resemble its responses to shock, the stimulation is defined as aversive or noxious and possibly painful; if the animal's responses to brain stimulation resemble those to a neutral stimulus, then the stimulation is called neutral. Unless the experimenter uses one of these approaches, it is impossible to determine whether the brain stimulation elicited primary or secondary aggression.

Neural Triggers or Centers for Aggression

As pointed out in the preceding discussion, there can be two kinds of neural triggers: aversive and nonaversive. When electrodes are attached to a monkey's skin and electric current applied, the monkey will attack another monkey. The attack is well organized and directed and cannot be differentiated from the animal's attack pattern under normal (nonshock) conditions. When these electrodes are removed from the skin and implanted into the brain, and electric current is applied, the monkey will attack another monkey. In the first example, it is easy to see that shock to the skin (pain) results in aggression. And, few would say that there is a trigger for aggression on the monkey's skin.

In the second case, because the electrodes are in the brain, some investigators have assumed that aggressive responses produced by brain stimulation are due to excitation of neurons directly involved in aggressive responses. Because these responses were elicited by brain stimulation, it is tempting to speak of neural triggers for aggression. But before they are called neural triggers, it is necessary to eliminate the possibility that these "neural triggers" are actually pain triggers. The latter is a very real possibility since painful or noxious sensations have been produced by brain stimulation in animals and humans (Delgado *et al.*, 1954; Bowsher, 1957; Olds and Olds, 1963; Kruger, 1966; Obrador and Dierseen, 1966; Hassler, 1968).

Peripheral Pain and Aggression

Before the effects of noxious brain stimulation are reviewed, it should be pointed out that there is a large body of research relating aggression to noxious stimulation to the body. Electric shock or other noxious stimuli have elicited well-directed and organized aggressive responses from many different species, both between animals of the same species, and against animals of different species or inanimate objects (Azrin *et al.*, 1963, 1964, 1965; Ulrich *et al.*, 1964, 1965; Ulrich, 1966). In these studies, it was the pain produced by noxious stimulation to the body which caused aggression. These aggressive responses produced by noxious stimulation to the body are similar to the aggressive responses elicited by noxious brain stimulation.

Central Pain and Aggression

If two squirrel monkeys are shocked on the feet, after a few seconds they will begin to fight one another. If a restrained monkey is shocked on the tail and there is a ball or tube within reach, the monkey will bite the ball or tube following each shock. In the first example, the aggression is elicited against the same species, while in the second example, it is elicited against an inanimate object. Aggression was defined as presentation of noxious stimuli to members of the same species. This definition of aggression would not include inanimate objects. However, since it had been shown that shock elicited aggression against the same species and, in addition, elicited a similar response against an inanimate object, there is good reason to call these responses to the tube aggressive. In other words, the behavior of biting a tube following tail shock has been called aggression because it would be directed against a member of the same species if a target animal were present. This kind of aggressive behavior, biting a tube, has also been shown to follow noxious brain stimulation, that is, brain stimulation the monkey works to avoid (Renfrew, 1969). Since this brain stimulation was aversive, the aggression elicited in this study was, most probably, mediated by pain or aversion and should therefore be called secondary aggression.

Secondary aggression has also been elicited with brain stimulation in unrestrained rhesus monkeys. Monkeys were tested for whether they would work for stimulation (positive or rewarding), work to avoid stimulation (negative or aversive), or work neither to obtain nor avoid stimulation (neutral). Of the 174 electrodes implanted in seven different monkeys, 35 were defined as aversive or negative, 22 as positive or rewarding, and 117 as neutral. None of the positive or neutral brain stimulation elicited aggression. Of the 35 aversive brain stimulation points, 14 elicited aggression while 21 did not. The aggression elicited was against a submissive monkey, never a dominant one, and the aggression always occurred *after* the stimulation. This study demonstrated that aversive brain stimulation elicited well-directed and organized aggression; that not all aversive brain stimulation produced aggression; and that aggression always occurred after the stimulation. Again, since this aggression was most probably mediated by some noxious sensation, this would be secondary aggression (Plotnik *et al.*, 1971).

There has been one study in which aversive brain stimulation in one unrestrained monkey elicited aggression during, rather than after, stimulation. The aggression was directed against a dominant monkey (Robinson *et al.*, 1969). One monkey with two different electrode points would attack a dominant partner when either of these points was stimulated. This monkey would also work to escape this stimulation, indicating that the stimulation was noxious and the aggression was secondary.

Secondary Aggression

There are a number of important points to be made about brain stimulation and secondary aggression. If the researcher states that brain stimulation elicits aggression, it is essential that the reinforcing properties of this stimulation be determined. If the stimulation properties are not determined, there is no way to differentiate between primary and secondary (pain-mediated) aggression. The fact that electrodes are implanted in the brain, rather than on the skin, does not rule out the possibility of pain's mediating the aggressive response! There is no way to tell from the aggressive response itself, that is, from observing the behavior, whether the aggression is primary or secondary. Aggressive responses elicited by aversive brain stimulation are well directed, organized, and responsive to the environmental cues. Brain stimulation that was shown to be aversive has elicited aggression in rats, cats, and monkeys. Since there is abundant evidence for secondary aggression (pain-mediated), there is compelling reason for future researchers to differentiate between primary and secondary aggression.

Comparison of Central and Peripheral Aversive Stimulation

If pain is mediating aggression, then similar aggressive responses should be elicited whether the pain is produced by stimulation of the brain or body. There has been only one study comparing aversive brain stimulation with aversive body shock in freely moving monkeys (Plotnik *et al.*, 1971). Immediately following any one of three aversive stimuli—foot shock, waist shock, or aversive brain stimulation—the same kind of aggressive responses were elicited; and, for all three stimuli, the occurrence of aggressive responses depended on the social rank of the stimulated animal. If a submissive animal was shocked on the skin in the presence of a dominant animal, the submissive animal did *not* make any aggressive responses. The same was true following aversive brain stimulation. If a dominant animal was shocked on the skin in the presence of a submissive animal, the dominant animal would attack or threaten the submissive animal. The same was true following aversive brain stimulation. Thus, whether the aversive stimulation was administered to the brain or body, the aggressive responses elicited were identical: well directed, organized, and related to the social hierarchy.

BRAIN STIMULATION AND AGGRESSION REEXAMINED

Monkeys

Since brain stimulation can elicit either primary or secondary aggression, it is necessary to reexamine previous research and determine, if possible, the kind of aggression reported. Brain stimulation in unrestrained squirrel monkeys has been reported to elicit aggression against other monkeys (Maurus and Ploog, 1969). These authors have not reported whether the stimulation was aversive so we do not know whether this was primary or secondary aggression. In a series of experiments on rhesus monkeys (Delgado, 1955, 1959, 1962, 1963, 1964, 1965a,b, 1966, 1967, 1968), brain stimulation was reported to elicit aggression against other monkeys. Since there was no test for the reinforcing properties of brain stimulation, it is not possible to determine whether this was primary or secondary aggression. In a study of unrestrained rhesus monkeys in which properties of brain stimulation were determined, only aversive brain stimulation elicited aggression or, in other words, only secondary aggression was found (Plotnik *et al.*, 1971).

There is one experiment reporting both primary and secondary aggression (Robinson *et al.*, 1969). In one monkey, two electrode points that had been shown to be noxious elicited aggression against another monkey, demonstrating secondary aggression. In another monkey, a single point when stimulated caused a submissive monkey to attack a dominant monkey. Since this animal would not work to avoid or escape this stimulation, it was considered not to be aversive. This is the only evidence for primary aggression in monkeys. Thus, because of the failure to define the reinforcing properties of brain stimulation, many previous studies are inconclusive as to whether the aggression produced was primary or secondary. Presently, there is a single study with a single electrode point in a single monkey that gives clear evidence for primary aggression.

Chimpanzees

Only a few attempts have been made to study brain stimulation and aggression in chimpanzees because of the expense of the animals and difficulty of handling (Delgado, 1969; Bradley, personal communication). In addition, the social structure of the chimpanzee is complicated compared with the stable social hierarchy of the rhesus monkey. For example, in one case in which it was possible to stimulate a chimpanzee, the brain stimulation, which was aversive, elicited a variety of responses. On some occasions the chimpanzee responded with loud vocalizations, distorted facial gestures, and violent banging on the cage (general tantrum). On other occasions, the stimulated chimp attacked its cagemate. In one case, after the cagemate had been attacked and vocalized loudly, the attacking chimp returned and put its arms around the attacked animal. It was very difficult to score and interpret these kinds of social interactions. Thus, it is probably premature to study the chimpanzee's social–emotional responses to brain stimulation until its social behavior is better understood and handling difficulties can be minimized.

Humans

Infrequently, humans have been implanted with electrodes for therapeutic reasons. For example, a neurosurgeon may be trying to determine if there are abnormal brain waves deep in the brain, and, in some cases, electrode implantation is the only method; or, a surgeon may have decided, on the basis of other neurological tests, to remove brain tissue and may therefore implant electrodes to help localize the area functioning abnormally. If a patient has medically uncontrollable seizures and a history of aggressive responses, the surgeon may implant electrodes to try to localize and destroy the area or areas involved in the epileptic and assaultive behavior; finally, electrodes may be implanted to relieve intractable pain by making tiny lesions in the brain.

For reasons not yet determined, aggressive or assaultive behavior sometimes accompanies temporal lobe epilepsy. Epilepsy is not so much a disease as a group of symptoms (Mark and Ervin, 1970) which might include abnormal brain waves with or without any behavioral manifestations. If there are behavioral manifestations they can include behavioral seizures, sometimes accompanied by visual or olfactory sensations; emotional feelings of fear, depression, or loss of control; and sometimes aggressive or assaultive behavior. Some patients with temporal lobe epilepsy may become aggressive with the slightest provocation or inconsequential annoyance (Gloor, 1967; Mark and Ervin, 1970). One might expect, therefore, that it would be easy to elicit aggression with brain stimulation in patients having a history of epilepsy and assaultive behavior. Just the opposite is the case. Chapman (1958) reported that brain stimulation elicited feelings of fear and other sensations, but never rage. Gloor (1967) has stated that at the Montreal Neurological Institute temporal lobe brain stimulation commonly elicited fear, but has never elicited anger or rage.

In another laboratory, the effects of brain stimulation have been reported for a girl with a history of assaultive behavior—she had stabbed two people (Delgado *et al.*, 1968; Mark and Ervin, 1970). This girl (aged 21) was implanted with electrodes in the amygdala (in the temporal lobe) and other brain areas. With a radio stimulator mounted on her head, she was free to move about during stimulation and recording. One hundred and fifty seconds after one stimulation to the amygdala, she interrupted her activity and began to beat on the wall; 90 sec after a second stimulation, the girl let loose of the guitar she was playing and hurled it across the room. Beating on the wall and letting loose of her guitar were interpreted by the authors as evidence for brain stimulation triggering aggression. That is certainly the most dramatic explanation.

Because this experiment, which has received widespread publicity in *Life* (Rosenfeld, 1968), is one of only two which report aggression elicited by brain stimulation in humans, and since it has been used as evidence for neural triggers for aggression (Mark and Ervin, 1970), the results should be analyzed in detail. The most serious problem with this study, reported both by Delgado *et al.* (1968) and Mark and Ervin (1970), is the lack of self-reports from the patient either during or after stimulation. The lack of self-report means that we cannot determine whether the stimulation produced sensations of anger, rage, or any emotion; whether the patient was intentionally directing her responses toward some object; or whether the stimulation was noxious. Without this information, we cannot determine whether the responses of beating the wall and hurling the guitar were motor movements without any emotional content, or whether these responses were directed and organized aggressive attacks.

The next problem is to decide whether this study presents evidence for neural triggers for aggression. As noted above, it is difficult to decide if these responses were aggressive, so the question of neural triggers for aggression may be moot. If there were evidence, besides that published by the two papers, which justified calling these responses aggressive, could we then consider these data evidence of neural triggers for aggression? As defined already, a neural trigger is some combination of neurons whose activation results in an organized and directed behavioral response. To ensure that these neurons are in fact eliciting the response, the behavior should either accompany or immediately follow (within 5–10 sec) activation of the neurons. In these reports, the patient beat against the wall 150 sec after brain stimulation and hurled the guitar some time over 90 sec after brain stimulation. The elicited behavior does not, by any stretch of the imagination, closely follow the brain stimulation. What is happening in the brain during the 90–150 sec between the time of stimulation and the onset of behavior? We do know from the EEG records of the patient that immediately before striking the wall or hurling the guitar, there were abnormal brain wave patterns which persisted into the behavioral response. Furthermore, these brain wave patterns were similar to those observed when a person has an epileptic seizure. Thus, the brain stimulation that lasted 5 sec appears to have resulted in abnormal brain functioning as shown by brain wave patterns usually seen during epileptic seizures. In the discussion of epilepsy, it was noted that, for undetermined reasons, assaultive behavior may accompany this abnormal brain wave activity. Thus, instead of evidence for a neural trigger of aggression, the authors found evidence for a neural trigger of epileptic seizure activity, which either accompanied or mediated the wall beating and guitar hurling.

It should also be pointed out that there are serious discrepancies in the two different authors' reports. Delgado *et al.* (1968) state that the patient struck the wall 7 sec after brain stimulation. But it is clear from the more detailed results, including EEG records, shown by Mark and Ervin (1970), that this response did not occur until 150 sec after stimulation. Also, Delgado *et al.* (1968) state that during the response of wall beating or guitar hurling, there was no brain wave pattern of seizure activity. Again, the EEG records (Mark and Ervin, 1970) indicate that there was brain wave pattern of seizure activity immediately before and during these responses. The data from the EEG records is very clear and specific as to time, so it would seem the errors lie in the Delgado *et al.* (1968) report. The reason that these discrepancies are pointed out is that one might conclude from the Delgado *et al.* paper that there is evidence for a neural trigger for the "aggressive" response. However, this conclusion (of a neural trigger) is precluded by the Mark and Ervin report in which the EEG recordings indicate a 150-sec delay between stimulation and response, as well as seizure activity during the wall-beating response.

Therefore, after detailed examination of the results reported by Delgado *et al.* (1968) and Mark and Ervin (1970), we must conclude that (*1*) there is not sufficient evidence to decide whether the elicited responses were well organized and directed aggressive responses or motor responses accompanying the epileptic seizure activity; and (2) instead of evidence for neural triggers for aggression, these results indicate, if anything, neural triggers for epileptic brain activity, which mediated the behavioral responses.

There is a report by Heath *et al.* (1955) of a woman stimulated in a brain structure, the amygdala, which is deep in the temporal lobe. This is the same patient as reported in another article (King, 1961). During stimulation, the patient sometimes reported rage or anger ("I'm going to hit you") and sometimes reported fear, with a desire to run away. The patient said that the stimulation was not painful. This is the only example of primary aggression elicited by brain stimulation in a human. Why was the response elicited by brain stimulation inconsistent, sometimes rage and sometimes fear? One explanation is a changing test environment—different doctors present during the stimulation sessions. It is absolutely necessary to have appropriate target stimuli present during brain stimulation. If there were different testing conditions for different stimulation sessions, the patient might express anger at one doctor and fear at another. This is true for monkeys, which, after aversive brain stimulation, show an aggressive response, no response, or a submissive response, depending upon whether there is a dominant monkey, no monkey, or submissive monkey present, respectively. In studies on humans, there are usually only doctors present and they are probably inappropriate targets (from the patient's viewpoint) for aggression.

Only a few conclusions can be drawn from human studies. Although some of the epileptic patients implanted with electrodes exhibited assaultive behavior prior to implantation, it has been extremely rare to elicit aggression with brain stimulation in these patients. This could have resulted from either a lack of appropriate targets during testing or could reflect the fact that the electrodes were not activating the brain areas involved in aggression. In humans, interpreting a response as aggressive and distinguishing between primary and secondary aggression depend on the patient's self-reports. There is a single example where a woman reported aggression, alternated with fear, during brain stimulation that was not painful. Because other studies did not include self-report data, the latter study is the only evidence for primary aggression in humans.

SUMMARY—CRITERIA FOR ELICITED AGGRESSION

Before one can conclude that brain stimulation elicits primary aggression, certain criteria must be met. (*1*) The rewarding or punishing effects of brain stimulation should be studied independently of tests for aggressive responses. Many studies do not meet this criterion. Although it is not difficult to test for reinforcing properties of stimulation, it can be very time-consuming. (2) If an animal is tested with a stimulus or target other than its own species, it should be tested, in addition, with one of its own species; also, responses to brain stimulation should be tested in unrestrained animals. Testing an unrestrained animal with its own species makes the interpretation of aggressive responses more valid since one of the animal's own species is, in effect, interpreting the response for the investigator. (3) When an animal is tested with its own species, the rank of the target animal is an important consideration since a submissive monkey will rarely, if ever, attack a dominant animal, particularly when the social hierarchy is well established. The earlier discussion of targets also applies to human studies, which generally employ doctors or psychiatrists as targets to aggress upon. They are probably inappropriate targets. (4) With animals, and humans if possible, the brain stimulation should be repeated until the experimenter is satisfied of the response reliability. This is more difficult to do with humans because there is always the possibility of danger to the patient or staff if an aggressive response is elicited by brain stimulation. (5) Only responses occurring during or immediately after the stimulation should be considered to have been elicited by brain stimulation. These responses can be termed stimulus bound, meaning the relationship between brain stimulation and elicited response is so close in time, and occurs so consistently, that there is little doubt that the cause of the response is the brain stimulation. If an aggressive response occurred 60 sec after stimulation, there would be grave doubt that the stimulation was actually eliciting the response.

Only a few studies to date have met these criteria. Because many studies have not met these criteria, we have limited knowledge about the neural organization of aggression from brain stimulation data. Only two brain structures, the hypothalamus and amygdala, have been implicated by brain stimulation. Because there is so little research satisfying the above criteria, it is probably premature to speculate about the brain structures involved in aggression.

NEURAL TRIGGERS FOR AGGRESSION REEXAMINED

If brain stimulation of a group of neurons results in some organized behavioral response, it is convenient to think of these neurons as a trigger or center or circuit for that particular behavior. Aggressive behavior may be elicited in at least two different ways by brain stimulation, either by activation of a neural center for aggression or by activation of a pain center. Previously, it was thought that if brain stimulation elicited aggression, that was evidence for neural triggers or centers for aggression.

But as we know now, the fact that brain stimulation elicits aggression does not necessarily mean a neural trigger for aggression; it could also mean a pain trigger.

Experiments which did not differentiate between primary and secondary aggression have not, in fact, differentiated between neural triggers for aggression or neural triggers for pain. If we eliminate those experiments that fail to differentiate and those experiments that reported only secondary aggression (pain triggers), we are left with essentially two studies—one study with one rhesus monkey (Robinson *et al.*, 1969) and one study with one human (Heath *et al.*, 1955; King, 1961)—that provide the total evidence for some kind of neural trigger or center for aggression in primates. This is not overwhelming evidence, but it does suggest the possibility that these circuits exist. If a writer insists that brain stimulation has shown that neural centers or triggers for aggression exist, this insistence is based on two subjects.

Aggression in Nonprimates

There is evidence for an innate neural circuit for predatory behavior in rats and cats. Experiments have shown that isolated cats will show predatory behavior when electrically stimulated in the brain (Roberts and Berquist, 1968). Laboratory rats that have been tested for killing of mice and do not kill, can be made to kill with electrical or chemical brain stimulation (Smith *et al.*, 1970; Panksepp, 1971). Apparently, the neural circuits for predation are present from birth in these animals.

There is no evidence that nonaversive brain stimulation elicited aggression between rats. Of course, rats will fight each other when stimulated with aversive brain stimulation. Thus, there is evidence for secondary, but not primary, aggression in rats.

There is evidence that brain stimulation will elicit aggression of one cat against another, but we do not have the data on whether this stimulation was aversive. A map of the cat brain showing various locations of flight, defense, and attack structures (Kaada, 1967) may reflect, in large part, pain or noxious neural circuits rather than aggression circuits. Thus, for the cat, there is evidence for brain stimulation eliciting aggression but there is no evidence for whether this is primary or secondary aggression.

Aggression in Primates

There is no evidence from brain stimulation data for *innate* neural centers in primates.

There is evidence for secondary aggression (pain-mediated) elicited by brain stimulation in primates. Secondary aggression is well organized, directed, and responsive to environmental cues.

There is evidence of primary aggression (nonpain-mediated) elicited by brain stimulation in one monkey and one human. These studies provide the very limited evidence from brain stimulation data for the existence of neural triggers for aggression. Most previous studies have not differentiated between primary and secondary aggression.

The environment and target for aggression elicited by brain stimulation are critical and, to a large extent, determine the kind of aggressive response elicited. The length of contact between target and stimulated animal has an effect on whether a submissive animal will attack a dominant animal.

There is evidence that certain temporal lobe diseases are associated with a high incidence of aggressive behavior, triggered in some cases by very trivial stimuli. Does this indicate that humans possess an innate aggressive circuit that is being triggered in the above patients by abnormal brain functioning? One investigator says yes (Moyer, 1971). It is equally possible that humans learn various aggressive patterns and that these patterns are somehow represented in the brain. Because of abnormal functioning, these neural patterns are triggered more frequently and inappropriately in the temporal lobe patients. It is certainly stretching the data to make the jump from a high incidence of assaultive behavior to the existence of innate aggressive circuits. Thus, from the high incidence of assaultive behavior in temporal lobe diseases, we can draw few conclusions except that the brain is functioning abnormally.

There is limited evidence that brain stimulation elicits primary aggression in humans. In the single human in which the properties of the stimulation were determined (patient said it was not painful), brain stimulation sometimes produced anger and sometimes fear (Heath *et al.*, 1955; King, 1961). In another human patient, reported by two investigators (Delgado *et al.*, 1968; Mark and Ervin, 1970), brain stimulation resulted in the patient's beating the wall or letting go of her guitar in a swinging movement (reported as hurling). Since these authors failed to report what the patient experienced during the stimulation or whether these responses were directed or were the result of seizure activity, there is no way to determine their aggressive content, much less whether it was primary or secondary aggression.

Theoretical Possibilities

The brains of monkeys, and to a lesser extent of humans, have been explored with electrical stimulation. The result has been one case of stimulation eliciting primary aggression in a monkey (Robinson *et al.*, 1969) and one not entirely convincing case of stimulation eliciting primary aggression in a human (Heath *et al.*, 1955; King, 1961). There are two possibilities for why more brain areas that elicit aggression have not been located. The first possibility is that researchers have not yet explored the right regions. Since entirely different behaviors can be elicited by electrodes no more than 2–3 mm apart, it may be necessary to explore the brain in millimeter steps, which is feasible in monkeys, but not humans.

Another possibility, and one that is more difficult to evaluate than the first, is that there are many different neural systems involved in aggression. It could be that more than one area must be stimulated simultaneously for aggression to occur; for example, there might be one area that evaluates the environmental stimuli to determine whether aggression should occur and another for the behavioral manifestation of aggression. If both of these areas are not stimulated simultaneously, aggression will not occur.

Finally, it is possible that the neural circuits for aggression are located in different parts of the brain for each individual. Thus, patient A, who shows a decrease in aggression following a lesion in the temporal lobe, may have had his circuits located in the temporal lobe area. Patient B, who did not show any reduction in aggression following temporal lobe lesion, may have had his neural circuits for aggression in a different area.

If we assume that in primates aggressive behavior is primarily learned, we might expect at least individual differences in the brain areas involved in elaboration of aggressive responses. Because of the complexity of brain connections and interrelatedness of environmental stimuli and brain stimulation, theorizing about why something does not happen in the brain is a limitless occupation. These few hypotheses illustrate the point that there are several viable alternatives to the notion that there are all-or-nothing circuits or triggers for aggression and that these circuits are innate.

ACKNOWLEDGMENT

The author wishes to thank Sandra Mollenauer for her critical reading of an earlier draft.

REFERENCES

Adams, D. B. (1968). Cells related to fighting behavior recorded from midbrain central gray neuropil of cat. *Science* **159**, 894–896.

Akerman, B. (1966). Behavioral effects of electrical stimulation in the fore-brain of the pigeon. II: protective behavior. *Behaviour* **26**, 339–349.
Ardrey, R. (1961). "African Genesis." Atheneum, New York.
Ardrey, R. (1966). "The Territorial Imperative." Atheneum, New York.
Azrin, N. H., Hutchinson, R. R., and Hake, D. F. (1963). Pain-induced fighting in the squirrel monkey. *J. Exp. Anal. Behav.* **6**, 620.
Azrin, N. H., Hutchinson, R. R., and Sallery, R. D. (1964). Pain-aggression toward inanimate objects. *J. Exp. Anal. Behav.* **7**, 223–227.
Azrin, N. H., Hake, D. F., and Hutchinson, R. R. (1965). Elicitation of aggression by a physical blow. *J. Exp. Anal. Behav.* **8**, 55–57.
Bowsher, D. (1957). Termination of the central pain pathway in man; the conscious appreciation of pain. *Brain* **80**, 606–622.
Buss, A. H. (1971). Aggression pays. *In* "The Control of Aggression and Violence" (J. L. Singer, ed.). Academic Press, New York.
Carthy, J. D., and Ebling, F. J. (1964). Prologue and epilogue. *In* "The Natural History of Aggression" (J. D. Carthy and F. J. Ebling, eds.), pp. 1–5. Academic Press, New York.
Chapman, W. P. (1958). Studies on the periamygdaloid area in relation to human behavior. *Res. Publ. Nerv. Dis.* **36**, 258–277.
Davis, D. E. (1964). The physiological analysis of aggressive behavior. *In* "Social Behavior and Organization among Vertebrates" (W. Etkin, ed.), pp. 53–74. Univ. of Chicago Press, Chicago, Illinois.
Delgado, J. M. R. (1955). Cerebral structures involved in transmission and elaboration of noxious stimulation. *J. Neurophysiol.* **18**, 261–275.
Delgado, J. M. R. (1959). Prolonged stimulation of brain in awake monkeys. *J. Neurophysiol.* **22**, 458–475.
Delgado, J. M. R. (1962). Pharmacological modifications of social behavior. *In* "Proc. Pharmacol. Congr. 1st." (W. D. M. Patton and P. Lindgren, eds.), pp. 265–292. Pergamon, Oxford.
Delgado, J. M. R. (1963). Cerebral heterostimulation in a monkey colony. *Science* **141**, 161–163.
Delgado, J. M. R. (1964). Free behavior and brain stimulation. *Int. Rev. Neurobiol.* **6**, 349–449.
Delgado, J. M. R. (1965a). Chronic radiostimulation of the brain in monkey colonies. *Excerpta Med. Int. Congr. Ser. No. 87; Proc. Congr. Physiol. Sci., 23rd* **4**, 365–371.
Delgado, J. M. R. (1965b). Pharmacology of spontaneous and conditioned behavior in the monkey. *In* "Proc. Pharmacol. Congr., 2nd." (M. Ya. Mikhel'son and V. G. Longo, eds.), pp. 133–156. Pergamon, Oxford.
Delgado, J. M. R. (1966). Aggressive behavior evoked by radio stimulation in monkey colonies. *Amer. Zool.* **6**, 669–681.
Delgado, J. M. R. (1967). Social rank and radio-stimulated aggressiveness in monkeys. *J. Nerv. Ment. Dis.* **144**, 383–390.
Delgado, J. M. R. (1968). Electrical stimulation of the limbic system. *Proc. Int. Congr. Physiol. Sci., 24th* **6**, 222–223.
Delgado, J. M. R. (1969). Offensive-defensive behaviour in free monkeys and chimpanzees induced by radio stimulation of the brain. *In* "Aggressive Behaviour" (S. Garattini and E. B. Sigg, eds.), pp. 109–119. Wiley, New York.
Delgado, J. M. R., Roberts, W. W., and Miller, N. E. (1954). Learning motivated by electrical stimulation of the brain. *Amer. J. Physiol.* **179**, 587–593.
Delgado, J. M. R., Mark, V., Sweet, W., Ervin, F., Weiss, G., Bach-y-rita, G., and

Hagiwara, R. (1968). Intracerebral radio stimulation and recording in completely free patients. *J. Nerv. Ment. Dis.* **147,** 329–340.
Eibl-Eibesfeldt, I. (1967). Ontogenetic and maturational studies of aggressive behavior. *In* "Aggression and Defense" (C. D. Clemente and D. B. Lindsley, eds.), pp. 57–94. Univ. of California Press, Los Angeles, California.
Eibl-Eibesfeldt, I. (1970). "Love and Hate." Holt, New York.
Flynn, J. P. (1967). The neural basis of aggression in cats. *In* "Neurophysiology and Emotion" (D. C. Glass, ed.), pp. 40–60. Rockefeller Univ. Press, New York.
Flynn, J. P., Edwards, S. B., and Bandler, R. J., Jr. (1971). Changes in sensory and motor systems during centrally elicited attack. *Behav. Sci.* **16,** 1–18.
Gloor, P. (1967). Discussion of brain mechanisms related to aggressive behavior by B. Kaada. *In* "Aggression and Defense" (C. D. Clemente and D. B. Lindsley, eds.), pp. 116–127. Univ. of California Press, Los Angeles, California.
Harlow, H. F., and Harlow, M. K. (1962). Social deprivation in monkeys, *Sci. Amer.* **207,** 136–146.
Hassler, R. (1968). Interrelationship of cortical and subcortical pain systems. *Int. Pharmacol. Meeting* **3,** 219–229.
Heath, R. G., Monroe, R. R., and Mickle, W. A. (1955). Stimulation of the amygdaloid nucleus in a schizophrenic patient. *Amer. J. Psychiat.* **73,** 127–129.
Johnson, R. N. (1972). "Aggression in Man and Animals." Saunders, Philadelphia, Pennsylvania.
Kaada, B. (1967). Brain mechanisms related to aggressive behavior. *In* "Aggression and Defense" (C. D. Clemente and D. B. Lindsley, eds.), pp. 95–116. Univ. of California Press, Los Angeles, California.
Karli, P., Vergnes, M., and Didiergeorges, F. (1969). Rat-mouse interspecific aggressive behaviour and its manipulation by brain ablation and by brain stimulation. *In* "Aggressive Behaviour" (S. Garattini and E. B. Sigg, eds.), pp. 47–55. Wiley, New York.
King, H. E. (1961). Psychological effects of excitation in the limbic system. *In* "Electrical Stimulation of the Brain" (D. E. Sheer, ed.), pp. 477–486. Univ. of Texas Press, Austin, Texas.
Kruger, L. (1966). The thalamic projection of pain. *In* "Henry Ford Hospital International Symposium on Pain" (R. S. Knighton and P. R. Dumke, eds.), pp. 67–81. Little, Brown, Boston, Massachusetts.
Lorenz, K. (1966). "On Aggression." Harcourt, New York.
MacDonnell, M. F., and Flynn, J. P. (1964). Attack elicited by stimulation of the thalamus of cats. *Science* **144,** 1249–1250.
Mark, V. H., and Ervin, F. R. (1970). "Violence and the Brain." Harper, New York.
Maurus, M. (1967). Nene Fernreizappuratur fur kleine Primaten. *Naturwissenschaften* **22,** 593.
Maurus, M., and Ploog, D. (1969). Motor and vocal interactions in groups of squirrel monkeys, elicited by remote-controlled electrical brain stimulation. *Proc. Int. Congr. Primatol., 2nd.* Basel, 1968, **3,** 59–63.
Montagu, M. F. A. (ed.) (1968). "Man and Aggression." Oxford Univ. Press, London and New York.
Morris, D. (1968). "The Naked Ape." McGraw-Hill, New York.
Moyer, K. E. (1968). Kinds of aggression and their physiological basis. *Commun. Behav. Biol.* **A2,** 65–87.
Moyer, K. E. (1971). A preliminary physiological model of aggressive behavior.

In "The Physiology of Aggression and Defeat" (B. E. Eleftheriou and J. P. Scott, eds.). Plenum Press, New York.

Obrador, S., and Dierseen, G. (1966). Sensory responses to subcortical stimulation and management of pain disorders by stereotaxic methods. *Confina Neurol.* **27**, 45–51.

Olds, M. E., and Olds, J. (1963). Approach-avoidance analysis of rat diencephalon. *J. Comp. Neurol.* **120**, 259–283.

Panksepp, J. (1971). Aggression elicited by electrical stimulation of the hypothalamus in albino rats. *Physiol. Behav.* **6**, 321–329.

Phillips, R. E. (1964). "Wildness" in the mallard duck: Effects of brain lesions and stimulation on "escape behavior" and reproduction. *J. Comp. Neurol.* **122**, 139–155.

Plotnik, R., Mir, D., and Delgado, J. M. R. (1971). Aggression, noxiousness, and brain stimulation in unrestrained rhesus monkeys. *In* "The Physiology of Aggression and Defeat" (B. E. Eleftheriou and J. P. Scott, eds.), pp. 143–221. Plenum Press, New York.

Plotnik, R., Mir, D., and Delgado, J. M. R. (1972). Map of reinforcing sites in the rhesus monkey brain. *Int. J. Psychobiol.* **2**, 1–21.

Renfrew, J. W. (1969). The intensity function and reinforcing properties of brain stimulation that elicits attack. *Physiol. Behav.* **4**, 509–515.

Roberts, W. W. and Berquist, E. H. (1968). Attack elicited by hypothalamic stimulation in cats raised in social isolation. *J. Comp. Physiol. Psychol.* **66**, 590–595.

Roberts, W. W., Steinberg, M. L., and Means, L. W. (1967). Hypothalamic mechanisms for sexual, aggressive, and other motivational behaviors in the opossum, *Didelphis Viriginiana. J. Comp. Physiol. Psychol.* **64**, 187–193.

Robinson, B. W. (1969). Brain telestimulation in primates. *Amer. Psychol.* **24**, 248–250.

Robinson, B. W., Alexander, M., and Bowne, G. (1969). Dominance reversal resulting from aggressive responses evoked by brain telestimulation. *Physiol. Behav.* **4**, 749–752.

Rosenfeld, A. (1968). The psycho-biology of violence. *Life*, June 21, 67–71.

Scott, J. P. (1971). Theoretical issues concerning the origin and causes of fighting. *In* "The Physiology of Aggression and Defeat" (B. E. Eleftheriou and J. P. Scott, eds.), pp. 11–41. Plenum Press, New York.

Smith, D. E., King, M. B., and Hoebel, B. C. (1970). Lateral hypothalamic control of killing evidence for a cholinoceptive mechanism. *Science* **167**, 900–901.

Storr, A. (1968). "Human Aggression." Athenum, New York.

Ulrich, R. (1966). Pain as a cause of aggression. *Amer. Zool.* **6**, 643–662.

Ulrich, R. E., Wolff, P. C., and Azrin, N. H. (1964). Shock as an elicitor of intra- and inter-species fighting behaviour. *Anim. Behav.* **12**, 14–15.

Ulrich, R. E., Hutchinson, R. R., and Azrin, N. H. (1965). Pain-elicited aggression. *Psycholog. Rec.* **15**, 111–126.

Ulrich, R., and Symannek, B. (1969). Pain as a stimulus for aggression. *In* "Aggressive Behaviour" (S. Garattini and E. B. Sigg, eds.), pp. 59–69. Wiley, New York.

von Holst, E., and von Paul, U. (1962). Electrically controlled behavior. *Sci. Amer.* **206**, 50–59.

Woodworth, C. H. (1971). Attack elicited in rats by electrical stimulation of the lateral hypothalamus. *Physiol. Behav.* **6**, 345–353.

AGONISTIC BEHAVIOR OF PRIMATES: A COMPARATIVE PERSPECTIVE

JOHN PAUL SCOTT

Bowling Green State University

INTRODUCTION

In this paper I shall try to relate the research on primate aggression to the broader findings from the study of other animals. One of the principal results will be to bring out those areas about which more information is needed. I will also attempt to place primates somewhere in the broad spectrum of behavior found in the rest of the animal kingdom and, more particularly, try to locate man within the somewhat lesser diversity of primate behavior.

A major clarifying concept is that of agonistic behavior, defined as behavior which is adaptive in situations involving conflict between members of the same species. In any given species that exhibits agonistic behavior—and there are many that do not, particularly in the lower invertebrates—such behavior is organized into a behavioral and motivational system made up of several alternative patterns of behavior such as fighting, escape, freezing, etc. The system involves motivations based on emotional and physiological processes. Some of these may be peculiar to the species or even to individuals within the species, and certainly in many cases peculiar to the class of individuals belonging to one sex or the other.

The concept of behavioral systems serves to separate phenomena that are biologically and experientially closely related to each other from those that are not so related. The more restricted concept of the agonistic behavioral system serves to cut out numerous surplus meanings that have become associated with the word "aggression" by eliminating other major behavioral–motivational systems. One of those so eliminated is predation, involving conflict between members of different species. In all of the predators that have been studied, predation and agonistic behavior form distinctly different systems, both behaviorally and motivationally. The notion that man is a "killer ape" and motivated to destroy his own kind because he is a hunter is therefore without foundation. Further, the concept of agonistic behavior differentiates social fighting from other systems such as sexual behavior or ingestive behavior, both of which have been either hypothetically related to aggression or included within it. For example, in common usage, "the aggressor" may mean simply a person who initiates sexual advances. Such crude and ill-founded analogies may lead to various sorts of peculiar thinking, both with respect to behavior and physiology. The relationship between two major behavioral systems, if any, is always an empirical problem to be solved by observation and experiment.

ADAPTIVE FUNCTIONS OF AGONISTIC BEHAVIOR

One of the major research problems in this area, and one which is too often neglected, can be expressed very simply as: "What are the animals fighting for?" Underlying this is the basic biological assumption that if behavior is commonly and repeatedly seen, it must serve some adaptive function for the species. As the following examples show, there are numerous functions of fighting, and any one species may exhibit one or many of these.

Defense

The most generally observed function of behavior adapted to conflict is that of a reaction against injury or the threat of injury. Such behavior can be found even in lower animals that show no true agonistic behavior and is obviously adaptive against predation or accidental injury from other species as well as from conspecifics. Among higher animals, behavior similar to that of the cornered rat or fear-biting dog is commonly found. Other examples are the threat display of the captured opossum who opens his mouth and shows all of his teeth, or the familiar "fear grimace" of the rhesus monkey.

Regulation of Space and Distance

One of the patterns of behavior in the agonistic system is to respond to injury or the threat of injury by moving away, thus increasing the distance between two individuals. In the case of an injured animal, this response removes it to a point where injury is no longer likely. This behavior in turn has often evolved into a mechanism for the regulation of distance between individuals (McBride, 1964; Esser, 1971) and takes two general forms. One of these is the phenomenon of personal space, common in the ungulate animals. As a flock of goats moves around, one will butt or threaten another if it comes too close, with the result that these animals rarely come closer than a foot or two from each other. An exception to the rule is that females do allow their young to come close to them, as is necessary for nursing.

Most primates seem to be highly tolerant of close contact and spend a great deal of time in mutual grooming. However, dominance relationships between male savannah baboons result in these individuals being spaced out from each other, at least under certain conditions. The regulation of space thus becomes much more complex than simple avoidance of close contact, related individuals permitting very close contact and unrelated individuals in the same social group maintaining distance from each other.

A second kind of space about which animals may fight is the actual possession of a particular piece of ground, usually called a territory. This behavior is generally typical of birds, being found in almost every species. The size of a territory may range from the distance around the nest that a sitting bird can reach with its beak, as in the nesting colonies of gulls, through territories as large as a quarter acre or so, in the case of most song birds, to the much larger hunting areas guarded by certain birds of prey. The usual function of the territory is to provide a defense of a breeding ground and, as a result, not only provides for more adequate food and hence survival of young in a particular area, but also limits the total number of broods raised in that area. Such a territorial system works best where the boundaries can be efficiently guarded, that is, in species with good eyesight who can keep the entire area under surveillance. Many birds, particularly song birds, use special territorial cries to advertise that a particular area is occupied. The result is that, except during a brief period at the outset of the breeding season when territories are being organized, there is very little overt fighting.

The occurrence of territoriality among mammals is much less general. There are a few cases like that of the prairie dog (King, 1955), in which specific boundaries are maintained under surveillance and guarded efficiently, but these are the exception rather than the rule. For nocturnal mammals, the efficient guarding of a territory is particularly difficult. Some species, such as the European rabbit (Mykytowycz, 1965), use scent-marking (in this case through the chin glands) as a means of advertising that a particular area is occupied, but surveillance of course is less efficient than that accomplished by vision. Many species of mammals have ritualized eliminative behavior to serve a communicatory function, and it is often assumed that any such scent-marking is territorial in nature. However, the communicatory function must always be established by experiment, as the message may mean nothing more than "Kilroy was here."

A much more fundamental and general phenomenon than territoriality is that of site attachment, which occurs in a great variety of invertebrates as well as in all vertebrates, even the aquatic ones. While many species may migrate, this is always between specific geographical locations, one of which is usually the area in which the individual grew up.

As among other mammals, the occurrence of territoriality among primates is spotty (Southwick, 1972). All species so far reported have specific home ranges but only a few guard territorial boundaries [such as the *Callicebus* monkeys, and even this species does not show territorial behavior in captivity (Mason, 1968)]. The most common arrangement is that of a central core area surrounded by a home range. The home range of one troop usually overlaps with those of other troops, but never completely so, the core area being occupied exclusively. In howling monkeys, males of different troops utter warning cries, with the result that the troops maintain considerable space between them. This phenomenon is similar to that of personal space except that it occurs between different social groups rather than individuals. Other species, such as the rhesus, may show actual fighting between troops if they accidentally come into close contact (Southwick, 1972), but in most cases group space is maintained and no fighting occurs. Thus, the net effect of a well-established spatial system is a reduction of agonistic behavior to low levels and innocuous forms.

Mating Rights

Among mammals, the best examples of this function are seen among the ungulates. Fighting between males in the season of rut is quite common among such animals as deer, antelope, and bison. In the red deer the result of such fighting is that a winning stag, usually an older one with large antlers, is able to round up and guard a few females and mate with them exclusively as each comes into estrus. However, the male is usually unable to maintain this position throughout the entire season and gives way to others that have not been exhausted by constant guarding (Darling, 1937; Lincoln *et al.*, 1970).

Carpenter's (1964) study of howling monkeys, in which he reported that females in estrus passed from male to male without overt fighting or other signs of jealousy between them, at first seemed to be merely an unusual quirk of evolution. However, such nonjealous behavior occurs in other species also, such as the savannah baboons (Hall and DeVore, 1965) and may be a common reaction among primates. That intermale fighting may sometimes have the function of providing exclusive mating rights in primates is shown in hamadryas baboons. Each male attempts to maintain constant contact with two or three females and keeps them close at hand by means of threats. However, violent fighting is quite rare under natural conditions, according to Kummer (1968).

Repression of Sexual Behavior

Closely allied to the above function is negative regulation of sexual behavior. Female rodents not in estrus usually respond to the advances of a male by avoidance and threat, and this sort of behavior is common among mammals generally. Also, males generally respond to sexual advances by other males with attack behavior. In these cases agonistic behavior may have the function of repressing homosexual behavior, which of course is nonadaptive for the survival of the species (Scott and Fredericson, 1951).

In groups of captive wolves the dominant female may prevent subordinate females from mating, which results in direct population control (Rabb *et al.*, 1967). Kuo (1967) reported that he could repress adult sexual behavior in dogs by punishing its immature expression. Aside from clinical reports of human behavior, I have read of no instances of this repressive function among primates.

Defense of Young

Among female mammals, the defense of young against potential predators is almost universal, and such behavior is often extended to members of the same species and thus becomes part of the agonistic behavioral system. The primate species are no exception. The studies of Koford (1963) and Sade (1967) in rhesus monkeys indicate that a mother's help in situations of conflict between her offspring and others helps determine their later relative positions in the dominance hierarchy of a troop.

In general, too little attention has been paid to the agonistic behavior of females, partially because it tends to be less violent and spectacular than that of males, and partially because male observers have been primarily interested in the behavior of their own sex.

Division of Food

Whether this function is important in any species usually depends on the nature of the food supply and its concentration. House mice ordinarily do not fight for the possession of food, but can be induced to struggle for the possession of a single pellet if they are hungry enough (Scott and Fredericson, 1951). Most grazing animals similarly do not fight over food, but goats again can be induced to do so by presenting a single handful of attractive food, such as grain, to a pair of hungry animals.

On the other hand, among social carnivores such as wolves, agonistic behavior organized into a dominance hierarchy has an important role in determining the division of prey obtained by group hunting. Among dogs, agonistic behavior can be induced even in very young puppies by presenting a single bone to the litter (Scott and Fuller, 1965). Thus, the importance of agonistic behavior with respect to food supplies is largely dependent on whether a species is herbivorous or carnivorous.

Primates in general tend to be anatomically and physiologically adapted for an omnivorous diet, including both vegetable and animal foods. A few, such as the gorilla, are entirely vegetarian, but most are facultative carnivores, eating the small amounts of animal food that they can kill or capture but not being dependent on it. An exception is the galago, which has become almost entirely insectivorous. However, such food as this is not likely to lead to competition.

In rhesus monkeys, Southwick (1972) has shown that reducing the food supply in a captive troop has opposite effects depending on its distribution. If the food is widely scattered, as is the case with a natural food supply, the troop spends more time hunting for it and fights less. On the other hand, if the troop is provided with an artificial supply concentrated in one place, troop members do fight to possess it, and the amount of agonistic behavior increases. The concentration of food supplies may also affect fighting between troops, as when two troops come together and fight for the possession of a food supply in a temple in an Indian city. The situation is similar in the semiwild rhesus colony in Puerto Rico, where artificial food is concentrated in feeding stations accessible to more than one troop.

Recent observations of chimpanzees indicate that competition for meat obtained by hunting leads only to mild disputes; the division of food is largely ritualized (Teleki, 1973).

Agonistic Behavior as a Tool

The most general cause of agonistic behavior, related to its defensive function, is to react to any sort of noxious stimulation or emotional discomfort as if a nearby animal had caused it and to attack the animal and drive it away. This reaction will take place whether or not the neighbor actually caused the discomfort, as can be shown experimentally with rats exposed to foot shock. The result is to drive the other animal away and, if the discomfort is in fact caused by its presence, thus to solve the problem.

Considered in more detail as operant behavior, the usual first reaction to an attack is to fight back, with resulting pain to the attacker. If this continues long enough, the effect will be continuous negative reinforcement, and the attacker desists. On the other hand, if the attacker inflicts enough pain, the other animal runs away, the attack becomes successful, and the attacker receives positive reinforcement. Thus the primary operant effect of fighting behavior is to produce escape behavior in another animal and thus increase distance. This in turn can result in the exclusive possession of various things such as a particular geographic spot, a female, or food. All of these things can occur in the nonhuman primates, but there is little evidence that they ever try the more complex uses of agonistic behavior sometimes attempted in human societies, such as forcing an individual to perform work. As is well known in human experience, such distortions of basic adaptive functions of agonistic behavior work very inefficiently.

Summary

Among nonhuman primates, various species show all of the above adaptive functions of agonistic behavior. However, there is a great deal of variation between species with respect to these functions, and in many cases the presence or absence of the functions depends on environmental conditions, as in the distribution of food. Thus there is no general rule to the effect that any given function must be present in every primate species. Primates have evolved in a number of different directions, and the evolution of their agonistic behavior is no exception.

EVOLUTION OF PRIMATE BEHAVIOR

As I have pointed out elsewhere (Scott, 1968), selection pressures affecting the evolution of behavior may act at any level of organization and thus be directed at genetic, physiological, organismic, social, and ecological levels. Selection pressures at different levels can produce either similar or opposite results, with all degrees of variation in between. Where the pressures are antagonistic to each other we would predict that they would eventually reach an equilibrium.

Agonistic behavior is strongly affected by social selection and in fact becomes a part of such selection pressures, especially where it modifies mating success or competition over food essential to survival. However, for evolutionary change to take place, the important event is not the survival of the individual but of the genes which he carries, and in order for these to survive, not only the individual but the mating population of which he is a part must survive. It is obvious that violent agonistic behavior resulting in the death or serious wounding of large numbers of individuals is nonadaptive and is therefore selected against on both the social and individual levels. Thus there are strong theoretical grounds for predicting the evolutionary reduction of violent behavior as well as considerable observational evidence supporting this expectation in species living under natural conditions.

On the other hand, this does not preclude the possibility that a species may evolve capacities for destructive behavior, provided these are kept in check under normal conditions of social and ecological organization. It is also possible that certain species have evolved destructive rather than adaptive behavior and are on the way to extinction.

If we examine the results of primate field studies, we find there is a wide range of expression of agonistic behavior. In some species, such as howling monkeys, it consists almost entirely of vocal threats, while at the opposite extreme serious wounds, at least under certain conditions, are common among species such as the Indian rhesus monkey. Some form of escape and defensive behavior is found in almost all primates, and we can hypothesize that, as with other major groups of animals, the evolution of agonistic behavior began with defensive responses and secondarily took on a variety of other adaptive functions, which now vary from species to species.

As an example of one evolutionary trend, the savannah baboons have to compete and protect themselves against large predators such as leopards and lions (Hall and DeVore, 1965). The relatively large size of males in this species and their huge canine teeth are adaptive in this respect. The behavior which goes along with these structural adaptations is one of group defense by the males against potential predators—a defense reported to be effective against any animals except lions. The possession of such strength and weapons makes male baboons potentially capable of inflicting severe injuries on each other, as well as on the females, but they do not. As young baboons grow up and get into quarrels they are quickly repressed by older males by means of threats. Thus, the young baboon learns the limits of acceptable agonistic behavior and his behavior becomes organized with respect to the rest of the troop. Similar kinds of organization are seen in rhesus monkeys.

ORGANIZATION OF AGONISTIC BEHAVIOR

Considered from the viewpoint of evolution, an animal species evolves capacities to develop agonistic behavior and express it in ways that are adaptive rather than harmful. Thus the species as a whole has an organized potential for agonistic behavior. This, however, is not the whole story. Organization also takes place in the development of the individual through the processes of learning in combination with the processes of growth and embryonic differentiation. The first important studies of such organization were done with chickens and resulted in the discovery of the well-known pecking order. As a result of fighting, two hens work out a relationship in which one threatens and the other submits or escapes, with resultant reduction of actual fighting to a low level. Such relationships can be developed by mammals, as well as chickens, and may take on complex forms. In organized dog groups, for example, puppies may work out relationships of mutual threat in which neither animal gives way in a competitive situation, or relationships of alternate dominance, in which the first animal to get possession of a bone remains dominant as long as he chooses to retain it, but will submit if the other dog gets it first. In addition, there may be all degrees of dominance, from complete control by one individual to very nearly complete absence of any control by either (Scott and Fuller, 1965).

It should be pointed out that dominance organization is not the only social order but that many other kinds of organization can exist in a social group, including among others the care-dependency relationships between adults and young, sexual relationships, and leader–follower relationships.

Among primates, dominance organization appears to be important in those animals living in large troops, and some of the more detailed studies of rhesus monkeys indicate that it may be far more complex than originally seemed the case. As stated above, the rank of mothers influences the development of dominance in their offspring so that there is relative dominance between families as well as individuals. Furthermore, dominance can exist between troops, as well as among the individuals composing them.

EFFECTS OF SOCIAL DISORGANIZATION ON AGONISTIC BEHAVIOR

If natural selection favors the reduction of agonistic behavior to non-violent ritualized and communicatory forms, why do we then see occasional outbreaks of severe violence among animal species? Most of these have occurred in conditions of captivity, but such outbreaks have also been observed under natural conditions. The most direct and obvious explanation is that the expression of agonistic behavior is developed under the influence of and regulated by social organization, and that this organization has in some way become degraded or destroyed. When I published my first studies of agonistic behavior (Scott, 1942) I was attracted by the laboratory mouse because it commonly showed violent fighting resulting in serious injury or death of some of the participants—i.e., the kind of fighting that becomes a serious social problem in humans. I now realize that I was dealing with mouse populations that had been disorganized, either by the introduction of strange individuals, or by the reduction of space to the point where the organization of behavior seen in a natural population could no longer occur.

In free-living mice an agonistic encounter between males usually consists of a brief attack and escape, or mutual avoidance without contact. Reduction of fighting to threat and submission, with animals remaining in close contact, is not part of the mouse's repertory of behavior, although mice will live together with little or no agonistic behavior, provided they have been reared in contact with each other. Consequently, the result of throwing two or more strange males together in a small area is that one male becomes dominant and continually harasses the others to the point where they are constantly wounded, if not eventually killed.

Social Disorganization in Primate Societies

Although the first experimental studies on social disorganization were done by Guhl and Allee (1944) on chickens, a classical example of social disorganization had already been reported by Zuckerman (1932) from his observations of hamadryas baboons in the London Zoo. This represented an ill-fated early attempt to exhibit animals under natural conditions. "Monkey Hill" was an enclosure 100 ft long by 60 ft wide, about one-seventh of an acre. Into this were placed 100 wild-caught baboons, mostly males, almost certainly not from the same troop. At the end of 2 years, the numbers had been reduced to 59 as a result of much fighting and other causes. At this point, the zoo officials introduced 30 females and 5 immature males. All males not only attacked each other, but in some cases actually pulled the females to pieces. Two years later there were only 39 males and 9 females left, and fighting was still going on. In all this time only one infant had survived. On the basis of information then available, Zuckerman concluded that such behavior was typical of the species in the wild, although he did express doubts concerning the severity of fighting there. His conclusions fitted the then-common neo-Darwinian concept of "Nature red in tooth and claw."

Seen from the vantage point of modern information, Zuckerman was observing a population that had been disorganized in three ways: one, by the introduction of strange individuals both at the outset and later; two, by the reversal of the normal sex ratio of more females than males; and three, by reduction of space and excessive crowding. The true picture of hamadryas agonistic behavior only emerged with Kummer's (1968) study under natural conditions in Ethiopia. The basic social group in this species is a male guarding and restraining a few females and their offspring. Such groups only coalesce into a larger troop when they come together at night in their sleeping places among rocky cliffs. In his extensive observations Kummer saw only one or two cases where one male actually bit another, and no serious injuries or deaths. Males bite their attendant females on the average about once a day, but almost never draw blood.

The effects of social disorganization have also been documented by field studies. Jay (1965) made an extensive study of the Indian langur monkey in the Orcha area and reported stable troops with very little agonistic behavior either within or between troops. In contrast, Sugiyama (1967) studied langurs in Dharwar and reported severe fighting, especially between troops, which sometimes resulted in strange adult males attacking and killing infants. These troops were obviously less well organized, and in one case disorganization was deliberately brought about by the removal of the dominant male in the troop. Later Yoshiba (1968) made a detailed ecological comparison of the two areas and their monkey inhabitants. There were two outstanding differences. The population density in the Orcha area was 7–16 a square mile, but that in Dharwar ranged between 220 and 349, over 30 times as great. Similarly, the average home range of the troops was reduced from 1.5 square miles to 0.07 square miles, differing by a factor of 20. We can conclude that the Dharwar population was disorganized by overpopulation, with the result that the troops could not maintain adequate distances between each other. There was no suggestion as to why the Dharwar population had grown, except for a probable decline in predators.

Similarly, in another species, Vessey (1971) has shown that removal or death of key individuals in the dominance organization of a rhesus monkey troop will result in increased fighting and wounding among its members.

Disorganization and Violence

In another paper (Scott, 1973), I have suggested that the concept of disorganization producing violence can be extended to systems on other levels than the social. For example, artificial selection in fox terrier dogs has produced individuals that no longer form dominance orders in early development in a harmless fashion, but rather kill some of the members at an early age. This is a distortion of the genetic system common to the species. This raises an empirical question yet to be answered —i.e., what are the kinds of disorganization that lead to violence? Even without such information it is obvious that this is a major theory for the understanding and prevention of harmful agonistic behavior. The results of disorganization and their prevention through appropriate positive organization should be studied in man as well as the other primates.

AGGRESSION IN PRIMATES: THE PLACE OF MAN

As knowledge concerning the biology and behavior of primates has accumulated, we have gone beyond the simplistic notions that nonhuman primate species represent stages in the development of humans and that a monkey species is really a subnormal human being with a long tail and a fur coat. The genetic theory of evolution implies that once species have separated, each sets out on a different path. Far from converging on the human goal, primates have been independently evolving in different directions for millions of years. Primate species are as divergent among themselves as any other order of mammals, and can be as different as bears, cats, wolves, and raccoons are among the members of the order Carnivora. Therefore, to look for man's past among other primates at best gives us no more than an inaccurate metaphor—a guess as to how ancestral human beings might have lived. The nonhuman primates are (to use another metaphor) not our grandfathers, but our distant cousins.

Seen in the perspective of the modern theory of evolution, the two great limitations of the primates are that they have never become aquatic and, with the exception of man, are confined to tropical and semitropical climates. A partial exception to the climatic rule is the Japanese macaque, and even this animal is able to maintain itself only in the relatively mild winters of Japan. The limiting evolutionary factor appears to be the development of hands and manipulative ability. Fingers are poor aids to swimming compared to the flippers of seals. Further, fingers are no longer sensitive instruments when stiff with cold, even if they do not become frozen. Carnivores with similar adaptations for manipulation have solved the problem by hibernating in the winter, as has the raccoon, but this has occurred in no primate species.

Man, with the use of fire, tools, and clothing, is an exception to this rule, and with such aids has been able to penetrate every land region of the world to become a truly world-wide species. Even before such tools were developed we have evidence that man evolved as a plains-living primate, in contrast to the vast majority of other primate species, which are either arboreal or semiarboreal. Part of this evidence is man's bipedal locomotion, an adaptation which has appeared independently in several other orders of mammals, typically in species living in desert areas—the kangaroo and similar related species among marsupials, and various species of kangaroo-like rodents.

Several reasons have been offered for this development, one of which is that it keeps the animal in minimum contact with the hot sand, but it is more likely that this kind of locomotion is only advantageous in areas where there are few obstacles and where it permits long-distance vision as well as specialization of the forelimbs for grasping and manipulation. Furthermore, dry semidesert habitats do not support large carnivores, against which a primitive ground-living primate without tools would be relatively helpless. Therefore it is probable that more may be learned about the social and behavioral adaptations of human beings from the study of other plains-living species, even if these are not as closely related morphologically, than from the forest-living species, including the great apes.

Even the dog and wolf, which are not at all closely related genetically to man and the other primates, but have evolved as plains-living animals living in social groups, are socially more like man than many closely related primates.

What was the agonistic behavior of primitive man? What are his basic biological capacities? Can we rate him on a scale of aggressiveness along with the nonhuman primates? Southwick (1972) has done an admirable job of comparison of other living primate species and finds that they vary in many dimensions, not only in the frequency of expression of violent acts, but whether these are expressed as ritualized threats or actual injuries.

To place man on such a scale is almost bound to be a largely hypothetical exercise. To do it fairly we should take away the two capacities that make man outstandingly different from other primates: his tools and his capacity for language. To do this, we would have to go back millions of years to a time when the human species must have been genetically considerably different from what it is today, inasmuch as man has since been evolving under the influence of these two factors. Consequently, trying to assess man's primitive nature is bound to be guesswork. Nevertheless, we cannot avoid the question of man's primitive nature, if for no other reason than that people are fascinated by their ancestors and frequently use bad guesses concerning them to excuse bad behavior—bad in the biological sense that it does not promote survival. To explain human behavior by saying that "I did it because my ancestors did it" is at the best an untested hypothesis and at the worst a poor excuse.

Our best lines of evidence are morphological, since, unlike behavior, we have remnants from the remote past in the form of a few actual skeletons, and from them can infer something of primitive man's physique. As more and more skeletons are discovered in Africa, it becomes apparent that our remote ancestors did not at all resemble the hulking caveman of fiction, with enormous jaws, long canines, and bulging biceps. Rather, he was a small, lightly boned creature built for speed and having jaws and teeth much like those of modern man.

With respect to physical weapons he was not equipped to act as a predator or even to defend himself from predators as the modern baboon does with his doglike teeth. Therefore, the argument that man kills man because he is a bloodthirsty carnivore has nothing to support it. Early man's teeth were those of an omnivore—an animal that can eat anything—even as they are today. He could have become an effective hunter only by developing tools, and his original protein diet must have come largely from robbing birds' nests, eating insects and grubs, and catching an occasional immature or helpless mammal.

How much man has evolved since he acquired the tools for hunting, and how much this activity has affected such evolutionary changes, is a matter of conjecture. Some authors have speculated that he became a pack hunter like the wolf, but pack hunting is efficient only when directed against prey animals large enough so that each kill provides food for all, and man's pursuit of large game depends on having efficient hunting tools.

In any case, predation has little to do with agonistic behavior. The most we can conclude is that a primitive man did not have the physical equipment to do much damage in combat with another individual of reasonably equal size. The injuries which modern men can inflict on each other in unarmed combat are the product of centuries of culture.

Another aspect of physique relevant to agonistic behavior is unequal size between the sexes. In our study of the development of dominance–subordination relationships in purebred dogs, we found that differences between breeds are expressed almost entirely in relationships between the sexes (Scott and Fuller, 1965). Males are usually about 20% bigger and heavier than females and even in highly aggressive breeds showed the most extreme dominance in this relationship rather than between members of the same sex. It follows that the differentiation of an agonistic relationship is the result of genetically-determined inequality. Among hybrid animals where variable genes affecting size and aggressiveness were assorting freely, more cases of dominance were expressed than in the parent pure strains, since there were now inequalities within sexes as well as between them.

Generalizing from these findings, we would expect to find that in species in which there is marked divergence of size, males would invariably dominate females, and of course adult animals would dominate the young, which are much smaller. In this respect man is intermediate between such species as the baboon, where males may weigh two or three times as much as females, and gibbons, in which males and females are very nearly equal in size. There appears to be no primate species in which the female is larger than the male, as in the case of a few rodents such as the hamster. In modern man the situation is complicated by the fact that there is so much individual variation within the sexes that there is a great deal of overlap, which would be conducive both to greater expression of agonistic behavior within sexes and to a fairly high frequency of female dominance over males. On physique alone (which is by no means the sole determining factor), modern man should be more quarrelsome than his ancestors.

Another line of evidence comes from the development of agonistic behavior in children. Although their behavior may have undergone considerable evolutionary change in the millions of years since our primitive ancestors lived, they have had relatively little time in which to be affected by culture and training. An immediately obvious characteristic of developing human behavior is that there are relatively few of the elaborate stereotyped patterns of behavior that ethologists like to call instinctive. Rather, the development of human behavior begins with a few very simple patterns that are capable of elaborate modification through experience. In the case of young children of nursery school age, defensive agonistic behavior is expressed sometimes by pushing or holding off the other individual, while attack seems to consist of raising the arms above the head and delivering an open-handed downward blow. Other than this, agonistic behavior is largely vocal, being expressed by crying and shouting. We can conclude that, like other species, man has evolved basic patterns of agonistic behavior that are mild in form.

From this point on, our search for man's primitive biological nature becomes more and more fanciful. If we look to the behavior of present-day primitive peoples, so-called, we find ourselves in the same difficulties that arise from analogies with present-day nonhuman primates. Modern Stone Age tribes have been evolving independently, both biologically and culturally, and do not necessarily represent stages on the way to civilized man. Furthermore, all of these peoples have gone far beyond the stage of using only nonfabricated tools.

On the basis of what little hard evidence there is, we can picture our remote ancestors on the warm, dry African Plains as confined to the semi-arid regions where large predators cannot make a living and making their own by wandering around searching for food within a few miles from shelter among rocks and cliffs. They must have lived in groups, as no single individual could have survived for long by himself, but only in the relatively small groups that the limited resources could support. Thousands of years later, as primitive hunting tools were developed, we can picture them entering more fertile regions and competing with the large predators such as leopards and hyenas.

The late Dr. Leakey once ran a test to see if man could have made a living with only stones and primitive flint knives as tools. He and a friend drove hyenas away from a kill by shouting and waving their arms, and during the 20 min or so before the hyenas came back were able to rush in and hack off a leg and then make their escape. We can therefore picture early man as making a living by means of bluff, arm waving, and shouting and, at the same time, being wary and timid but putting on a show of bravery. If there is any heritage in our biological nature that is injurious under modern conditions, it is far more likely that it consists of inappropriate timidity, fear, and anxiety rather than a boiling reservoir of rage. Further, the one occasion on which desperate attack is adaptive is that of defense against annihilation. In short, instead of the picture of man as a proud and bloodthirsty savage, I am suggesting the alternate picture of the fear-biter.

The amount of useful information that can be gained by these speculations is indeed limited. In contrast, we can apply with considerable profit a general idea based not only on studies of nonhuman primates but on many other social species—that any species will evolve toward a condition in which agonistic behavior serves one or many useful functions in the life of the species. The corollary of this principle is that harmful and destructive agonistic behavior results when social organization breaks down, for any of a number of reasons. This concept has immediate applications to human affairs on any level of organization, whether it be that of family, relationships between small social groups, or between nations. The solution to such disorganization is, of course, to build up and maintain adequate organization, a process which, unlike any dream of a static Utopia, is something that requires constant effort. At the present time it is obvious that one of the ways in which *Homo sapiens* is completely off the scale with respect to other primates is that of population. Human beings now exist in numbers and densities that exceed those of any other primate species that has ever lived. Can we develop organization appropriate for such numbers, or will sheer numbers cause our present society to break down, with the inevitable result of violence?

REFERENCES

Carpenter, C. R. (1964). "Naturalistic Behavior of Nonhuman Primates." Pennsylvania State Univ. Press, University Park, Pennsylvania.

Darling, F. F. (1937). "A Herd of Red Deer." Oxford Univ. Press (Clarendon), London and New York.

Esser, A. H. (1971). "The Use of Space by Animals and Men." Plenum Press, New York.

Guhl, A. M., and Allee, W. C. (1944). Some measurable effects of social organization in flocks of hens. *Physiol. Zool.* **17**, 320–347.

Hall, K. R. L., and DeVore, I. (1965). Baboon ecology. *In* "Primate Behavior, Field Studies of Monkeys and Apes" (I. DeVore, ed.). Holt, New York.

Jay, P. (1965). The common langur of North India. *In* "Primate Behavior, Field Studies of Monkeys and Apes" (I. DeVore, ed.). Holt, New York.

King, J. A. (1955). Social Behavior, Social Organization, and Population Dynamics in a Black-tailed Prairie Dog Town in the Black Hills of South Dakota. Contrib. from the Lab. Vertebrate Biol., No. 67. Univ. of Michigan Press, Ann Arbor, Michigan.

Koford, C. (1963). Ranks of mothers and sons in bands of rhesus monkeys. *Science* **141**, 356–357.

Kummer, H. (1968). "Social Organization of Hamadryas Baboons." Univ. of Chicago Press, Chicago, Illinois.

Kuo, Z. (1967). "The Dynamics of Behavior Development." Random House, New York.

Lincoln, G. A., Youngson, R. W., and Short, R. V. (1970). The social and sexual behaviour of the red deer stag. *J. Reprod. Fert. Suppl.* **11**, 71–103.

Mason, W. (1968). Use of space by callicebus groups. *In* "Primates: Studies in Adaptation and Variability" (P. C. Jay, ed.). Holt, New York.

McBride, G. (1964). A general theory of social organization and behaviour. *Univ. Queensland Papers, Faculty Vet. Sci.* **1**(2), 75–110.

Mykytowycz, R. (1965). Further observations on the territorial function and histology of the submandibular cutaneous (chin) glands in the rabbit *Oryctolagus cuniculus* (*L.*). *Anim. Behav.* **13**, 400–412.

Rabb, G. B., Woolpy, J. H., and Ginsburg, B. E. (1967). Social relations in a group of captive wolves. *Amer. Zool.* **7**, 305–311.

Sade, D. (1967). Determinants of dominance in a group of free-ranging rhesus monkeys. *In* "Social Communication among Primates" (S. A. Altmann, ed.). Univ. Chicago Press, Chicago, Illinois.

Scott, J. P. (1942). Genetic differences in the social behavior of inbred mice. *J. Hered.* **33**, 11–15.

Scott, J. P. (1968). Evolution and the domestication of the dog. *Evol. Biol.* **2**, 243–275.

Scott, J. P. (1973). Biology and the Control of Violence (In press).

Scott, J. P., and Fredericson, E. (1951). The causes of fighting in rats and mice. *Physiol. Zool.* **24**, 273–309.

Scott, J. P., and Fuller, J. L. (1965). "Genetics and the Social Behavior of the Dog." Univ. Chicago Press, Chicago, Illinois.

Southwick, C. H. (1972). "Aggression among Nonhuman Primates." Anthropology Module No. 23, pp. 1–23. Addison-Wesley, Reading, Massachusetts.

Sugiyama, Y. (1967). Social organization of Hanuman langurs. *In* "Social Communication among Primates" (S. Altmann, ed.), pp. 221–237. Univ. Chicago Press, Chicago, Illinois.

Teleki, G. (1973). The omnivorous chimpanzees. *Sci. Amer.* **228**, 33–42.

Vessey, S. H. (1971). Free-ranging rhesus monkeys: behavioural effects of removal, separation and reintroduction of group members. *Behaviour* **40**, 216–227.

Yoshiba, K. (1968). Local and intertroop variability in ecology and social behavior of common Indian langurs. *In* "Primates, Studies in Adaptation and Variability" (P. Jay, ed.), pp. 217–243. Holt, New York.

Zuckerman, S. (1932). "The Social Life of Monkeys and Apes." Harcourt, New York.

THE MYTH OF THE AGGRESSION-FREE HUNTER AND GATHERER SOCIETY

IRENÄUS EIBL-EIBESFELDT

Max-Planck-Institut für Verhaltensphysiologie

INTRODUCTION

In a number of recent publications hunters and gatherers have been depicted as lacking aggression and territoriality (Helmuth, 1967; Sahlins, 1960; Lee, 1968; Woodburn, 1968; Schmidbauer, 1971). And since hunters and gatherers represent a stage in which man has lived through most of his history, it has been alleged that man by nature is not territorial or aggressive at all, but that these traits developed as cultural adaptations hand-in-hand with agriculture, and found their climax in the competitive achievement societies of modern industrialized civilizations.

If hunters and gatherers were indeed nonaggressive, nonterritorial people, the theory about the primarily peaceful nature of man would have some probability. The statement about the peaceful nature of hunters and gatherers does not stand critical examination, however. The allegedly peaceful Eskimos actually perform a rich variety of aggressive acts, although most tribes settle their disputes in a fashion that does not result in bloodshed. The tribes of Siberia, Alaska, Baffinland, and northwestern Greenland settle their disputes by wrestling; occasionally one gets killed. The Eskimos in central Greenland slap one another's faces; in western and eastern Greenland, song duels are a favorite means of settling disputes. The Kwakiutl Indians, quoted by Schjelderup (1963) as people lacking the "instinct" to fight, are well known for potlach feasts, during which the chiefs compete fiercely to outdo the guests by destroying valuables. Their songs are aggressive and they indeed call the performance a fight:

"Furthermore, such is my pride," announced one host on such an occasion, "that I will kill in this fire my copper Dandalayu, which is groaning in my house. You all know how much I paid for it. I bought it for 4000 blankets. Now I will break it in order to vanquish my rival. I will make my house a fighting place for you, my tribe . . . [Benedict, 1934, p. 195]. It is indeed difficult to see how one could ever come to the conclusion that the Kwakiutl are particularly nonaggressive. Schmidbauer, an eloquent promotor of the neo-Rousseauan myth, argued recently (1972) that, after all, the Kwakiutl were only destroying their property. Yes indeed—including their slaves, as every one familiar with the literature knows. One of the ceremonial killing clubs is on exhibit in a Washington museum.

The myth of the aggressionless society is not new—it actually dates back to Rousseau—and it has popped up again and again. Nansen, by creating the myth of the aggression-free Eskimos, wanted to create a favorable picture of his beloved people. König pointed to this fact as early as 1925:

> To focus on the latter [the alleged peacefulness], the only source that this opinion is based on is Nansen. However, he learned very little about the Eskimos in their natural state, and his moral point of view is tendentious in order to arouse sympathy—as he stated in his "Eskimo Life" [p. 294].

Although this correction was published in 1925, the myth is not dead. On the contrary, in blaming our industrial achievement society for all aggression, the members of hunting and gathering societies are depicted as the noble, noncompetitive, and therefore nonaggressive savage. The argument runs like this:

> If human aggressiveness is a phylogenetically determined disposition, rooted in man's inherited genetic disposition, then it must be especially evident in those cultures which characterized the stage occupying 99% of the human evolutionary timetable. The stage here in question is designated as the Stone Age, characterized by the patterns of the roaming hunters and gatherers. If such groups of hunters and gatherers had existed for approximately one million years, one can assume that the psychological and physical properties of (present) man became deeply ingrained at this time. The mere 10,000 years in which man became a farmer could have hardly changed the genetic biological basis of humans. If, therefore, there exists any reason to assume that the life patterns of hunters and gatherers

> were stamped into our heredity, then the theory of an inborn, aggressive drive would have some support. If it could be proved to the contrary that hunters and gatherers in general are more peaceful than the later, more highly developed cultural types, then the hypothesis of an aggressive drive would be totally unbelievable [Schmidbauer, 1971b, p. 41].

After this introduction Schmidbauer concludes, on the basis of the data of cultural anthropology, that the *majority* of hunters and gatherers are notably unaggressive and most notably that they would not defend territories. The statement is not at all based, however, on a thorough count of the known hunters and gatherers and an examination of their aggressiveness. It is based on a selection of some allegedly nonaggressive hunters and gatherers. In addition to the Eskimos, already mentioned, the Hadza and the Bushmen of South Africa are quoted as examples. In doing so, Schmidbauer primarily relied on the articles by Lee and DeVore (1968). Woodburn (1968) claims that the Hadza defend no territories, show no aggression, and also live in open groups. Woodburn, however, studied this group in 1958 and the following years—when the group which had formerly occupied about 5000 km^2 of land was pressed into an area of 2000 km^2. As a result, the group was certainly uprooted, and consequent changes of their original behavior and social structure are to be expected. Indeed, Kohl-Larsen (1958), who visited the Hadza between 1934 and 1936 and later between 1937 and 1939, recorded with respect to this a very noteworthy story told to him by a Hadza friend. The protocol of August 7, 1938 reads as follows:

> In the old times the Hadzapi fought amongst one another. One tribe which was located in Mangola went to another tribe over in Lubiro. When they arrived there, a man was picked out. He went to the Lubiro band and said: "We have come to you today to fight you!" Now someone from the Lubiro tribe said: "Yes, if you have come here to wage a war with us, we are satisfied." The people in Lubiro come together and confer among each other. They pick a man who is to fight with the man from Mangola. He is chosen: each of the men gets two sticks. With them they beat each other. If no one is victor over the other, both tribes begin to battle each other with arrows and spears. As they are battling each other, an old lady steps out of one crowd and out of the other crowd an old man. Both of them place themselves in the middle of the two fighting tribes and say: "Sit down and rest a bit!" Having rested a bit, they begin hitting each other again. They fight each other for a long time. When one tribe is beaten, they run away. The others, the victors, follow them a stretch, then go back again to their camp and sleep. The next day the tribe that won goes to the losers. They stay overnight with them. In the morning all the strong men and youths go hunting. When they kill a few animals, they take its meat, which is fatty, and sit down and eat it. Only the men may

> be around then. If a woman goes to the men, she can be killed. However, if the woman has a child which she carries on her back, she pinches him so he cries. When the men hear a crying child they cannot hit the woman. When they (all) eat the meat the friendship between them is reestablished. Having lived many days like this and seeing that they are living for no good purpose (no work and nothing to do), they look for another tribe to fight with. They hit each other so badly that often a few people are killed. Whoever loses goes then into the big crowd of the victor [p. 35].

The protocol is not only noteworthy because of its numerous interesting details. It also demonstrates that it is a grossly false generalization to conclude, on the basis of the current behavior of the Hadzas, that they, or hunters and gatherers in general, were originally peaceful peoples.

THE CONTROVERSY ON BUSHMEN TERRITORIALITY

In recent publications (Sahlins, 1960; Lee, 1968; Schmidbauer, 1972) the Bushmen have been referred to as living in open nonexclusive bands and as not expressing territoriality in the sense that it is usually defined, namely as intolerance confined to space. For example, this is explicitly stated by Lee (1968) concerning the !Kung:

> The camp is an open aggregate of persons which changes in size and composition from day to day. Therefore, I have avoided the term "band" in describing the !Kung-Bushmen living groups. Each waterhole has a hinterland lying within a six mile radius which is regularly exploited for vegetables and animal food. These areas are not territories in a zoological sense, since they are not defended against outsiders [p. 31].

Such statements are puzzling in view of the overwhelming evidence of territoriality among the same Bushmen, as published by older authors. Thus Passarge (1907) describes the !Kung-Bushmen as belligerent and he emphasizes that not only the bands, but every family owns their particular collecting grounds. He writes in this context:

> The classification of the Bushmen in families has been known for a long time . . . even though I have found no report indicating that property and ground are distributed according to law. However, this is a fact of enormous importance. Only when considering this fact can a clear view of the social organization in the Bushmen be achieved [p. 31, translated from the German].

Zastrow and Vedder (1930) report that the Bushmen are not allowed to hunt or collect food in the land of another band:

> Where the Bushman's ground has not yet been divided into farms, but kinship-areas follow kinship-areas, every Bushman knows that he must

not hunt or collect "veld cos." Being caught at poaching is grounds for forfeiting his life. This does not mean that he necessarily will be killed; vendettas may prevent this . . . [p. 425, translated from the German].

Lebzelter (1934) reports of the great distrust the !Kung show when meeting members of foreign bands:

> Every armed man is considered as an enemy. The Bushman must not enter other tribal territory, except unarmed. When a Bushman is sent as a messenger to another farm, the mutual hostility will not permit him to leave the path that is recognized as some kind of a neutral zone, even at the boundaries of the farm zone. When two Bushmen approach each other, their weapons will be put down within range of sight [p. 21, translated from the German].

Similar reports can be found by Brownlee (1943), Vedder (1952), and Wilhelm (1953). Marshall (1965) also reports on territoriality:

> The !Kung say that one cannot eat the ground itself, so it does not matter to whom it belongs. It is these patches of veldkos that are clearly and jealously owned and the territories are shaped in a general way around these patches. . . . The strange concept of ownership of veldkos by the band operates almost like a taboo. No external force is established to prevent one band from encroaching in another's veldkos or to prevent individuals from raiding veldkos patches to which they have no right. This is just not done [p. 248].

Tobias (1964) emphasizes that the Bushmen strictly move within their territory:

> Territoriality applies among bands of the same tribe and between different tribes. Tribal bounds are sometimes reinforced by social attitudes, such as the traditional enmity between the Auen and Naron. Under special conditions such as an abundance of food these bounds and the accompanying enmity are forgotten [p. 206 of the offprint in David].

Silberbauer (1973) found territoriality to be typical for the G/wi. After referring to the ethological definition of Willis of a territory as "a space in which one animal or group generally dominates others, which become dominant elsewhere" Silberbauer writes:

> Willis' description aptly describes the relationships between G/wi bands with regard to their territories; a visiting band, or a single visitor submits to the dominance of the host band by either waiting for an invitation or by seeking permission to enter and occupy the territory [p. 117].

He furthermore emphasizes that visitors who cross a territory en route to another destination and those who are headed for the occupying band both call at the band encampment and ask permission "to stay in your country and drink your water." This wording, by the way, is a standing phrase, since they employ it even though the waterholes are usually dry at the times the visits are made. Continuing (p. 212), Silberbauer writes: "In each band there are individuals known as !u:ma (owner) or !u:sa in the case of a woman. In the vernacular account the !u:ma is the original founder of a band or his male or female descendant."

Strangely enough he still refers to the band as being open, referring to the fact that visitors, after asking for permission, may come and stay for a while and that occasional adoption into a band occurs; although he emphasizes that the change of group composition is actually a rare event. It seems that Silberbauer misunderstood the term "closed groups" as used by ethologists, as being absolutely closed. This is not the common definition, however. A group is closed—in contrast to open—if members of other groups are not freely allowed to move in, except under certain circumstances. In a number of monkey species (macaques) it has been repeatedly described how peripheral males seek acceptance in other bands, by slowly associating themselves first as peripheral males with the other group, appeasing the aggression of the others against the stranger by a number of rituals. But adoption can occur and a closed group is never absolutely closed, either in nonhumans or in man.

In view of all these reports, we have to assume that the group studies by Lee no longer show the typical pattern, probably due to acculturation. In addition, they may not have been aware of the larger unit in the !Ko-Bushmen.

TERRITORIALITY AND AGGRESSION IN THE !KO-BUSHMEN

The !Ko-Bushmen belong to the central Kalahari. Bushmen live in the area south of Ghanzi (see map on p. 441). Many of the bands are still living as hunters and gatherers. Heinz (1972) gives a detailed report on this group, and I had the opportunity to repeatedly visit one of his groups in 1970, 1971, 1972, 1973 and 1974 and live with these people, documenting unstaged social interactions by film (see Eibl-Eibesfeldt 1971, 1972). These studies not only shed light on Bushmen aggression and aggression control, but also contribute to the more general problem of phylogenetic adaptations in human aggressive behavior.

Territoriality

This subject has been dealt with extensively by Heinz (1966, 1972). I will therefore review his findings. My observations are in complete

agreement. Heinz distinguishes three levels of social organization: (*1*) the family and extended family; (*2*) the band; (*3*) the band nexus.

All these units have a definite pattern of bonding *and* spacing. The sitting position of family members around the fire is less formalized than in the !Kung (Marshall, 1960). The wife can sit anywhere at the house side of the fire. But the proper place according to Heinz is the right side of her husband. Parents settle at least 12 m away from their married children and the entrance is always arranged in a way as not to allow them to watch their married children sleeping.

> Though all band territory is accessible to everyone in the band, nevertheless the family's area of activity is recognized. When hunting alone, a man is expected to hunt on the side of the village on which his house is built, a

> rule applying to women collecting veldkos (wild food) or firewood as well. Only collective activity—which is most prevalent—breaks this rule [Heinz, 1972, p. 407].

Heinz further reports that the bands periodically split into family groups, each family then moving to a family place which is respected by the others.

While families have no direct territorial claims, the band definitely has. A band considers a piece of land as its territory. The control over it is exercised by the "headman" (Heinz, 1966, 1972) on behalf of the band, a group of old men and women acting as consultants. The band hunts and collects wood and veldkos within its territory. In cases of emergency they may ask for permission to hunt and collect in another band's territory. To members of the same nexus permission is normally readily granted.

I consider the nexus system as the most important discovery of Heinz, since it may explain at least to a certain extent how some of the controversial statements on Bushmen arose. The nexus constitutes a group of bands. Members refer to themselves as "our people." The people are bonded by friendship and kinship ties, by ritual bonds[1] and they share slight peculiarities in their dialect. There is extensive intermarriage within the nexus. The band nexus is a territorial group which is more exclusive than the band territories. The nexus territory is demarcated from those of another nexus by a strip of no-man's land, which is generally avoided by the members of both sides. It would not occur to a !Ko-Bushman to seek permission to hunt in the land of another nexus.

Access to territory is acquired by birth, admission to a band, or by marriage. If the parents come from different bands, dual band membership is the result. When marrying, the groom resides for a certain period of time with the bride's band and has access to this territory. Thereafter the couple moves to the man's band, where the bride receives access to the band's territory. "It is this phenomenon which gives parents rights in each other's land, and which is transmitted to the children and which might in certain cases blur band territoriality [Heinz, 1972, p. 409]."

Patterns of Aggressive Behavior

Being interested in the universals of human behavior, I spent many hours studying and filming unstaged social interactions. In total, approxi-

[1] For example, by coming together for trance dances.

mately 12,000 m of 16-mm film were taken on the !Ko-Bushmen, about 2000 m on the !Kung, and 1000 m on the G/wi. On the basis of this documentation it can be proved that: (*1*) Aggressive behavior patterns are fairly frequently observed. (*2*) Many of the patterns are identical in form with those observed by people of other cultures in the same context.

Aggressive acts in this context are defined as all those which lead to spacing or to the establishment of a dominance–subordination relationship. Whether the person involved hurts another person physically or not does not enter the definition, nor do we incorporate the "intent." If one were to speak of aggression only when damage results, one would have to omit all the patterns of aggressive threat and other ritualized patterns of aggression. I see no reason to do this.

SIBLING RIVALRY

Intersibling rivalry can be observed even at a very early age. I documented a most dramatic example during my stay with the !Kung. The parties involved were two brothers, the younger about 1 year, the older perhaps 5 years old. The older was evidently seeking bodily contact with the mother but her attention was concentrated on her smaller son, whom she nursed exclusively. The older son tried fairly frequently to harm his younger brother. He attempted to scratch and beat him on several occasions, tried to poke him with sticks (Fig. 1A–C), and the mother had to be on the alert to keep the brothers from fighting. The older brother also tried to interfere with his younger brother's play, teasing him, such as by taking his toys and throwing them away. But the younger brother also was aggressive. While his brother was drinking he was observed to give him a well-aimed kick. The older brother certainly was suffering quite a lot, and in his frustrated efforts to achieve contact with the mother, he often cried. His mother did not respond very eagerly to his efforts to seek contact. She allowed contact when her small son was away playing, but she did not invite it; rather she seemed fed up with him. Less dramatic cases of sibling rivalry were observed in the !Ko-Bushmen. But here too sibling rivalry occurs. The bushmen refer to it as a normal inevitable occurrence that has to be accepted. The rivalry is not restricted to a particular sex. Girls rival with their newborn brothers as well as sisters, and vice versa. Nonetheless, a strong and lasting friendly bond develops later on. In view of the fact that hunters and gatherers are so often reported as growing up without any frustrations, these observations are of importance in correcting such ideas. In general, babies show aggressive behavior at a very early age. A 10-month-old boy used to attack other babies, e.g., by pushing them over and by scratching them (Fig. 2; for further examples see Eibl-Eibesfeldt, 1972).

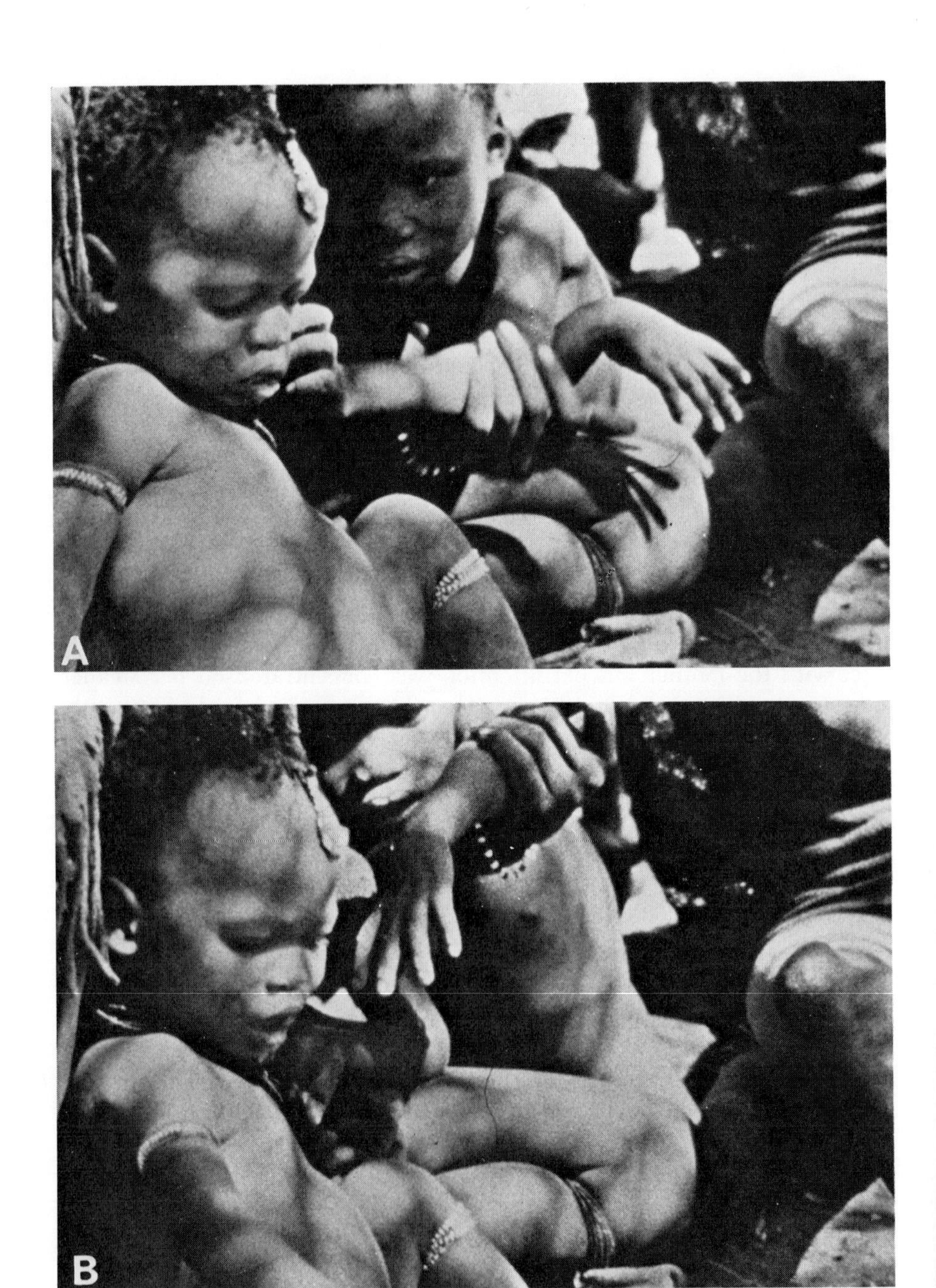

Fig. 1 Legend and part C on facing page.

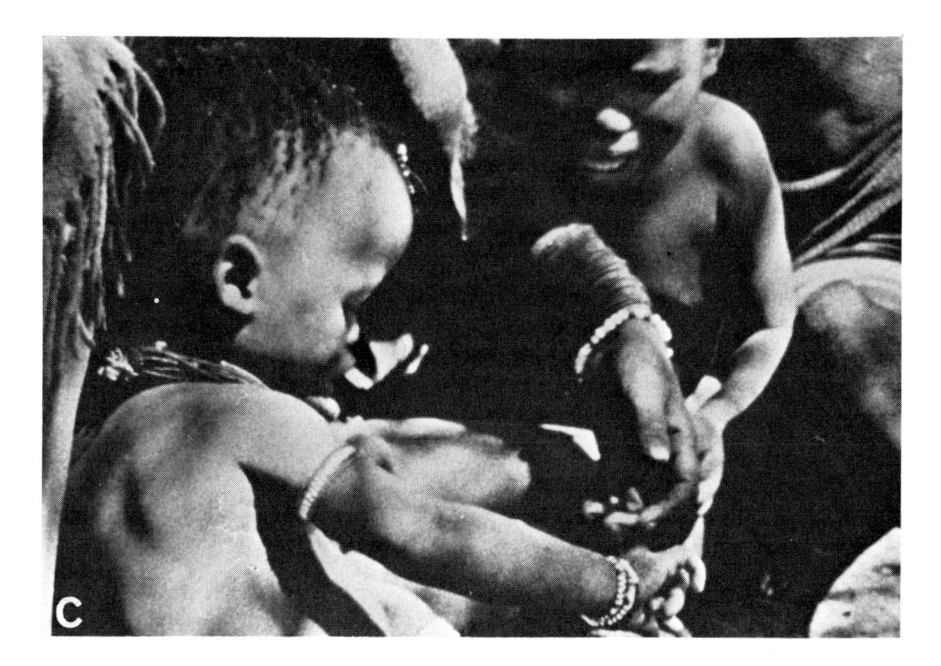

Fig. 1 Sibling rivalry in the !Kung-Bushmen. A: The older brother (in the background) tries to scratch the younger. B: The mother interferes, pulling the attacker's hand away and shielding off further attacks with her hand. C: The older brother cries in evident frustration. (From a 16-mm film, HF 41, by the author.)

AGGRESSION IN PLAYING GROUPS OF CHILDREN

Playful Aggression. Patterns that normally lead to spacing subordination and the cutoff of contact are often performed during play. Although the motor patterns in such a case are often difficult to distinguish from real aggressive acts, additional signals such as laughing and smiling allow us to recognize that the aggressive interaction is actually play. And so does the fact that the roles of attacker and defender or pursuer and pursued change freely. I do not want to deal here with play aggression, since this will be dealt with elsewhere by one of my students.

Serious Aggressive Interactions. Serious aggressive interactions are fairly common within playing groups of children. The acts of aggression are slapping with the palm, beating with a stick or other object, throwing objects toward a child, throwing sand, punching with the fist, kicking with the foot, pushing the other with one or both hands, ramming a child with the shoulder, pushing with the hip, pinching, biting, pulling the hair, scratching, wrestling, spitting, stealing (Fig. 3). All these patterns have been described in detail in my monograph (Eibl-Eibesfeldt,

Fig. 2 Attack of a boy, approximately 11 months old, against a baby girl who tried to take an object from his hand. The boy pushes the girl over and scratches her at the same time. (From a 16-mm film, HF 4, by the author.)

1972). There are some remarkable patterns of threat and submission which I want to describe. When threatening, the child frowns, clenches his teeth, often exposing them at the same time, and stares at the opponent. Sometimes a hand with or without an implement is raised. Both opponents may engage in such a threat–stare duel. Such display can lead to the submission of one. Then the loser lowers his head, tilts it slightly, turns sideways, and pouts. This behavior strongly inhibits further aggression—but not only that. One can observe that the aggressor quite often tries to comfort his victim, seeking friendly contact. The victim normally responds to this only after a certain lapse of time. At first he responds to the effort by clear cutoff behavior, e.g., by turning away. Another aggression-inhibiting behavior is crying. Both submissive behaviors are to be found in all cultures I have visited so far (Figs. 4, 5A and B).

Aggressive interactions are fairly common in groups of playing children. Within 191 min I counted, in a group of seven girls and two boys, 166 aggressive and defensive acts: slapping, punching with the fists, or beating with an object, 96 times; kicking with the foot, 23 times; throw-

Fig. 3 Girl about to throw a stick toward a boy. (From a 16-mm film, HF 2, by the author.)

ing sand, 8 times; and a number of other acts. Ten times a child cried loudly during this observation period, which indicates clearly that many of the aggressive acts lead to serious conflicts. Only about one third of the interactions were clearly playful aggression by the criteria mentioned above. Not all play sessions are disturbed by so many aggressive acts. Another time 12 children played for 88 min together, and only 7 aggressive acts were counted.

There were a number of typical situations in which aggression did occur.

Quarrelling about the possession of objects. Children like to play ball with melons, and the "ball" is often the object of a quarrel. In particular, boys try to rob others of the ball. Pursuits and fights for, and in defense of, the objects develop. The loser often shows all the signs of serious anger, and sometimes even cries. Children rob each other also of other objects, sometimes even food, which is considered a more serious offense.[2] Robbing another child of an object is often a means to tease a playmate and entice him into a fight.

[2] Refusing to share food is an offense which invites punishing attack. Once, a girl gave a morsel to an 8-month-old baby, expecting that it would return part of it. The baby crawled away with all of it and immediately was beaten by the girl.

Fig. 4 Boy threat–staring at a girl. She is pouting. (From a 16-mm film, HF 2 and HF 3, by the author.)

Punishment for offenses. Older children often interfere with the quarrels of their younger playmates, such as by punishing the attacker. In one play group the oldest prepuberal girl regularly acted in such a way that she practically controlled the play activity. She initiated many of the play activities and demanded obedience. If a child broke a rule of the game, or sometimes even if he behaved in an unskilled way, she attacked. We called her the "Spielleiterin" for this reason.

Demonstrative aggression. The "Spielleiterin" sometimes attacked another child without evident reason. In the morning, when she came to the already assembled group of the other children, she kicked the melon out of the hand of a child, or boxed another, and indeed all children showed signs of respect to her by approaching and showing their melons, even at the danger of being robbed of this prized object. It seems to me that the function of this aggression is to achieve and keep rank and respect, which is a prerequisite for functioning as a mediator and soothing quarrels between the other playmates.

Unprovoked spontaneous attacks. Children often attacked others without apparent reason. The patterns were of particular interest, since it was evident that a child is inhibited against starting a full attack upon

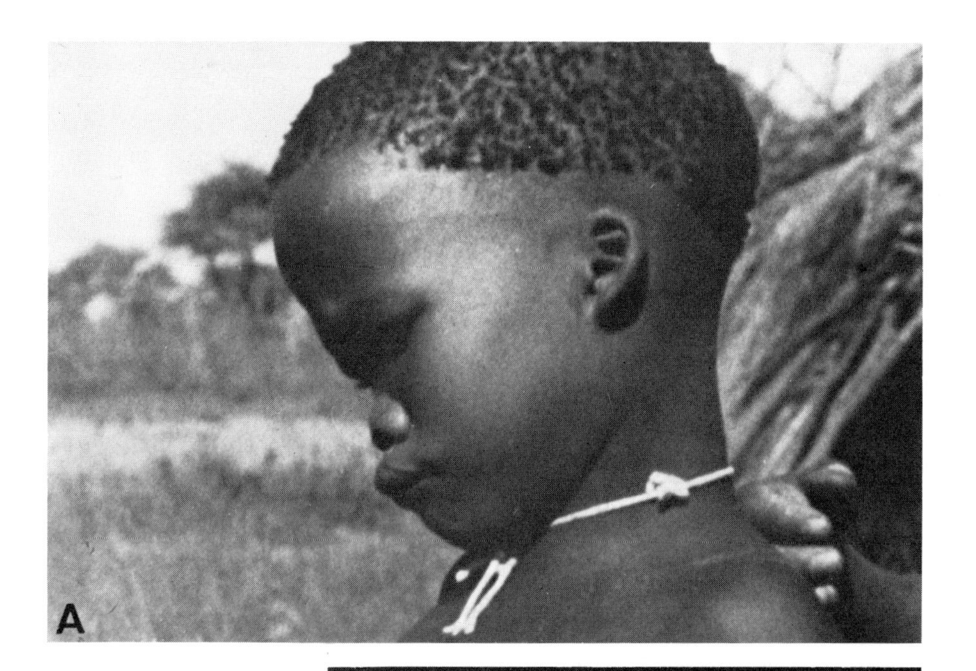

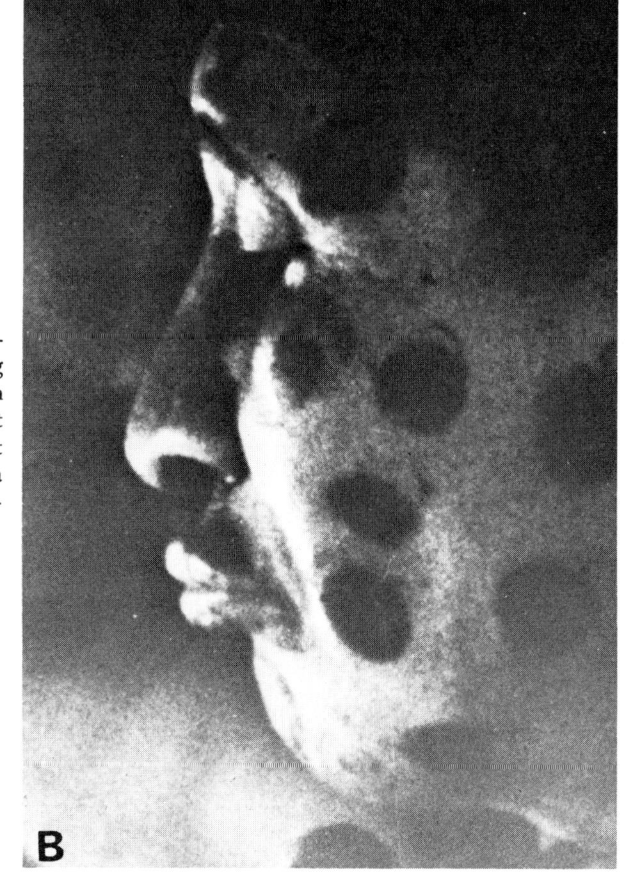

Fig. 5 A: Pouting Bushmen girl. B: Pouting Waika Indian. The man wanted to join our boat for a ride. His request was refused. (From a 16-mm film by the author.)

an innocent bystander. If a child seeks a fight he incites it by a pattern of provocation. Teasing, by offering something and withdrawing it when the child tries to grasp it, is a common way. A slight slap, a kick, showing the tongue, or robbing an object and throwing it away, these are all typical small insults that provoke the other. And once he attacks, a heavy retaliation follows. It looks as if the provoking child seeks an excuse to attack by the provocation of aggression which allows him then to retaliate. Unprovoked aggression often serves to test out the social "Handlungsspielraum" as it does in our culture. Through social encounters, the child expects an answer as to what is allowed and what no longer is. By experimenting it learns about freedom and limits for behaving. If not finally meeting resistance, the child's social explorative aggression tends to escalate, as was recently discussed by Hassenstein (1973) in connection with a critique of the so-called "antiauthoritarian" education.

Retaliation. Retaliation is the regular answer to an attack. This retaliation may not necessarily follow immediately. I observed once that an attacked boy went into the bush, broke and adapted a branch from a small tree and returned after about 10 min to retaliate.

Escalation of play. Rough-and-tumble play occasionally develops into a fight. If one accidentally or by intent hits another a little more strongly than expected, he invites retaliation and a fight may start.

QUARRELS BETWEEN CHILDREN AND ADULTS

Children sometimes respond violently when parents admonish them. Once a girl was scolded by her father for carelessly kicking against a pot, causing part of its contents to spill. Upon this admonishment the girl grasped the pot and threw it against the floor, spilling all its contents. The father did not say anything further. On another occasion a girl kicked her mother with her foot, after she had been scolded for her greedy begging for food. Once a girl robbed a small boy of a piece of meat. The boy cried out and his father came, took the meat back to the boy, slapping the girl slightly on the head. The girl retaliated by throwing sand at the man, who in turn slapped her a second time. This did not subdue her at all, for she answered by throwing sand after him again when he left. A final example: A man hid a little object to prevent a baby from swallowing it. The baby and another boy, approximately 6 years of age, then searched for the object. They were searching the man with much laughter when finally the man and the little boy started in a playful fashion to exchange slappings. This escalated, however, and when the man slapped harder, the boy started to cry out. He ran away and returned with large bones and horns, threatening to throw them at the man. At this moment the boy's father intervened and soothed his son's excitement.

AGGRESSION IN THE ADULT

Within a band, aggressive behavior of the adult is under good control. The cultural ideal of the Bushmen is indeed to live a peaceful life. Should tension arise between two families of one band, the problem is solved by splitting up. One family moves for a while to another place and this relieves the tension (Heinz, 1966, 1967). I myself had the opportunity to observe mostly only playful aggression between adults. Young men wrestle; married couples engage in teasing and mock fighting. An element of aggression is certainly to be observed in the numerous games (see Sbrzesny, in preparation). Heinz described males as teasing and maltreating "underdogs," that is, low-ranking members of the group. Verbal threats of murder are often uttered (I will kill you with my medicine). Males, after insulting each other, occasionally use their weapons and Heinz reports a case of manslaughter. He also describes temper fits, which occasionally overcome the Bushmen: "An angry Bushman finally settles down with a face that shows an unbelievable degree of anger. It takes very little for this anger to cause a wrestling and punching encounter with sticks and knobkerries. If the reasons are serious, the fight will deteriorate into one in which knives and spears are used . . . [Heinz, 1967, p. 6]." Married couples fight for reasons of jealousy, and so do women. Adultery may lead to bloodshed among the males.

Recently a clash occurred between the band that I regularly visit, and another band of the same nexus. It started with a man approaching a married woman whom he had courted many years ago. The brother of the woman interferred. The woman and her brother were from the other band and so it happened that others took sides according to band membership. A brawl started, which resulted in one open wound on the head, a seriously bitten hand, a broken rib, and numerous bruises and smaller wounds from bites—all among the men. The women only verbally joined in, their main concern being to separate the fighting men.

SORCERY

During my recent visit to the !Ko I was told by them that a member of the band who had been angered by a Bantu had thrown bones to cause his death (which failed to occur). The technique of bone-throwing is the same used for an oracle. The !Kung were well known for their small magical bows and arrows, which they carried hidden on their bodies. According to Vedder (1937) they used to shoot the tiny arrows in the direction of their enemies' homes, at the same time cursing them.

TEASING AND MOCKING

By teasing and mocking, group homogeneity is enforced and an outlet for aggression provided. Heinz describes the existing joking relationships in detail and I shall not pursue the subject here. I want, however, to discuss some patterns of mocking. These patterns are released by behaviors of individuals that deviate from the group's norm. By this mocking the outsiders are under pressure to conform. Mocking certainly has an educational function. Mocking is done by imitating the patterns which provoked the hostility of the group and thus ridiculing it. I furthermore found that female genital display, showing the tongue, and showing the rear are used during mocking. Several ways of female genital display can be observed. Children repeatedly mocked me, while I was filming with my mirror lens. They imitated my behavior, and if I did not pay attention, they approached dancing and singing and at short distance lifted their genital aprons (Fig. 6). Often the girls posed with

Fig. 6 Three girls mocking the author by lifting the genital apron. The author was filming with a mirror lens. When he looked up, the girls stopped their mocking. (From a 16-mm film, HF 1, by the author.)

their hands on the hip, posturing in a characteristic way with the legs (Fig. 7). In addition to this frontal genital display, a genital display from the rear was observed. The mocking girl turned the rear toward the person and bent deeply (Fig. 8). The pubic region was exposed in a

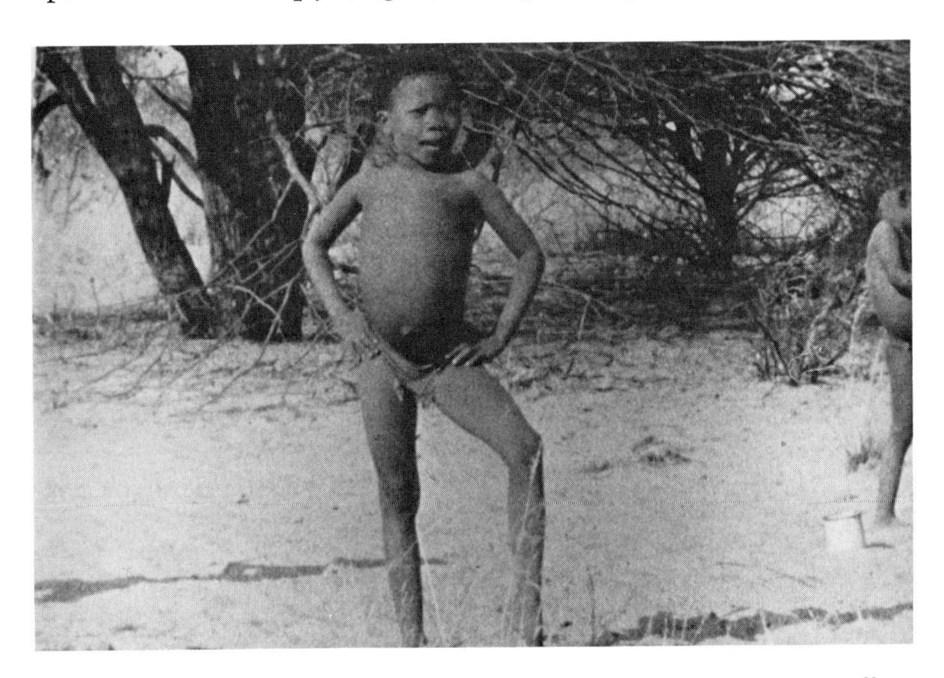

Fig. 7 Posturing of a mocking girl. Situation as in Fig. 6. (From a 16-mm film, HF 1, by the author.)

very conspicuous way, particularly since this race as a characteristic feature shows a strong lordosis. It is noteworthy that the enlarged labia minora contribute to the marked visibility of the vulva. I assume that this posturing is indeed homologous to primate sexual presenting: In fact the Bushmen copulate from the rear while lying on their sides.

The widespread use of patterns of sexual presentation (see Eibl-Eibesfeldt, 1972) in mocking displays may be explained by the fact that sexual inciting is taboo, except in certain circumstances. To use it outside its original context is therefore a demonstration of disrespect. Not to be confused with the pattern of genital presentation are patterns of showing the rear. This pattern is more often used as an aggressive threat. I observed that girls mocked boys who were teasing them in this way: They held sand between their buttocks, then approached the boys, turned in front of them, and bending let the sand go as if defecating.

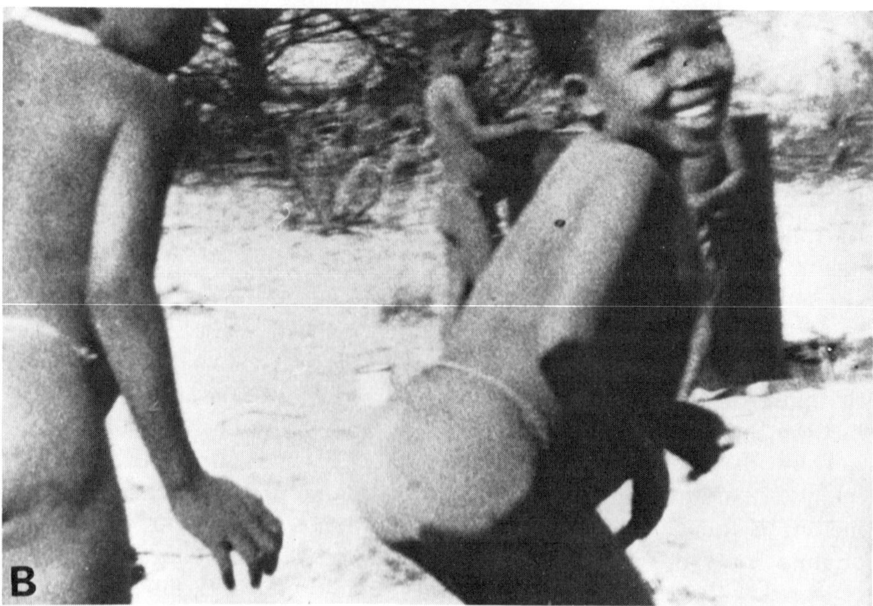

Fig. 8 A, B: Genital presenting by a girl. Situation as in Fig. 6. (From a 16-mm film, HF 1, by the author.)

Sometimes wind is passed. Both clearly indicate that this pattern is a ritualized form of defecation.

Tongue showing is another pattern used in mocking. It is difficult to interpret, since it occurs in various forms. In derogatory tongue showing the tongue is stuck out and turned down, as if the person were about to vomit. Spitting can even occur in this context. There exists, furthermore, a sexual tongue flicking, and to complicate matters further, there is also a friendly tongue showing derived from licking, which I will not discuss here. I want to mention, however, that tongue flicking is often used in a teasing fashion.

DISCUSSION

The much-quoted cross-cultural "evidence" for man's primarily peaceful and nonaggressive nature proves on examination to be a weak point in the environmentalists' argument. Among others, the oft-mentioned Bushmen certainly do not lack territories nor do they fail to be aggressive. Even within the band, numerous aggressive acts can be observed. The situations and most of the movement patterns are identical with the aggressive patterns in other cultures. They are universals and although the similarity of many of the movements can be explained on a functional basis (hitting, kicking, etc.) and thus be acquired independently in a similar way, this certainly does not hold for the more elaborate aggressive displays, such as the facial expressions of threat, the threat stare, the patterns of submission (pouting, head lowering, gaze avoidance, crying) or some of the patterns of mocking (genital presenting, tongue showing). We can assume that many of these patterns are phylogenetic adaptations ritualized in the service of spacing and aggression control.

It is true, however, that Bushmen are not belligerent. Their cultural ideal is peaceful coexistence, and they achieve this by avoiding conflict, e.g., by splitting up and by emphasizing and encouraging the numerous patterns of bonding. This way they achieve a peaceful life. Most of the socialization of aggression takes place within the playing groups of children. In interaction with others, children learn to control their aggression. It should be mentioned that many of the patterns of bonding and the urge to bond are inborn in man. Man is, so to speak, by nature a bonding as well as a spacing creature.

Culture puts emphasis on one or the other and may unbalance man for better or worse. The Bushmen certainly belong to those people whose culture shapes them according to a peaceful ideal. What is striking when observing the Bushmen is not their lack of aggression, but their efficient way of coping with it. Friendly bonding behavior predominates in the interaction of adults, and these people spend many hours a day grooming each other, chatting, sharing a pipe, and playing with their children, to mention but a few of these social interactions. Since wood and food gathering consume only 2–3 hr a day for the women and since men hunt only occasionally, the Bushmen have plenty of time for intimate social interactions. One could say that these people have more time at their disposal to be "human" in a friendly way, while we are losing this capacity to a greater and greater extent.

REFERENCES

Benedict, R. (1934). "Patterns of Culture." Houghton Mifflin, Boston.

Brownlee, F. (1943). The social organization of the Kung (!Un) Bushmen of the North-western Kalahari. *Africa* **14,** 124–129.

Eibl-Eibesfeldt, I. (1970). "Ethology, Biology of Behavior." Holt, New York.

Eibl-Eibesfeldt, I. (1972). Die !Ko-Buschmanngesellschaft: Aggressionskontrolle und Gruppenbindung. "Monographien zur Humanethologie," Vol. 1. Piper, München.

Hassenstein, B. (1973). "Verhaltensbiologie des Kindes." Piper Verlag, München.

Heinz, H. J. (1966). The social organization of the !Ko Bushmen. Masters Thesis, Dept. Anthropol., Univ. of South Africa, Johannesburg.

Heinz, H. J. (1967). Conflicts, Tensions and Release of Tensions in a Bushmen Society. Inst. for the Study of Man in Africa, Isma Papers No. 23.

Heinz, H. J. (1972). Territoriality among the Bushmen in general and the !Ko in particular. *Anthropos* **67,** 405–416.

Helmuth, H. (1967). Zum Verhalten des Menschen: die Aggression. *Z. Ethnol.* **92,** 2, 265–273.

König, H. (1925). Der Rechtsbruch und sein Ausgleich bei den Eskimo. *Anthropos* **20,** 276–315.

Kohl-Larsen, L. (1958). Wildbeuter in Ostafrika. Die Tindiga, ein Jäger- und Sammlervolk. Berlin (Reimer).

Lebzelter, V. (1934). "Eingeborenenkulturen von Süd- und Südwestafrika." Hiersemann, Leipzig.

Lee, R. B. (1968). What hunters do for a living. *In* "Man the Hunter." (R. B. Lee and I. DeVore, eds.), pp. 30–48. Aldine. Chicago, Illinois.

Lee, R. B., and DeVore, I. (1968). "Man the Hunter." Aldine, Chicago, Illinois.

Marshall, L. (1960). !Kung Bushmen bands. *Africa* **30** (4), 325–355.

Marshall, L. (1965). The !Kung Bushmen of the Kalahari Desert. *In* "Peoples of Africa" (J. L. Gibs, ed.) pp. 241–278. Holt, New York.

Montagu, M.F.A. (1968). "Man and Aggression." Oxford Univ. Press, New York and London.

Passarge, S. (1907). "Die Buschmänner der Kalahari." Reimer, Berlin.

Sahlins, M. D. (1960). The origin of society. *Sci. Amer.* **204,** 76–87.

Schjelderup, H. (1963). "Einführung in die Psychologie." Huber, Bern.

Schmidbauer, W. (1971a). Methodenprobleme der Human-Ethologie. *Stud. Gen.* **24,** 462–522.

Schmidbauer, W. (1971b). Zur Anthropologie der Aggression. *Dyn. Psychiat.* **4,** 36–50.

Schmidbauer, W. (1972). "Die sogenannte Aggression." Hoffman and Campe, Hamburg.
Silberbauer, G. B. (1973). Socio-Ecology of the G/wi Bushmen. Theses, Department of Anthropology and Sociology, Monash University, July, 1973.
Tobias, Ph. v. (1964). Bushmen—hunter—gatherers. A study in human ecology. *In* "Ecological Studies in Southern Africa," (D. H. S. Davis, ed.), W. Junk, The Hague, reproduced in Cohen, Y. A. "Man in Adaptation" (Y. A. Cohen, ed.), pp. 196–208. Aldine, Chicago (1968).
Vedder, H. (1937). Die Buschmänner Südwestafrikas und ihre Weltanschaung. *South African Journal of Science* **34**, 416–436.
Vedder, H. (1952/53). Über die Vorgeschichte der Völkerschaften von Südwestafrika. *J. South West Afr. Sci. Soc.* **9**, 45–56.
Wilhelm, J. H. (1953): Die !Kung Buschleute. *Jahrb. Mus. Völkerk. Leipzig* **12**, 91–189.
Woodburn, J. (1968). Stability and flexibility in Hadza residential groupings. *In* "Man the Hunter" (R. B. Lee, and I. DeVore, eds.), pp. 103–110. Aldine, Chicago, Illinois.
Zastrow, B.v., and Vedder, H. (1930). Die Buschmänner. *In* Schultzewerth, E. and Adam, L. (eds.), "Das Eingeborenenrecht: Togo, Kamerun, Südwestafrika, die Südseekolonien." Strecker u. Schröder, Stuttgart.

Films Quoted

Eibl-Eibesfeldt, I. (1971). HF 1 !Ko-Buschleute (Kalahari)—Schamweisen und Spotten. !Ko-Bushmen—Mocking and Genital Display in Females. Homo, **4**, 261–266.
———, HF 2 !Ko-Buschleute (Kalahari)—Aggressives Verhalten von Kindern im vorpubertären Alter, Teil I. !Ko-Bushmen—Aggressive Behavior of Children, Part I. Homo, **4**, 267–278.
———, HF 3 !Ko-Buschleute (Kalahari)—Aggressives Verhalten von Kindern im vorpubertären Alter, Teil II. !Ko-Bushmen—Aggressive Behavior of Children, Part II. Homo, **4**, 267–287.
———, HF 41 !Kung-Buschleute (Kungveld, Sudwestafrika)—Geschwister Rivalität. !Kung-Bushmen—Sibling Rivalry. Homo (in press).

AGGRESSIVE BEHAVIORAL SYSTEMS

C. R. CARPENTER

The Pennsylvania State University
and University of Georgia

INTRODUCTION

Aggressive behavior covers a broad stage, with action in many keys of excitement and levels of biologics. On the stage, too, are varied performers, animals and man, taking different roles and using many languages and dialects. It is the difficult, self-assumed, and unfinished task of man to order existing knowledge of aggressivity, sketch its changing boundaries and limitations, evaluate its consequences, and point to pathways of future inquiry.

There are further challenges for research in determining significance of what is observed and known, however restricted, and in using this information to achieve and better human adaptations and adjustments.

The achievement of understanding and managing all levels of organic and social aggressivity is based on pivotal and crucial values. These essential demands are being made with urgency on modern man. Significant accomplishments can and must discover and use realizable possibilities of many alternative strategies and lines of development. Failure to select the right scenarios, or error in their execution, can lead instantly to irreversible disaster now that the horrendous power of nuclear and hydrogen weapons lies at finger-tip control of competitive and aggressive man.

Scholarly symposia and volumes (Berkowitz, 1962; Bliss, 1962; Carthy and Ebling, 1964; Clemente and Lindsley, 1967; Fried, Harris, and Murphy, 1968; Garattini and Sigg, 1969; Howell and Clark, 1963; Lee and DeVore, 1968; Otten, 1973; Washburn and Jay, 1968) have scanned repeatedly the growing information and the results of scientific inquiries on characteristics, bases, "causes," contingencies, and biosocial consequences of *aggressivity*. Inquiries have ranged over all kinds of living organisms from insects and fish through birds to mammals, including nonhuman primates and man himself. These symposia and volumes have reported what is known of phylogenetic and ontogenetic development of aggressivity, its behavioral components and patterns, its anatomy and physiology, and its biosocial utilities and pathologies.

The authors of the chapters in this book are continuing these interdisciplinary efforts to review what is known with fair certainty, and to add new information and viewpoints, in the hope that there may be in due time a synthesis of the diverse lines and patterns of facts, observations, experimental results, questions, propositions, hypotheses, concepts, and speculations. After this book the tasks will still be unfinished.

Among the relatively limited number of certainties is the conclusion that aggressivity is a quality of many modalities of behavior, and if aggressive behavior is to be understood and used fruitfully, regulated and controlled as a set of dynamic biotic forces, its facets must be investigated in the varied ecosystems. Analytical research on genetic factors, neurophysiology, biochemistry, and ethology will need to be organized into themes and plots on a *broad* stage with many kinds of actors.

It is also necessary to learn to what degree the observations and the data collected in one environment, or on one kind of organism, can be generalized to other environments, species, and genera. There is the difficult problem of knowing what abstractions and deductions can be validly generalized to apply to man in his living context.

The problem of abstracting and forming generalizations about aggressivity can be partially resolved by making inferences from episodes of anchoring observations and levels of symbolic language that apply across different environmental contexts, different genera, and that include different sets of contingencies which yield and form aggressive behavior. It is proposed, furthermore, that realistic, contextual observed episodes of aggressive behavior can serve as the basis for abstractions that are descriptive of functions, solutions, and conditions for comparative studies of organismal aggressivity.

Against the background of the observations and experimentations of this volume, this chapter will use a syncratic method of generalization. The procedure will consist of describing—photographing would be better —limited space–time frames of biotic contexts with defined animals, especially primate subjects that complexly interact with each other. The episodes are selected to show contingencies of environment and behavior that have varied elements and different kinds of functional and pathological aggressivity. Space–time frames, organisms, behavioral elements, and kinds of aggressivity are selected to have high dependability or surety, repeatability, and predictability. The episodes can be restaged or will reoccur when the set of observed contingencies operate; since the episodes have a high probability of occurring repeatedly, they provide confirming observations and validation. Inferences are abstracted from each anchoring episode and many of these deductions are supported by the summarized evidence of this volume on the biologics of aggressive behavior.

EPISODE ONE: EARLY SPRING AGGRESSIVE BEHAVIOR OF A BIRD COMMUNITY

The space frame, a rural area in the southeastern United States, was limited by a window through which could be seen a 50-ft^2 area of grass, the lower sections of the brown-barked trunks of six large pine trees, a holly, and a cedar tree. A backdrop was formed of honeysuckle vines, blackberry canes, and a variety of low-growing bushes. The time-frame was 30 min in the morning of March 28, 1973, early spring. Beside the window a television screen showed the last of the U.S. prisoners of war returning from the longest, bloodiest, most expensive, and least rationalizable war in this country's war-born history.

The grassy area had been seeded with milo grain and sunflower seed, and the typical mobile spring community of 25 birds flew into the feeding place, and in the bright sunlight reflected their new breeding coats of brown, gray, blue, white, yellow, and red.

Five blue jays flashed down on the grain-strewn grass; two were close together, but three were widely separated. Where the preferred sunflower seeds were, one jay and then another hopped with assertive priority, and the others flew. Eight cardinal males and four of their brownish-red females filtered over the grass. The male cardinals superseded the females, but withdrew before the aggressive blue jays of both sexes. Towhees competed on the same levels of the same social order, and in the same spaces with the cardinals. Two red-headed woodpeckers dived and took one sunflower seed at a time and returned to a vertical pine tree trunk to shell the seed and eat the kernel. They made many trips, each time eluding the confident blue jay and performing in spaces and on surfaces for which their feet, wings, and beaks were especially adapted but those of the jays were not.

A single mockingbird occupied the holly tree where he had been for several mornings. Each time as another of the larger birds alighted in his tree the mockingbird would dive at them and drive them away. Defending his domain fully occupied him as he flashed his white-eyed wings in frequent threats and flights toward intruders while the seeds were being eaten by other birds.

Small wrens, juncos, chickadees, and sparrows twittered in among the larger birds without obvious conflict but with delicate selective movements and special orders of priority among themselves. Three squirrels came aggressively on the scene, ate sunflower seeds, and repeatedly with short, quick runs and with flagging tails briefly drove all of the birds from limited parts of the feeding area. Three times during this morning's half-hour a squirrel with white ears drove at another smaller squirrel and sent him scurrying a short way up a tree trunk where he sat and flagged his tail.

Suddenly two bright spring red cardinals clashed on the ground, then engaged each other in a contest of strength and assertions. They fought vigorously while climbing straight upward for 30 ft, creating in the sun a red fountain of rhythmic movements and brilliant color. A large dog entered the feeding area. All the birds flew. The squirrels climbed the trees.

What generalities can be abstracted from this time-and-space limited ecobehavioral episode which is reiterative and predictable? For several weeks, and deep into April until nesting sites are selected, pairs are formed, copulations occur, and nests are built; this early morning drama is repeatable, given the same contingencies with only slight variations. From the events embedded in this limited biotic episode, what can be abstracted and stated about the contingencies, characteristics, and functions of aggressivity?

INFERENCES

1. Threats and signaling precede and, if effective, supersede, actual fighting, but fighting is imminently in reserve, and its possible effects shape much aggressive and related modalities of behavior.
2. There are clear interspecies differences in aggressiveness, and this behavioral quality results in previously established priorities for preferred food.
3. Threats, signals, conflict behavior, and fighting operate to give order

and structure to the dominance interactions and roles of both males and females of a species.

4. In species of the bird population which have sexual dimorphism, cock birds dominate hens in a limited-feeding situation.
5. Relative size is contingent to having priority to preferred food for interspecies and intersex control behavior, for movement alternatives, and for access to incentives for appropriately aroused kinds of motivation.
6. The defense or guarding of preferred places can so completely occupy the efforts of a territorial bird that other kinds of behavior may be inhibited, diverted, or prevented.
7. Aggressivity regulates and orders spacing and dispersal interactions in microspace among different species of birds and mammals, although it does not prevent most of the individuals from feeding if food is dispersed.

Restatement of the Problem

It clearly requires great leaps of imagination to understand or conceive of homologies and analogies between the molecular behavior of birds on a limited feeding ground and the determining contingencies and consequences of the human war in Vietnam as reflected on the peripheral electronic tube showing the return of prisoners from that political, legalistic involvement of the United States. Is it, or is it not, possible to employ the processes of linguistic abstractions and symbolizing as a conceptual bridging operation from percept (observation), to concept, to generalization from one genera and from one biome to another? More specifically, how can knowledge about aggressivity on the levels of animals, including nonhuman primates, be applied in understanding and regulating the nonadaptive or maladaptive aggressive behavior of man?

EPISODE TWO: INTERGROUP AGGRESSIVITY IN NEW WORLD PRIMATES

The last day and hours had arrived of the July–August 1959 field study of the howling monkeys of Barro Colorado Island, Panama Canal Zone (Field Notebook, Barro Colorado, 1959). Three groups of black howlers (*Alouatta palliata*) were triangulating each other's location where the Barbour and Snyder–Molino trails intersect and where the howlers' ranges overlap in the deep rain forest valley toward Donato Trail and the laboratory clearing. There were more than 50 animals in three distinct groups closely interacting within an area of 15 to 20 acres. The morning's heavy downpour, typical of the rainy season, had stopped suddenly and the sun's rays reflected from water dripping from leaves and limbs.

As two groups of howlers approached each other, the males responded with repeated huffs, then, after reaching a high level of arousal, with full opened-throated roars and reverberating hyoid cavities they blasted the dense air and the ears of any creatures within a thousand yards of surrounding jungle. First one group roared and then the other, both with several males projected in concert but with irregular alternation of their respective groups' barrages of aggressive–defensive sounds. As the groups came closer together, both the frequency and volume of sound increased and females and subadults added their barks and whines and blended them with the male-led choruses. The intervening distance between group males was reduced to 300 yd as Group 1 maneuvered to checkmate the other from further approaching the center of its range, which lay like a half moon around the laboratory clearing.

Thus confronting males of both groups were now in visual and sound contact. They clutched with both hands and pulled strongly on branches, some of which broke off and fell crashing to the water-swept forest floor, not too far from where I sat partially hidden between the high-ribboned buttress roots of a great tree. Some of the males, as the level of excitement continued to increase, swung conspicuously but briefly by their powerful tails, scratched themselves, and continued to give their panting roars. The contest lasted for a long half-hour, and then the encroaching group began to change its course slowly and move near the ridge along the gentle hillcrest along which ran Barbour Trail.

The third group came over the same ridge, hesitated, and then with radarlike vocalizations it probed the forest ahead, received responses from, and began to react to howlers in the already crowded little valley. The group males added their voices to the threatening and blustering waves of howler roars. The third group stopped its advance in a large tree filled with ripe figs, while the second and smallest group reversed its course over arboreal pathways. Group 1 moved into a hedge of cecropia trees in the center of its range and fed.

As the sounds faded, an unusual event began to unfold, signaled at a distance by the intermittent "caw" calls of white-faced capuchin monkeys (*Cebus capucinus*). Along an advancing front of 300 or more yards, agile *Cebus* monkeys in numbers I had never seen before filtered through the middle altitudes of the rain forest and deftly searched for favored foods down to the ground level.

Two heavy-shouldered males saw me and, while displaying angry faces, broke off and dropped pieces of dead branches, once pulling a section across the supporting branch from the far side to my side to drop it with an awkward aim in my direction.

A half-dozen capuchins moved at leisure and without much excitement through the fig tree where the third group of howlers was feeding. But, between these two species of New World monkeys there were no agonistic interactions; only avoidance and indifference. A juvenile howler and a like-age capuchin briefly and playfully chased each other over and back along a large wet slippery fig limb. Most of the howlers continued feeding, resting or turning on their backs against limbs, thus pressing the rain water from their coats. The 40 or 50 capuchins zigzagged their travel course through the second and then the first howler group as males from many places gave their pleasant, coordinating directional cues of "caws," searched for insects in bark cavities and under leaves, for grubs and beetles, and for other food items too small to see at a distance with 7 × 30 binoculars.

The subcommunity of three howler groups, excepting the playful young, became relatively silent; the primate assembly with the capuchins dissolved without conflict as they moved through distant trees and their guiding "caws" blended with somewhat similar caws of parrots.

What abstractions can be drawn from this limited episode that may have implications for other species and genera of primates, man included?

INFERENCES

1. Buffering vocalizations (as well as other regulatory functions of produced sounds) substitute for actual fighting. More specifically, vocal buffering adjusts spacing of groupings of howler monkeys, and this use of space reduces conflictful interactions.
2. Howler males cooperatively produce sounds that serve the function of place marking and perhaps even group identification. Discriminations among frequently encountered neighboring groups are possible because of differences in number of males that roar and the qualities of sound produced. The latter may provide information which could be decoded and translated by the howlers into estimates of the strength of opposition to be accepted or avoided in group encounters.
3. Special patterns of sounds produced and exchanged by some primates are needed to supplement vision, especially in dense rain forests, in detecting the location and movements of conspecifics. Thus, these sounds operate like social radar. There is reflection of sound in the sense that the projected and probing sound stimulates the receiving conspecifics to reply with appropriate vocal patterns.
4. Sexual dimorphism in aggressive–defensive vocalizations and related

behavior of adult males and females, and perhaps of subadults, makes distinctly differentiated contributions to the aroused group choruses.

5. The aggressivity of howlers is differentially varied, contingent upon the location in ranges and separation of distances among the approaching and defending groups, and relative to biotically significant locations in learned ranges or territories. These locations are the most frequented areas with trees for sleeping, resting, and feeding.
6. Groupings of different species of primates, even though the species have shared ecological niches, are more compatible than the organized groups of the same species. The former condition leads to little direct competition and conflict, but the latter sympatric state often leads to avoidance behaviors or conflict, but generally stops short of fighting and injury since the behavioral signs and signals buffer aggressive activity.

EPISODE THREE, A AND B: OCELOT ATTACK AND PERIPHERAL MALE GROUP DEFENSE

Episode A: Ocelot Attack and Howler Males' Defensive Reactions

The time was January 2, 1933 and the place was near Shannon Trail, Barro Colorado Island, between 100-yd markers 6 and 10. I was searching for Group 2 when I heard nearby the distress cries of a young howler which were immediately followed by unique, thunderous, and continuous roars of several adult males (Carpenter, 1964a). I ran toward the sound and glimpsed a catlike form jumping from a corozo palm to the ground and running away. A juvenile howler moved out of leaf clusters to where it was fully observable. It seemed to be stunned. Large drops of blood fell from the young animal and splattered on the dry leaves on the ground. At least two and possibly three adult male howlers rushed toward the juvenile in the corozo palm tree from which the cat had just leapt. The howlers roared as I had rarely heard them roar. The main body of the howler group was several tree-top distances away. The males traveled over this distance unusually rapidly for relatively slow-moving howlers and arrived near where the juvenile sat bleeding.

The roars gradually were replaced by angry grunts and huffs as the males became calmer and, followed by the juvenile, returned slowly to the disturbed group.

INFERENCES

1. Howler males and not females produce concertedly a blasting roar of sound immediately when a group member, especially a young one, is attacked by a predator. The males but not the females rush toward the place of attack, producing a blanket of protective sound.
2. The sound travels rapidly, more rapidly than the defenders could move through trees to the action spot, and *the vocalizations operate at a distance to repel the attacker.* The sound is continued as the defenders move toward the attacker. They charge to where cooperative fighting might be needed if the vocal defense should fail. Therefore, for the relatively slow-moving arboreal primates, the combined sustained protective roars of the males are effective both quickly and at a distance and thus serve defensive functions short of fighting behavior.
3. Adult males acting together as a subgroup of an organized group are the front-line force for defense using the adaptations of loud aversive sound reinforced if necessary by actual fighting.

Episode B: Peripheral Male and Group Defense

The place was Lutz Ravine near the laboratory clearing (Carpenter, 1964a). The laboratory group or Group 1 of the Barro Colorado Island howlers was moving into an espave tree to the left of Lutz Trail 1 when, after a brief period of increasing excitement, the males formed a front line facing Donato Trail 1–2 and began a vigorous series of intermittent howls and roars. These continued from 7:15 A.M. to after 11:00 A.M., or almost 4 hr. The source of provocation was observed to be a young peripheral adult male. He was also excited and displayed from one limb to another and barked feebly. His excitement increased as he responded to the group males' rage displays which consisted of shaking branches along with frequent loud and sustained roars. The peripheral male from time to time tried to hide himself from the directed vocal barrage behind tree trunks or dense clusters of leaves. Finally, the lone male withdrew and other forest sounds replaced the irritating 4 hr of multimale roars, each male giving on the average one full roar per minute for a total of 240 countable vocal series.

In undisturbed howler populations it is known that males of different ages average one male to two or three females. This socionomic sex ratio within organized groups determines that there shall be extra-group males. These howler males live separately among integral heterosexual groups. Also, it is known that howler groups, like many other primate groupings, are permeable and when specifiable conditions prevail, peripheral males from the excess-male pool can gradually penetrate the group boundaries and join a group. Generally, group males resist the approaches and entry of foreign individuals by vocal and, if necessary judging from observable injuries, by other defensive attacks including biting.

This additional information helps to complete the description of what was happening in Lutz Ravine on February 8, 1932 and what happens consistently when unidentified strange or peripheral males approach and attempt to enter cohesive organized groups of howlers. The same reactions would characterize the responses of macaques and gibbons, and other genera of free-ranging nonhuman primates.

INFERENCES

1. Adult group males clearly exercise concerted and rapid defense against predators and peripheral conspecific males prior to the defensive behavior of females. Therefore, males are more exposed to risks which may result in injuries and death. In many nonhuman primate population samples, organized groups are known to have a predominance of adult breeding females over the number of adult breeding males.
2. Excess males in a nonhuman primate population that are not recognized as group members, the extra-group or peripheral males, when approaching organized cohesive groups testing their permeability to entry meet with defensive–aggressive behavior including vocalizations and other coercive displays of group males. The extra-group male may withdraw, or persistently follow the group, and he may eventually achieve, in short or long term, acceptance into that group if it has enough openness and other favorable contingencies to accept another male. The group that was the object of observations for the above episode had an increase of three to four adult males in 1932 and from four to five the next year. Some primate males may grow to adulthood within the group but others from nonnatal groups become members against the resistance of group males and by means of persistent following efforts resulting in *familiarity conditioning*.
3. Defensive–aggressive behavior of the adult group males against the entry of lone males is only one of many contingencies which regulate adult male group membership and determine the socionomic sex ratio. Therefore, these factors regulate male access to breeding

females. This, in turn, determines the possibilities for males to reproduce.

EPISODE FOUR: INTERGROUP AGGRESSIVE BEHAVIOR

The place was Santiago Island off the southeast coast of Puerto Rico and the year was 1940 (Carpenter, 1964b). By then, 1 year after about 400 macaques (*Macaca mulatta*) had been transported to the island, six groups of varied sizes had formed. These groups were learning to adjust to ranges and to establish territories on different but overlapping parts of the 37-acre island. On the island, which had one large and one small part connected by an isthmus of sea gravel and sand, Group 1 with about 150 animals focused its activities on the large part in a 3-acre coconut grove. Group 2 with about 50 animals lived mainly on the small tear-drop part of the island across the isthmus from that coconut grove. While the colony was being organized, aggressive actions often led to injuries and death, especially among the males, and also caused some to be driven into the sea and drowned. Aggressive behavior was a prominent kind of behavior during the first year of the establishment of the Santiago colony. This was a time when groups were being organized and were becoming adapted to special areas of the island. It was during this first year that the colony lost rather than gained population by new births. Of the hundred infants that were brought to the island from India with their mothers, about one-fourth were killed by fights, as were other young, adult females, and a few prime males. Two-thirds of the 15 old males, especially selected because of their age for possible special study of cancer of the prostate, were killed or fatally injured by group male attacks that first year. During the second year when the colony became organized, the following limited episode occurred which demonstrated important elements of intergroup fighting, a frequent occurrence as the colony became adjusted to its new western hemisphere home.

Group 1 had been feeding along the crest of the large part of the island and in areas fairly well overgrown with wind-formed trees and tough grasses growing among rocky outcroppings. Group 1 was returning one afternoon to the focus of its range and the place where it would spend the night on the edge of the coconut grove in a dense hedge of sea grape bushes.

Group 2 had left the small part of the island, a rare occurrence, crossed the isthmus, and moved through the coconut grove to explore new territory beyond. The two groups clashed on the grassy hillside among outcropping rocks above the coconut grove. The attack was led by the third male (number 150)[1] in the eight-male dominance hierarchy, and he was supported in the driving attacks on Group 2 by the second male (number 170). Diablo (number 160), clearly and consistently the most dominant male as judged by all observable behavioral characteristics, held himself aloof from the battle but in alert reserve. The other adult males, and of these especially the fourth (number 159) and fifth (number 173), gave close and loud-voiced support to the second and third males. There were three remaining males, of which number 171 and number 149 stayed together and by summated dominance were able to maintain their group status, including the eighth male (number 180) who was gradually changing from a peripheral to a low group-accepted status. These three gave added but even more reduced support to the conflict, performing much circling and showing wide-ranging conspicuous, but evasive, tactics. The three low dominance males were not formed into a peripheral male subgroup. Females, contributing more calls than force, deployed themselves behind the front lines and along an irregular formation of threatening, driving, and charging males. Occasionally the male attackers of Group 1 would catch an individual, most often a prime male or female, "neck bite" and shake it briefly, throw it aside, and then drive again at the principal animals of Group 2. Diablo (number 160) with appropriate dignity took the general's role. He watched from the rocky outcroppings, sometimes standing up the better to scan the field of conflict. Excitement mounted and the volume and number of kinds of calls and cries increased. The second and third males in the dominance order drove deeper and more swiftly into Group 2 animals which were massing, and then they rapidly retreated to the position where they had closer support from their own Group 1 members. Minutes before the conflict reached a climax, Diablo (number 160) charged directly at the leading male cluster of Group 2, caught one of the animals, and slashed him severely with a canine bite. Group 2 animals began simultaneously to contract their scatter pattern, to become more compact, and to retreat by way of the rugged coast and through the dense sea grapes toward the isthmus. The third male of Group 1 (number 150) drove several females, one carrying an infant, out into the rough water, where they had to swim circling through the surf and jumping from rock to rock in order to rejoin Group 2 farther down the coast. The speed of the retreat increased as the 50 macaques of Group 2 bunched close together and fled across the isthmus to their own range of the small section of the island. The lead animals of Group 1, males and females, but not Diablo, followed the

[1] Before the macaques were released, they were marked with an identifying number.

retreating group to the boundary isthmus, then circled back, calmed down, ate, and selected sleeping places for the night in the sea grape bushes and coconut trees. Diablo suddenly and swiftly climbed a 60-ft coconut tree, surveyed his world, walked down backward to the ground, and rejoined his estrous female consort (Carpenter, 1947).

INFERENCES

The style of aggressive behavior of the terrestrial macaques during intergroup encounters differs markedly from the patterns of aggressiveness of arboreal howlers, but the biotic functions are similar. Thus the following generalization can be made to extend and reinforce previous ones.

1. The place and area or places in the territorial topography affect the patterns, strength, direction, and targeting of intergroup conspecific conflicts involving offensive–defensive behavior.
2. There are genera and species differences in signaling and buffering and in the patterns of fighting behavior.
3. Male adults at the center in macaque groups take different roles in agonistic encounters between groups, and males differ greatly as fighters, with one contingency being the status of a male in the dominance hierarchy.
4. The direction and targeting of aggressivity are very selective and may be reoriented at times to the least resistant among a number of alternatives. They are selective, also, in response to the threat levels of available target organisms (displaced aggression), hence counterattacks may be made on weaker individuals and deflected from more threatening individuals.
5. Aggressive fighting behavior on the social level is dichotomous and involves always two sides; two or more organisms, two or more groups, two or more parts of groups and, unless inverted or directed to the aggressor himself, fighting has the general form of social action and reaction cycles. These fighting episodes often occur in extended sequences that have definable climaxes, predictable terminations, and observable consequences.
6. Aggressive behavior that is directed outward from a group toward another group reduces aggressivity within the group.
7. The amounts and kinds of fighting are reduced by the formation of organized cohesive groups; fighting, injuries, and killing occur in the highest frequencies in unstructured groupings of strange individuals that do not have territorial ranges to which they are adapted.
8. The contingencies to maladaptive aggressive behavior can be man-

aged and, therefore, aggressive behavior itself can be regulated in nonhuman primate populations. The following are procedures that reduce and control levels of aggression in free-ranging primate colonies:

a. Reducing the relative number of adult socialized (group) breeding males; for example, establishing a socionomic sex ratio for adult rhesus macaques of 1 male to about 30 females.
b. Providing optimum space and ecological conditions which permit dispersion and conditions for escape and protections; for example, for macaques, trees for climbing offer escape routes and protection for individuals being driven or attacked.
c. Dispersing food and putting food in established ranges greatly reduce fighting and injuries in primate colonies.
d. Avoiding overpopulation and overcrowding.

EPISODE FIVE: AGGRESSIVE BEHAVIOR ON A FEEDING GROUND

Takasakiyama lies between Beppu and Oita in Kyushu Prefecture, Japan, and on this mountain reservation is located one of the world's largest colonies of macaques. In 1972 the colony of 1500 animals had three groups, one of which had about 1000 individuals (Carpenter, personal observation, 1972). This group is one of the largest known naturalistic groups of nonhuman primates.

The structure of this sample population and colony is very complex, as the primatologists of the Primate Research Institute, Kyoto University have shown in many excellent articles and monographs. The three-group colony is fed sweet potatoes, wheat, and soy beans on one large feeding area. This procedure creates an almost constant dense flow of macaques through the feeding area among several Buddhist temples. Of special interest is the regulation and biotic and biosocial functions of different kinds of aggressive behavior in this large dense colony. Related is the complex communication system which coordinates the movements of groups and group components, such as subgroups and segments as they move on and off the area where food has been spread.

Wednesday, November 15, 1972, after 3 frustrating days of rain the *Wild Kingdom* television filming crew of five men arrived at 8 A.M. in sunshine at the entrance to the Takasakiyama reservation. Tourists, already arriving by busloads, were streaming up the clean, gravel pathways and past two beautifully executed bronze statues of macaques, one in memoriam of Jupiter, a group leader for many years, and the other a male and female consort pair.

Monkeys of Group A were dispersed along pathways, under and in trees. One subgroup consisted of eight vigorous adult males; another of several females and their female juveniles of the birth season 1 year removed, and male and female infants of the June, July, and August 1972 birth season. Throughout the parklike wooded area, along paths to the feeding ground there were scattered peaceful macaques, singles, dyads, and clusters of three to five monkeys. All of these were members of Group A.

As our crew approached the feeding area, two kinds of aggressive attacks were observed. First, there were small group attacks directed toward single monkeys, both adult male and female and by both males and females. These attacks often were led by several peripheral males. They were driving attacks over distances up to 100 yd, and the longest were sustained for brief periods lasting no more than a minute. The attack resulted in fleeing and escape, some of the animals being caught and bitten. The sequences of behavior were always associated with voices of aggressiveness and voices of escape, the cries of being caught and punished. The chains of actions and reactions were sometimes interrupted briefly and then resumed. The objects of attacks were sometimes occasional intergroup animals but most often were intragroup individuals moving from one part of Group A to another, and the attacks operated to regulate membership in group components or in the group as a whole. Boundaries are not observably clear in organizations as large and complex as Group A in this location.

As we approached close to the last terraced area where many hundreds of macaques of both sexes and all ages (the "central part" of the group including the most dominant males) were feeding, another pattern of aggressivity at first could be heard, and then both heard and seen. This behavior consisted of threats of quick, short dashes of one animal toward another, including threat faces and harsh voices. Simultaneously over the main feeding ground and surrounding it, there were many of these flashing eruptions of raucous drives, flights, and escapes, of catching and being caught and bitten. Many of the encounters led to counterattacks. Others were sequential and involved, for example, animal A attacking B, with B escaping, then the counterattack of C on A. Sometimes both B and C would jointly counterattack A, who might in turn attack D. There were many other permutations and combinations, such as one to several and several to one with attack relays when A attacked B, who attacked C, and then C attacked either D or E. The swiftness of onset was matched by the suddenness of ending of the disturbance as truces for peaceful eating were arranged.

The pattern of spacing or dispersion of the several hundred animals on the feeding ground was like a map of that group's variegated structure.

Some animals had envelopes of space around them, and the space was especially large around the central males. Some individuals were permitted to enter these areas of individual space while others were prevented from doing so, and, if they did, they were attacked. Which animals associated with which other animals were arrangements resulting from sensitive selectivity involving both inclusion or nearness and exclusion or greater separation distances. Favored individuals such as estrous females and young could enter and pass through the envelopes of spaces around the feeding central males but many females and almost all other males, if they entered the spaces, would be attacked, not always by the occupying male but sometimes by one of his supportive peers.

The supreme or alpha male of Group A had lost vision in one eye and was clearly identifiable both in appearance and behavior. He and the beta male executed about half of all the attacks on the crowded feeding ground. The alpha male had pulled all the hair from his chest, inner arms, and upper abdominal areas. The breeding season was beginning but this stressed alpha was reported to engage less in primary sexual behavior than other males of the "central part" who were lower in the dominance-control hierarchy. As the alpha and beta males moved through the feeding ground and the other monkeys avoided them, a wake of open space moved like a wave along with them.

INFERENCES

1. Groupings of primates which form large multigroup colonies have semiclosed or semipermeable boundaries. Some animals, both individually and in groups, pass into the group and out from it across these mobile boundaries. Most frequently these are males and not females. Males and especially subgroupings of young aggressive adult peripheral males move across group boundaries into noncentral segments, associate temporarily with a group, and then change to another group or they may for a time remain peripheral to groups. These males constitute reserves from which males can be recruited for the "central parts" of the group.
   The important points of this inference are
   a. The boundaries of naturalistic groups are semiclosed or semipermeable, and not closed.
   b. The passage of individuals and subgroups across group boundaries is very selective and this selectivity indicates individual and/or group member identification.
   c. The regulatory behavior of this selectivity consists of pervasive driving aggressive attacks made by group members acting special roles, resulting in escape or being caught and bitten on the part

of the individual that attempts to enter the organized group. Aggressive behavior based on perceptual discrimination and identification is the implementing regulator of selectivity.

2. The microspacing of individuals and clusters in compressed groupings where there is crowding provokes aggressive behavior which regulates behavioral norms established by genealogical relationships; that is, a female and her succession of young. The place and status of each principal male and female have the support of conditioning and learning effected through reinforcements, both positive and negative, with aggressivity being especially prominent as negative reinforcement.

EPISODE SIX: MACAQUE ATTACK ON HUMANS, FEMALE DEFENSE OF YOUNG, AGGRESSIVE PLAY

Macaque Aggressivity Displaced to Humans

Four physically mature and fully adult but transitional males had moved temporarily from Group B to Group A of the Takasakiyama colony in late September 1966 (Carpenter and Nishimura, 1969). They were on the feeding ground in the afternoon with other loosely integrated members of a Group A segment that were collecting sweet potatoes, eating and carrying them away as this last group segment moved from the feeding ground through the high temple grove up the mountain and toward the regularly used sleeping areas. Akisato Nishimura and I had spent an hour studying and photographing the socially mobile male subgroup on the feeding ground, working at times within 5 ft of the animals.

The four physically mature males climbed on top of a large boulder outcrop which stood higher than my head and groomed each other reciprocally. I took advantage of this situation to move cautiously to the side of the great boulder and to hold a directional microphone to within a few feet of a male to record his close-in throaty threat vocalizations. This disturbed the males slightly and they vocalized as I had expected. Soon the males continued moving up the mountain. Suddenly a female threatened a small juvenile that ran out on a dead tree limb, which broke, crackled, and crashed to the ground. The female threatened again, the juvenile screamed, and the four males dashed to its defense. I moved into the melee to record the complex sequences of threat vocalizations of all the monkeys but especially of the combined aggressive sounds of the four males. Quickly, only seconds later, the four males charged together toward Dr. Nishimura and me. We backed off, but they pressed the charge forcefully as a gang attack, and in two's they jumped on each of our shoulders in attempts to bite our necks and heads. A male bit Dr. Nishimura on the upper arm. I protected my head and neck by swinging the live microphone around and over my head. The concerted attacks continued to be pressed strongly as a massed gang effort with increasing levels of excitement and aggressive vocalizations. We ran down the steep mountainside, jumped over a 4-ft wall, and thus escaped the vicious attack. The subgroup of transitional males changed directions, calmed down, and followed the other animals along a trail toward the sleeping places.

INFERENCES

1. Approaches by a human beyond the limits of the operative tolerance distance provoke aggressive behavior.
2. The place where primates interact, e.g., the feeding ground as contrasted with the mountainside, operates differentially in interaction with the social stimuli which provoke aggressive actions of both attack and defense.
3. Subgroups of transitional males are especially sensitive and reactive to threats and distress cries of infant and juvenile classes, and they react by fast concerted attacks on the perceived target whether it is an animal or a human. The concerted defensive aggressivity is similar in character to the howler male attack on the predator ocelot.
4. The directionality of an attack can be changed or "displaced" rapidly as primate attackers perceive other and different targets "to be responsible" for the threat disturbance, and they react aggressively to the alternative instigating target.

Female Defense of an Infant Nursery

It was during the third week in November 1972 and the breeding season was just beginning in the Takasakiyama colony (Carpenter, personal observation). The last year's crop of 3-, 4-, and 5-month-old infants was becoming more and more independent of its mothers as females were beginning to redden in their faces and perineal areas and become sexually receptive. While most of the monkeys of the central segment were feeding, a dozen infants formed, or were deposited in, a nursery-like cluster where a fallen tree invited play. Young females of the previous year, 15 to 18 months of age, continued to associate with their mothers, while young males of the same ages had formed unisexual young subgroups which were independent of their mothers and separate from the natal Group A.

For many hours daily in places suitable for play the infants wrestled, chased each other, manipulated objects, and tested their perceptual motor skills of climbing, clinging, and jumping. While the adult females were feeding or beginning to approach males as they became sexually active and their sex drive intensified, a few of the juvenile female sisters of infants stayed near where they clustered and played. Close by several mothers of the infants maintained a constant visual reference and alert guarding behavior for protection of the playing infants. When I moved close to the playing cluster to try to photograph the play behavior better, I was strongly threatened and charged by several females. Immediately, they dashed into the playing cluster or nursery area, and each selectively and correctly retrieved its own infant.

INFERENCES

1. In large naturalistic groups of primates infants may be deposited or assembled in fairly large clusters where they are guarded.
2. When the breeding season begins and maternal behavior phases out as sexual behavior phases in, females still guard clusters of infants from encroachment by threat targets, and they use attacks short of contact to defend their young against perceived threat targets.
3. Females specifically select, identify, and retrieve their own infants and quickly carry them to places and distances of safety.

Aggressive Play and Threats by Subadult Male Macaques

There was a grass-covered area near the feeding ground of the Takasakiyama colony of monkeys in September, 1966 (Carpenter and Nishimura, 1969). This was a favorite place for play of subgroupings of young male macaques in the 2- to 3-year-old age range. Patterns of play consisted of charges, attacks, avoidance, escape, catching, being caught, play-biting (without damage) and being bitten, running, chasing, and following over and over circuitous routes. The playing activities of these young males were very vigorous and often verged on fighting. For an observer, fine discrimination is required to differentiate a short sequence of the most energetic play from fighting. The accompanying vocalizations are different. However, there are contingencies which signal and stimulate play, for example, the play face, and there are other action patterns which prevent the vigorous play from sparking over into fighting.

I moved close to the subgroup of about 10 young males at play to make motion pictures of them while they were excited and intensively engaged in play. When I was seen minutes later by some of the playing monkeys, 5 or 6 of them ceased playing and charged me in a concerted gang attack. For safety I climbed to the top of a large rock outcropping from which I completed photographing the action sequence.

INFERENCES

1. Vigorous play occurs most prominently in young subadult primate males; and during intense activity sustained for many hours each day, this level of aggressivity begins to establish the identity and role of individuals. Play also establishes orders of control status and priorities for each individual relative to others of the subgroup. Furthermore, during play there are formed cohesive and persisting subgroupings of young males. These young male groupings have identity in their natal groups, but also they have social mobility for joining other groups.
2. Patterns of facial expressions and other actions operate to signal play and not aggressivity, or perhaps to stimulate play and to inhibit aggression.
3. Subgroups of young male macaques defend their temporarily possessed areas by concerted aggressive attacks on strange human intruders.

EPISODE SEVEN: COORDINATION OF INTERGROUP INTERACTIONS BY SIGNALING AND AGGRESSIVITY

The place was Takasakiyama. The time was noon of November 18, 1972 (see also, Carpenter, 1969). The main actors were the central males of Group C and the final departing segment of Group A that had been eating sweet potatoes on the temple feeding ground. They were being observed by several hundred visitors to this world-famous primate colony. Of the three groups, A, B, and C, B was still in its range to the left of the temple area. Group C members were coming in from their range from the opposite direction to B's range, and they were crowding below a cliff, the refuse dump, and food-storage and service houses.

The five or six principal males, along with a few females and their young, approached to within 50 yd of the feeding area and deployed themselves irregularly along a cliff-top fence and on the food and service sheds. On the roof tops several males invited grooming and were groomed. Clusters of adult females and their young, some females showing the first blushes of estrus, reciprocally groomed each other.

Aperiodically as members of Group A and Group B interacted, there were sharp, swift, driving attacks, rapid avoidance, and quick follow-throughs but rarely were there aggressive contacts which resulted in biting and being bitten. Sometimes two or more males would seesaw back and forth as one side would drive and retreat and the other side would intersperse equal but opposite drives and attacks. There was formed a line of expressed tensions as Group C members filtered slowly but steadily up to the receding boundaries where there were aggressive actions and reactions between Group A and Group C members.

Dramatically and conspicuously, the alpha male of Group C climbed swiftly a 70-ft tree with a damaged top. On arriving at the bare top, where he was clearly in view of the feeding ground and his silhouette was etched against the blue sky, he vigorously shook the dead polelike top of the *Cryptomaria* tree. A minute later he repeated the shaking sequence. He rested and watched both the feeding ground where Group A members were, and the area where his own group members were massing at the foot of the rock-lined cliff. Then, while looking to first one side and then the other, he shook the tree top in methodical rhythm. This series, too, was repeated several times, and then this alpha male bounded down the tree from limb to limb and to the fence and onto a sheet-metal roof where he bounced aggressively.

With movements covering yards of ground at a time and with alert rests between his moves, the Group C alpha male cautiously approached the feeding area. He moved with that erect style and positive pattern of locomotion which so clearly marked and signed his dominant alpha role for an observer or another macaque who can read the behavior. Members of his group massed densely at his rear where there were other central group males that supported the alpha. However, the alpha male alone spearheaded the cautious advancing wedge onto the feeding area. There were a few sharp aggressive skirmishes associated with aggressive vocalizations, exchanges by females and subordinate males of the two groups. More and more of the Group C monkeys crowded toward the newly spread, clean sweet potatoes. The last monkeys of Group A snatched hands and mouthfuls of potatoes and left the feeding area. The continuing contact calls of the peripheral monkeys of Group A faded as they followed the sound trails of Group A up the mountain. Just as quickly, seconds later and now with reduced tension, Group C monkeys came to satiate their hunger with sweet potatoes. The usual intragroup driving attacks began to supersede and replace all the former clashes between Group A and Group C members. About 200 hungry monkeys from Group C crowded on the feeding ground and mapped the structural pattern of that organization.

INFERENCES

1. Frequent contacts between naturalistic groups of nonhuman primates result in the development through learning, conditioning, and accommodations of complex patterns of adjusted behavior which are markedly different from the hostile conflicts which occur when relatively strange groups meet and compete prior to the development of these learned accommodations.
2. The learned interactions are characterized by both near and far visual and auditory signaling. The signaling is reinforced by threat and mild attack, or the imminent possibility of such attacks and the consequent punishment.
3. The developed and complexly subtle intergroup coordinating behavioral patterns, using communicating processes including identifications of individuals and group members, effect displacements from areas of competition and access to places and incentives. This is done with reduced energy expenditure as compared to actual fighting, and the risks of injuries and killing are reduced by signal communication.
4. Distant signaling, in this specific case from a tree top, provides a threat plus other information in a visual strategy which buffers or reduces the possibility of serious intergroup fighting by indicating a need (hunger) and intention movements toward the feeding ground. The signaled actions given in advance permit the target group to withdraw without a fight. A close unanticipated or sudden encounter of strange monkeys usually provokes a sequence of attacks and counterattacks in alternating sequences. In brief, signing, signaling, and communication processes in open space can greatly reduce or eliminate serious energy-demanding aggressive behavior.

EPISODE EIGHT: AGGRESSIVE BEHAVIOR AT THE BEGINNING OF THE BREEDING SEASON OF MACAQUES

The place was Choshi Valley between mountains on an island in the Inland Sea, Japan (Carpenter, 1969). It was the first week in October 1966. The one-group colony of about 300 Japanese macaques was well into the fall breeding season. In the shaded areas of the pine forest with a laurellike ground cover, the faces of adult females and males were ripe-cherry red and shone like mobile lanterns. The whole colony was compacted on the feeding ground and an adjacent area with a combined space of about 100 yd^2. Here they ate and competed for limited quantities of grain. Almost continuously the fast, loud, driving attacks were made mainly by males on and near the crowded feeding ground. The vigor, intensity, and frequency of the intragroup attacks were at least five times as great as the intragroup attacks on the feeding ground of the much larger Takasakiyama colony 2 months before its breeding season. A high percentage of the sexually active males and females showed injuries from being canine bitten. The partially blind alpha male had a deep, long incision through his lower lip which hung as a flap and uncovered his incisor teeth.

In addition to the attacks, threats, drives, gruff sound threats, and cries from animals being attacked and fleeing into trees as places of escape and transient safety in the central feeding areas, aggressive behavior of other characteristics and kinds and intensities was occurring around the perimeter of the concentration of monkeys. Extended observations showed that these attacks and fights involved peripheral males with red faces and scrota, attracted by receptive females, which were attempting to penetrate the group and the feeding area. The peripheral males especially were being driven away and chased up trees by a dozen central males of the main group.

A subordinate male and an estrous female of the group engaged in a typical copulation sequence to climax in a tall tree. The beta male saw the final mount and climbed the tree, directing his threat at the fleeing female. He performed a limb-shaking display, looked toward the female repeatedly, and interacted briefly but nonhostilely with the cringing subordinate male. After his threats toward the female and the limb-shaking display assertive of his special status, he returned to the ground and joined the half-blind alpha male and two closely associated estrous females.

INFERENCES

1. During the breeding season when sexual and other levels of excitement and arousal are high, intra- and intergroup aggressive behavior is increased in frequency and intensity and oftentimes attacks lead to injuries by biting during contact attacks.
2. Intragroup displays, threats, attacks, and serious fighting are targeted to females entering estrus and to subordinate males prior to consort formation and copulatory activity. These sequences of aggressivity act selectively on fertilization and reproduction.
3. High levels of intragroup aggressivity during breeding seasons result from, or are contingent to, changes in social structure, group niche organization, and modified temporary roles of females that are coming into estrus and are approaching males. Involved is the penetration of the individual niches and spaces of breeding males.

Therefore, aggressive behavior is a regulator of which females breed with, and are fertilized by, which males. Also, clearly by inference physiological and biochemical regulators that have been defined by experimentation take over from behavioral regulators to complete the selectivity involved in the whole system of reproduction.

EPISODE NINE: AGGRESSIVE BEHAVIOR IN HANUMAN LANGUR TROOPS

The observer was Yukimaru Sugiyama of the University of Kyoto's Institute of Anthropology, Japan, and the time was June 1961 (Sugiyama, 1965). The study was made during the Japan–India Joint Project in Primate Investigation. The location was a tropical dry deciduous forest 15 km west of Dharwar, Mysore State, India.

Troop Number 30, with 24 members but only *one* adult male, had a range of about 65 acres, including trees lining road sides and river banks. There were cultivated fields interspersed with grasslands. The range of Troop 30 was bordered by but did not overlap with the ranges of four other single-male langur troops that had 7, 10, 21, and 19 members respectively. In addition, a unisexual male troop of six adults and one subadult lived in a range that was partially shared by another heterosexual troop to the south.

On the first day of the 2-week episode, the all-male group approached Troop 31. The leader male of that group became very excited and singly attacked the seven males of the group and drove them away. Following its repulse the male group moved swiftly into the range of Troop 30. Three adult females of this troop, one with an infant, came close to the males. They surrounded the females, and, as excitement increased, individual males attacked the clustered females. Conflicts developed in the group of seven males. The most dominant male, "L," displayed by grinding his teeth and by making convulsive threats. Two hours later one of the females showed early signs of estrus and sexual excitement by shaking her head and sexually presenting to the males. However, when males approached her, they were driven away by a female peer. Meanwhile, the leader of Troop 30, male "Z," was attacked and severely wounded, receiving a long slash on his leg. Nevertheless, he continued to grind his teeth and make convulsive aggressive displays toward the all-male group.

The next day several individuals of the male group were bloodstained. Some of the males repeatedly made driving attacks toward two of the females and a juvenile of Troop 30, while other males attacked two different females and a juvenile. Troop 30 was dispersed by repeated attacks made especially by males "L," "M," and "N" of the male group. No defense was made by male "Z." The female who had shown signs of estrus did so again, but none of the males responded.

On the third day "Z" and "L" of the all-male group exchanged aggressive displays, chasing and counterchasing each other. The other males and Troop 30 females watched the conflict intently but did not take part in it.

Male "L" went close to the females of Troop 30, and they exchanged vocalizations. He also gave alarm barks, and "acting like a leader," he attacked a dog. Then "L" went directly to the center of Troop 30. The other six males of his group stayed 20 m away. About 2 hr later, "L" made violent attacks on Troop 30 females. Two females counterattacked. The estrous female continued to approach males without getting any response from them. Troop 30 members and the all-male group formed a temporary cluster of 18 langurs. "Z" and the other 13 langurs of Troop 30 were separated into another association at this time; all of the langurs were disturbed and dispersed.

On the fourth day, after the all-male group had been with a part of Troop 30 for the night, "L" attacked the six males of his own former unisexual group and drove them away. Again "L" acted like a leader and directed threats and display jumps toward a dog. The six males returned and some females began to follow them, but "L" dashed to the space between the females and males and, with threatening barks, herded the females back to Troop 30. "Z," the deposed Troop 30 leader, watched the events from a tall tree from which he displayed threats. He then approached but stopped 100 m short of the group now led by "L." Three juvenile males of Troop 30 stayed close by male "Z," along with a few other of that troop's members.

By the fifth day, "L" had consolidated his new troop, and he copulated with the estrous female. By the seventh day, an infant, who had been seriously injured, disappeared, and another infant had received a long, deep canine slash on its hip. Since this infant walked with difficulty, it clung to its mother for assistance. Later it was missing. "L" severely attacked the mother, who fought back, then directed a vocal attack toward several deserting females and drove them densely together into a part of the riverine forest.

"Z" returned to Troop 30 on the eighth day and fought severely with "L," who afterward left the troop. However, on the morning of the ninth day, "L" returned once more to his all-male group. There was intense fighting with Troop 30, especially between "Z" and "L." "L" had a slash on his head which severed the right ear lobe. "Z" ran away and was followed by three females and their infants. "L" again fought with his former peers of the all-male group which fled, then he was followed by the females that had been intercepted. On the tenth day, also, "L" fought off the male group, and on the eleventh day he fought with "Z" in a tree for about 2 hr. "Z" received a second deep wound on his leg, fell from the tree, and fled in defeat. He returned for another fight, and then a new subgroup from Troop 30, consisting of six juvenile males, was formed around him. This male subgroup concertedly voiced aggressive threats toward "L." Thereafter "L" repeatedly made brief attacks and drove at the juvenile males of the new troop. The mothers did not defend their juvenile sons.

"Z" and the male juveniles of Troop 30 left the area and moved to the north, and the six-male group went to the south. Another infant was injured and was then left behind by its mother who, within hours, became sexually receptive. A female and the new troop leader "L" fought each other vigorously. The displaced leader "Z" with the six juvenile males formed a stable group and occupied a territory north of Troop 30's range.

Three new groups were formed during the time of this episode. First, there was the displaced leader "Z" and six juvenile male langurs, perhaps his sons. Second, there was the new Troop 30 with leader "L," nine adult females, and two female juveniles about 1 year old. This 12-member group continued to occupy its former range. Third, there was the former all-male group of six individuals that repossessed its former range.

INFERENCES

1. By means of fighting, a mature male from a unisexual male group displaces the leader of a single-male langur troop.
2. The displaced male leaves or is driven away and forms a new unisexual group of juvenile males, perhaps his sons.
3. There are high levels of antagonism among adult langur males that compete for and possess females.
4. An intruding male leader attacks females, injures their infants, and many of them die.
5. When a female's infant is killed she becomes sexually receptive.
6. Females are attacked and injured during early estrus.
7. In langur troops the duration of control by a single group male is regulated by his aggressive intrusive–defensive and display behavior, motivated in part by seasonal and sexual contingencies.
8. Aggressive behavior results in new groups being formed and in groups being restructured or reorganized and reduced in size.
9. Individual recognition and selective discrimination of individual

characteristics operate as basic preconditions to aggressive behavior, group formation and organization, and social change.

10. The new single-male, heterosexual breeding group remains in the territory or range to which the females are conditioned, even though the new leader has changed his territory.
11. Females follow the male that wins fights.

EPISODE TEN: INTERGROUP AGGRESSIVITY OF GIBBONS

The place was on the tree-covered cliff face above Camp 2 of the Asiatic Primate Expedition at Doi Dao Temples, Thailand, and the time was noon, May 12, 1937 (Carpenter, 1964c). The action took place where the behavioral ranges of gibbon Groups 1 and 6 overlapped in their shared border area.

Group 1 with five gibbons, all black with white face ruffs, white hands, and feet, had typically an adult male and female and three young, an infant, and exceptionally two young adults of unidentified sex.

Group 6 had an adult black male, a buff adult female closely associated with and sometimes carrying a 1-year-old juvenile, and an intermediate-sized black juvenile, a total group of four members.

The actions of Groups 1 and 6 swirled around a tree that was vine covered, and the vines were heavy with ripening grapes, a highly preferred food of gibbons. The tree was located more in Group 6's range than in that of Group 1. Also, the calling and actions included responses to the vocalizations of six captive gibbons tethered to the center of horizontal bamboo poles located in clearings around my pool-side shed.

For more than an hour the gibbons of Group 1 brachiated widely and vigorously through trees and bamboo clumps which covered the mountainside above the temple grounds. With flashes of speed rarely exhibited, they swung with superb brachiation downward and in a series of long leaps through trees and clanking arched-over bamboo fronds to reach the tree with the vines filled with purple grapes. All the while one or more of the male gibbons gave calls with high levels of excitement.

At least two and maybe more Group 6 animals also called, using the male vocal patterns loaded with excitement, and swung into the vine-covered tree. As Group 6 members entered the tree, Group 1 fled rapidly and the female buff adult of Group 6 followed them until the procession was deep into the territory of Group 1.

INFERENCES

1. Group 6 had for some days before and during the morning of the 12th fed heavily on the ripening grapes. A reasonable interpreta-

tion of the observations would be that Group 6 was guarding the grapes as a food source and reacting defensively against Group 1 with calls and aggressive targeted actions. Clearly, Group 1 was driven back toward the center of its own territory. Also, Group 1's excited behavior and calling for an hour reflected behavioral characteristics that were an invasion and intrusion into the other group's range. Furthermore, this episode is supported by many other observations that intergroup conflicts are contingent on where groups are in their well-known ranges, and that animals have advantages when in their own ranges. In brief, gibbons defend and guard highly preferred incentives in their territorial ranges.

2. A gibbon group that invades the range of another group evidences stressful, defensive, exploratory behavior which can be observed and differentiated from the behavior of the occupying group.
3. Approach threats which are rapid and clearly directional are associated with aggressive–defensive type calls, most characteristic of males, which serve to repel an invading group of gibbons. Actual attacks may not be necessary, except as an ultimate reserve alternative of reinforcements for behavioral signs and signals. It is known from collected scarred specimens that gibbons in natural habitats do attack and injure each other in fights.
4. The ordinal positions of group dominance in communities and colonies, or priority behavior for preferred incentives, are contingent upon where the groups are in a central-peripheral axis across their ranges.

EPISODE ELEVEN: ANTAGONISTIC BEHAVIOR BETWEEN A GROUP AND A PERIPHERAL MALE

The place was Doi Dao, Thailand, around a fig tree near Clearing 3 (Carpenter, 1964c). It was midmorning of April 13, 1937. The episode began with a pair of beautiful adult buff male and female gibbons feeding on figs while I observed them from a blind 50 ft away.

A black young adult that was either from Group 11 which had six individuals or was an individual-living peripheral male that I had seen at a nearby spring of this forest came into view. As he approached and entered the fig tree, the buff female of Group 3 continued to eat figs while the male of that pair fed intermittently between swift swinging sallies from the fig tree and back again, showing threat displays. The black individual, which I judged to be a young adult male, approached the feeding buff female when her mate swung away. This was followed by the buff male swinging quickly back near her. With these intervening rushes, the black gibbon would leave the tree swiftly, only to return when the buff male was again off guard. The approaches and fleeing of the buff and black males alternated in and out of the tree four times, with the female as one focus of their activity. As the buff male made driving charges and displays, the black gibbon effectively evaded them.

The female reacted more positively to the buff male than to the black gibbon, but she did not display agonistic reactions to him. She was relatively indifferent to the conflict and after eating her fill she swung from the tree and disappeared in the dense canopies, with the buff male closely following her. The black gibbon trailed them at a distance of 50 to 100 yd.

I could not be certain whether the black gibbon was attracted to the buff female or figs, or both. Clearly the rivalry was keen between the two males. I believed that the female may have been sexually receptive. When she left the tree, both males followed her.

INFERENCES

1. There is keen competition expressed by swift, driving, sorties, displays, and interventions of a male gibbon with his bonded female. As this aggressivity is expressed toward other young males, the group male is dominant over an intruder.
2. Competition and aggressive guarding behavior occur between a group male and a lone male that is attempting to enter the group, and this behavior regulates the family unit typical of gibbon social organization. The same behavior is regulatory of maturing males' leaving the natal family.
3. The target of aggressivity is repulsed without overt contact fighting. Threat behavior and conditions which permit charges and avoidance in open arboreal space are contingencies which operate to prevent actual fighting and to make adequate regulation for the "species-specific" signs and signals of arboreal gibbons.
4. The guarding behavior in gibbons has reference to food sources, a bonded female in estrus, and other incentives that may summate when they occur in the same situation.

EPISODE TWELVE: CHIMPANZEE PREDATION AND BABOON KILL

This episode's basic observations were made available by Geza Teleki whose dramatic and pace-setting book *The Predatory Behavior of Wild Chimpanzees* is published in a series on primatology by the Bucknell University Press (Teleki, 1973).

The place was the Gombe Stream Research Center which has been established and made known throughout the world by Jane van Lawick-Goodall, her husband Hugo van Lawick, Cambridge University's Subdepartment of Animal Behavior, and the National Geographic Society. The center lies close by Lake Tanganyika, Tanzania, and in the Gombe National Park of that rapidly developing country.

The action occurred near camp in the feeding area where four observers recorded in speech and photography the live drama which can be viewed as a replay of the behavior and social interactions reflective of hunting events that may have characterized food getting for 99% of human history.

The time was March 19, 1968, about 7 A.M. on a rainy morning. Four of the many feeding boxes stocked with bananas were opened and two adult chimpanzee males, Hugh and Mike, with several other chimpanzees ate and mingled with the camp's troop of about 60 baboons.

At 8 A.M. a female baboon, Arwen, with her infant Amber carried on her belly, sat on a food box below the tree in which male chimpanzees Mike, Charlie, and Hugh were resting and waiting to feed on bananas. By then 18 chimpanzees were mixed with the baboons in the camp clearing.

The remaining banana-filled boxes were kept closed to avoid the expected chaos of aggressive competition, both among the chimpanzees and with the baboons, as the hungry primates anticipated eating their choice exotic food.

Tension mounted sharply, and the primates' behavior became agitated, especially that of the adult male chimpanzees. As the tension increased further with hunger and frustrating delay, the excitement erupted as chimpanzee Leakey attacked Figan. During the noisy encounter the adult Mike moved quickly away from Charlie and Hugh to behind Arwen and "—with a flick of his right hand" caught Amber. Mike ran bipedally, swinging Amber over his head, and twice struck her against the grass-covered ground. This action provoked excited screaming from the chimpanzees and raucous barks from the baboons. All treed baboons jumped to the ground as Mike, swinging the screeching Amber, dashed away for 15 yd and, as the baboons converged on him, took Amber in his mouth. He continued his arm-flailing and bipedal flight. Alive, Amber waved her arms and kicked her legs while perhaps screeching, although her distress cries were blanketed in the uproar of chimpanzee screams and baboon barks. Mike closed his teeth and jaws across her back.

All activity and about 12 baboons now converged aggressively around Mike, slowing his travel. Chimpanzees swaggered bipedally, swung their arms, striking out at the baboons, which in turn performed canine threats and barked. Excitement continued to increase and interspecific attacks and counterattacks rose in frequency.

Three male baboons charged Mike as the battling primates moved into head-high grass. A male baboon leapt on Mike's back and while gripping his hair with hands and feet, the baboon raked his teeth across Mike's shoulders without observable injury. Mike took Amber in his left hand and, while again walking bipedally, shook his shoulders to dislodge the mounted baboon. Mike waved his arms and hands violently and by accident hit the back of the hand in which he held Amber against a small tree trunk. The impact released his grip on the infant who fell free in the deep grass and was lost to the observers' sight.

The interspecific aggressive behavior, along with quickly reduced vocalizations and incitement, was displaced by searching and exploring behavior by both chimpanzees and baboons. Although during the searching the primates were closely associated, persisting tension was expressed in irregular attacks and counterattacks, and by varied vocalizations. The primates criss-crossed the area where Amber was lost. Hugh, Leakey, and the other chimpanzees came out of the grass, and then Mike appeared without his prey, Amber.

A male baboon lunged at and then caught the adult male chimpanzee Rix. The baboon barked loudly; Rix turned quickly, slapped the attacker, swaggered bipedally with hair erect and arms spread. Again the baboon lunged, yawned, gave the eyelid threat display, and withdrew. The chimpanzees and baboons slowly dispersed. The excited behavior decreased. Hugh briefly chased Figan. Mike disappeared. A few chimpanzees and baboons crossed and recrossed the grassy south slope where Amber had disappeared. Arwen was the last to leave that slope.

Within half an hour, as the rain stopped, a cluster of chimpanzees appeared 200 yd south of the melee ground. There were screams and six male chimpanzees, along with Mike, in determined travel left the feeding area and went toward the cluster of tree-mounted chimpanzees.

Humphrey, an adult male who had been peripheral to the conflict and capture, sat on a limb holding a limp infant baboon. Quickly Mike, Charlie, Leakey, and Hugh arrived near Humphrey, and the separating, sharing, and eating of the prey procedures, characteristic of wild chimpanzees, began and were completed about an hour later. A grooming session followed.

INFERENCES

1. Prey capture may occur quickly in the context of close interspecies spacing, competition for food, and increasing levels of excitement

with expressive vocal behavior being a prominent part of the varied activity.

2. Adult males are the principal actors in chimpanzee prey episodes on young baboons, and adult male baboons are the principal defenders and counterattackers.
3. The capture and killing of prey by chimpanzees are actions associated with excitement which rises sharply to a crescendo and which quickly subsides when the objects or targets of attack and defense are lost, when searching behavior displaces aggressive conflict, or when the prey is killed and eaten by the males.
4. Searching and exploratory behavior persist longest in the aroused mother of the infant prey.
5. Excited and vigorous interspecific group battles between chimpanzees and baboons which occupy the same ranges prominently consist of vocalizations, hand and arm actions, charges and countercharges, evasive movements, and aggressive signing and signaling like the eyelid displays of the baboons and the firm, set-lipped faces of the preying chimpanzees. Limited physical injury occurs.
6. After prey episodes in which chimpanzees make a capture, and following brief aggressive encounters and the shared consumption of the baboon prey object by male chimpanzees, the two species continue nonaggressive associations on and around the feeding ground.

EPISODE THIRTEEN: AGGRESSIVITY OF THE DANI IN NEW GUINEA

Normal Battle

The basic information for this episode was the report of the Harvard–Peabody New Guinea Expedition (Gardner and Heider, 1968). The observed aggressive behavior has been eloquently described in text and portrayed in poignant photographs by Robert Gardner and Karl G. Heider in their magnificent book *Gardens of War.* The authors duly accredited the unique contributions of Jan Broekhuijse, Abututi, a Dani himself, Peter Matthiessen, and especially Michael Rockefeller, each of whom in special ways provided keys to unlock the portals for observing yet another variation of human evolutionary adaptation, and the aggressivity of man.

The location was the 400-square-mile Grand Valley of the Baliem River which lies under the mountain wall in Dutch New Guinea. Here 50,000 Dani lived in a dozen alliances, each in perpetuating hostility with all the other alliances. The place was Kurelu, Northeast, which resisted until the first half of 1961 spearheads of intrusion by the contemporary world, missionary and military alike, as the territorial Dani practiced their ritualistic war games and delayed pacification by Dutch patrols.

Each alliance had its own range which was bounded with other territories by frontier zones. On each side of these border zones were watch towers that were manned each day from dawn to dark. From the pole-top towers observations were made of the enemy, other Dani across the border, of their activities, of preparation for attacks, or the sudden beginnings of a climactic attack. Events that portended the attack stimulated the sentinels to shout signals that called to immediate battle warriors who, since childhood play, had been trained to fight with spears and arrows.

Small groups of Dani warrior-leaders decided when and where to fight the next round of perpetual cycles of battles. After the divined decision, the challenge was made by shouts projected across the zone of conflict between tower lines. Usually the challenge was accepted.

Dawn and dusk bracketed the battle. Early in the morning bands of men from each side approached the zone of conflict and with calls and cries given and returned sent waves of signals that reached all villages.

The leaders in each village decided for themselves whether or not to fight at this time and place. Most will prefer immediate battle, wrote Gardner and Heider (1968). However, men who lived most distantly and those formerly wounded might refrain from fighting. Those men who would fight ate large quantities of sweet potatoes in preparation for the long day's battle. Some warriors greased themselves with pig fat, owned or borrowed. The plumes of birds-of-paradise and feathers of egrets, cassowaries, and parrots as well as colored clays were used by warriors to prepare themselves for the conspicuous aggressive display of their formal battle.

Assemblies of men awaited action in the shadows of watch towers lining the zone of conflict. Midday approached. The leaders who decided for battle consulted as their roles required about various strategies. They performed ceremonies and rituals. The magic rites were varied like different cures prescribed for the same disease of man since this makes the leaders seem important. The Doctors of War conspicuously displayed their wisdom. Nevertheless, the leaders attempted to divine what they could not surely know—the final results of the decisive hours of that battle day.

Nonfighting young and old men caught small creatures—bugs, grasshoppers, birds, and rats—and bound them tightly with grass and leaves. They gave these immobilized creatures to the leaders in conference who were pleased to have these symbols of dead enemies. In addition, old men drew maps and pathways in wind-blown ashes to guide ghosts to the zone of conflict.

By noon the warriors had made their battle formations along the two lines of towers on each side of the 500-yd wide ground of conflict. Expectant and alert silence reigned. In this open space the afternoon "would bring the pleasure of fighting to several hundred on both sides—terror to a few who would feel the pain of a piercing arrow or spear, and rarely the shock of death to that warrior who acted stupidly or clumsily [p. 138]."

Warriors of both sides initiated ceremonial thrusts to test for the opponents' readiness for battle. Thirty men plus or minus ten or so advanced over the open zone of encounter. The stylized advances from both sides moved, stopped, ran, stopped, advanced, and retreated, but not entirely back to where the last action started. In brief, there were thrusts and counter-thrusts. Arrows flew, signaling readiness to engage in formal conflict. There were several more advances and retreats with symbolic releases of arrows. Tension mounted. The excitement and action tempo increased during the first half hour of probing from both sides into the zone of conflict.

The open space filled with cavorting men in motion who threw and dodged flights of spears and arrows. Massed men in groups of 30 to 50 charged near together in the mid-ground to the rhythmic sound of several hundred feet on hard ground and clashed in action. Spears and arrows were discharged, flew in waves, and were avoided for an exhausting hot half hour.

Fresh men replaced the weary as all continued to exhibit agile motions; each warrior exercising his long-practiced skills of alert observing and reacting, of throwing unfeathered missiles into wobbly trajectories, and of warning his companions to avoid approaching shafts. The finest spears were reserved and risked only in the fiercest fighting. Trophies were sought and taken.

As the sequential battle continued, the duration of attacks and counterattacks shortened and the intervals between attacks lengthened. A few men were wounded, grazed by passing missiles or hit by ones that became embedded in their skin and muscle. The wounded moved well to the rear where wounds were bathed and bound with leaves and grass. Spears and arrowheads were removed dramatically, along with infectious orchid root fibers. The seriously injured were carried away by kinsmen and friendly villagers and were given incantations of praise by words, murmurings, and magic wands.

After 10 to 20 clashes the war of the day began to abate as men from the most distant villages left the field in order to arrive home before the frightening darkness. The remaining men on both sides shifted the battle to vocal taunts and strong threats. These tirades hurled across the 100-yd-wide zone of conflict gave the weary warriors pleasure, released their tension, and were amusing. Personal insults were shouted and laughter resulted.

Slowly and reluctantly both sides withdrew, leaving eager rear guards for protection against a surprise attack. Later these guards ran for home ". . . racing the night and the ghost they cannot see in the dark."

INFERENCES

1. The human Dani is socially organized to increase and perpetuate the basic bioecological patterns of aggressive behavior.
2. Myths, beliefs, and fears which are woven into sanctioned tradition both motivate and socially normalize aggressive behavior.
3. Balances of rewards and reinforcements for aggressivity perpetuate the behavior when the positive rewards and satisfactions outweigh the negative penalties and costs.
4. Levels of excitement contingent to fighting are both necessary for its occurrence and are rewards for the risks, fighting, and injuries.
5. Different kinds of symbolic language and vocal behavior both instigate fighting and serve functions similar to fighting.
6. Territorial aggressive behavior, having served to disperse a conspecific population in space, then functions to maintain separation of component groupings.
7. The male hunter (predator) is also the principal fighter in war, and the roles that are derived from biological conditions are strengthened and extended by social–cultural factors. Different roles are formed for other members of a society.
8. Viewed as part of an ecosystem, aggressive behavior is regulatory at normal levels of energy intake and expenditure, but, when aggressivity reaches high levels, it can become pathological for the community of organisms.
9. Leaders play, with obvious satisfaction, their socially defined and socially reinforced roles in formed war.
10. A highly closed and long-isolated society perpetuates its aggressive traditions until forces external to that society intervene and change the social structure.
11. Aggressive behavior that is polarized for different semiclosed groups (assemblies) requires high levels of cooperation within the kindred organizations.

The Raid

The raid was another expression of Dani hostility and pugnacity (Gardner and Heider, 1968). A typical raid was made by about 20 warriors selected for their aggressive characteristics and led by a veteran fighter. The focused and consuming purpose of the raid was to kill one of the male enemy in reciprocal revenge for the killing of the boy, Weaké.

The day was cloudy and cold, with a dense mist that hung at tree-top levels in Balcin Valley beneath the mountain wall, obscuring the tops of sentinel towers. The village from which the raiders came learned quickly of the imminent foray. Reserve support groups of regular warriors were formed and readied for action to support the raiders.

The silent raiders moved quietly and separately by hidden pathways to a rendezvous point near the zone of conflict. From there the band traveled to that garden area where a kill was anticipated. The party passed low blinds manned by "alert ghosts" who sometimes shook tree tops and alerted enemy sentinels. The raiding party arrived at the garden objective. However, the party had been detected and was allowed to filter through a trap formed by enemy warriors. At the right moment, the raiders were attacked from ambush and several were wounded. Yonokma was killed and speared repeatedly, and 13 men retreated to the rear in defeat and frustration. Behind the lines the escaped raiders stood in cold rain, humiliated and dejected, while regular reserve warriors staged a short and formal postraid battle.

The much-speared Yonokma was ceremoniously surrendered by the enemy back to his people under a protective shower of arrows. A counterfight of these shielded the retrievers of Yonokma from the zone of conflict. They built a litter and carried him quickly home. "The only sounds were those of feet sucking in the muck of the swampy trail and the sobbing dirge of Yonokma's forsaken women [p. 144]."

Yonokma's mother and two sisters had learned swiftly of his killing in the dread fog, and they ran to the battleground, a privilege extended to Dani women when a family member is killed. The slaughtered youth was displayed in a chair, decorated with shells and a net. A dense encirclement of women mourned in their socially channeled expression of sorrows.

A pyre was built and lit for Yonokma. His kinsmen wafted four grass arrows toward the enemy to certify future revenge and to ensure another cycle of hostility and aggressive behavior. There were both Weaké and Yonokma who must be revenged by yet two other enemy killings the delicate scales governing life and death among the Dani continued to adjust to alternating fortunes.

> War is one of the paramount institutions of Dani life. With agriculture and pig raising, it constitutes one of the few major focuses of all people's interest and energy. Without it the culture would be entirely different; indeed, perhaps it could not find sufficient meaning to survive . . . [p. 144].

INFERENCES

The reader is earnestly invited to draw his own conclusions and generalizations based on this and other chapters of this open-ended book.

REFERENCES

Berkowitz, L. (1962). "Aggression: A Social Psychological Analysis." McGraw-Hill, New York.

Bliss, E. L. (ed.) (1962). "Roots of Behavior: Genetics, Instinct, and Socialization in Animal Behavior." Harper, New York.

Carpenter, C. R. (1947). *Social Behavior of Rhesus Monkeys* (film), PCR-2012. Psychological Cinema Register, Audio-Visual Serv., Pennsylvania State Univ., Pennsylvania.

Carpenter, C. R. (1964a). A field study of the behavior and social relations of howling monkeys (1934). *In* "Naturalistic Behavior of Nonhuman Primates" (C. R. Carpenter, ed.), pp. 3–92. Pennsylvania State Univ. Press, University Park, Pennsylvania.

Carpenter, C. R. (1964b). Sexual behavior of free ranging rhesus monkeys (1942). *In* "Naturalistic Behavior of Nonhuman Primates" (C. R. Carpenter, ed.), pp. 289–341. Pennsylvania State Univ. Press, University Park, Pensylvania.

Carpenter, C. R. (1964c). A field study in Siam of the behavior and social relations of the gibbon (1940). *In* "Naturalistic Behavior of Nonhuman Primates" (C. R. Carpenter, ed.), pp. 145–271. Pennsylvania State Univ. Press, University Park, Pennsylvania.

Carpenter, C. R. (1969). Behavior of the Macaques of Japan (film). PCR-2184K. Psychological Cinema Register, Audio-Visual Serv., Pennsylvania State Univ., University Park, Pennsylvania.

Carpenter, C. R., and Nishimura, A. (1969). The Takasakiyama colony of Japanese macaques (*Macaca fuscata*). *Proc. Int. Congr. Primatol., 2nd,* Atlanta, Georgia, 1968 (C. R. Carpenter, ed.), Vol. 1, Behavior, pp. 16–30. Karger, Basel.

Carthy, J. D., and Ebling, F. J. (eds.) (1964). "The Natural History of Aggression." Academic Press, New York.

Clemente, C. D., and Lindsley, D. B. (eds.) (1967). "Aggression and Defense: Neural Mechanisms and Social Patterns." Univ. of California Press, Berkeley, California.

Fried, M., Harris, M., and Murphy, R. (eds.) (1968). "War: The Anthropology of Armed Conflict and Aggression." Natural History Press, Garden City, New York.

Garattini, S., and Sigg, E. B. (eds.) (1969). "Aggressive Behaviour." Wiley, New York.

Gardner, R., and Heider, K. G. (1968). "Gardens of War: Life and Death in the New Guinea Stone Age." Random House, New York.

Howell, R. C., and Clark, J. D. (eds.) (1963). "African Ecology and Human Evolution." Aldine, Chicago, Illinois.

Lee, R. B., and DeVore, I. (eds.) (1968). "Man the Hunter." Aldine, Chicago, Illinois.

Otten, C. M. (ed.) (1973). "Aggression and Evolution." Xerox College Publ. Lexington, Massachusetts.

Sugiyama, Y. (1965). On the social change of Hanuman langurs (*Presbytis entellus*) in their natural condition. *Primates* (6), 3–4.

Teleki, G. (1973). "The Predatory Behavior of Wild Chimpanzees." Associated Univ. Presses, Cranbury, New Jersey.

Washburn, S. L., and Jay, P. C. (eds.) (1968). "Perspectives on Human Evolution." Holt, New York.

INDEX

D

E